AF585131

THE BOTANICAL ART OF

WILLIAM T. COOPER

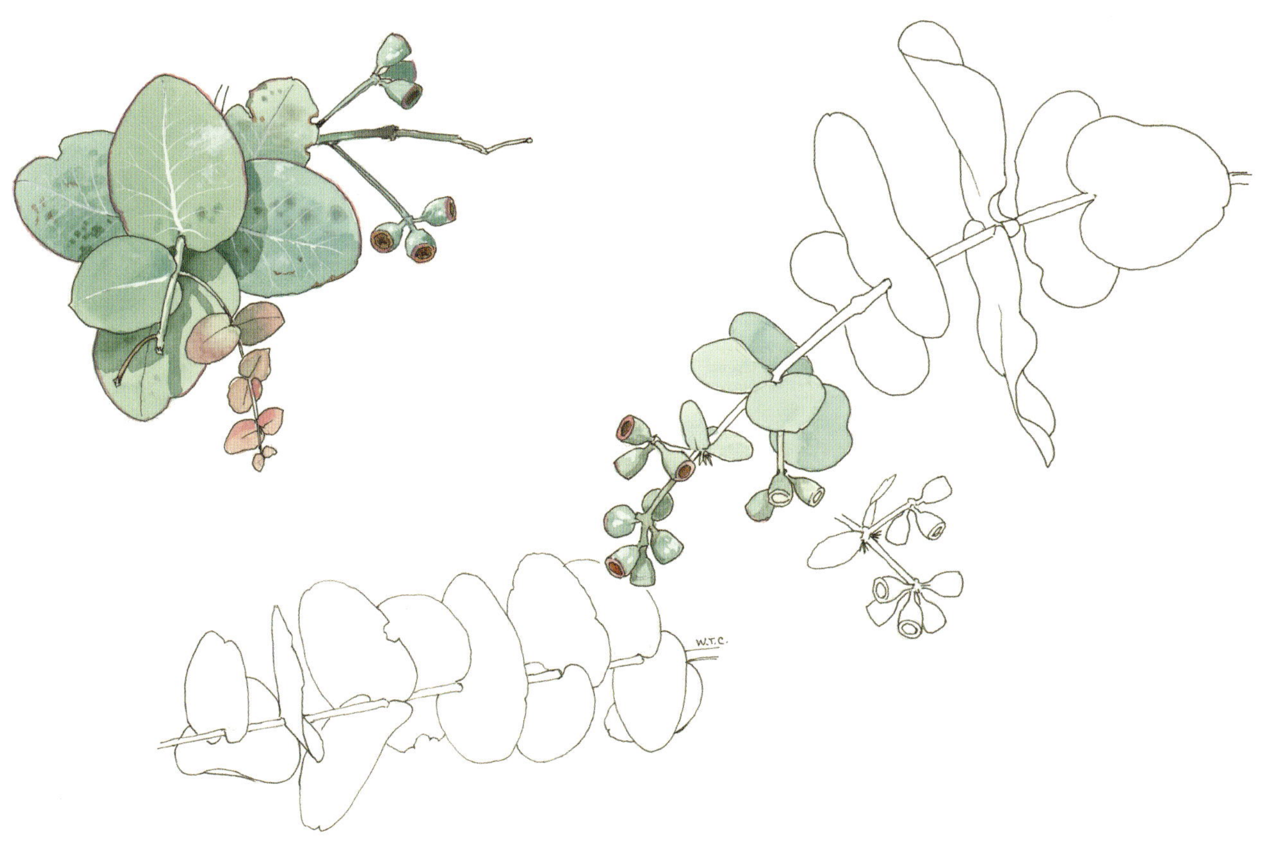

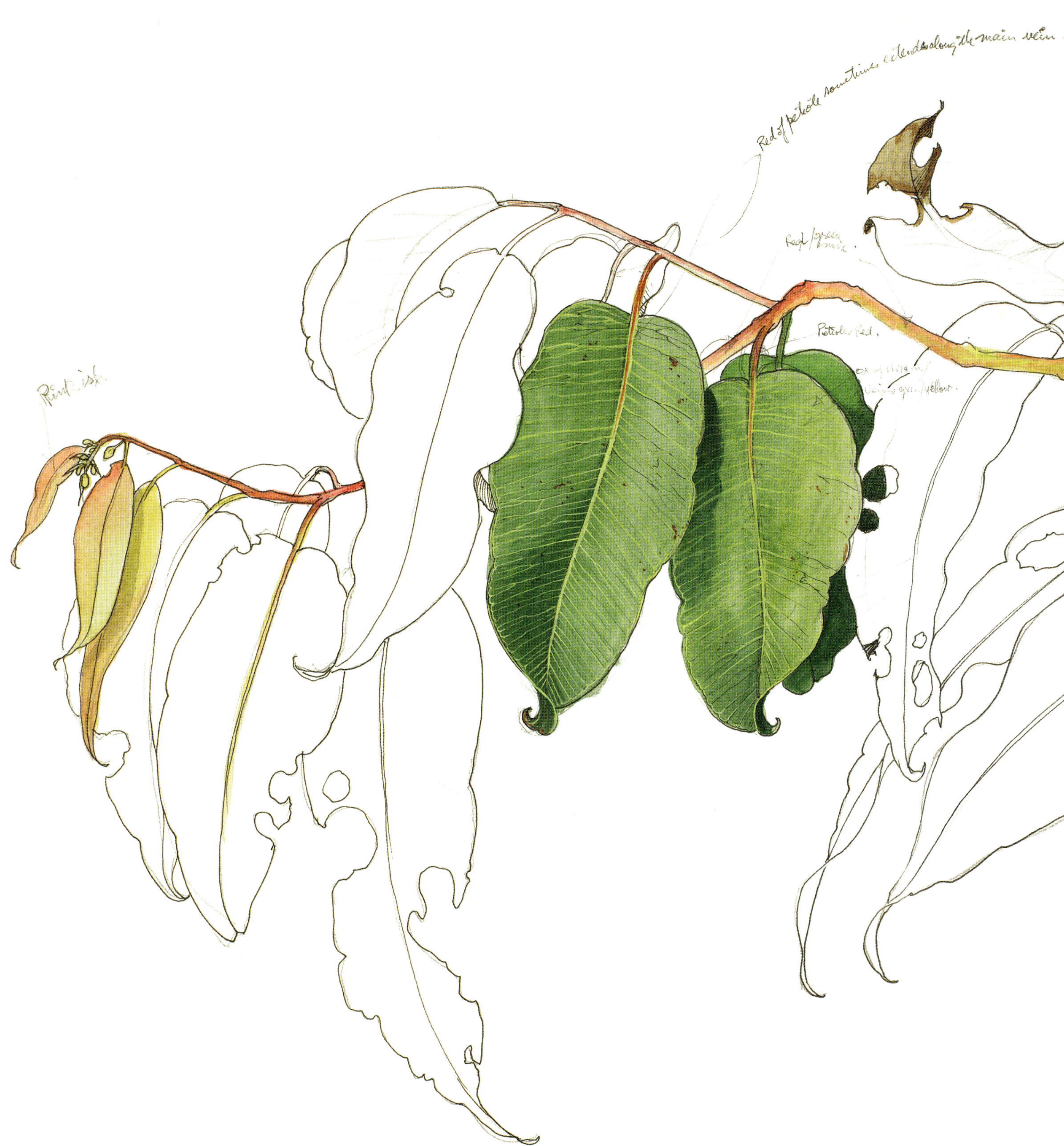
Red of petiole sometimes extends along the main vein.
Red/green mix.
Petioles Red.
Pinkish

THE BOTANICAL ART OF
WILLIAM T. COOPER

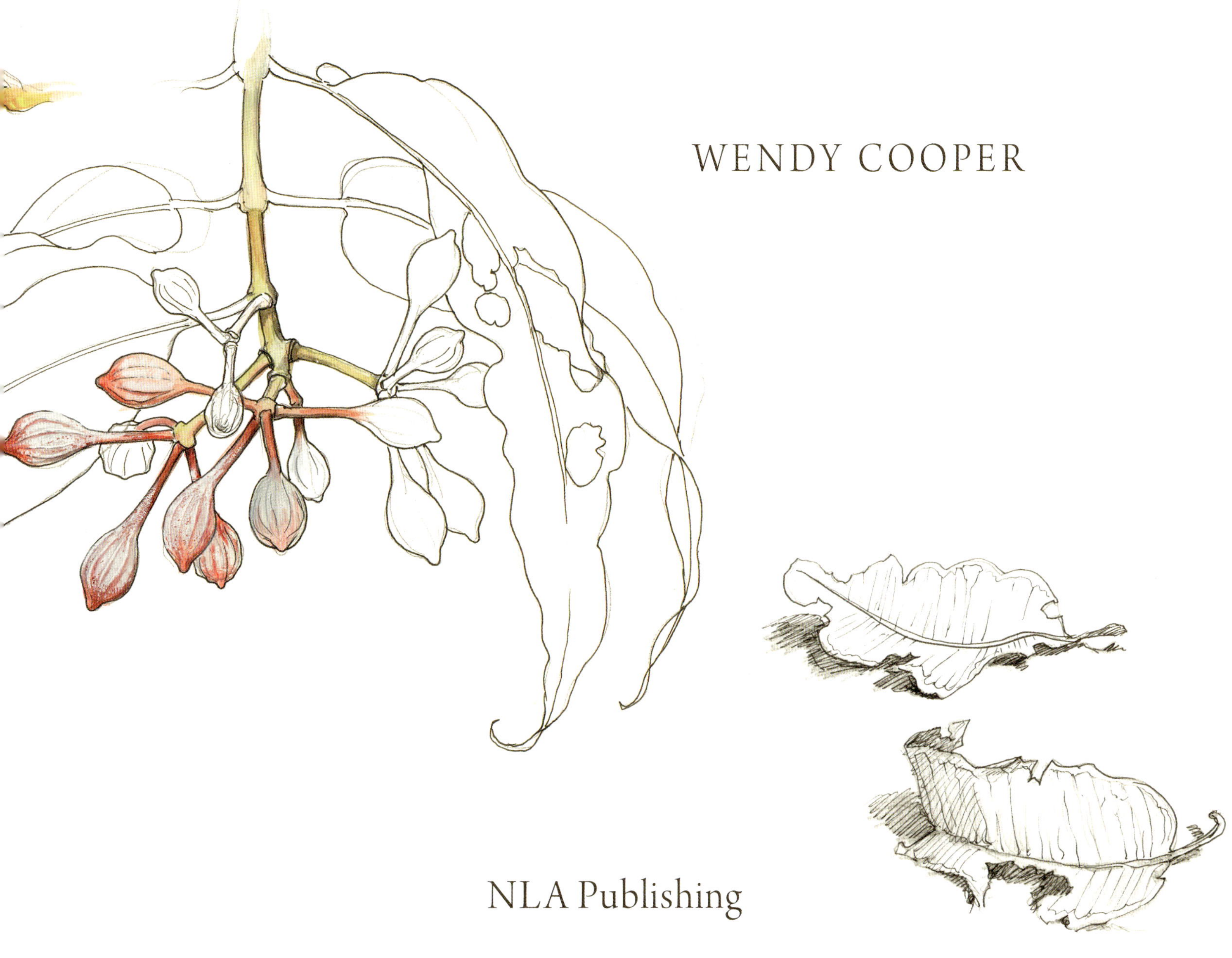

WENDY COOPER

NLA Publishing

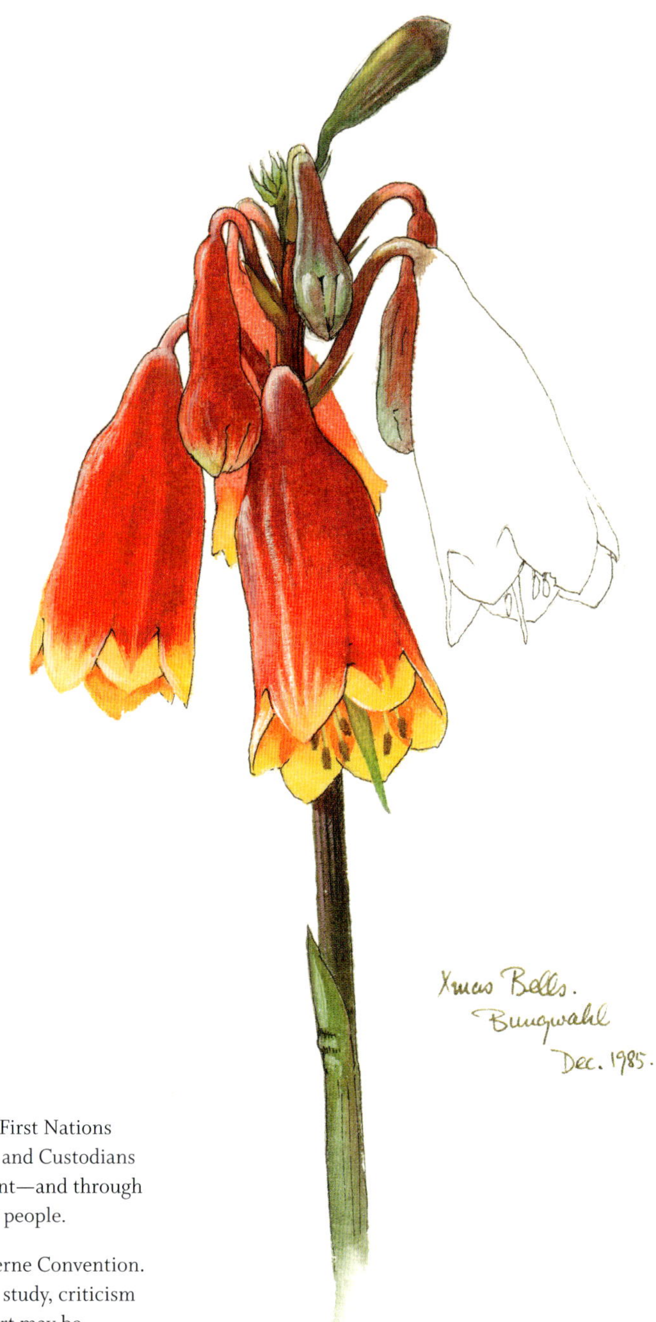

→ *Blandfordia grandiflora* (Christmas Bells)
1985; pen and watercolour; 300 x 250mm
State Library of New South Wales

Published by NLA Publishing
Canberra ACT 2600

ISBN: 9780642279712

The National Library of Australia acknowledges Australia's First Nations Peoples— the First Australians—as the Traditional Owners and Custodians of this land and gives respect to the Elders—past and present—and through them to all Australian Aboriginal and Torres Strait Islander people.

Designer: Stan Lamond
Printed in China by Asia Pacific Offset

The National Library of Australia would like to thank Simone Alcorso for her help with this book.

Find out more about NLA Publishing at
nla.gov.au/stories/national-library-publishing

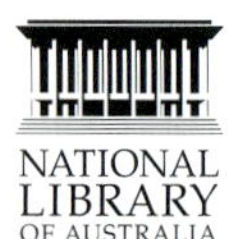

A catalogue record for this book is available from the National Library of Australia

Contents

Introduction
vi

Chapter 1
Rainforest Trees & Shrubs
2

Chapter 2
Vines, Ferns & Epiphytes
106

Chapter 3
Dry Country
166

Chapter 4
Swamps, Mangroves & Beach Forest
222

Chapter 5
Beyond Australia
252

Glossary 312

Index of Families, Genera and Species 313

Introduction

William T. Cooper was a prolific artist throughout his life, most notably as a painter of birds, but also of mammals, landscapes, seascapes and plants—numerous plants.

Bill always lived in the bush. He spent his childhood in the bushy outskirts of Newcastle, New South Wales, and his early adulthood on the coast to that city's north. The latter half of his life was spent at Topaz in north Queensland. A keen naturalist from childhood, he became exceptionally knowledgeable in many fields of natural history, including birds, reptiles and mammals, in Australia and worldwide.

I was the very lucky girl who won Bill Cooper's heart in 1976. From that time, beginning at Bungwahl in New South Wales, our lives were shared as we lived and worked together for 39 years.

Initially, Bill's knowledge of plants was limited to the familiar coastal plants of his younger years in the greater Newcastle area. However, when we bought our first property at Bungwahl, which had a couple of rainforested gullies, we learned those plants together. Then in 1987 we moved to a new property completely covered with rainforest, at Topaz, in the wettest part of tropical north Queensland. Learning the plants was daunting. Bill decided to paint every fruit we found in the forest so that we might one day be able to use the illustrations to find a name for each plant. These paintings were the beginnings of the 'fruit books'. From that point on, I seemed to learn the plant names more easily than Bill; he stepped back and mostly stuck to illustration. This was at a time when he was keen to develop his painting skills, particularly by adding more plants to his bird paintings.

We were soulmates from the beginning but our highly co-dependent professional relationship bloomed when Bill, after many attempts, finally convinced me to write our first fruit book, *Fruits of the Rain Forest*. For me, this was a revelation and from there we published *Fruits of the Australian Tropical Rainforest* and *Australian Rainforest Fruits—A Field Guide*. Soon after that I became a taxonomist, naming and describing new species of plants. Both of us had grown up with scarce resources, and neither of us completed high school, yet despite that we were lucky enough to make a living in fields we relished. We were in utter disbelief when, in 2014, the Australian National University bestowed on us honorary doctorates for our work.

Over the many years of our life together, we shared a well-lit studio/library/office that always had a special aroma of books, paints, plant specimens and bird specimens. We often worked on the same project but with different and complementary roles. Bill relied on my botanical knowledge but did not always need it. He was a tireless worker and when he was 'on a roll' his excitement was almost frantic to gather all of the material required to compose a picture and begin the actual painting. If the current piece was causing problems, we would head out into the forest for some time to 'smell the roses'.

Although Bill did not think of himself as a botanical artist, he was long considered one by those who appreciated his artistic skill in rendering the plants in his bird paintings. His keen eye, precise draftsmanship and extensive botanical knowledge allowed him to play with light and shade to give his paintings realism and a sense of perspective. A healthy artistic confidence enabled him to include blemishes and imperfections to add authenticity. His memory and attention to detail were truly remarkable.

Bill always wanted to tell a story in his paintings, with the botanical component as part of that story. This would often begin as a suggestion from him that we needed to go west to the dry country to collect a plant appropriate for a bird that occurred in that habitat. Or he might want to use a plant from somewhere further afield. If he was after a plant from the habitat of, for example, an Indonesian bird, I would scroll through plant lists and descriptions looking for a species that occurred both in Indonesia and in north Queensland. Then it was a matter of finding a location where the plant occurred, mostly within a day's drive, collecting a specimen and bringing it home for a detailed illustration, as seen on pages 68–69 in *Pleiogynium timorense*, which was used with the Umbrella Cockatoo. Bill was well known for always having the correct plants with his subjects.

Bill rarely painted plants as completed botanical illustrations, other than those for the fruit books. However, hundreds of working drawings of plants were made opportunistically for potential use in his paintings. Others were painted because they were especially attractive and might be able to be worked into a painting. It was hoped that these stored drawings might reduce the need to dash out urgently to find plants relevant to a proposed painting.

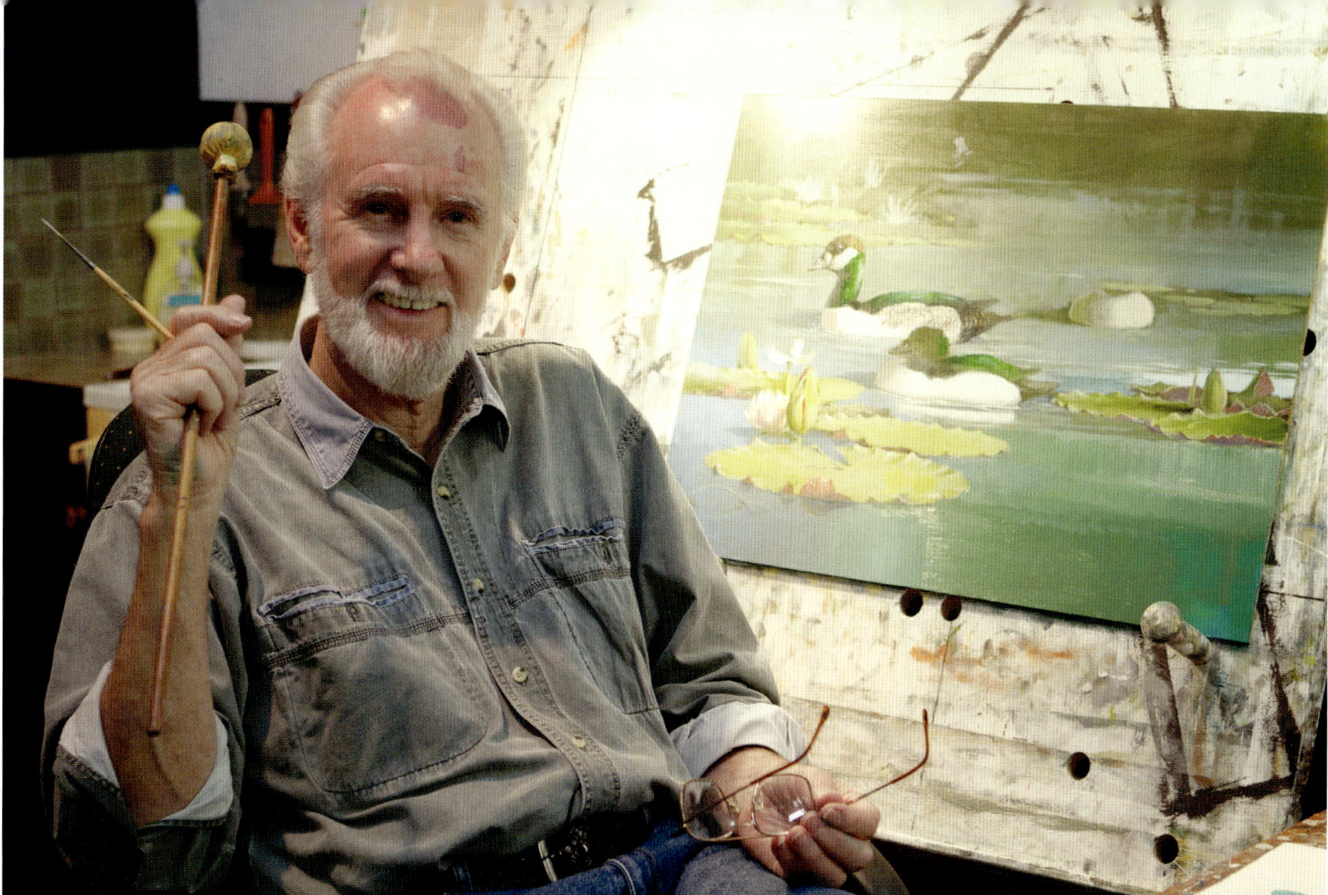

↑ William T. Cooper in His Studio, Topaz, 2013

Bill frequently kept diaries for recording his observations of wildlife behaviour. Most of these records are not related to plants, but where possible I have included extracts relevant to some of the botanical illustrations. They document Bill's interest in and compassion for all things wild.

The idea for this book was suggested by Bill's biographer, ornithologist and author Dr Penny Olsen, who believed that Bill's abundant and lovely botanical work (complete and incomplete) might have broad appeal, not simply for its artistic merit but also for those interested in plant illustration. Many observers find half-finished works interesting, the artist's process sometimes more obvious than in a completed painting. Most of the pieces chosen here are from Bill's archives, now largely held in the collections of the National Library of Australia and the State Library of New South Wales. The majority of these exquisite and scientifically accurate drawings and paintings are on public display in this book for the first time.

Plant names

Plant names are those given by taxonomic botanists as part of the formal and scientific description for each species. The names in Botanical Latin, which is derived from Latin and Greek, frequently give some insight into the plant's appearance, for example *longiflorum* translates to long-flowered. In some other instances, plants have been named for persons of historical fame.

Acknowledgements

I am sincerely grateful for the generosity of friends who assisted with creating this book: Nan Nicholson, for frankness as can only come from a dear friend to both Bill and me, as she accompanied me on this journey from the beginning, advising, improving and rearranging the stories throughout; Rupert Russell, also a long-time friend, for editorial suggestions, and especially for his sensitive way with words, his respect for Bill and his shared love of the environment, evident in his essays in this book; and Bryony Barnett, Mary Hoban, Hugh Nicholson and Juliana Russell, for kindly and generously taking time to proofread the manuscript and provide thoughtful comments and suggestions. Peter and Marilyn Chapman, Ashley Field, Rigel Jensen, Robyn Lewis, Penny Olsen, Sarah Scragg, Peter and Trixie Tuck and Frank Zich helped in various ways, from scanning images to supplying names for unidentified illustrations and offering editorial suggestions. It has been a pleasure to work with Susan Hall, Amelia Hartney, Jemma Posch, Katherine Crane and Stan Lamond from NLA Publishing. I sincerely thank them and their team.

The National Library of Australia and the State Library of New South Wales generously supplied scans for the majority of the artwork.

To those above, you have contributed enormously to making this a better book for Bill's legacy.

Chapter I

Rainforest Trees & Shrubs

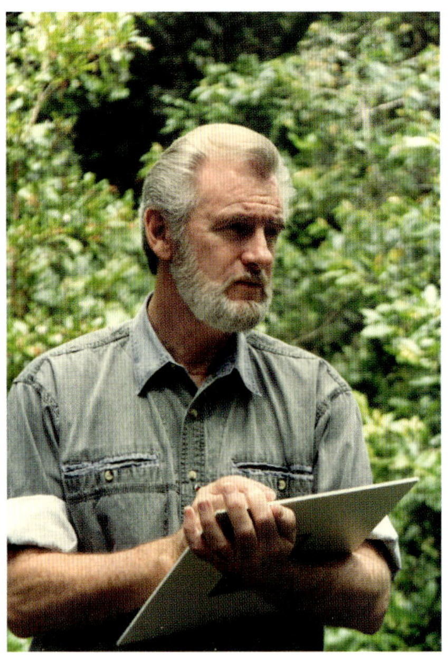

Study light and shade as if you are about to paint it and you will learn to see as an artist does.

Bill Cooper in *Capturing the Essence*

Rainforest is sometimes described as closed-canopy forest, with trees so densely spaced that their intermingled branches obstruct much of the light. For artist William T. Cooper, it was precisely the patches of light that drew his gaze. When walking with a companion, he might halt progress to point up to a flare of light striking the fronds of a fern on a high branch, or to bright rays silhouetting a cluster of foliage still glistening after an overnight shower. 'Look at the light on that,' he might say. 'Look at those fern leaves against that dark tree.' All such images Bill absorbed and stored, as another might store verses from a poem or the words of a song. Bill could call up a remembered image for use as part of a painting, whenever needed.

Sounds from the surrounding forest, chiefly birdsong, were an undertone, but welcome, with Bill paying fond attention to any favourite call. Scents carried up from damp litter or from lichen on a leaf brushed in passing added to the experience, but it was how the light treated every surface that most involved him. Mist in the forest brought added drama, turning the light almost green as it shafted between tree trunks. Endless interest came from the huge diversity of plants in any established rainforest, with north Queensland offering the greatest variety within any small area.

Bill, an accomplished painter already gaining fame for his portrayals of Australian birds, had developed an interest in plants while exploring the first property he owned together with his wife Wendy. This land at Bungwahl, on the Mid North Coast of New South Wales, comprised open forest along the ridges and enticing rainforest in the gullies between. Some years later, a move to Topaz in the Atherton Tablelands region of Far North Queensland brought a patch of richly diverse rainforest to their back door.

In this Topaz forest, each time either Bill or Wendy discovered a fruit it was carried home to be recorded. Painting fruit was an easy and enjoyable way to compile a register of the forest around them. Bill's work was so accurate and realistic that when the paintings were seen by Geoff Tracey, a renowned botanist and ecologist, putting a name to each fruit was simple. Throughout this time, Wendy's ability as a botanist was rapidly advancing: recognising groupings of plants, the intricacies of foliage and flowers and the structures of fruit while memorising it all so any new plant could be compared to what she already knew. As naturalists, both the artist and the botanist also recorded which birds were attracted to which fruit, and whether they fed on the fruit while it was still on the tree or vine, or waited until it dropped to the forest floor.

A lovely working partnership brought about their first book of rainforest fruits in 1994, followed by a second much expanded work, *Fruits of the Australian Tropical Rainforest*

← Bill Sketching in Rainforest at Chowchilla, Topaz, c.1989

↓ *Podocarpus grayae* (Once Known as *Podocarpus neriifolius*) with Leaves in Clusters, Bill's Preferred Leaf Orientation
pen and pencil; 365 x 480mm
State Library of New South Wales

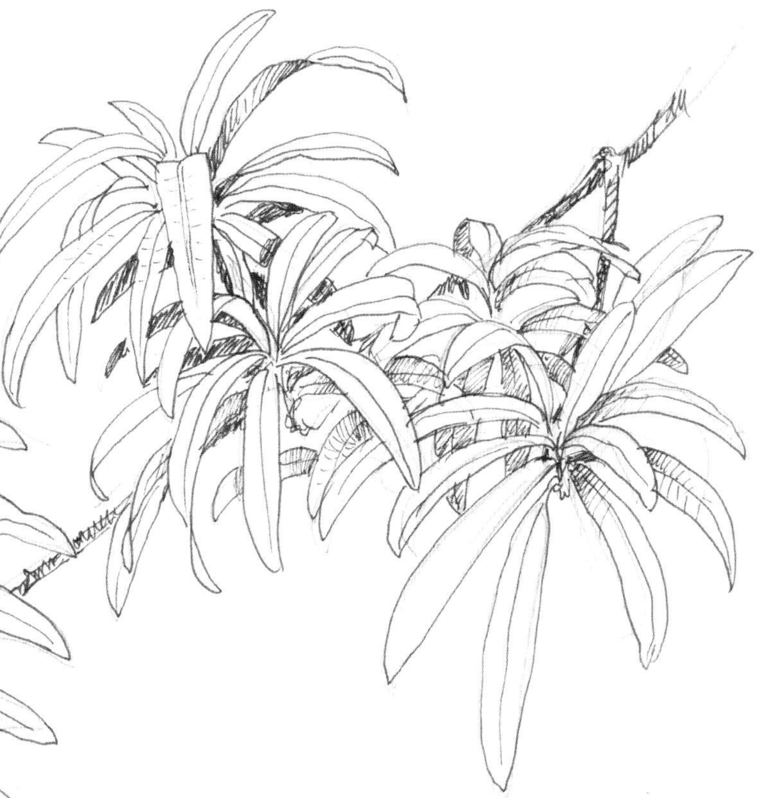

in 2004, which is a magnificent reference for all who are enthusiastic about Australia's northern rainforests. In a work that beautifully illustrates more than a thousand fruits, the exuberance of the artist can be seen in the decorations on quite a few pages. Here a Rose-crowned Fruit-dove plucks a ripe berry from a *Cissus* vine, and on another page a Metallic Starling eyes the ripest fruit on a Weeping Fig (*Ficus benjamina*). Yet there was no room in the *Fruits* book for paintings done just to show the beauty of leaf, flower and fruit. It is some of these illustrations that are presented in this volume.

Most of Bill's botanical paintings were done when he was planning a composition for his wondrous large-format images of birds. Paper used for watercolours does not fare well in the humidity of a rainforest, so nearly all foliage selected for trial in a composition would be collected from a forest and carried to the home studio. If foliage was required to suit a bird restricted to some distant forest such as Iron Range, well north of Topaz, it would be picked near the end of a field trip to be stored in a car fridge for the journey home. In his studio, Bill would set the specimen at the angle it occupied in the forest before beginning to draw. Every painting was preceded with a careful drawing. While drawing was in process, there would usually be silence from the studio unless passing clouds changed the light, which could bring out cross mutterings. Once he was satisfied with a drawing and the way the birds were to be placed, painting could commence. Now silence in the studio might be replaced with music of many styles but chiefly classical.

Bill enjoyed painting foliage that grows clustered on a stem, whorls of leaves on a *Terminalia* or a *Sloanea*, rather than leaves of plants that sprout evenly along a stem. When possible, he would avoid painting leaves with toothed margins because of the extra time required for such detail—detail that would not necessarily enhance the painting. And he was heard to complain about the accuracy needed to depict subjects such as a Bunya cone or the fruit of a *Pandanus*, in which small erratic variations interrupted the already complex pattern. Many of the plant specimens collected by Bill or Wendy or by botanist friends were painted in great detail, only for the paintings to be filed away because the result did not suit the composition of the bird and its surrounds that Bill finally decided upon.

For an observer, there is astonishment at Bill's absolute attention to the light and shade on every surface in every painting. Any leaf might cast a shadow on a leaf lower down, across part of a bird's tail feathers or its shoulder. Every detail illuminated by light or suggested in shadow accentuates the effect of a three-dimensional subject on a flat page. Working in his studio, Bill did not need to illuminate his specimens from one angle or another; his understanding of light and shadow was so complete that it could be worked into a painting just as he chose. He often drew a circle with an arrow at the top corner of a painting to remind himself of his chosen light angle for that particular painting.

Bill realised that most people did not see nature as he could. Through his paintings, he offered what he saw to any viewer who gazed with care; when someone discovered and remarked—usually with surprise—on a detail, the artist was pleased. To him, the work of most botanical painters looked flat, rendered as though the subject were a pressed specimen readied for preservation in a herbarium. Commenting on a painting of foliage by a third party, an artist friend said it looked like 'a cabbage run over by a truck'. Bill chuckled in agreement.

Rupert Russell

Alloxylon flammeum

Red Silky Oak

PROTEACEAE

Alloxylon is from *allo-* (other) and *oxylon* (wood), referring to the unique timber.

flammeum is from *flammeus* (scarlet), referring to the flower colour.

Red Silky Oak is a canopy tree of the Wet Tropics in medium- to high-altitude rainforest in north Queensland. The majority of Australian rainforest trees have insignificant white flowers, unlike Red Silky Oak, whose large Waratah-like flowers bloom sometime between June and December. Slender red petals cover the ground beneath trees that have been visited by honeyeaters or Rainbow Lorikeets foraging for nectar.

This painting was used on the cover of volume 1 of *Australian Tropical Rain Forest Trees: An Interactive Identification System*. Its use was requested by the author, rainforest botanist Bernie Hyland (b.1937), who was of great assistance to us during the compilation of the fruit books.

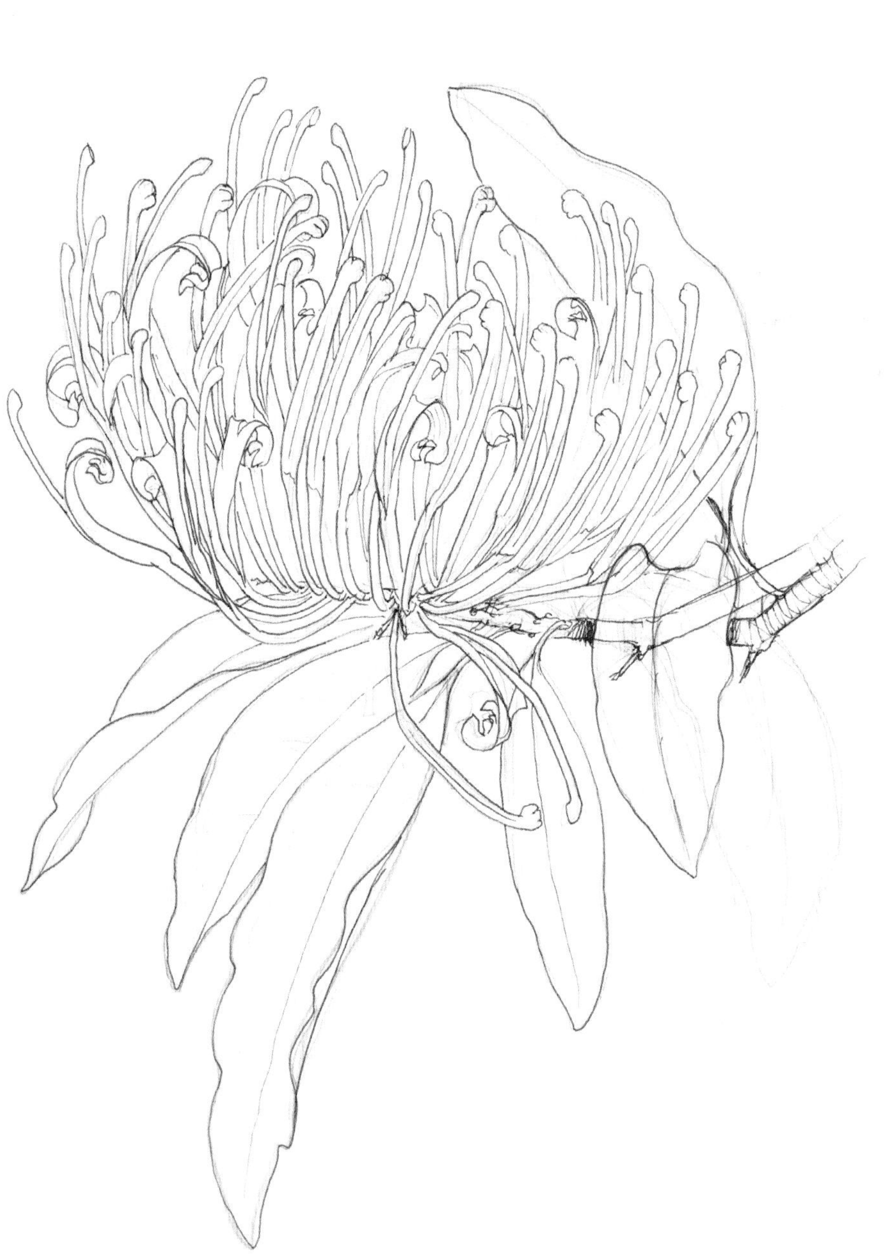

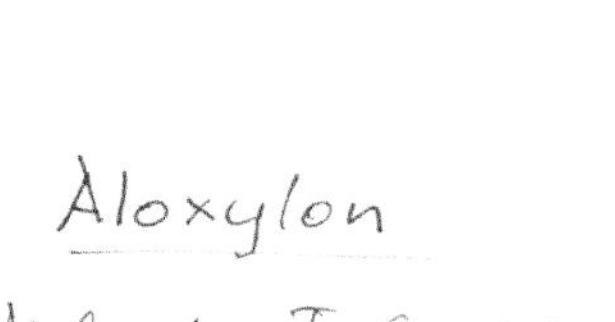

← Red Silky Oak
1992; pen and pencil; 345 x 390mm
State Library of New South Wales

→ Red Silky Oak
1993; watercolour; 510 x 420mm
Courtesy Bruce and Joy Gray

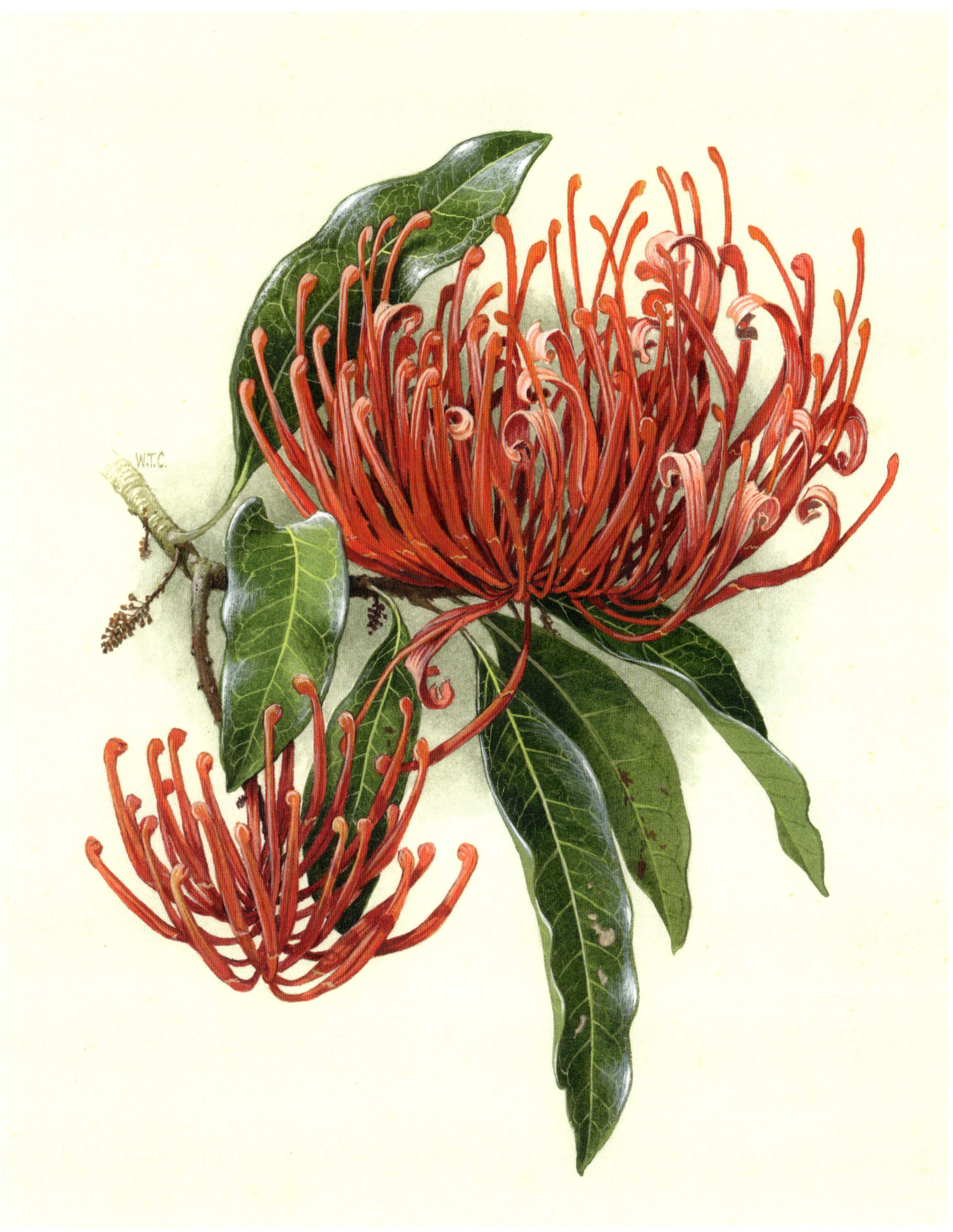
W.T.C.

Alphitonia petriei

Sarsaparilla

RHAMNACEAE

Alphitonia is from *alphiton* (barley meal), referring to the dry mealy fruit.

petriei is after Walter Rollo Petrie (b.1875), a forester who alerted botanists to the fact that this plant was a new species.

As a common regrowth tree, Sarsaparilla can be seen along forest edges and roadsides. It is easily recognised by the layered branching. Sarsaparilla is named for the smell of the freshly cut timber.

While we were quietly walking in the rainforest at Paluma, near Townsville, we became aware of a female Victoria's Riflebird in a Sarsaparilla tree. Then, nearby, a male riflebird suddenly began displaying to the female. We had not seen this display very often and Bill was keen to draw it, but soon realised he had forgotten his sketchbook. He did have a pen, however, so I stretched my pale-coloured shirt tightly across my shoulder blades and suggested he use this as a canvas. The drawings were later transferred to paper.

The plant drawing was done at Chowchilla, the name of our property at Topaz.

Paluma - North Queensland
Sept. 83.

15.30 hrs. — 5½ km from Paluma Village
Saw a VICTORIA RIFLEBIRD searching for food on the underside of the long horizontal branches of an Alphitonia by hanging over the branch in the same manner as Birds of Paradise (Lophorina & Parotia)

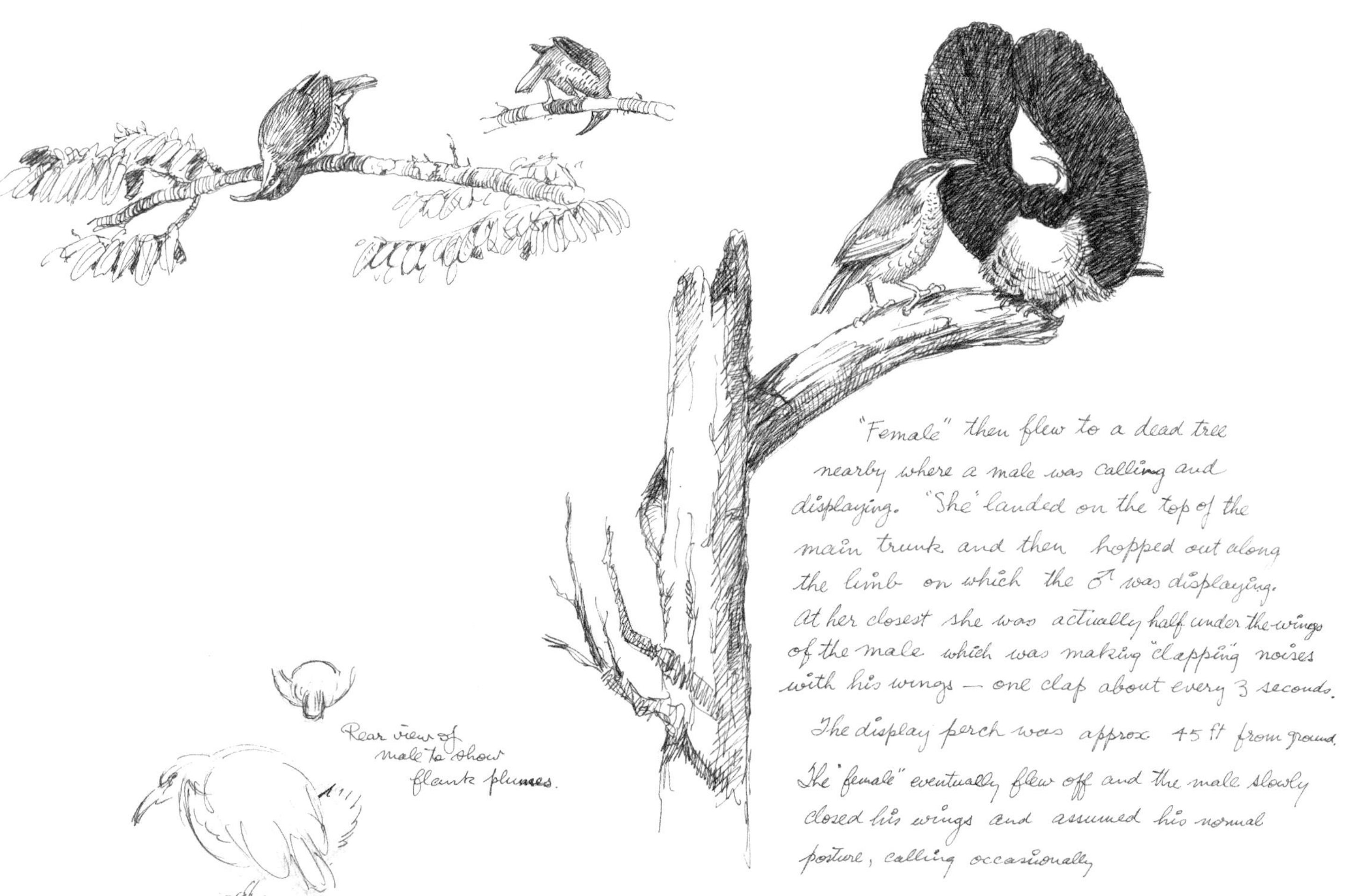

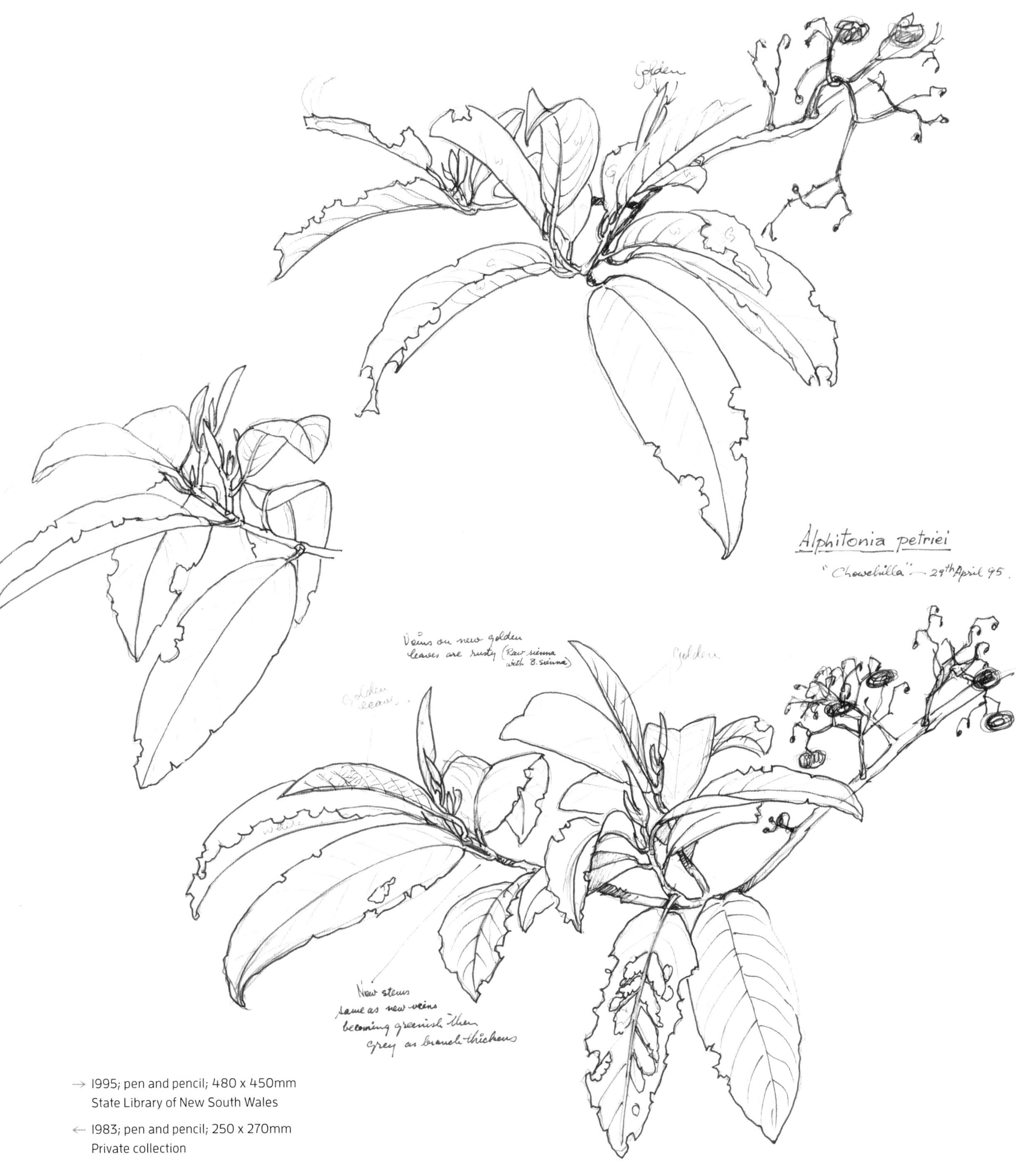

→ 1995; pen and pencil; 480 x 450mm
State Library of New South Wales

← 1983; pen and pencil; 250 x 270mm
Private collection

Antirhea tenuiflora

Crimson Berry

RUBIACEAE

Antirhea is from *anti-* (against) and *rheo* (flow), referring to the plant's use against dysentery and haemorrhaging.

tenuiflora is from *tenuis* (slender or thin) and *-florus* (-flowered).

Crimson Berry is a sub-canopy tree in north Queensland's lusher rainforests, occurring from Cape York to the Conway Range near Mackay.

The ribbed red fruits are a favourite of many birds and would have been an appropriate plant for a bird painting. Bill noted in his diary on 8 March 1992: 'Saw King Parrots eating *Antirhea tenuiflora*. Actually, they were extracting the seeds and dropping the flesh'.

It is a common tree on our rainforested property, where we walked most days. Bill especially liked to walk in the morning because of the warm light that dappled leaves, leaf-litter, trunks, everything. The afternoon light, especially in winter with its long shadows, he found grey and miserable, preferring to retreat to the studio at that time of day.

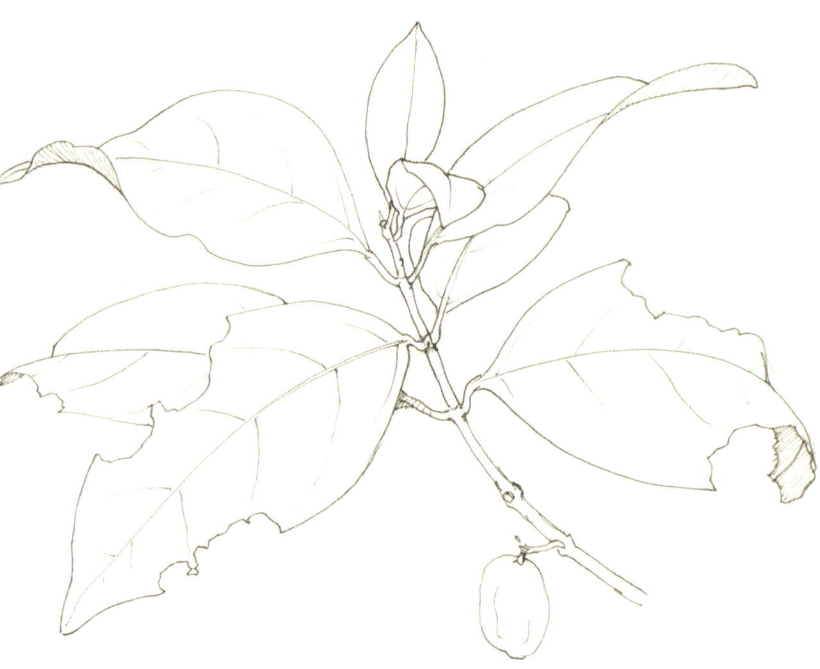

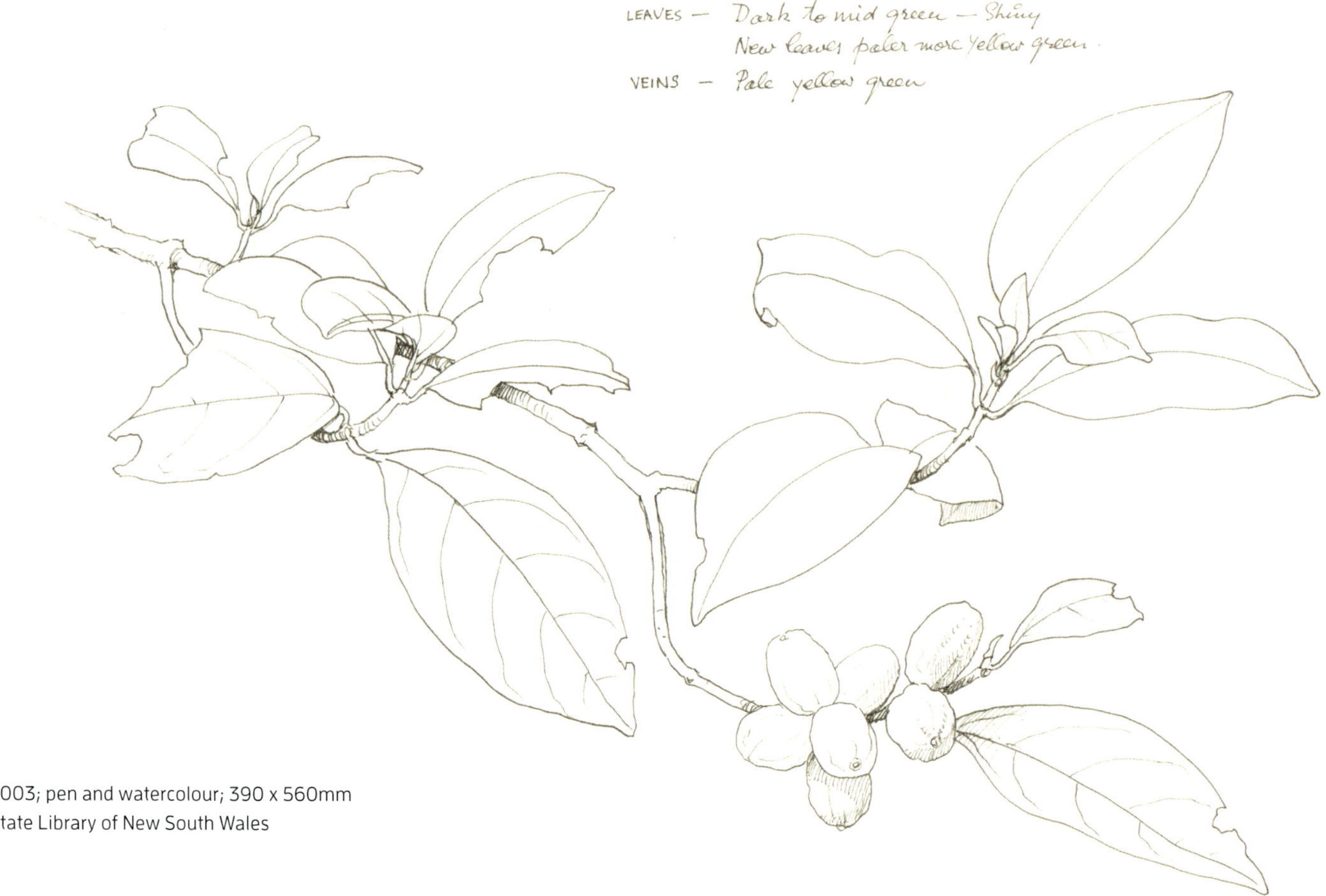

→ 2003; pen and watercolour; 390 x 560mm
State Library of New South Wales

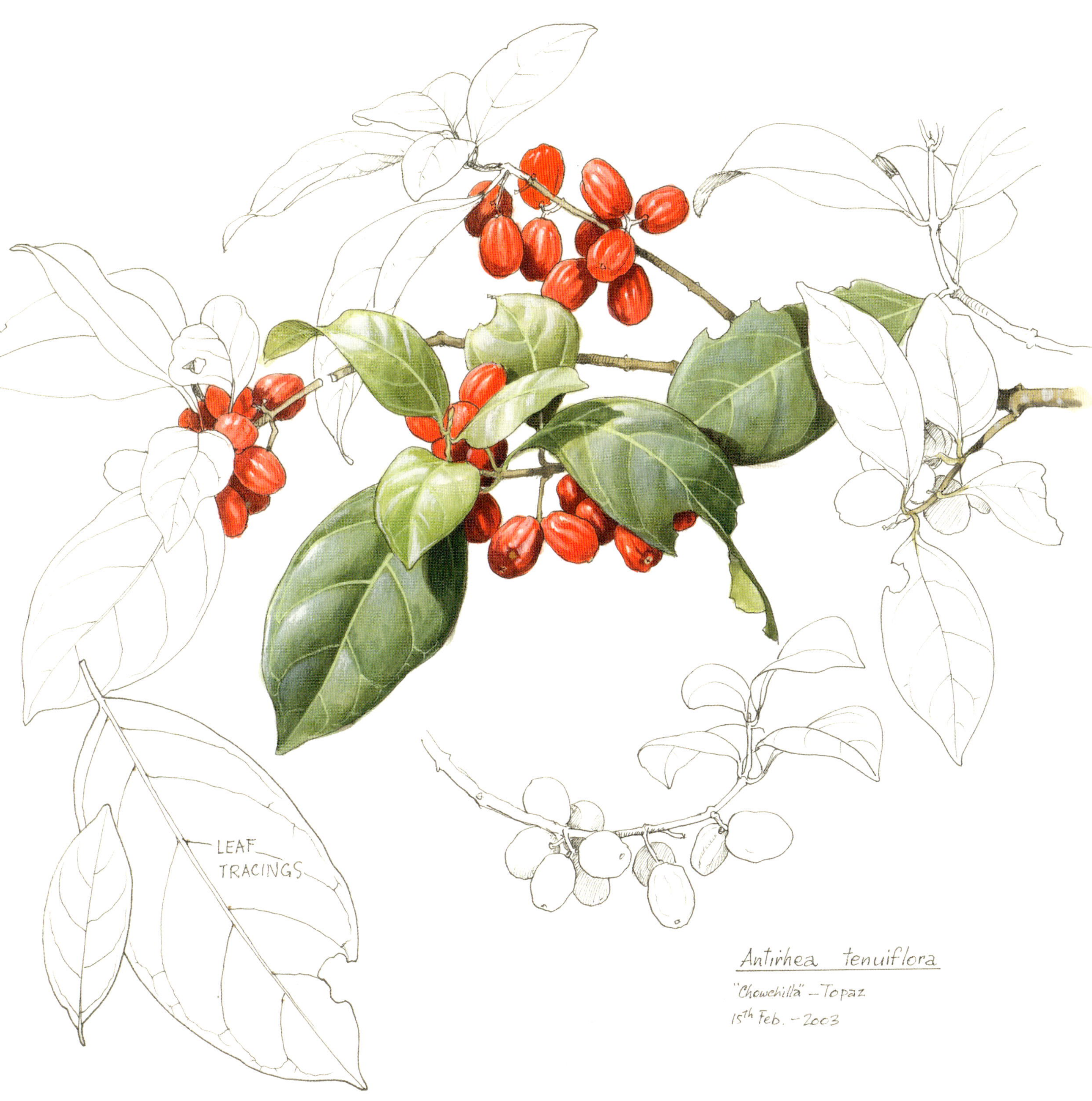
LEAF
TRACINGS
Antirhea tenuiflora
"Chowchilla" – Topaz
15th Feb. – 2003

Castanospermum australe

Black Bean

FABACEAE

Castanospermum is from *castanos* (Chestnut Tree) and *-spermus* (-seeded), referring to the Chestnut-like seeds.

australe is from *australis* (southern).

Black Beans are large trees of tropical and subtropical rainforests. The woody fruit is big, plump and bean-like and the seeds are poisonous to humans.

The dense dark foliage of Black Bean trees can be seen commonly along rainforest streams from northern New South Wales to Iron Range on Cape York Peninsula. Their prolific nectar-filled flowers age from yellow through to bright red. Trees in full flower are a riot of colour and movement, alive with screeching and jostling parrots.

Bill sketched the flowers near Malanda on the Atherton Tableland for use in a large acrylic painting of Rainbow Lorikeets feeding.

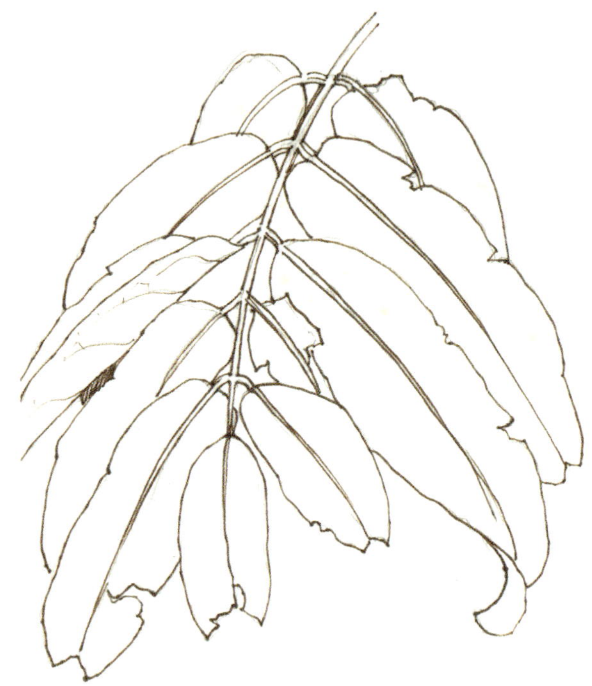

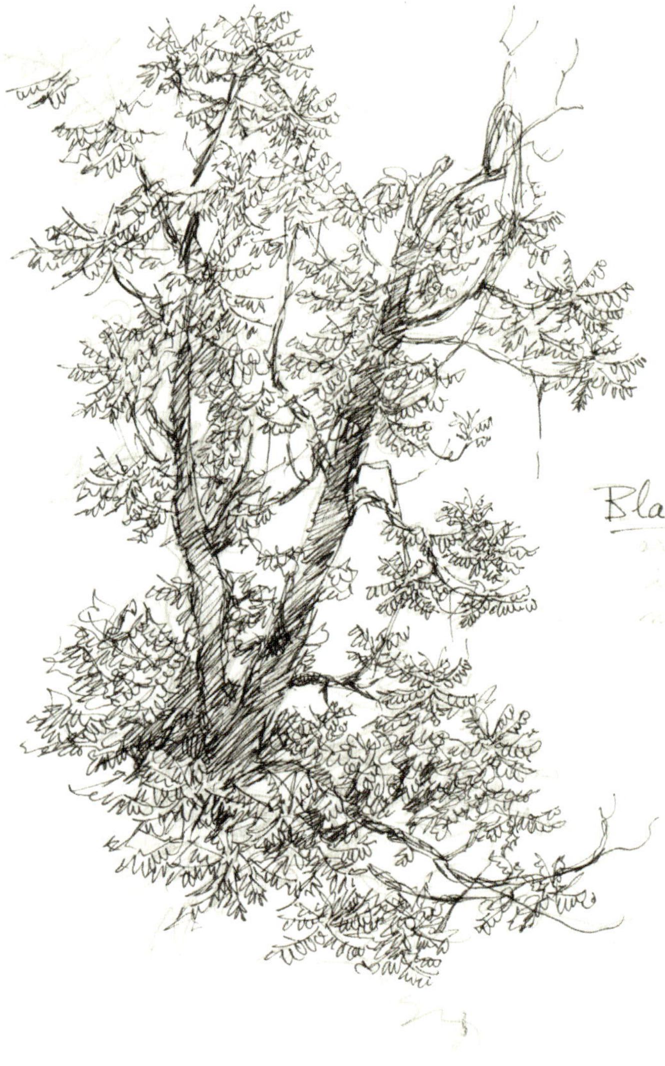

McILWRAITH RANGE

Rocky River

Blackbeans

← undated; pen and pencil
(in part) 230 x 325mm
Private collection

→ 2006; pen, pencil and watercolour
420 x 300mm
State Library of New South Wales

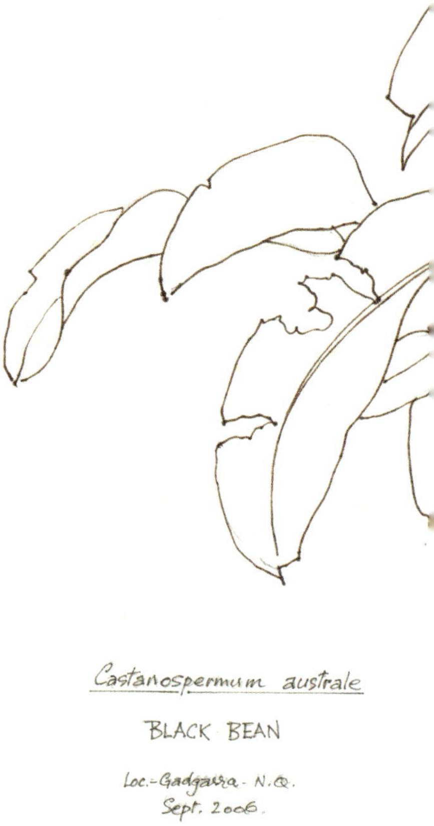

Castanospermum australe

BLACK BEAN

Loc.-Gadgarra - N.Q.
Sept. 2006.

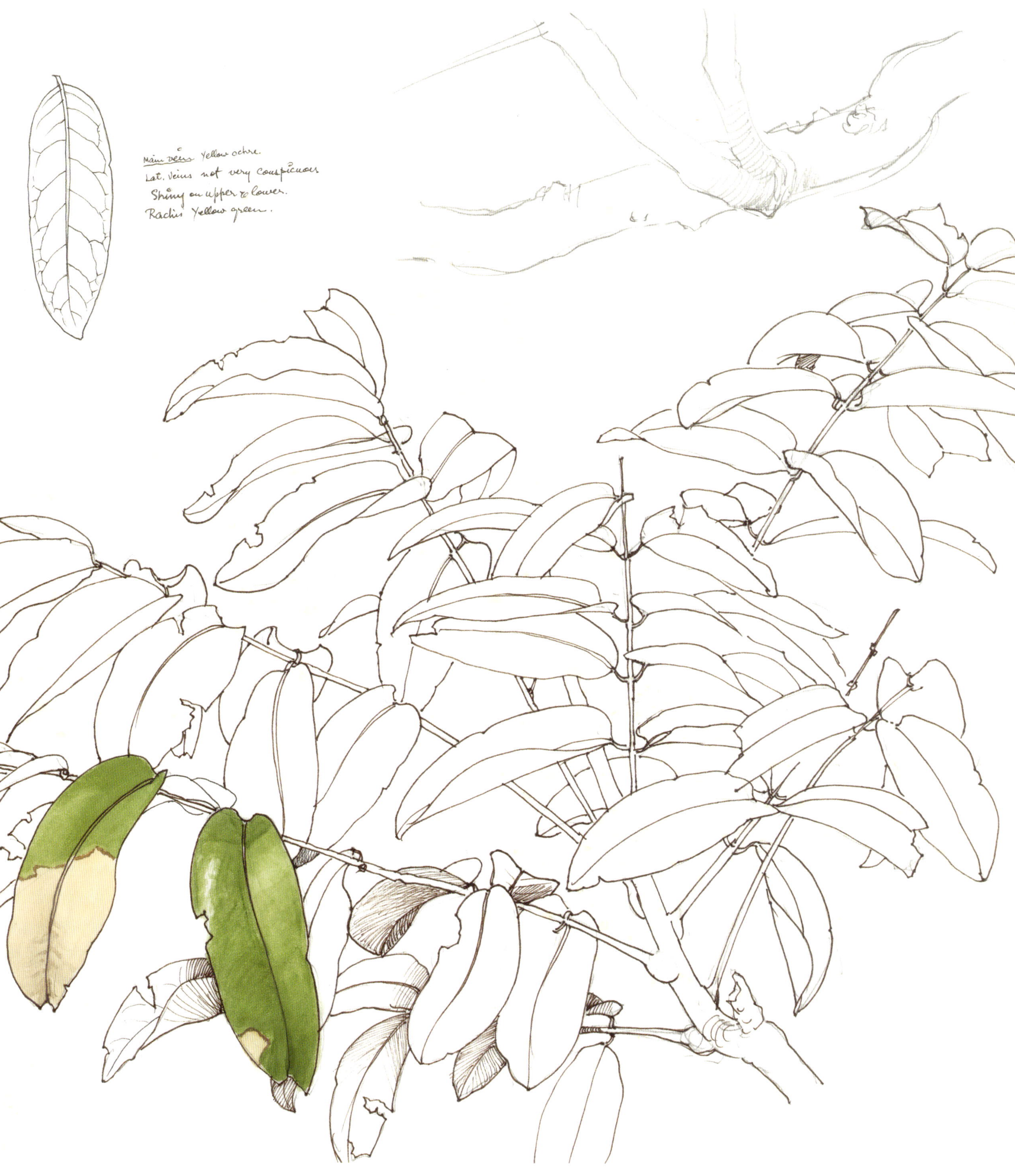
Main veins Yellow ochre.
Lat. veins not very conspicuous
Shiny on upper & lower.
Rachis Yellow green.

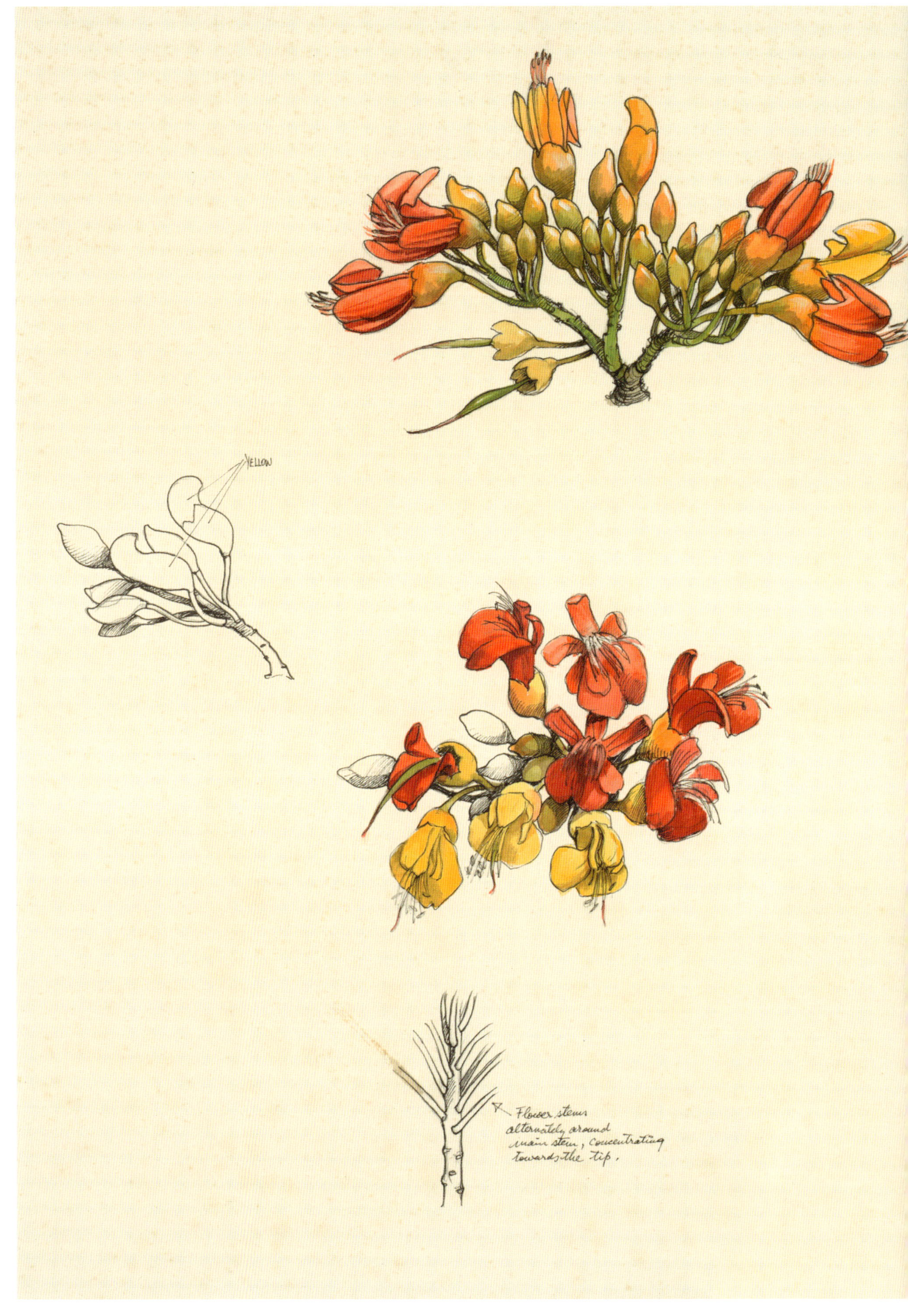

→ 1980; pen, pencil and watercolour; 412 x 584mm
National Library of Australia, 8068806

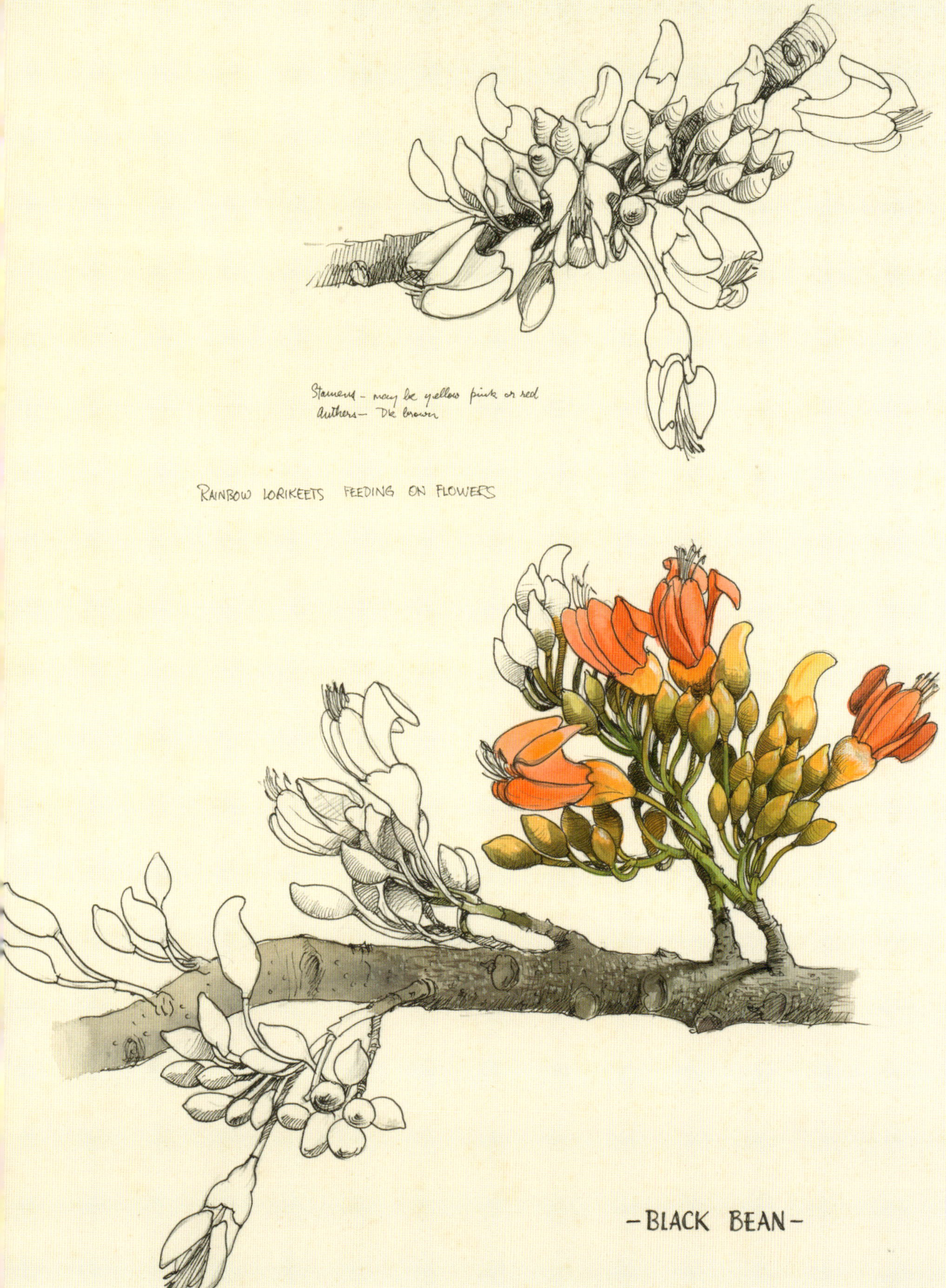

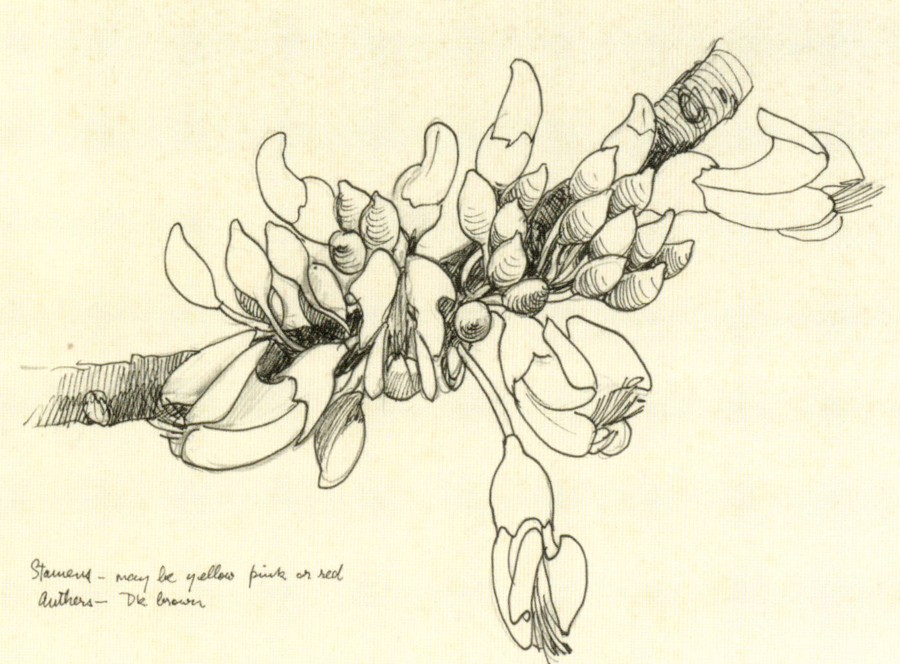

RAINBOW LORIKEETS FEEDING ON FLOWERS

-BLACK BEAN-

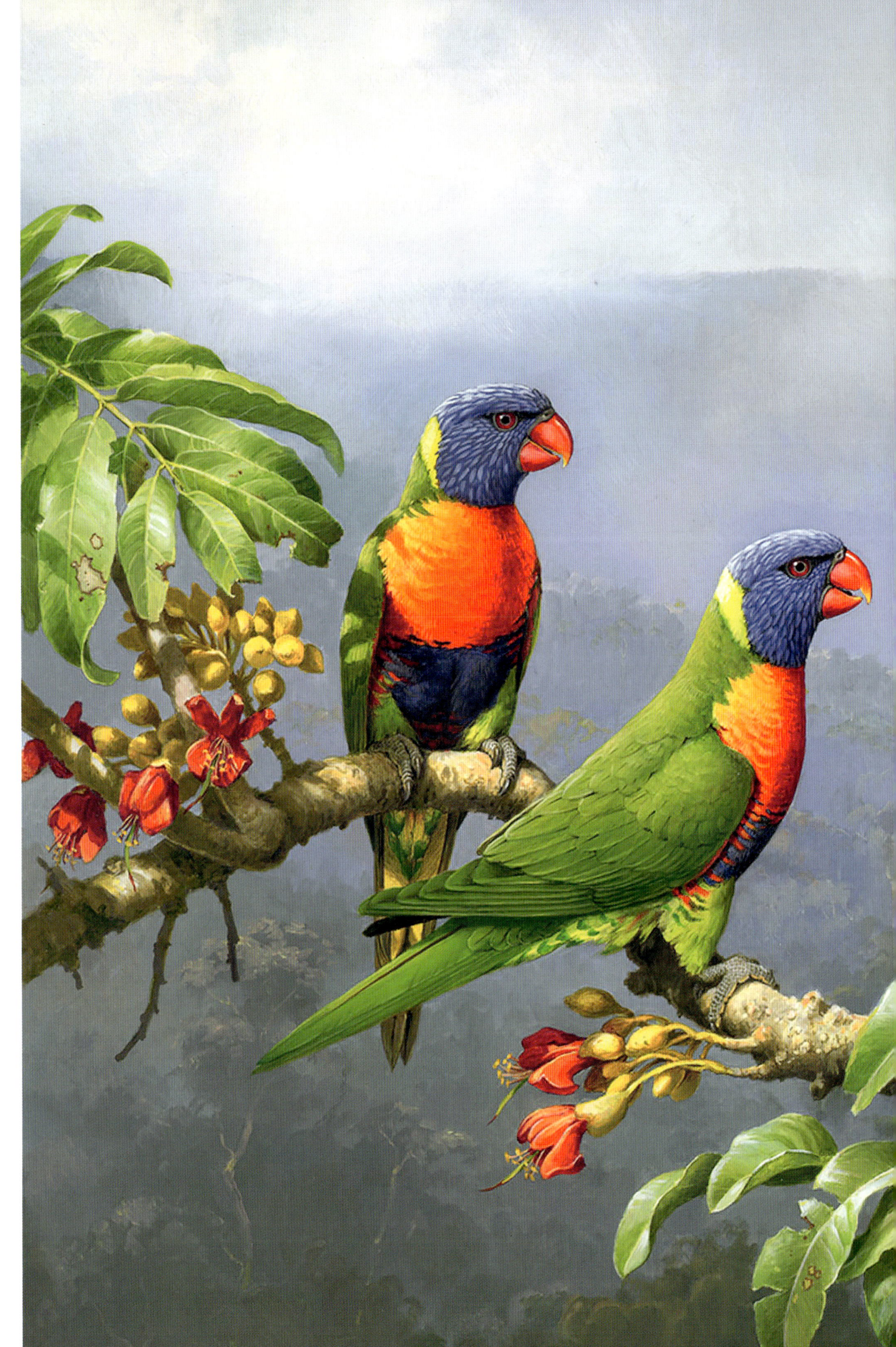

→ 2006; acrylic
on canvas
c.750 x 515mm
Private collection

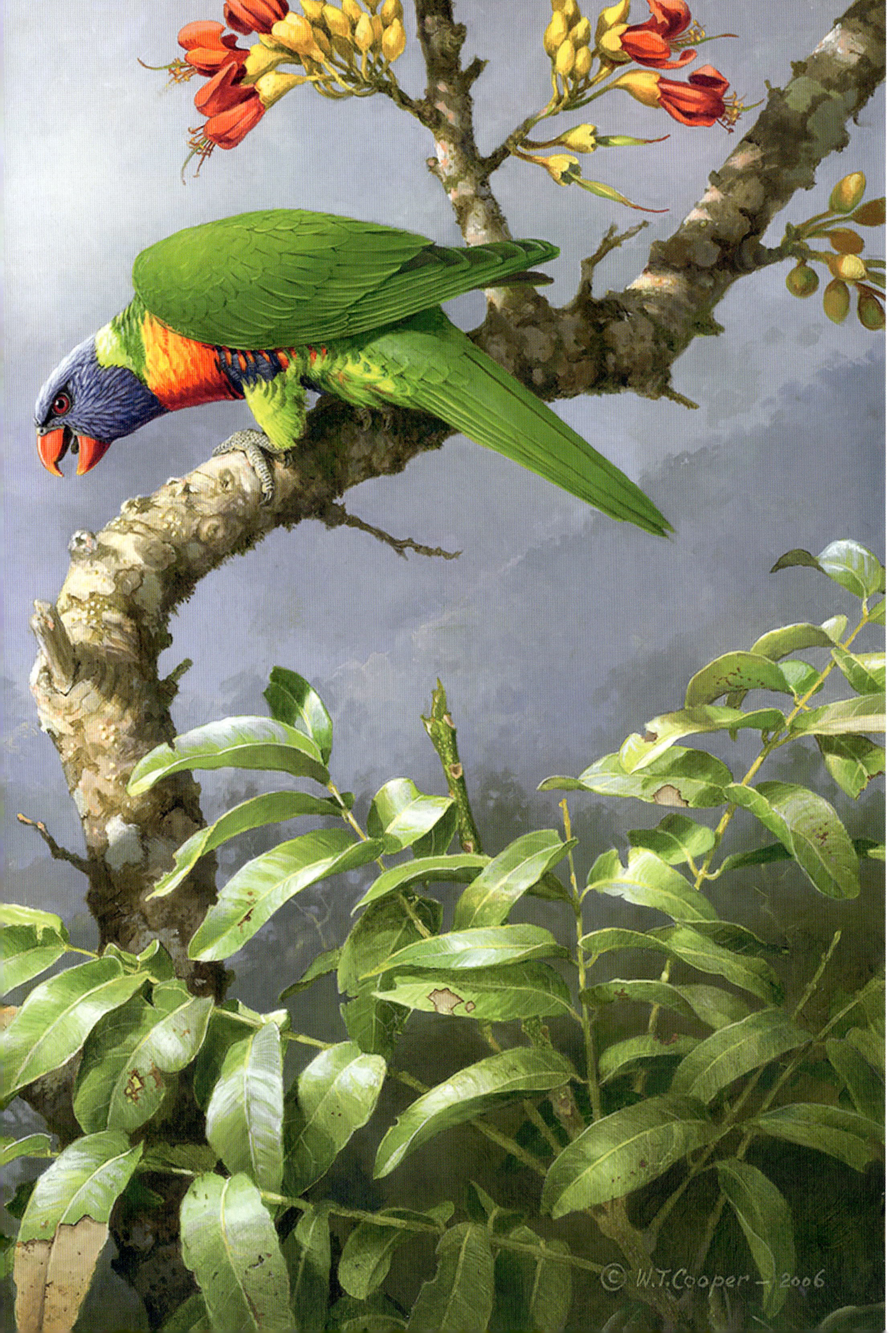
© W.T.Cooper – 2006

Commersonia bartramia

Brown Kurrajong

MALVACEAE

→ Black Hornbill, colour plate from *Kingfishers and Related Birds*, 1992
National Library of Australia, 2494045

↓ 1992; pen, pencil and watercolour
460 x 500mm
State Library of New South Wales

Commersonia is named in honour of Philibert Commerson (1727–1773), French botanist and naturalist.

bartramia is after John Bartram (1699–1777), American botanist, horticulturist and explorer.

Brown Kurrajong is mostly a regrowth or rainforest-edge tree, occurring from Cape York to New South Wales, as well as in New Guinea and South-East Asia. It has velvety pink growth, which gave a nice touch of soft colour to the plate of the Black Hornbill.

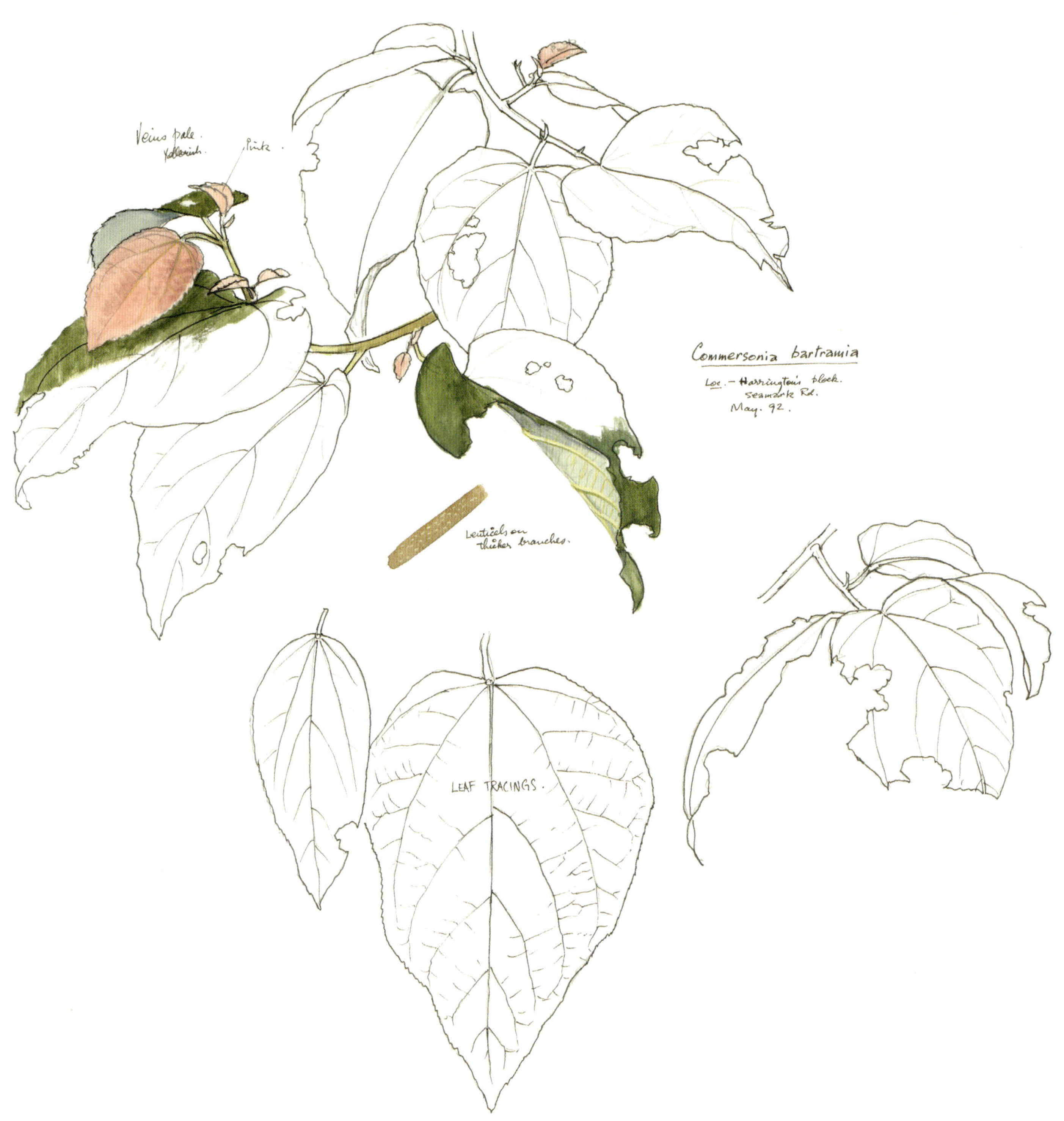

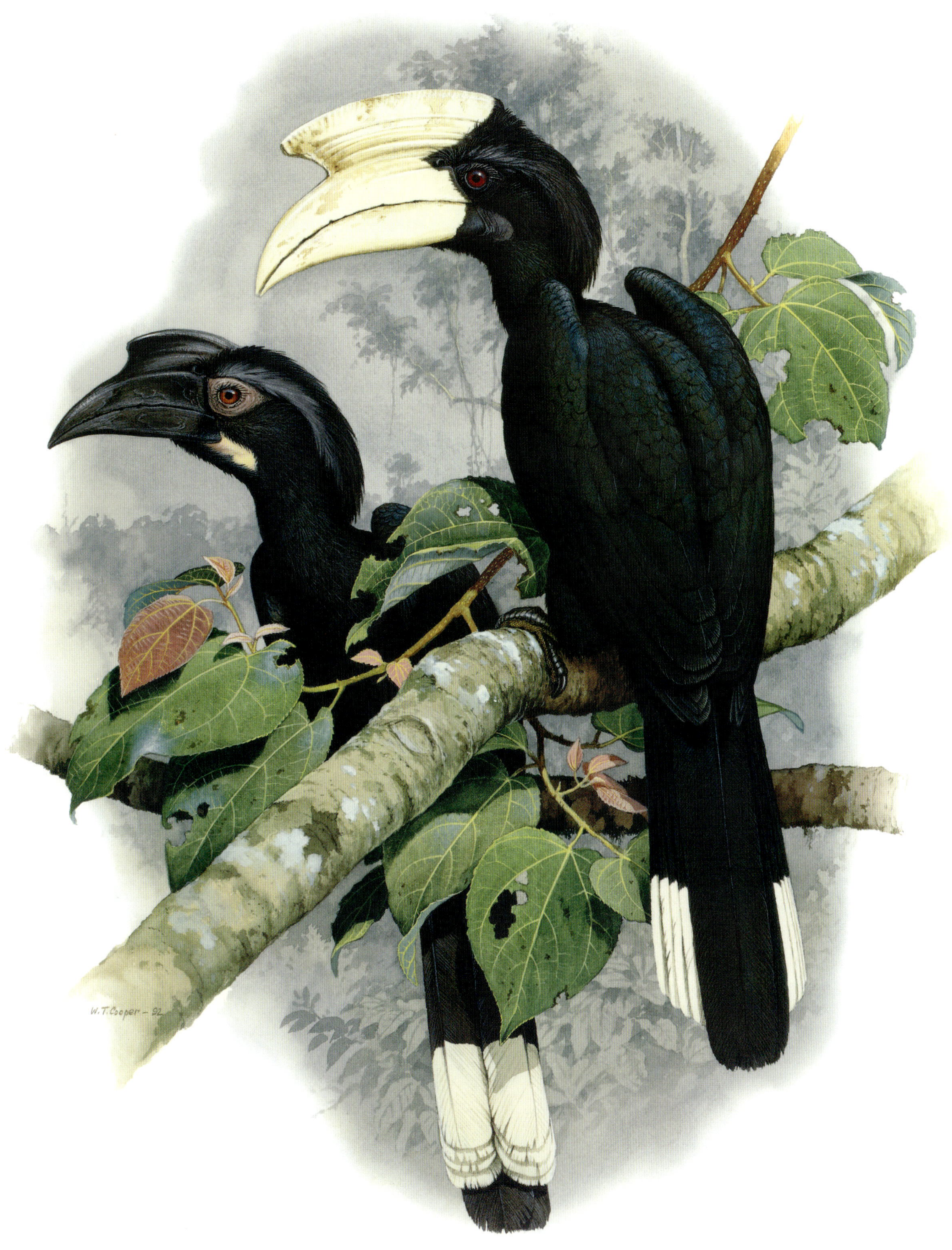
W.T.Cooper - 92

Cordia subcordata

Sea Trumpet

BORAGINACEAE

Cordia is named in honour of Euricius Cordus (1486–1535), German botanist and professor of medicine.

subcordata is from *sub-* (almost or not completely) and *cordatus* (heart-shaped), referring to the leaves.

Sea Trumpet is a small tree of monsoon and littoral rainforest, occurring from the Pacific through South-East Asia to India and East Africa.

Forever on the lookout for plants to be used as backgrounds in bird paintings, we spotted this street tree in the business area of Cairns. A small flowering specimen was surreptitiously taken for use in the painting of the Tarictic Hornbill. The bird and plant both occur in the Philippines.

→ Tarictic Hornbill, colour plate from *Kingfishers and Related Birds*, 1994
National Library of Australia, 2494045

↓ undated; pen, pencil and watercolour
410 x 570mm
State Library of New South Wales

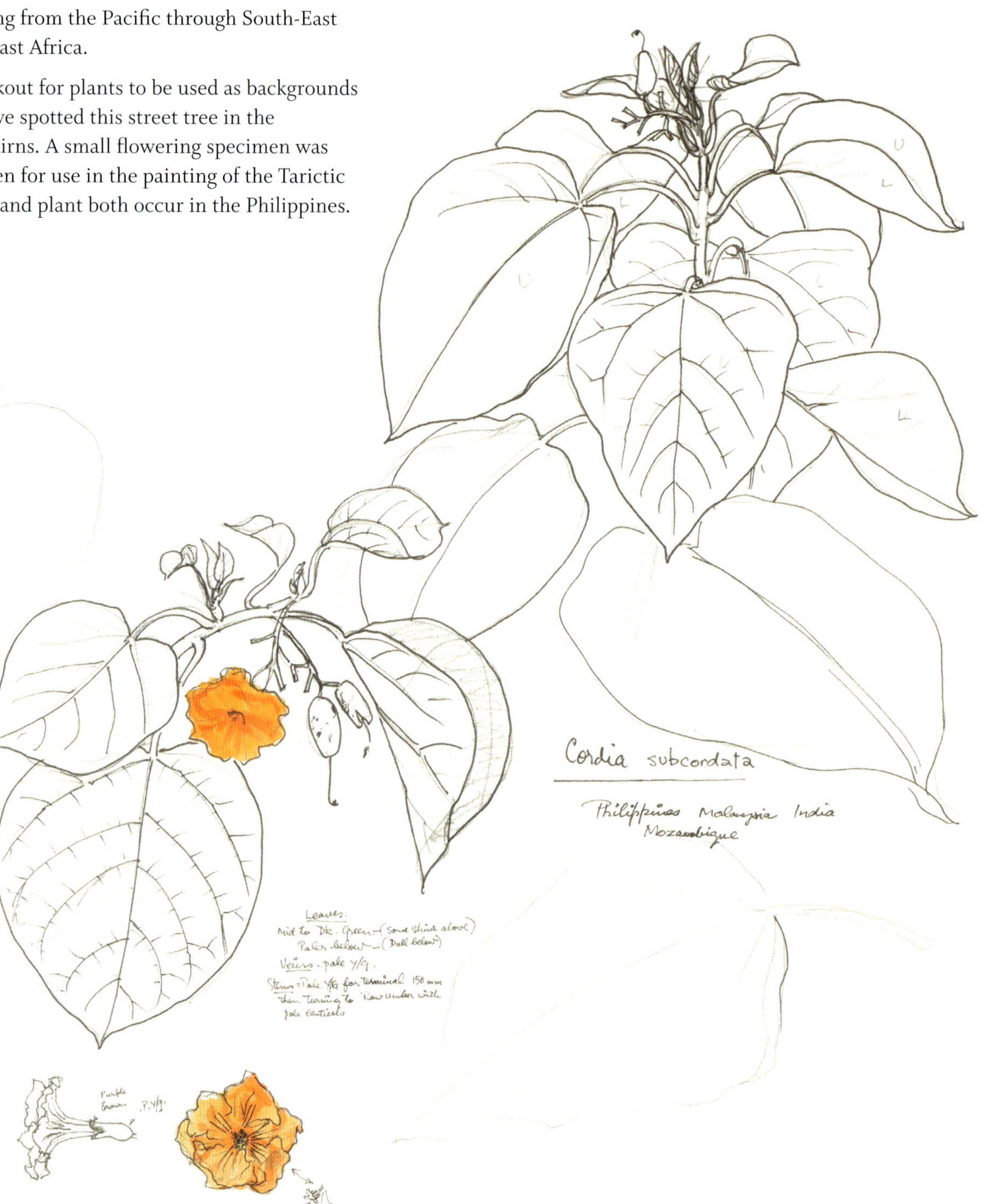

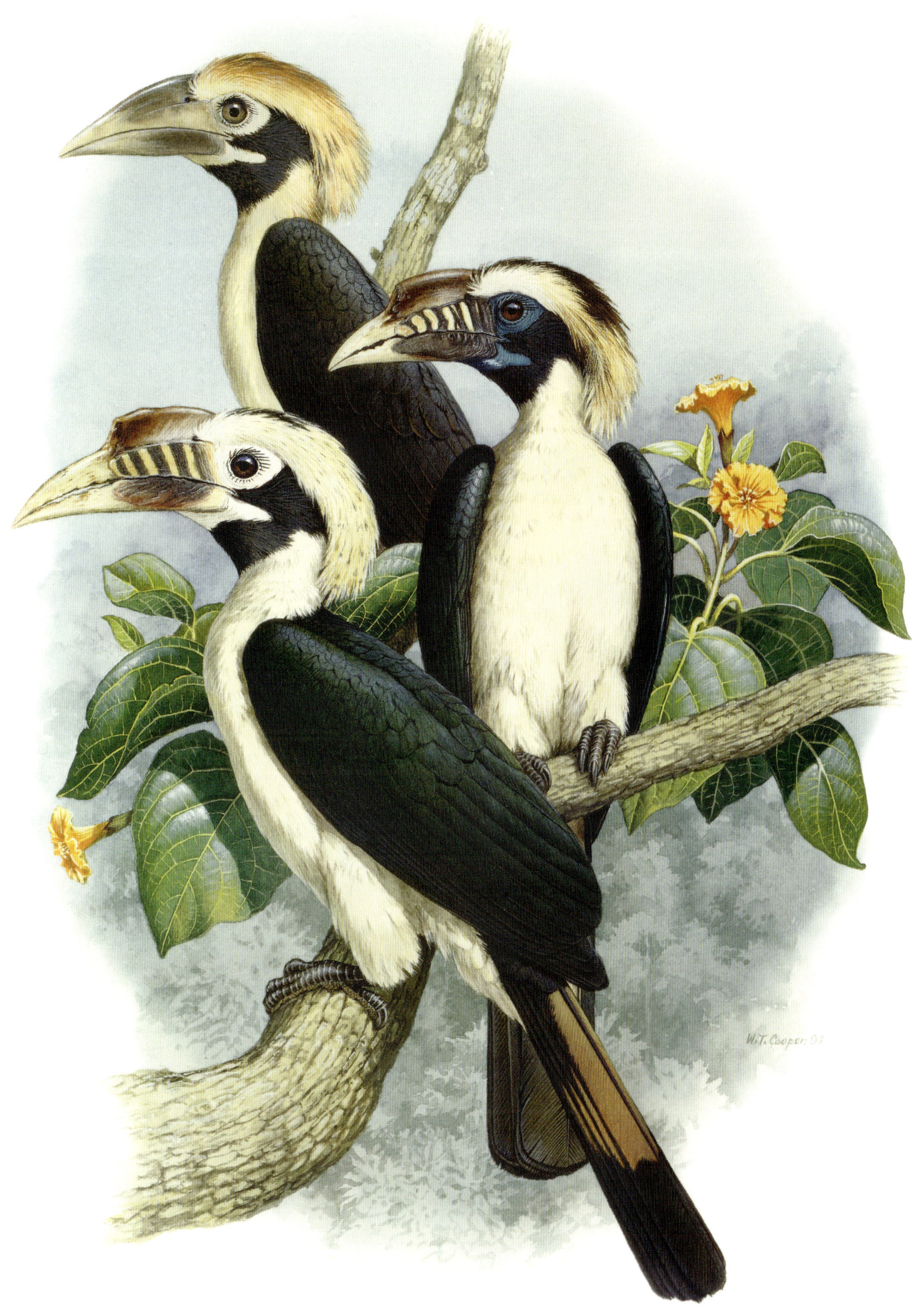
W.T. Cooper

Cordyline stricta

Narrow-leaved Palm Lily

ASPARAGACEAE

Cordyline is from *kordyle* (club), referring to either the club-like roots or stems in some species.

stricta is from *strictus* (tight, straight or upright), either referring to the leaves or the plant's habit.

Narrow-leaved Palm Lily is not a palm but it is related to lilies. The plant is mostly single-stemmed and up to 5 metres tall. A dense cluster of strap-shaped leaves crowns the top of the scarred trunk (scars are left by old leaves).

The plant grew in the wet sclerophyll forest and rainforest at our home at Bungwahl. It was common and fruited regularly but this was the first time that we observed King Parrots feeding on the juicy black berries. The parrots appeared to be extracting the seeds and dropping the flesh. Bill, the avid naturalist, was always keen to note animal behaviour previously unknown to him. This reference sketch was never used in a painting.

→ 1982; pen, pencil and watercolour
400 x 320mm
State Library of New South Wales

↓ 1970; pen and watercolour
310 x 290mm
State Library of New South Wales

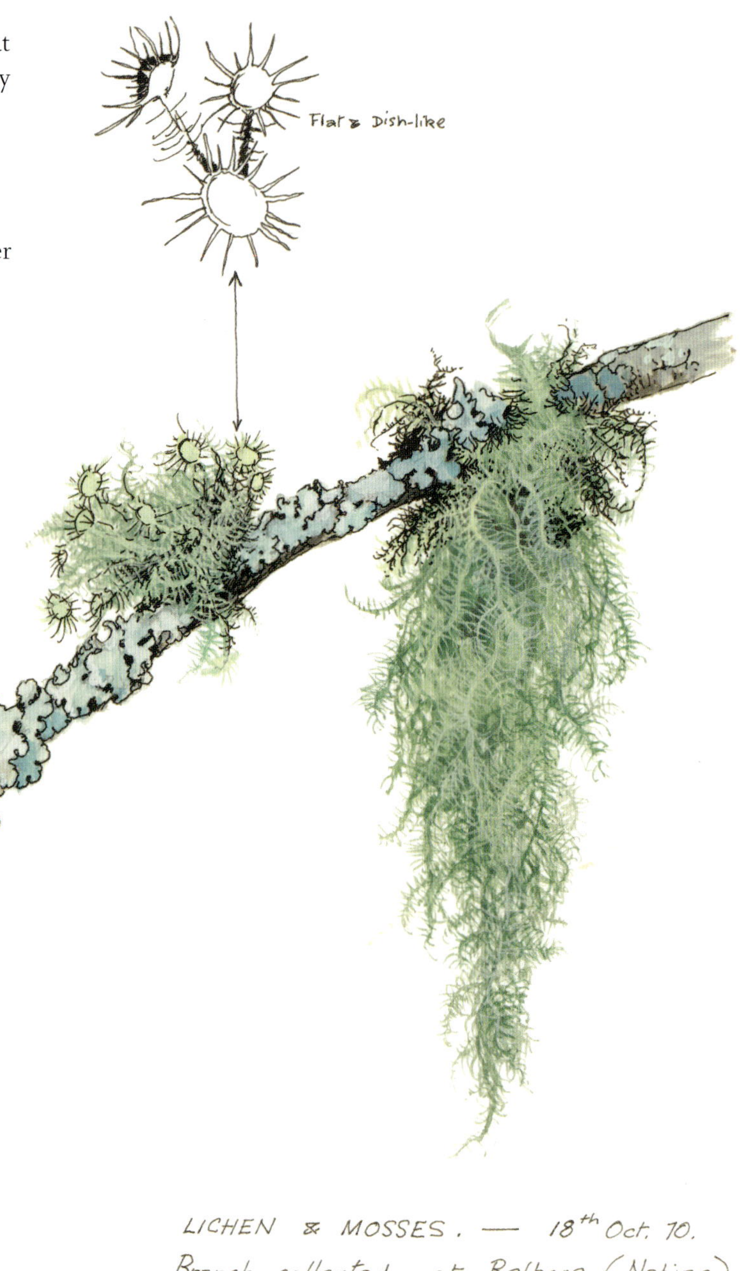

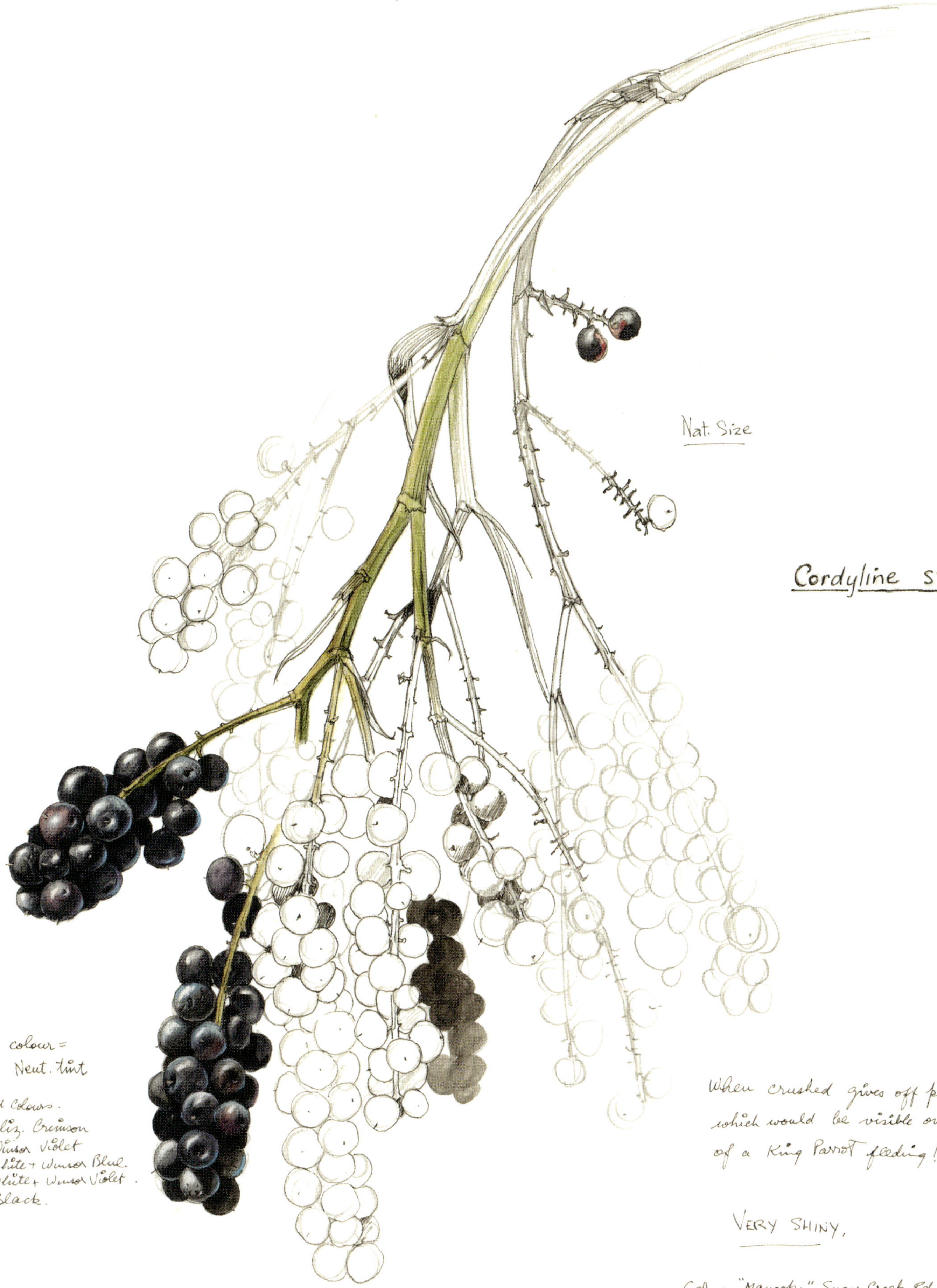

Nat. Size

Cordyline stricta

"Base colour =
W&N. Neut. tint

added colours.
Aliz. Crimson
Winsor Violet
White + Winsor Blue.
White + Winsor Violet.
Black.

When crushed gives off purple ink which would be visible on the bill of a King Parrot feeding!

VERY SHINY.

Coll. – "Manooka" Sugar Creek Rd.
March 1982.
Fruit almost all finished – had trouble finding this specimen.

Cupaniopsis anacardioides

Beach Tamarind or Tuckeroo

SAPINDACEAE

Cupaniopsis is after *Cupania* (a related South American genus) and *-opsis* (resemblance).

anacardioides is after the genus *Anacardium* and *-oides* (resembling).

Tuckeroo is a small tree of littoral rainforest, monsoon forests and vine thickets, occurring from the Torres Strait Islands to Illawarra in New South Wales. The fleshy red arils that surround the seeds are a favourite food of many birds.

This painting was done from plants at Treachery Head near Seal Rocks, in New South Wales, and the sketch below shows the rainforest canopy at Seal Rocks, near where the Tuckeroo opposite was collected. These beaches were the closest to our Bungwahl home and we often spent time there in the littoral rainforest. It harboured different plants and birds from our own place just a few kilometres away and provided an excuse to visit the headlands and beaches. Bill was a keen rock fisherman, so we often needed to check whether the seas would allow access to the best places, which were often precarious ledges.

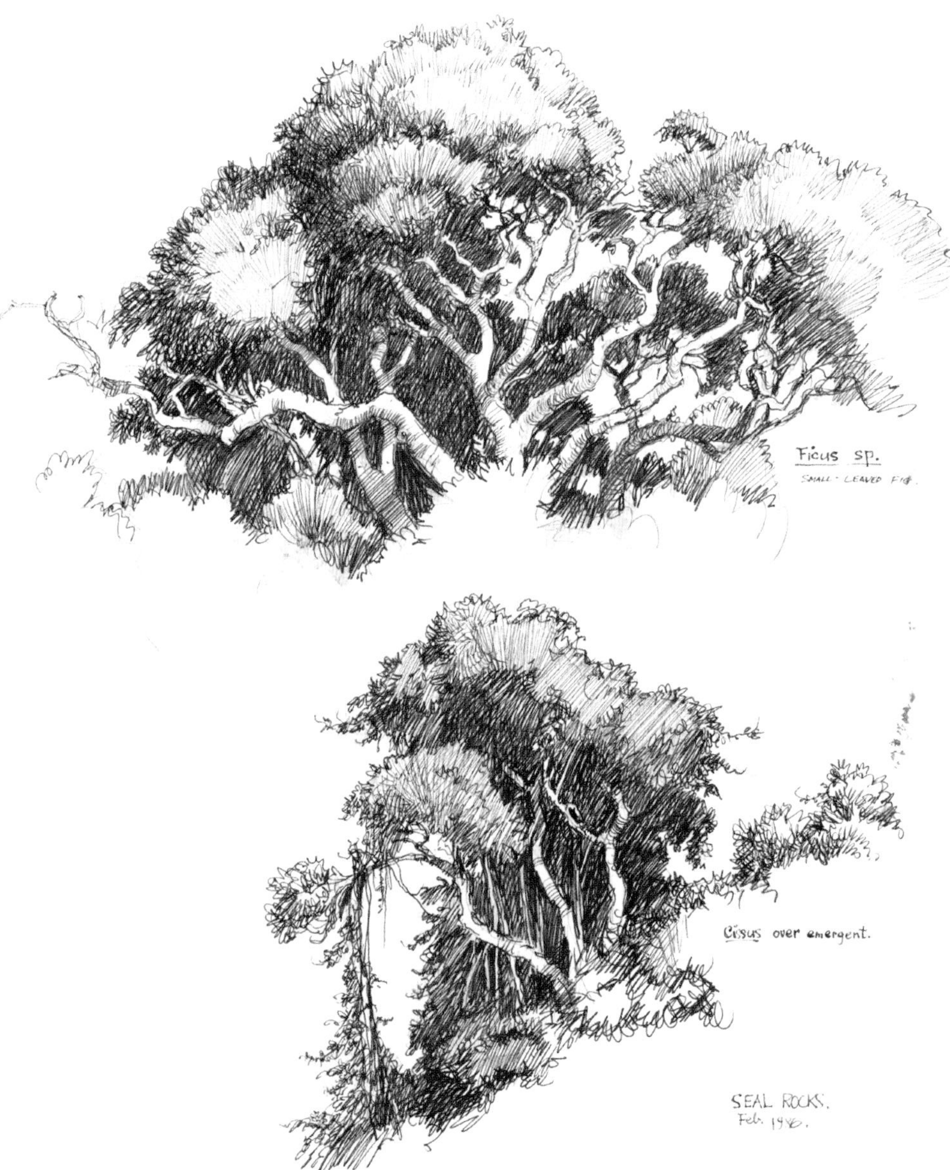

→ 1981; pen, pencil and watercolour
500 x 365mm
State Library of New South Wales

← 1986; pen and pencil; 375 x 325mm
State Library of New South Wales

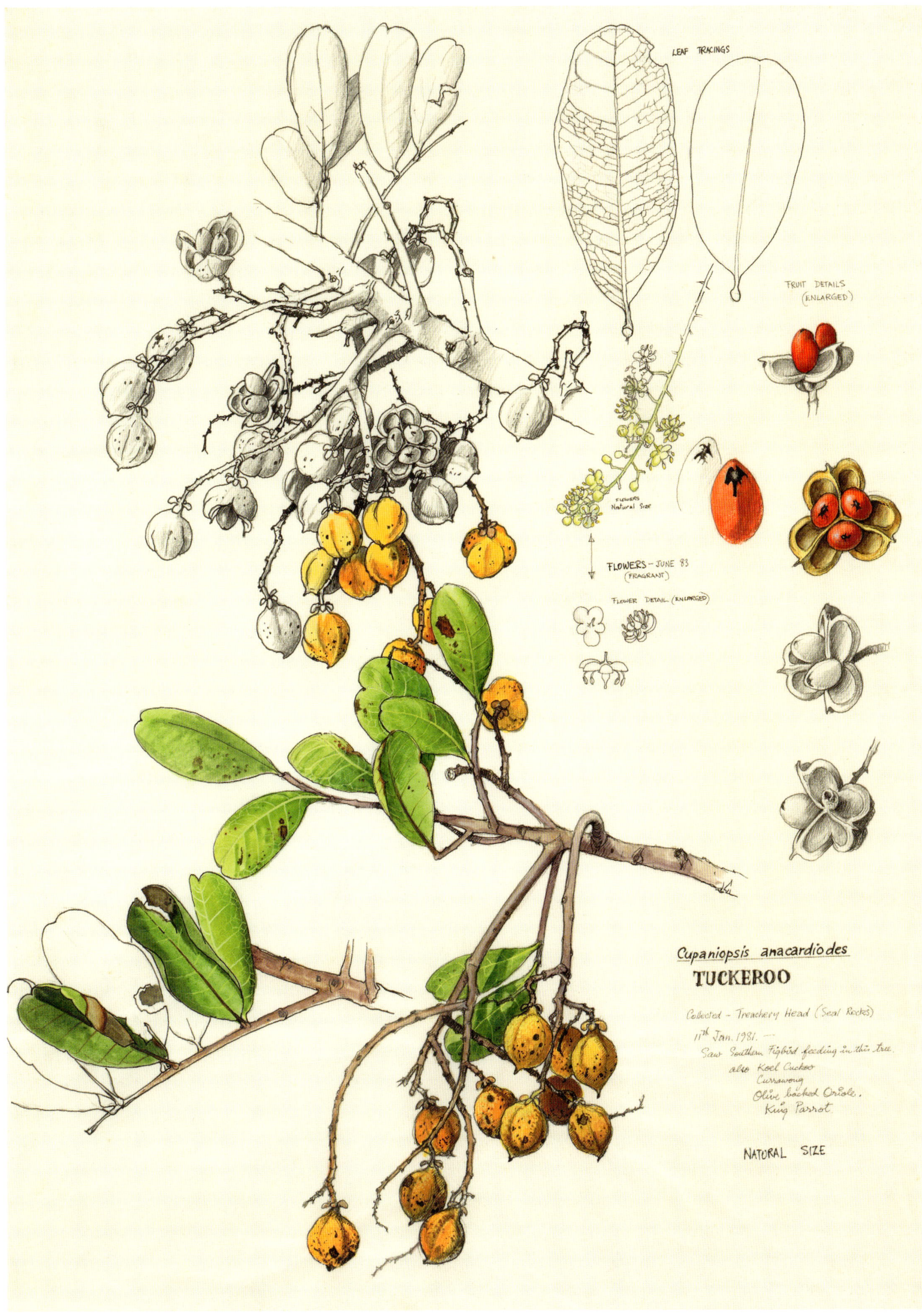
LEAF TRACINGS
FRUIT DETAILS (ENLARGED)
FLOWERS Natural Size
FLOWERS – JUNE 83 (FRAGRANT)
FLOWER DETAIL (ENLARGED)
Cupaniopsis anacardiodes
TUCKEROO
Collected – Treachery Head (Seal Rocks)
11th Jan. 1981. —
Saw Southern Figbird feeding in this tree.
also Koel Cuckoo
Currawong
Olive backed Oriole.
King Parrot
NATURAL SIZE

Dysoxylum arborescens

Mossman Mahogany

MELIACEAE

Dysoxylum is from *dys-* (bad) and *-xylon* (wood), referring to the malodorous wood in some species.

arborescens means tree-like.

Mossman Mahogany is a large tree that reaches the canopy of tropical forests in Queensland, New Guinea, the Pacific and Asia.

Both of these sketches were done while we were holidaying on Dunk Island in 1983, well before we lived in tropical Queensland. Bill noted in his diary:

> *Monday 15th Aug. 83 … Fabulous lowland rainforest right up to the house—primary forest. The island is way above my expectations. There is a resort at one end and a cleared farm area but most of it is national park and apart from the walk track which encircles about ½ the island it is virtually untouched, and to a great degree (mostly) beautiful primary rainforest—beautiful little beaches with heavy rainforest actually leaning out over the sand. Junglefowl calling at night. Pittas common.*

The Mossman Mahogany was very appealing as the colourful fruit was much bigger than anything we were familiar with in the forests of New South Wales. It was collected from a freshly fallen branch near the house that we rented in the forest not far from the beach.

I remember Bill being fascinated by the draping vine growth on the forest wall (pictured below), a common sight along this stretch of coast. In later years, we realised this growth habit was created by the massive disturbance caused by cyclones prevalent in the greater Mission Beach area.

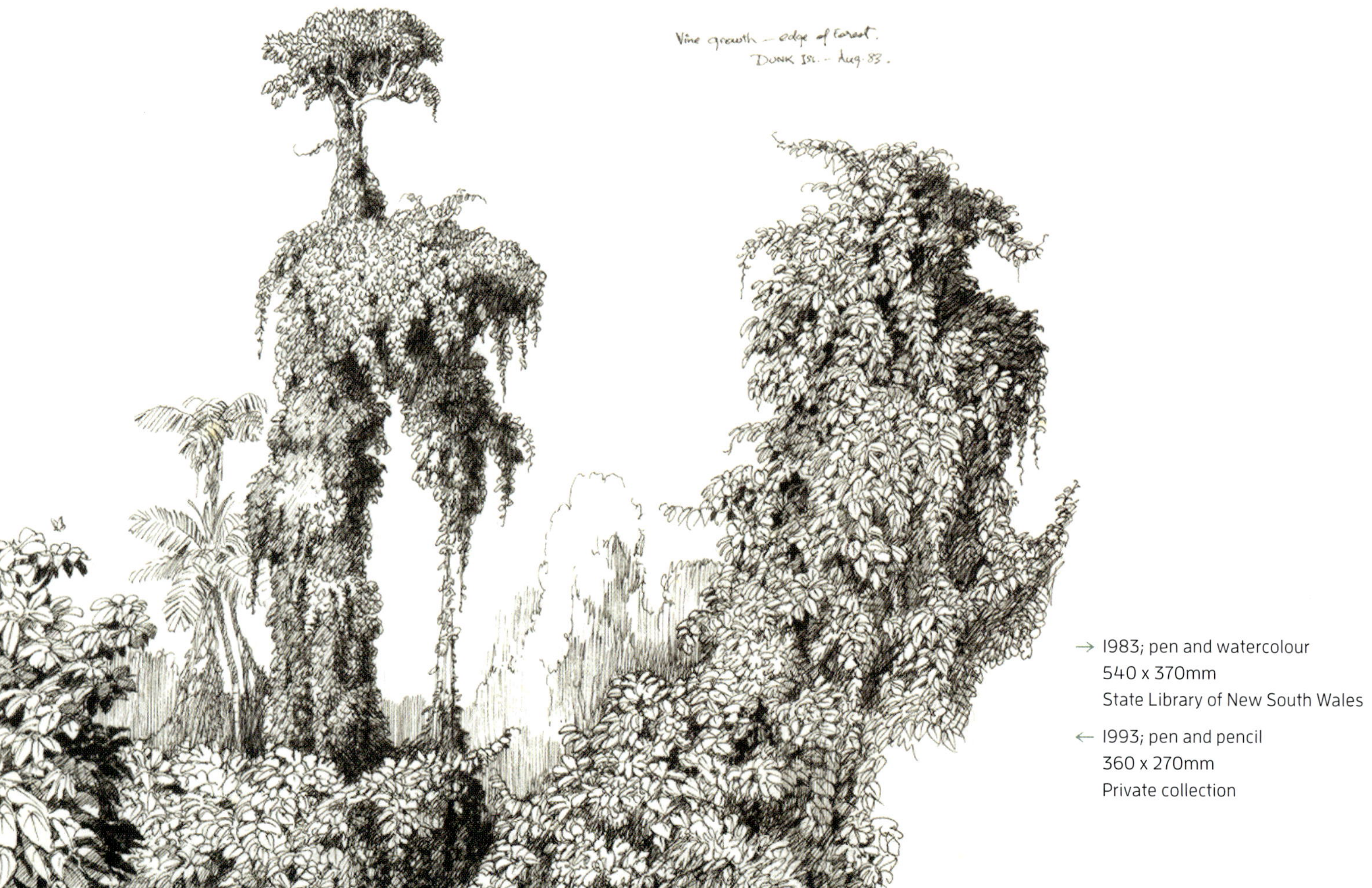

→ 1983; pen and watercolour
540 x 370mm
State Library of New South Wales

← 1993; pen and pencil
360 x 270mm
Private collection

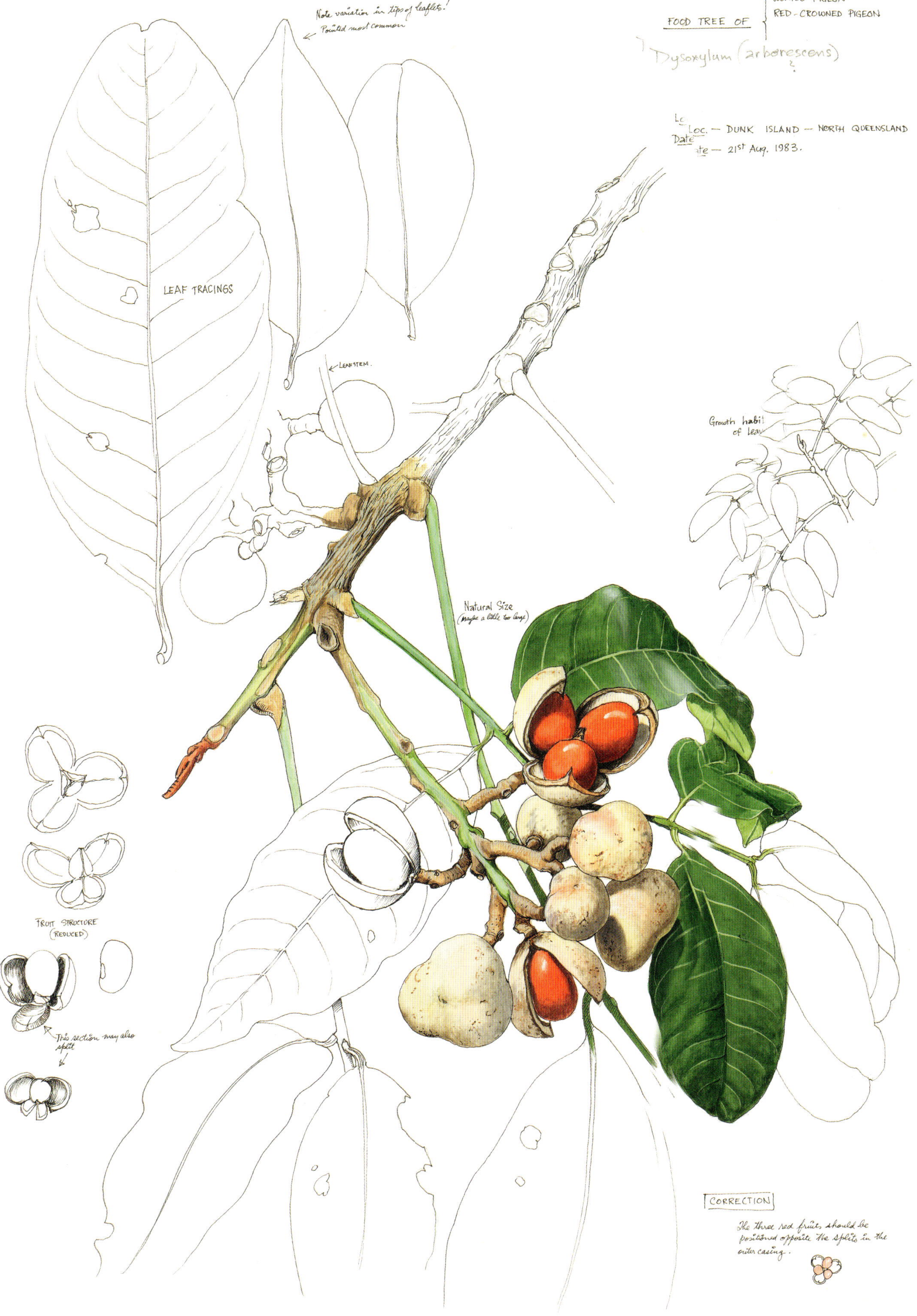
FOOD TREE OF
WOMPOO PIGEON
RED-CROWNED PIGEON
Dysoxylum (arborescens)
Loc. — DUNK ISLAND — NORTH QUEENSLAND
Date — 21st Aug. 1983.
Note variation in tips of leaflets!
Pointed most common
LEAF TRACINGS
LEAFSTEM.
Growth habit of leav
Natural Size
(maybe a little too large)
FRUIT STRUCTURE
(REDUCED)
This section may also split
CORRECTION
The three red fruits should be positioned opposite the splits in the outer casing.

Eupomatia laurina

Bolwarra

EUPOMATIACEAE

Eupomatia is from *eu-* (good) and *pomatos* (lid or cover), referring to the cap or operculum that covers the unopened flower buds and is shed as the flower unfurls.

laurina is after the genus *Laurus*, referring to the similar leaves.

Bolwarra is a shrub or small tree, which grows to about 10 metres and belongs to an archaic plant family endemic to Australia and New Guinea.

This painting depicts the waxy Bolwarra flowers being pollinated by *Elleschodes* weevils, which are attracted to the sticky, strong-smelling exudate produced by the staminodes.

Dunk Island, opposite, is one place where Bolwarra occurs, as recorded by E.J. Banfield in *The Confessions of a Beachcomber*.

→ Lowland Rainforest, Dunk Island
1983; pen; 360 x 270mm
State Library of New South Wales

↓ c.1991; watercolour; 203 x 261mm
National Library of Australia, 8069081

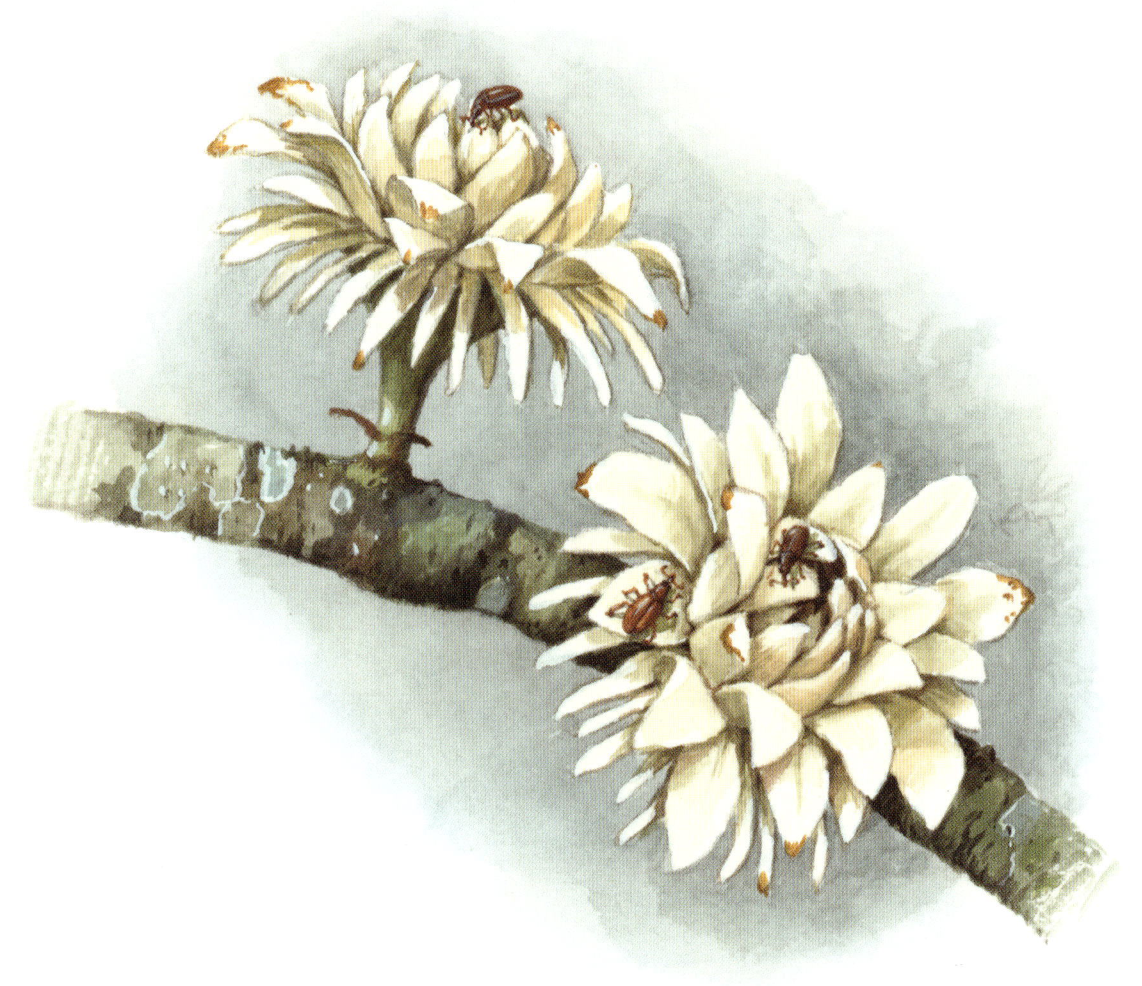

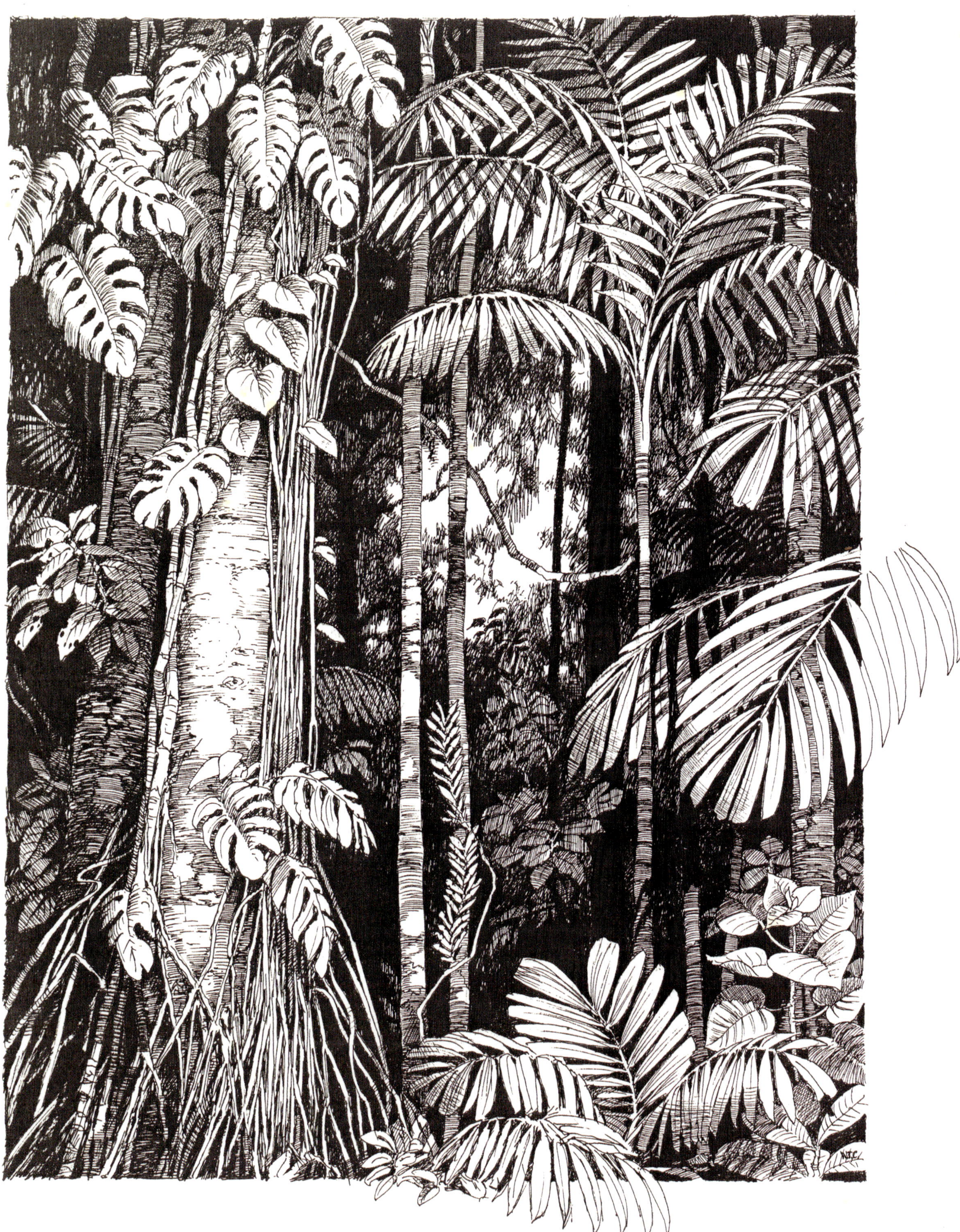

Ficus

Fig

MORACEAE

Ficus is from the Greek name for the edible fig *Ficus carica*.

There are about 1,000 *Ficus* species throughout the world, with about 41 species in Australia. Fig trees are enormously important for fruit-eating birds and flying-foxes. With their adventitious roots, bark texture, fruit colour and leaves, they are an interesting botanical part of various bird paintings.

Most, if not all, fig species have a particular wasp that fertilises the flowers. The flowers emerge on the inner wall of unripe figs and are the precursors of what are commonly thought of as fig seeds.

→ 1981; pen; 360 x 260mm
State Library of New South Wales

↓ 1981; pen and pencil; 370 x 520mm
State Library of New South Wales

Ficus copiosa

Plentiful Fig

copiosa is from *copiosus* (plentiful), referring to the abundant fruit.

Plentiful Fig is a small rainforest tree that grows to 10 metres, with figs mostly clustered on the trunk and larger branches. This particular specimen was appealing to Bill for its large, tightly packed leaves. That it was also destined to be bird food made it a worthwhile drawing for reference material.

The black and white drawings of a strangler fig and a *Polyscias* tree were done at Iron Range close to where the Plentiful Fig specimen was found. Bill wrote in his diary on 16 July 1981: 'I went for a walk in the morning and did some drawings of a strangler Fig—epiphytes—and *Polyscias* trees. Wendy organising the truck for tomorrow's drive out'.

Although Bill could draw from memory (and frequently did), he always found it useful to have new sketches from the field so that he didn't repeat familiar patterns.

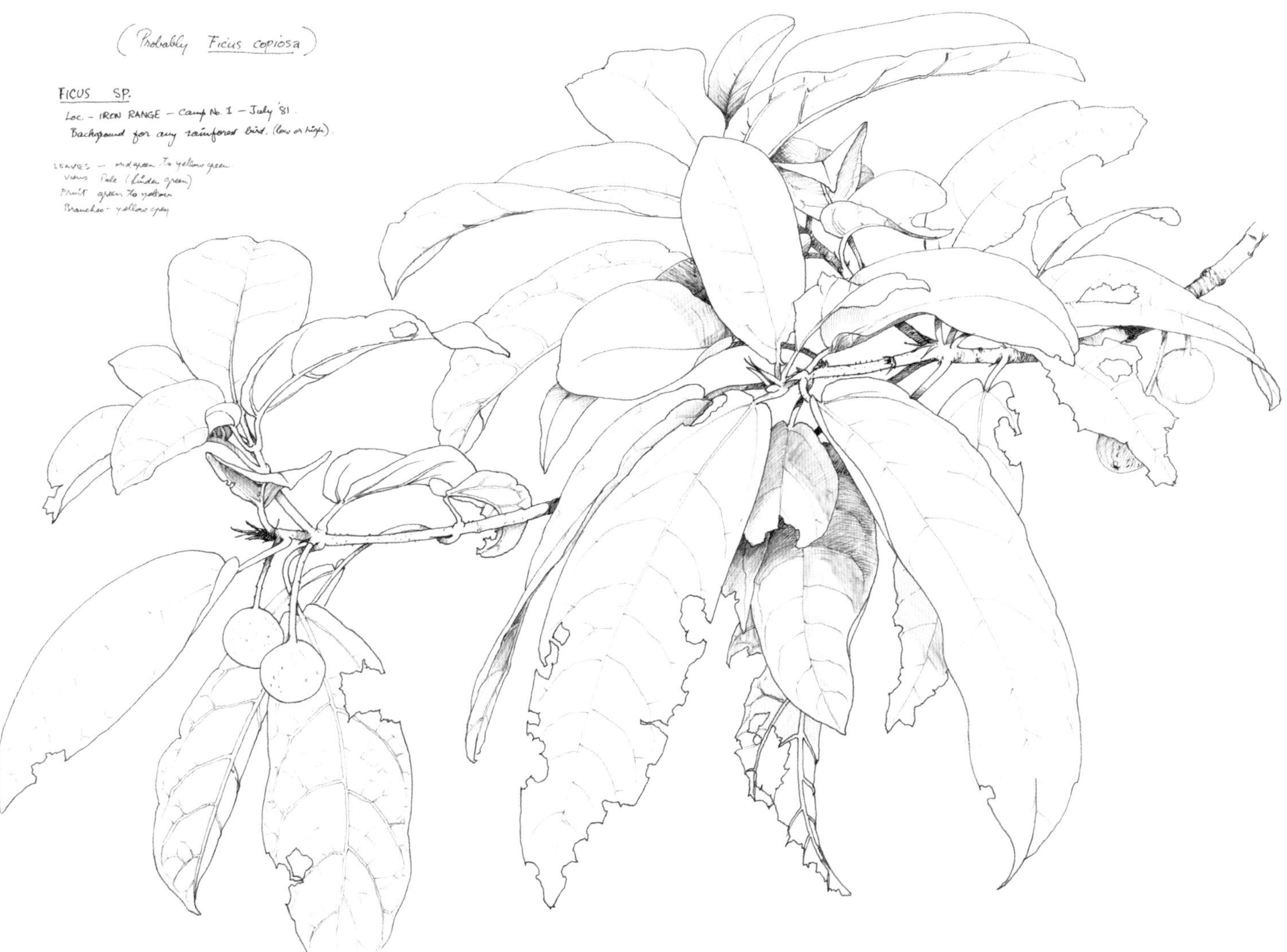

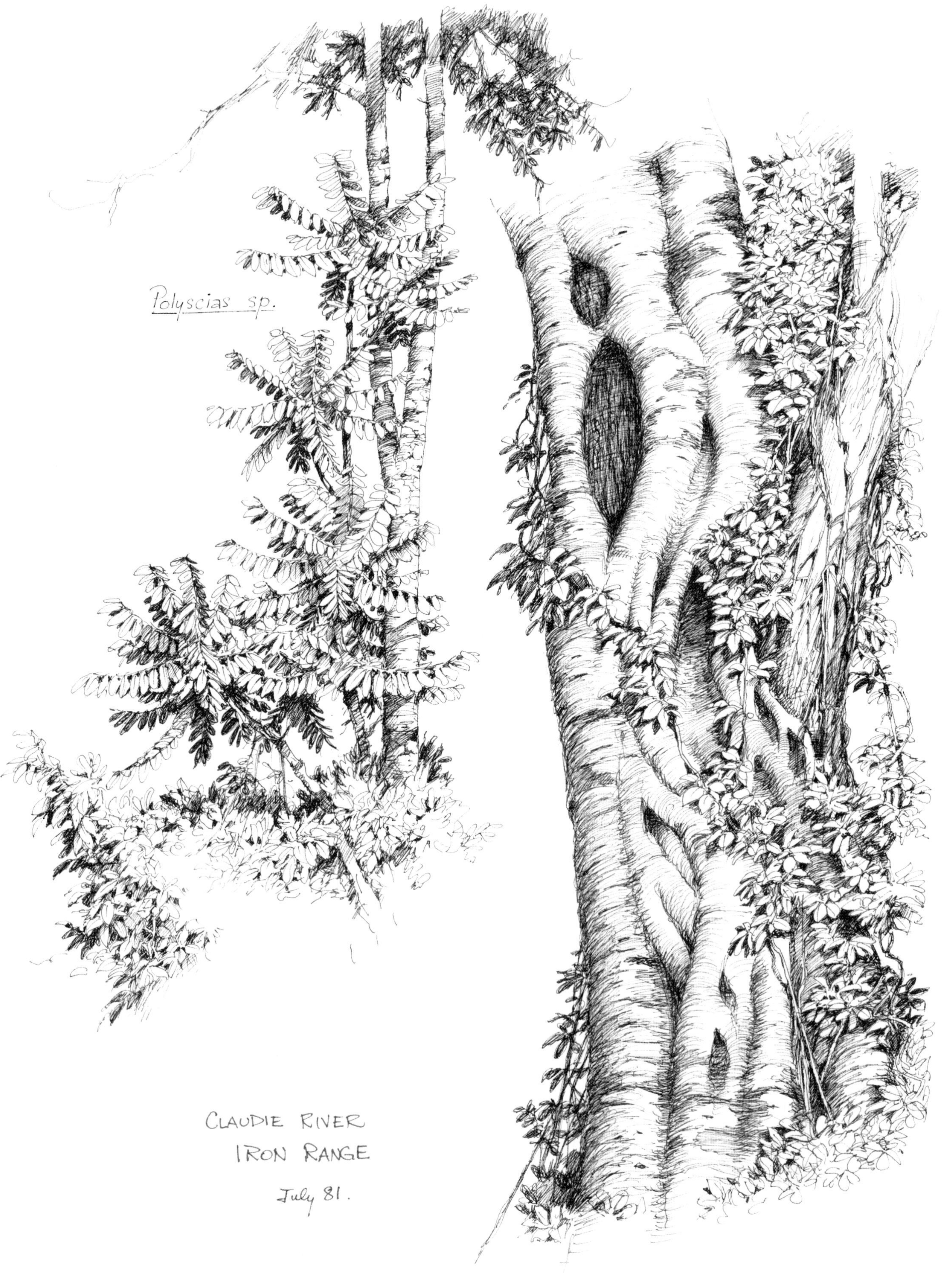
Polyscias sp.
CLAUDIE RIVER
IRON RANGE
July 81.

Ficus fraseri

Sandpaper Fig

fraseri is named in honour of Charles Fraser (1788–1831), colonial botanist in New South Wales from 1821 to 1831. Fraser created controversy when he advised that the soils of the Swan River area were fertile, when in fact they are low-nutrient sand.

Sandpaper Fig is found along the Australian east coast north from Sydney to the Torres Strait Islands. It is also in the Northern Territory and the Pacific Islands. It is sometimes deciduous and can grow to the height of the forest canopy.

Its variously coloured fruits and bunched leaves make it a perfect plant for a bird painting. The sectioned 'fruit' in the painting shows the fleshy receptacle, or fig, which envelops the tiny flowers and seeds (technically the fruit). Bill noted that Currawongs, Figbirds and Orioles were feeding on this tree at Seal Rocks—just some of the many bird and bat species that depend on this fruit.

→ 1981; pen, pencil and watercolour; 370 x 295mm
State Library of New South Wales

↓ 2003; pen and watercolour; 200 x 325mm
State Library of New South Wales

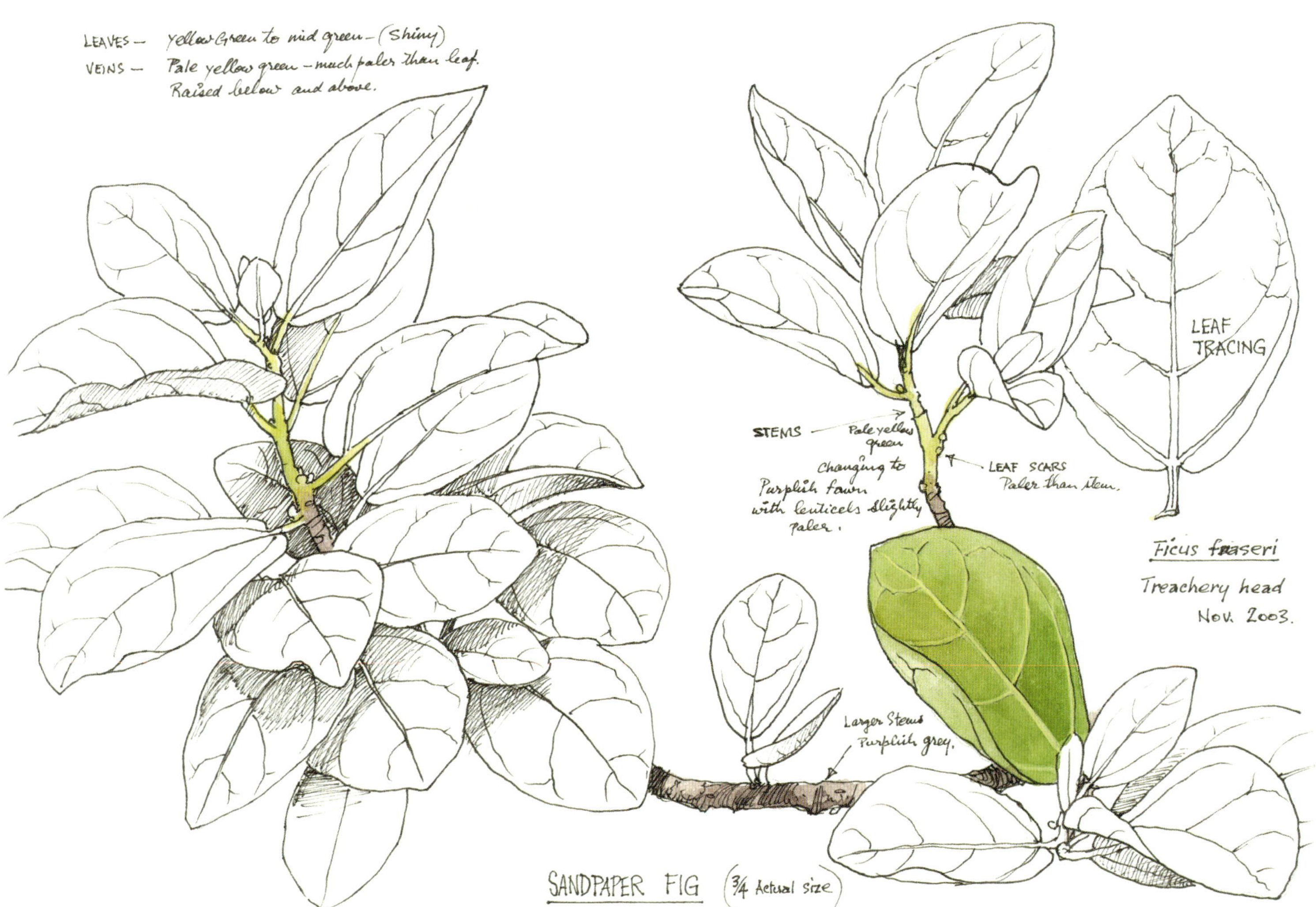

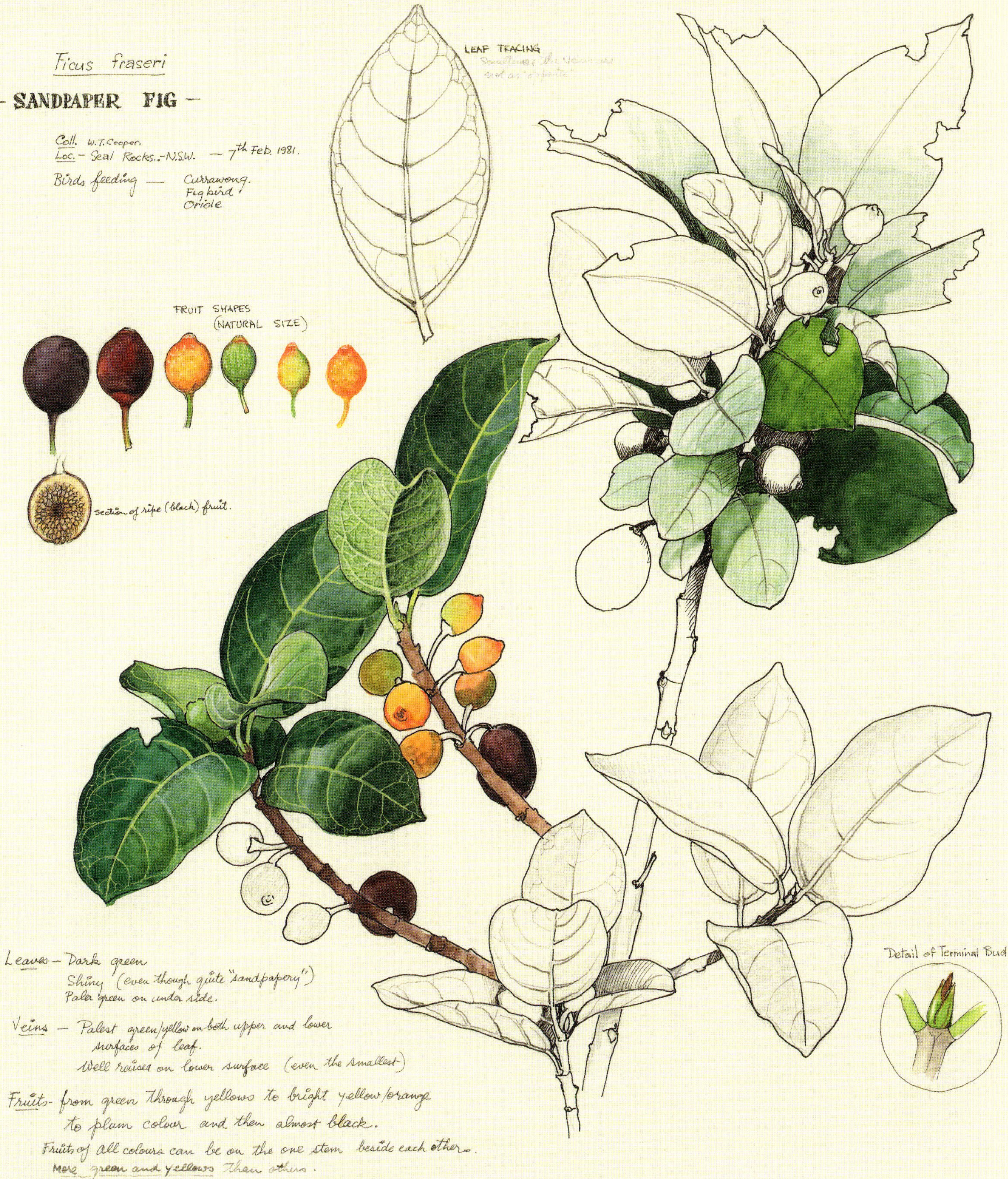
Ficus fraseri
– SANDPAPER FIG –
Coll. W.T. Cooper.
Loc. – Seal Rocks. – N.S.W. – 7th Feb. 1981.
Birds feeding – Currawong. Figbird Oriole
LEAF TRACING
Sometimes the veins are not as "opposite"
FRUIT SHAPES (NATURAL SIZE)
section of ripe (black) fruit.
Detail of Terminal Bud
Leaves – Dark green
Shiny (even though quite "sandpapery")
Paler green on under side.
Veins – Palest green/yellow on both upper and lower surfaces of leaf.
Well raised on lower surface (even the smallest)
Fruits – from green through yellows to bright yellow/orange to plum colour and then almost black.
Fruits of all colours can be on the one stem beside each other.
More green and yellows than others.

Ficus pleurocarpa

Banana Fig

pleurocarpa is from *pleuro-* (ribbed) and *-carpos* (-fruited).

Banana Figs are stranglers that begin life when the seeds germinate on the branch of a suitable tree. They send roots down to the ground, eventually enclosing the host's trunk and crowding out its crown. This species is limited to the Wet Tropics.

The figs are large—up to 65 millimetres long—and keenly sought by Spectacled Flying-foxes, Green Possums and several fruit-eating bird species. Because this tree was on the edge of our garden, we frequently watched the fruit being taken.

From Bill's diary:

> *19th Nov. 1991. Wendy found a Green Possum with a small young on its back. It was walking along a horizontal limb … leading into the large Banana Fig which it climbed—sometimes on vines and sometimes on the trunk itself. When it got to the limbs it went out along them to bunches of fruit which it examined (smelled), tried another bunch and so on rejecting at least a dozen bunches before picking a fig and clambering back to a place to sit and eat it. The little one was now sitting beside it and was eating some of the fig that the mother had picked. We couldn't actually see the female eating as she had her back towards us, but when the little one ran out of fig it leaned over and got some more from the mother and sat up holding it in both front paws … It was really lovely to see, as it climbed it gave the appearance of 'snaking' its way up the tree, and when it was going out to give a bunch of figs the smell test it went right out to the ends of the branches where they are quite thin. Once the young came right off or appeared to get wiped off by a limb above. The mother made no effort to wait. The young one soon hurried up and jumped on her back again.*

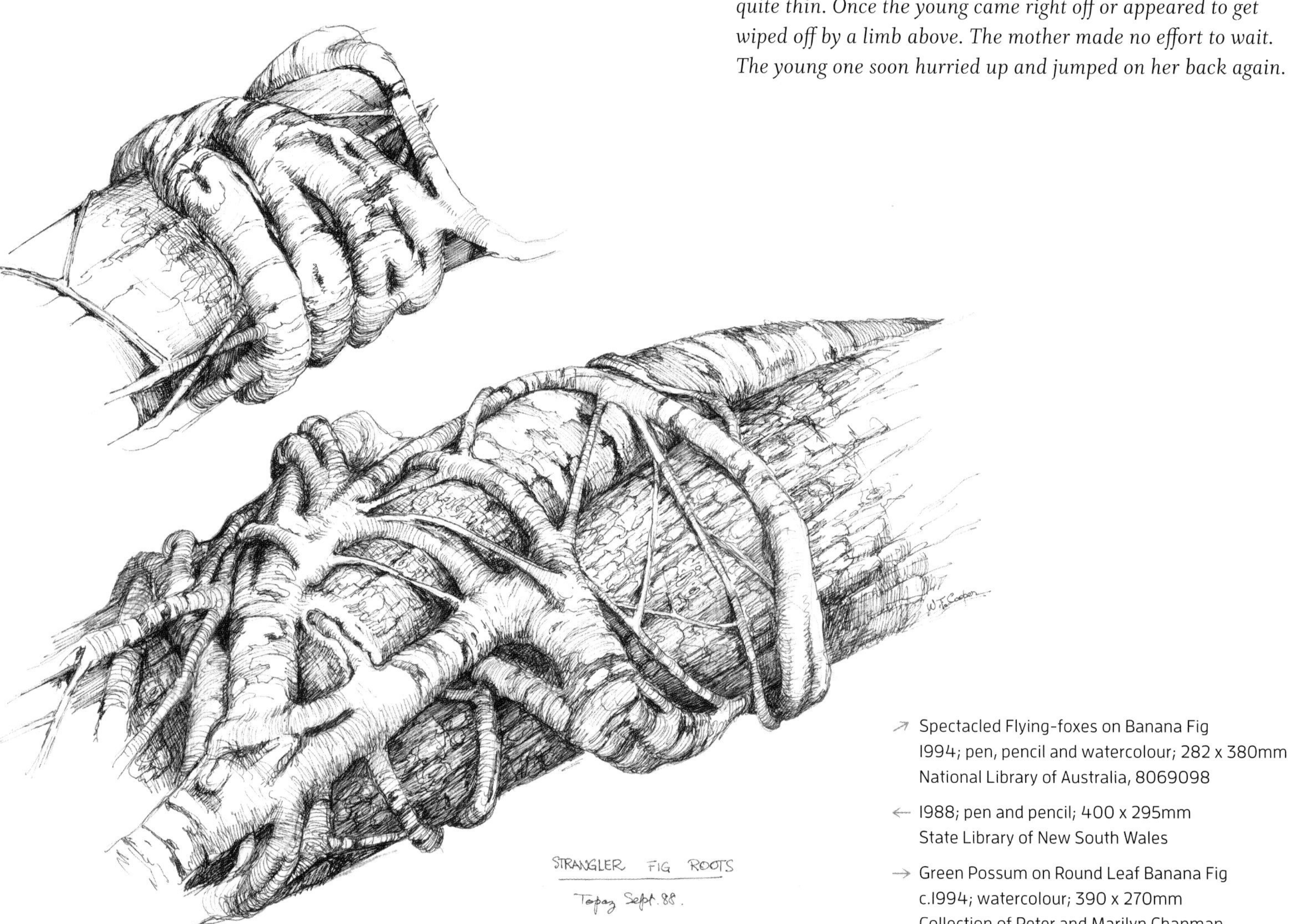

↗ Spectacled Flying-foxes on Banana Fig
1994; pen, pencil and watercolour; 282 x 380mm
National Library of Australia, 8069098

← 1988; pen and pencil; 400 x 295mm
State Library of New South Wales

→ Green Possum on Round Leaf Banana Fig
c.1994; watercolour; 390 x 270mm
Collection of Peter and Marilyn Chapman

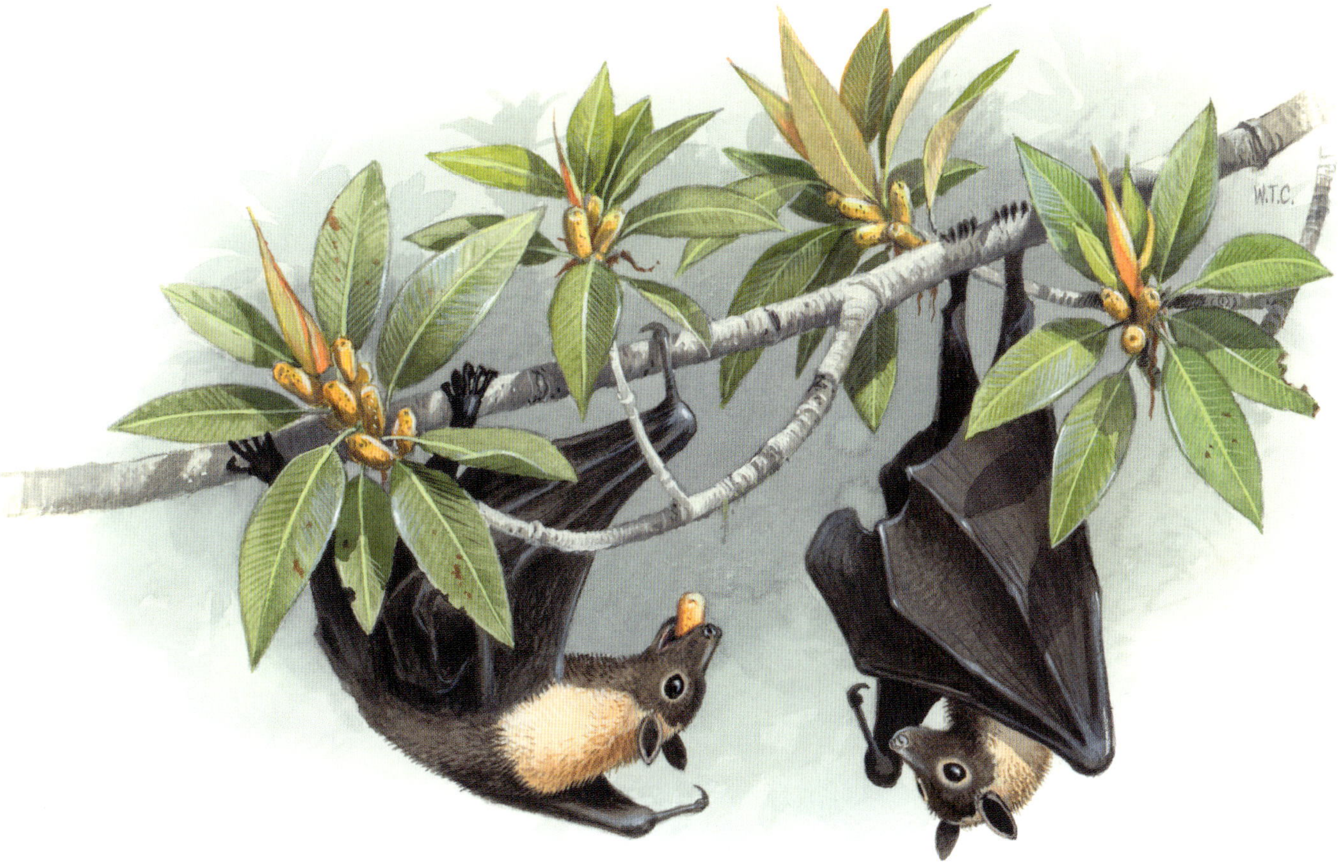

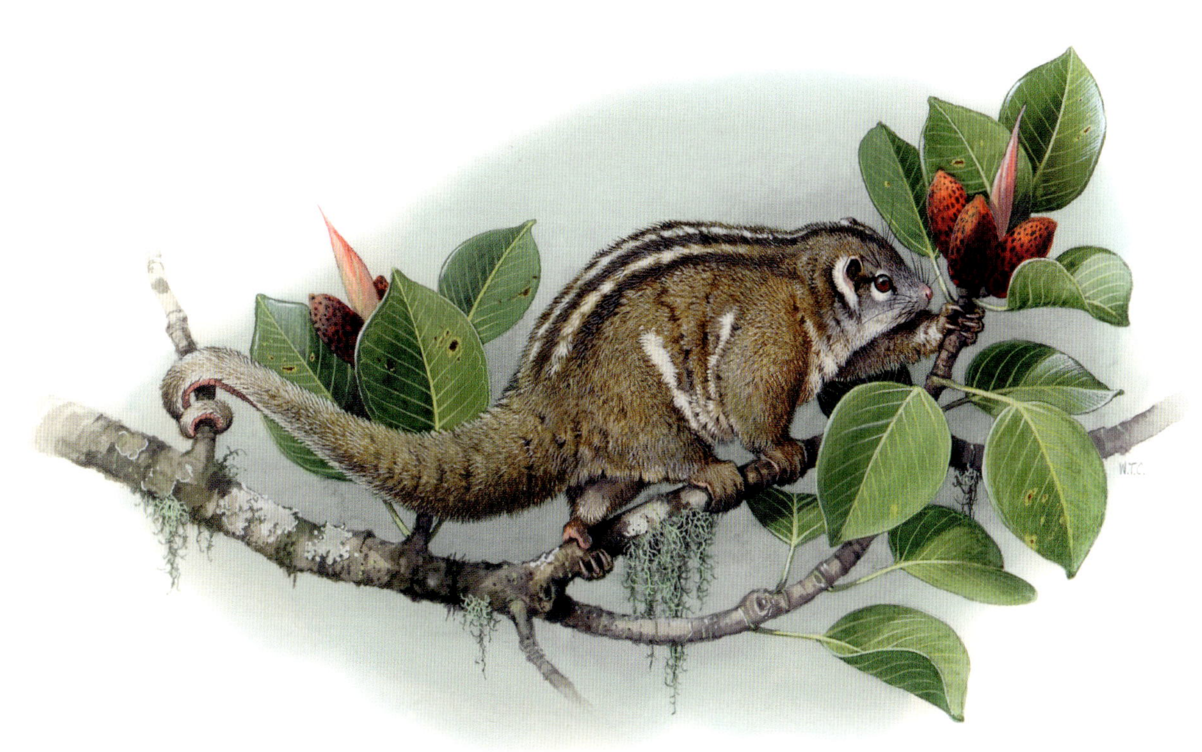
W.T.C.

Ficus variegata

Variegated Fig

variegata is from *variegatus* (variegated), referring to the fruit.

Variegated Fig is a tall buttressed fig occurring from Paluma north to Cape York, Asia and the Pacific. The figs, which mostly grow in clusters on the main branches and trunk, are an important food for Spectacled Flying-foxes, Cassowaries and Double-eyed Fig-parrots. This painting was especially created for use in *Fruits of the Rain Forest* (1994) as a double-page spread to add variety to the book's design.

→ Double-eyed Fig-parrots on Variegated Fig
1994; watercolour; 320 x 420mm
Collection of Peter and Marilyn Chapman

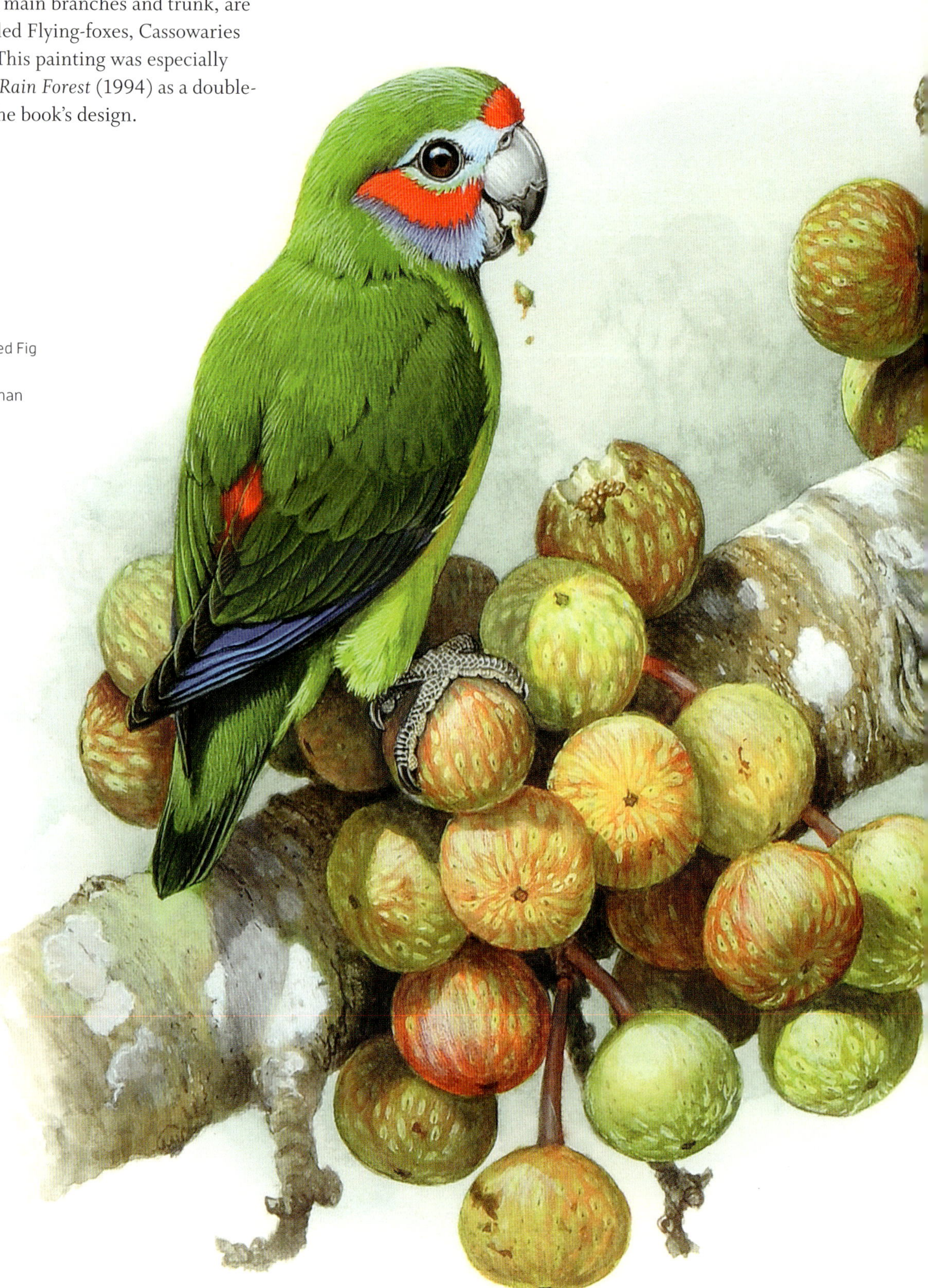

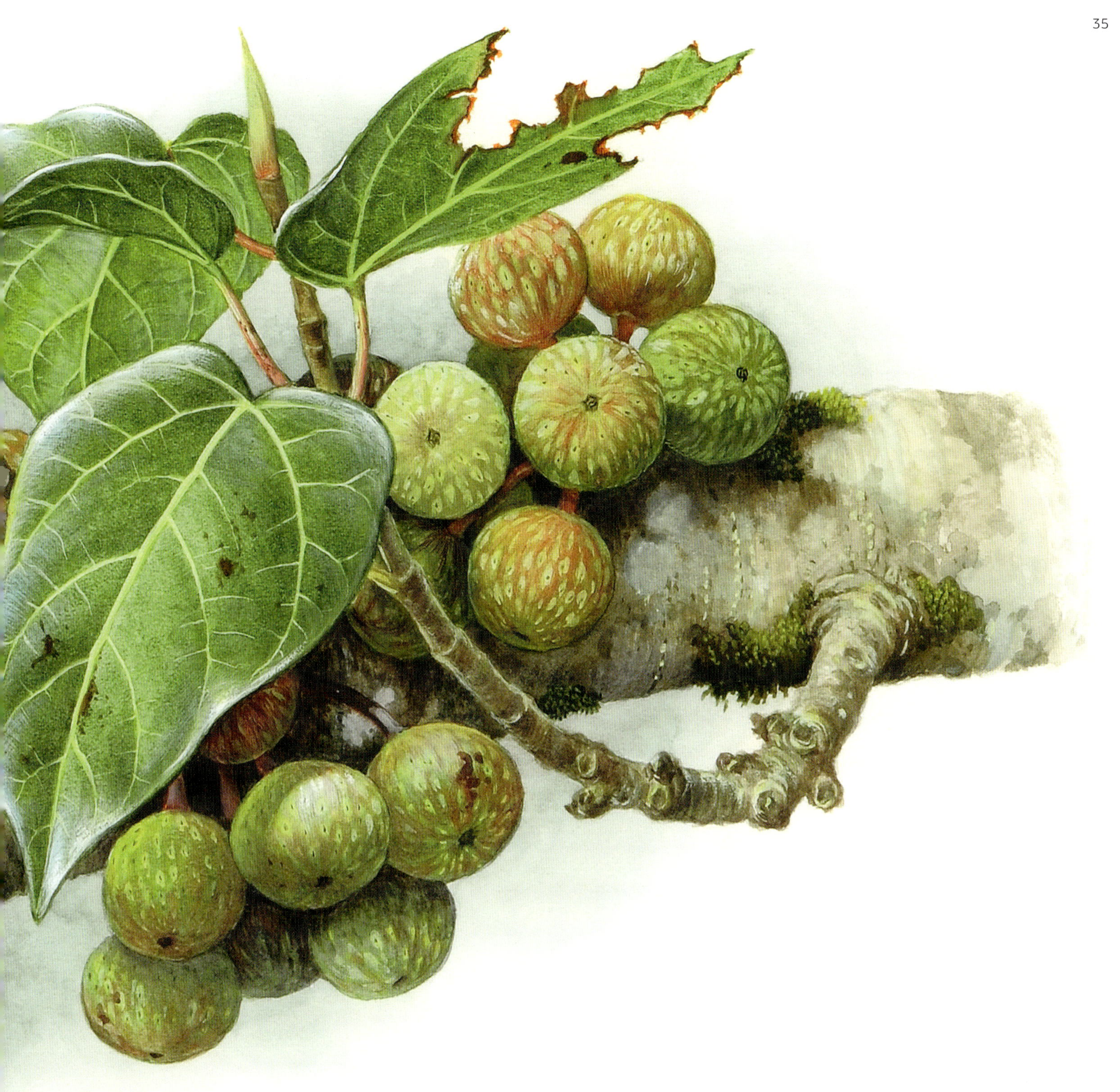

RED - BROWED FIG PARROT
Cyclopsitta diopthalma macleayana
on
VARIEGATED FIG
Ficus variegata

Ficus virens

White Fig or Green Fig

→ 2004; pen and pencil; 420 x 290mm
State Library of New South Wales

↓ 1986; pen and pencil; 400 x 285mm
State Library of New South Wales

virens means green, referring to the leaves.

White Fig is mostly a tree of deciduous rainforest, often attached to rock faces in sheltered gorges. Both of these drawings were done at Kakadu in the Northern Territory, as Bill noted in his diary:

> *Wednesday 30th April. Wendy and I up at daybreak, went back to rocks we visited yesterday, took photographs, saw Rock Pigeons. Collected some Ficus virens. Did drawings back at house.*

Our Kakadu trip was full of excitement, with highlights including sandstone escarpments, vast billabongs and waterfowl, rock art and the opportunity to meet an Aboriginal artist working in the X-ray style. Bill and Bluey Ilkirr compared techniques. Bill was fascinated by Bluey's ability to do such fine and intricate painting with a simple brush on bark. Bluey gave Bill a brush cut from a grass stem that was more like a slender and flexible palette knife. It was treasured by Bill for many years.

Ficus virens

KAKADU – N.T. 29th 4-86.

Grows in sandstone crevices like Ficus platypoda

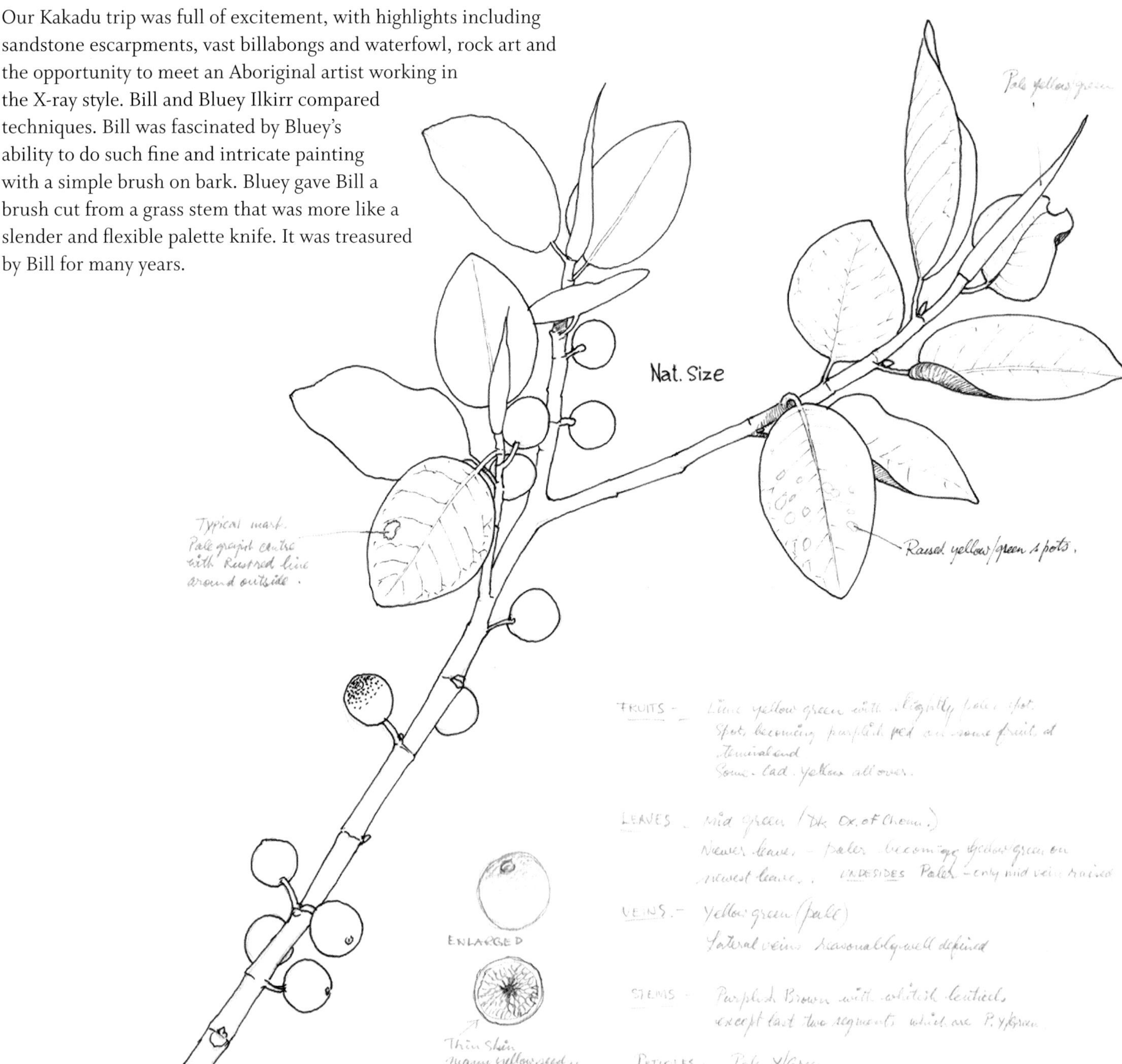

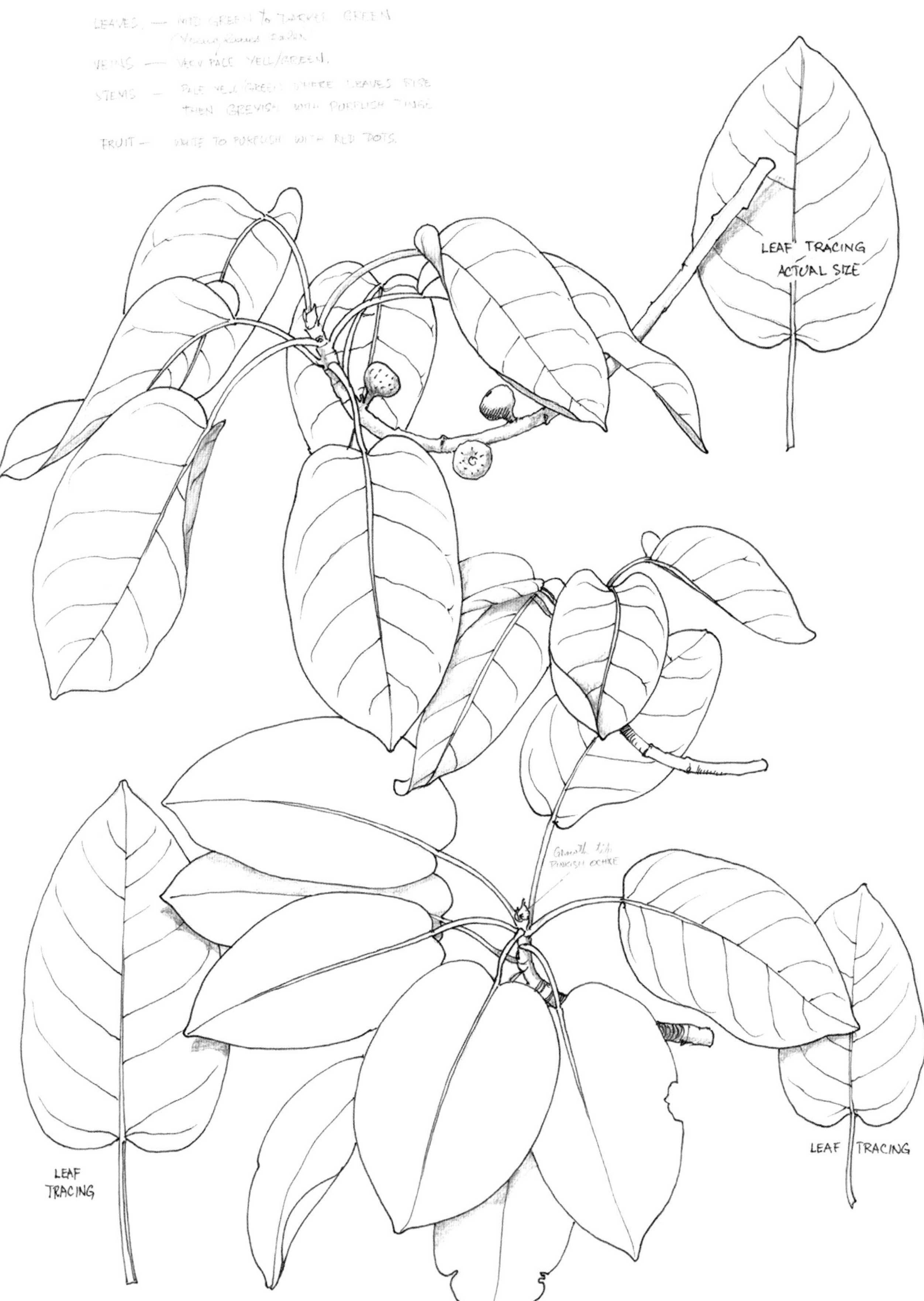
FICUS VIRENS
KAKADU N.T. – July 2004
Food plant of Banded Pigeon
LEAVES — MID GREEN to DARKER GREEN
(Young leaves paler
VEINS — VERY PALE YELL/GREEN.
STEMS — PALE YELL/GREEN WHERE LEAVES RISE
THEN GREYISH WITH PURPLISH TINGE
FRUIT — WHITE TO PURPLISH WITH RED DOTS.
LEAF TRACING
ACTUAL SIZE
Growth tip
PINKISH OCHRE
LEAF TRACING
LEAF TRACING

Ficus watkinsiana

Watkins' Fig

→ 1977; pen, pencil and watercolour
470 x 340mm
State Library of New South Wales

↓ undated (c.1978); pen and watercolour
360 x 320mm
State Library of New South Wales

watkinsiana is named in honour of George Watkins (1848–1916), plant collector, naturalist and pharmacist.

Watkins' Fig is a large strangler fig that grows to 45 metres. The distribution is disjunct, with a population from Dungog in New South Wales to south-east Queensland and another from Proserpine in central east Queensland to an area west of the Daintree in north Queensland.

The specimen drawn below was shot down from the forest canopy in north Queensland and posted to Bill in New South Wales to be used for the fig parrot plate in *Australian Parrots*. At the time, it was known as a food source for fig parrots, but our botanical knowledge was limited, so we didn't realise that a specimen of this tree species could have been collected from nearby forests. The sketch opposite was of a then unidentified plant from New South Wales, which I can now confirm is Watkins' Fig.

There are a few different techniques for getting plant specimens down from the rainforest canopy. Catapults or shanghais are often used, as well as firearms. Our preference was a .22 rifle, for which, because of the small calibre, you needed to have still conditions, preferably in the early morning. The target could be no thicker than a pencil and a slight breeze in the canopy would mean missing it.

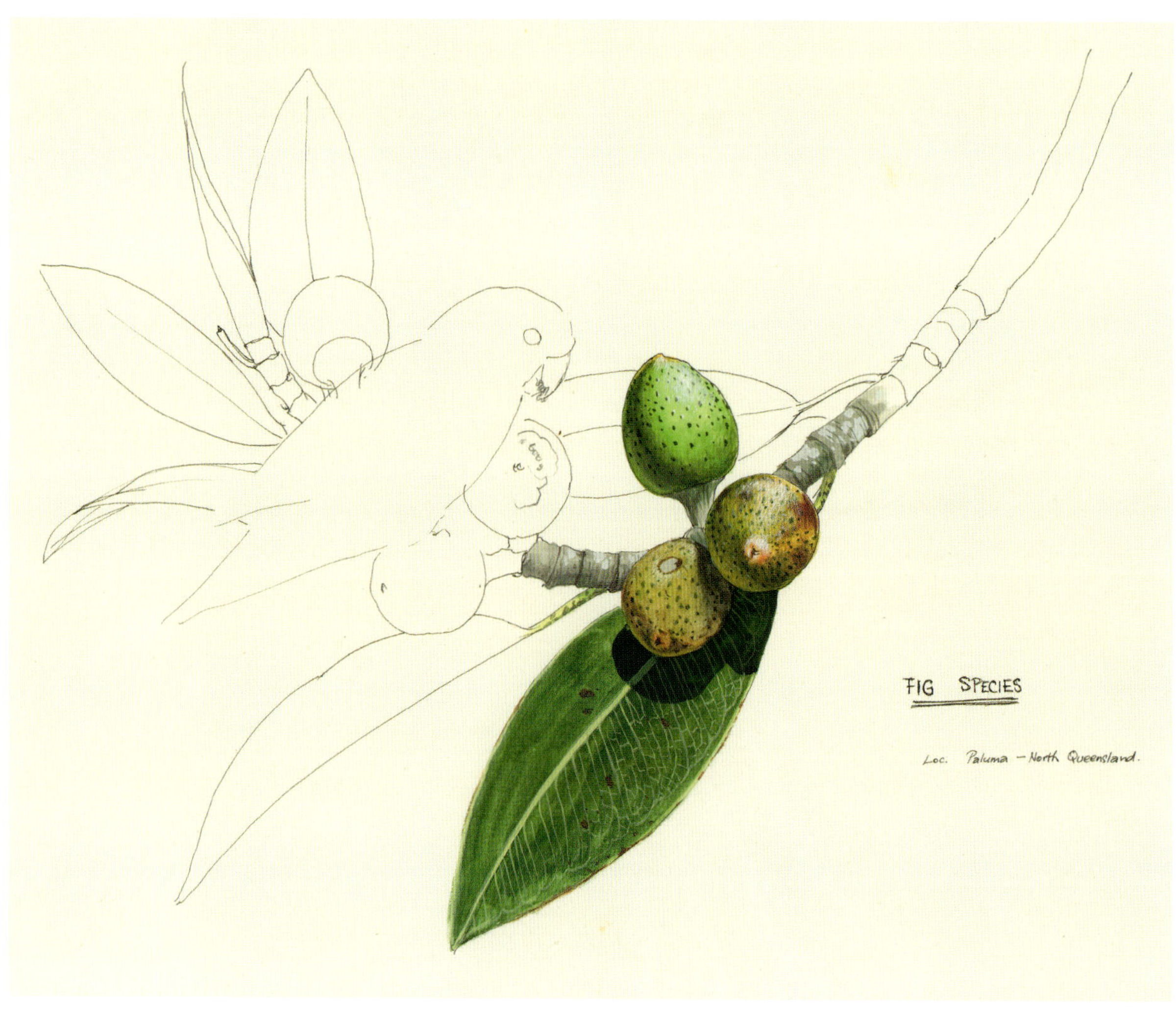

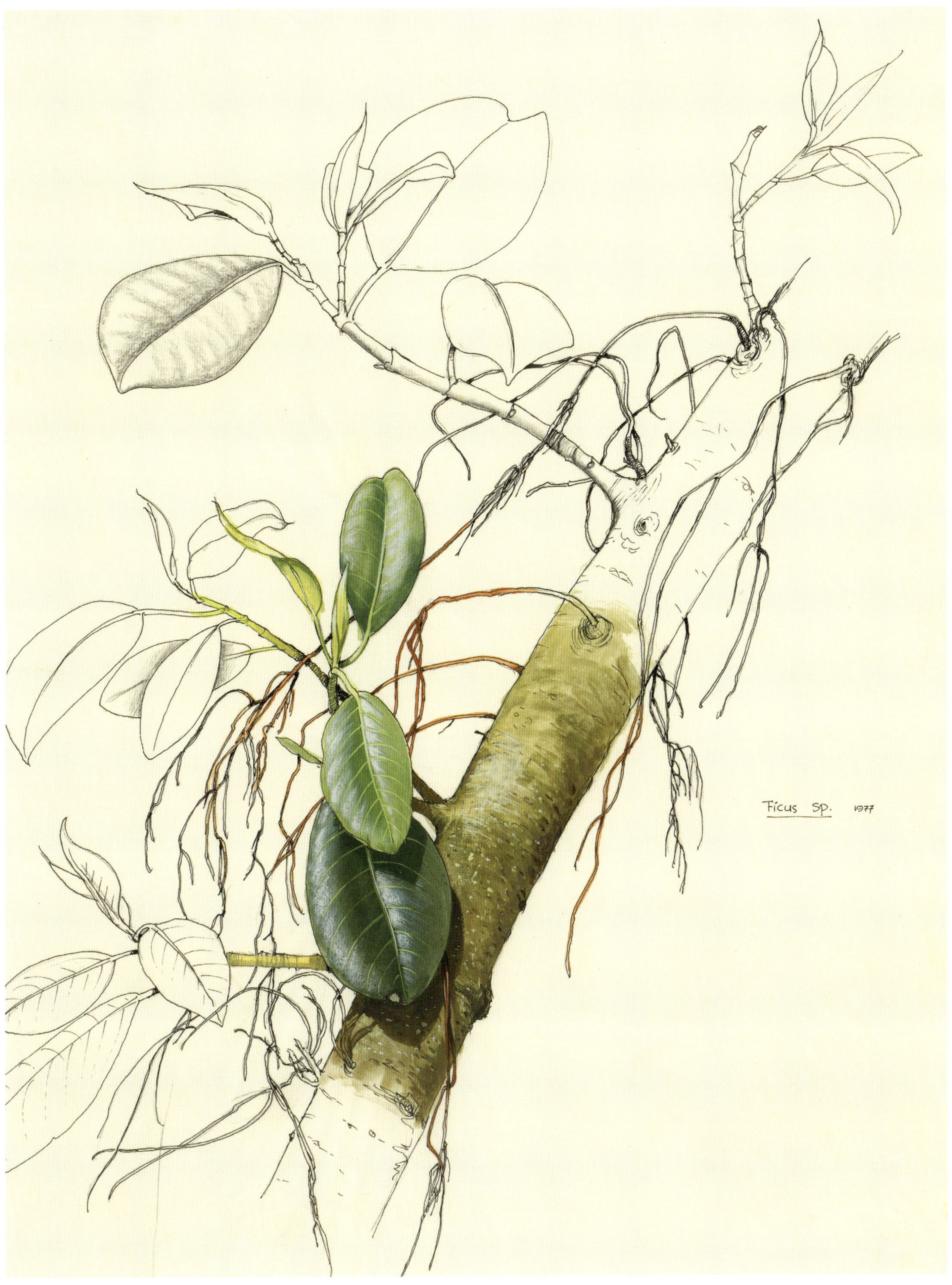
Ficus sp. 1977

Ganophyllum falcatum

Daintree Hickory

SAPINDACEAE

→ Yellow-billed Turaco
1997; watercolour; c.600 x 500mm
Collection of Peter and Marilyn Chapman

↓ undated; pen, pencil and watercolour
590 x 430mm
State Library of New South Wales

Ganophyllum is from *ganos* (lustre) and *phyllon* (leaf), referring to the shiny leaflets.

falcatum is from *falcatus* (sickle-shaped), referring to the leaflet shape.

Daintree Hickory is a wonderful lowland rainforest tree growing to more than 30 metres and occurring in Queensland, New Guinea, Asia and West Africa. The leaves usually have entire margins but this sample appears to have been chewed by an insect to create a couple of serrated leaflets, as annotated by Bill on the drawing.

As botanical records show that this plant occurs in West Africa, Bill was able to use it in a painting of the Yellow-billed Turaco from West and west Central Africa.

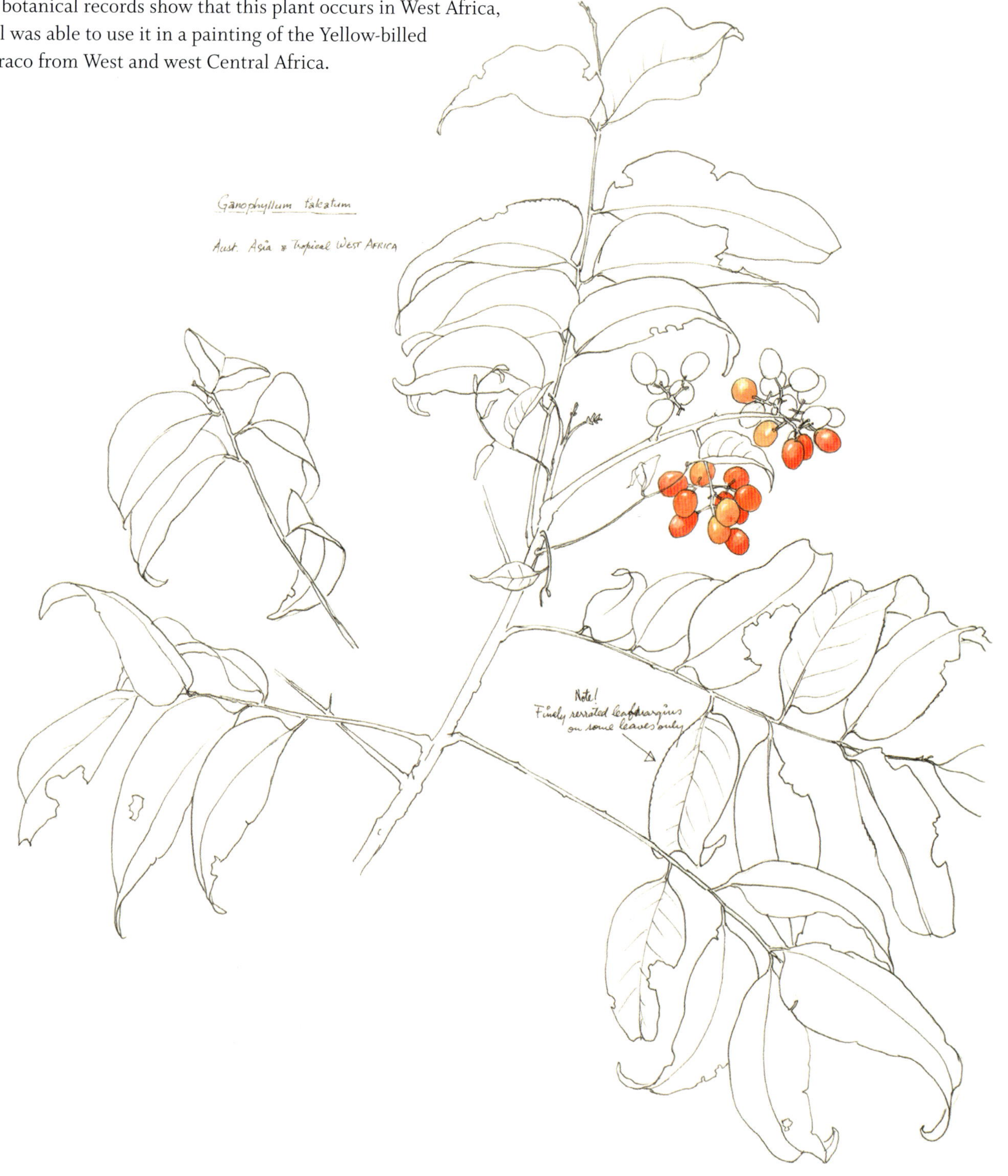

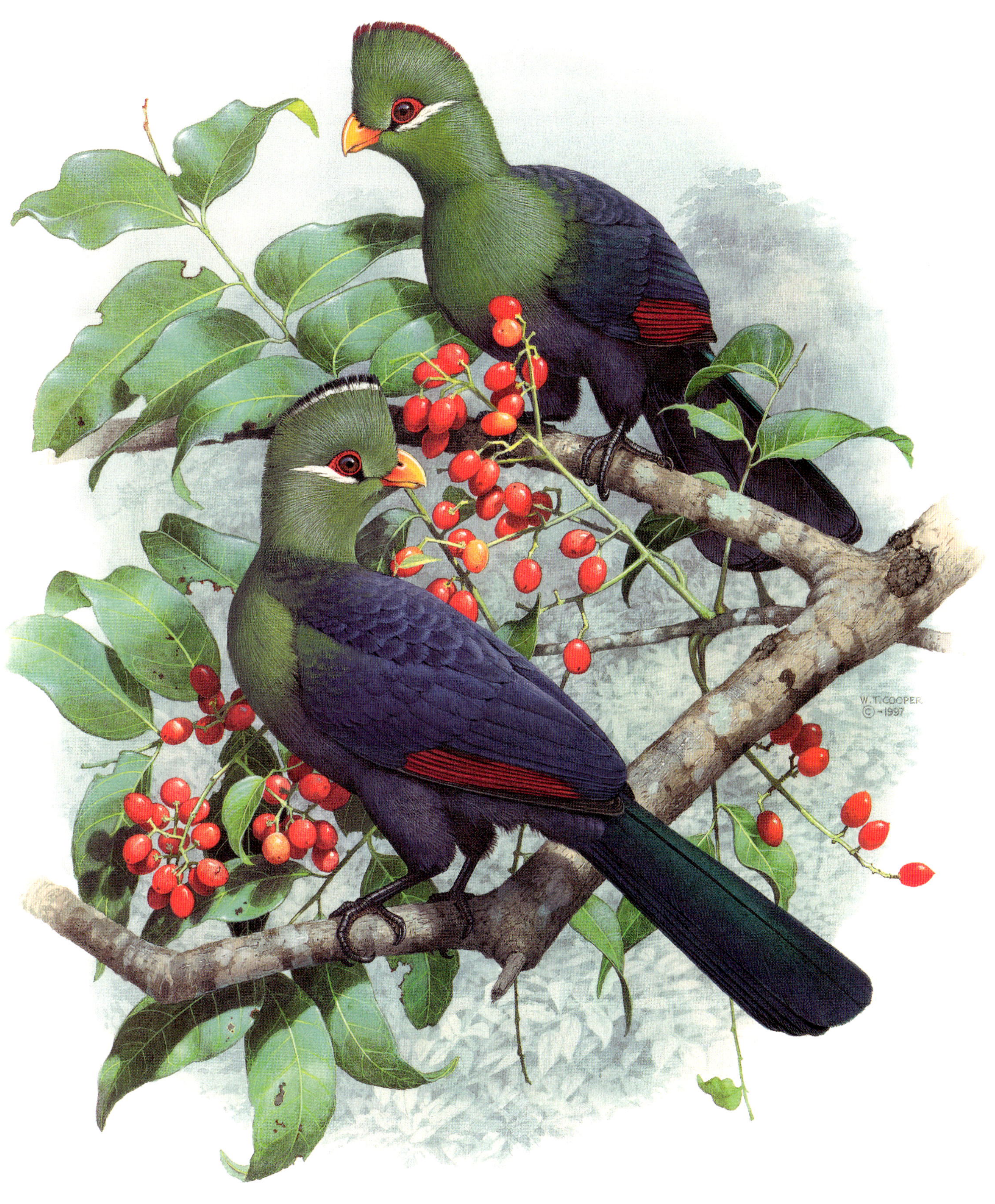
W.T. COOPER
© 1997

Gardenia scabrella

Star Flower

RUBIACEAE

Gardenia is named in honour of English botanist Alexander Garden (1730–1791).

scabrella is from *scabrellus* (slightly sandpapery), referring to the leaves.

Star Flower is a shrub growing to about 4 metres, which occurs in various types of rainforest on northern Cape York Peninsula. The flowers are large and fragrant, with a diameter of up to 80 millimetres.

The authors of *Australian Tropical Rain Forest Trees and Shrubs* requested the use of this painting for the cover of their publication. It was later presented to Peter Stanton on his retirement as Senior Conservation Officer of Queensland National Parks and Wildlife Service.

→ undated; watercolour; 340 x 300mm
Collection of Peter Stanton

↓ Tree Shapes
undated; pen and pencil; 210 x 305mm
State Library of New South Wales

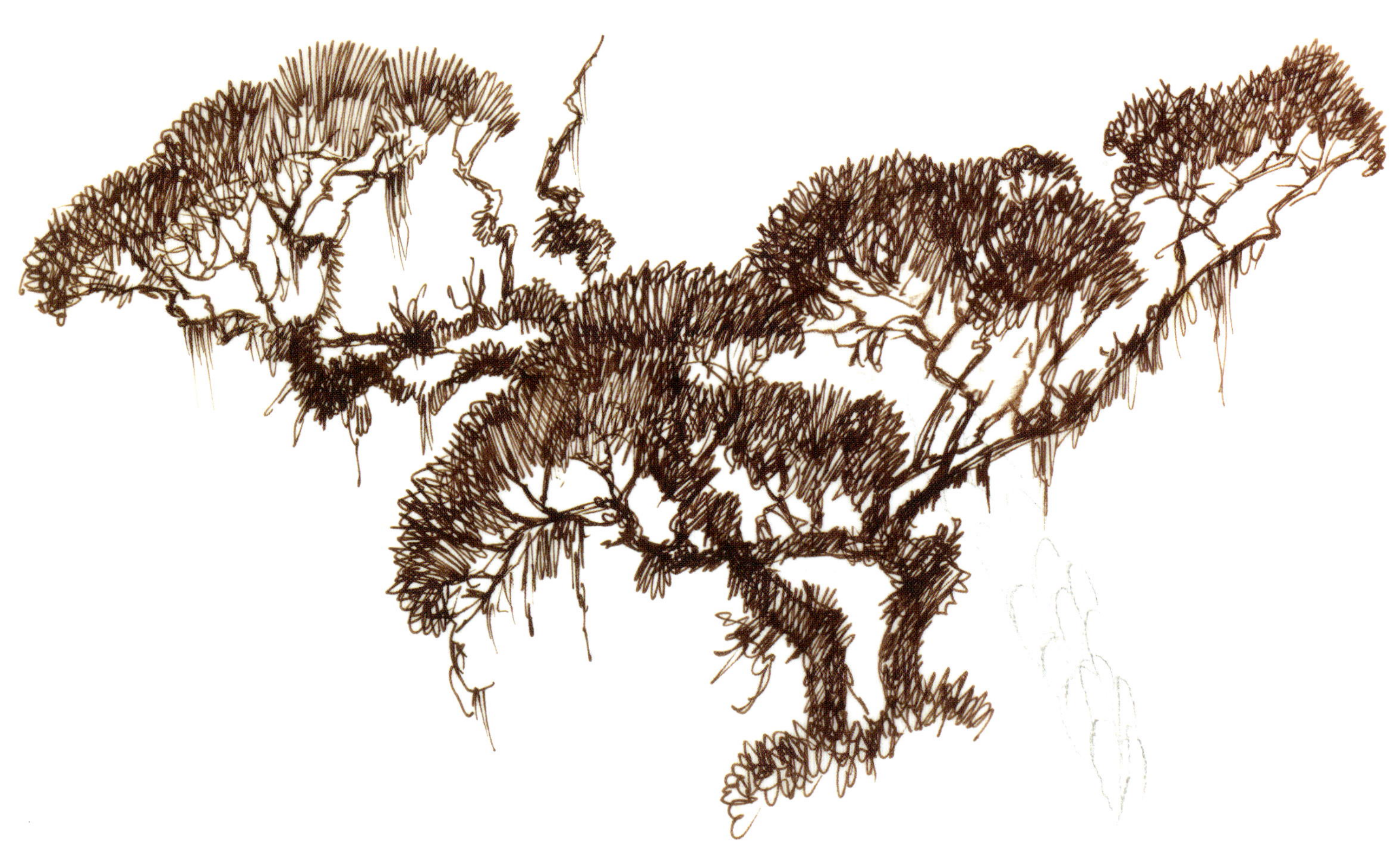

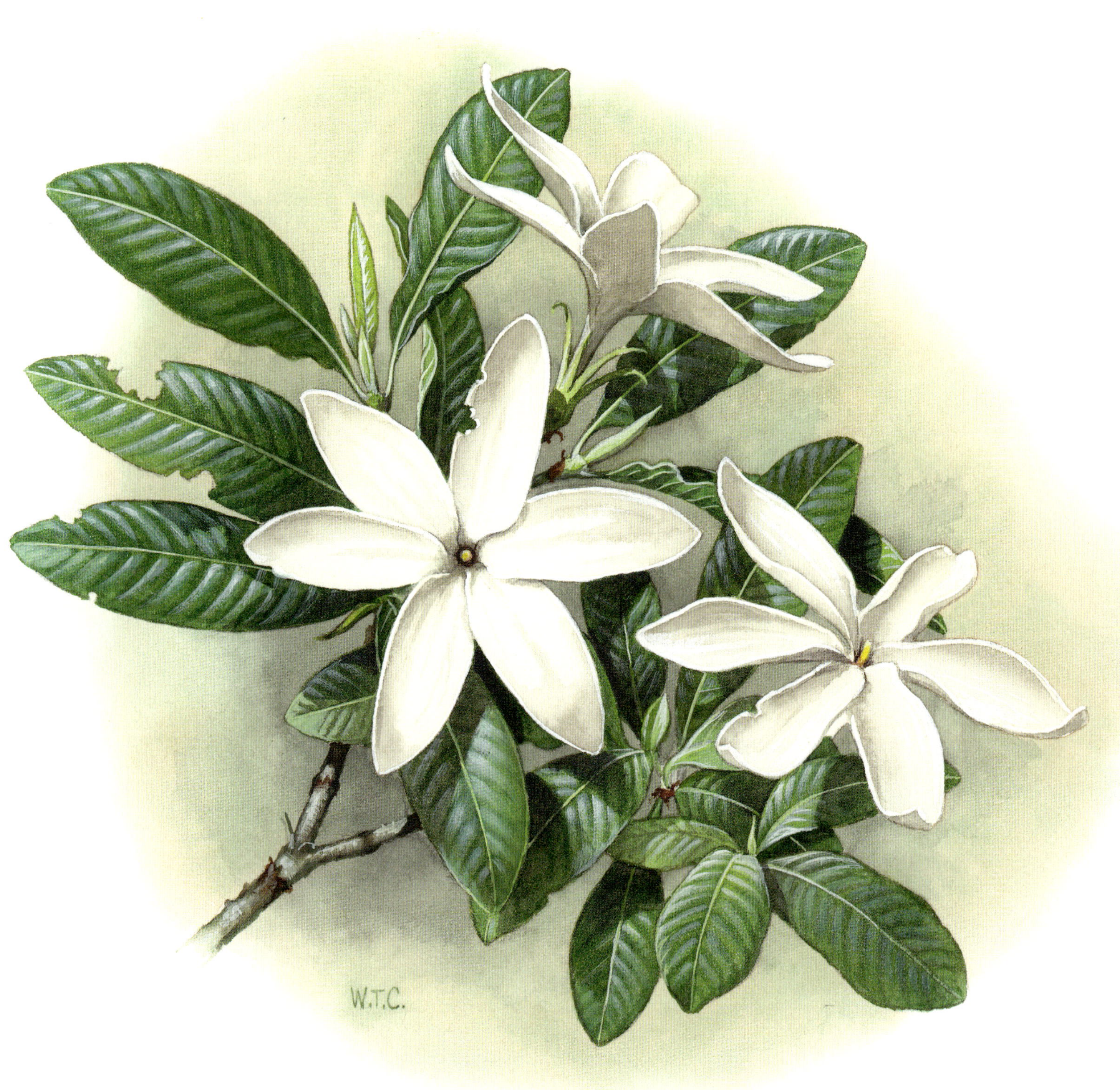
W.T.C.

Glochidion harveyanum

Harvey's Buttonwood

PHYLLANTHACEAE

Glochidion is from *glochin* (a projecting point), referring to the prominent style of the flowers.

harveyanum is after Harvey Creek (south of Cairns), where the type specimen was collected in 1910 by Czech botanist Karel Domin (1882–1953).

Harvey's Buttonwood is a shrub or small tree occurring in a wide variety of habitats from rainforest to dry open forest. The outer pink casing on the fruit splits and falls to reveal red-coated seeds that remain attached to the plant for some time.

We had gone fishing at Running River west of Townsville but, alas, the fish were not biting, so Bill took the opportunity to make this plant sketch.

It has been labelled *Glochidion sumatranum* but is actually *Glochidion harveyanum*.

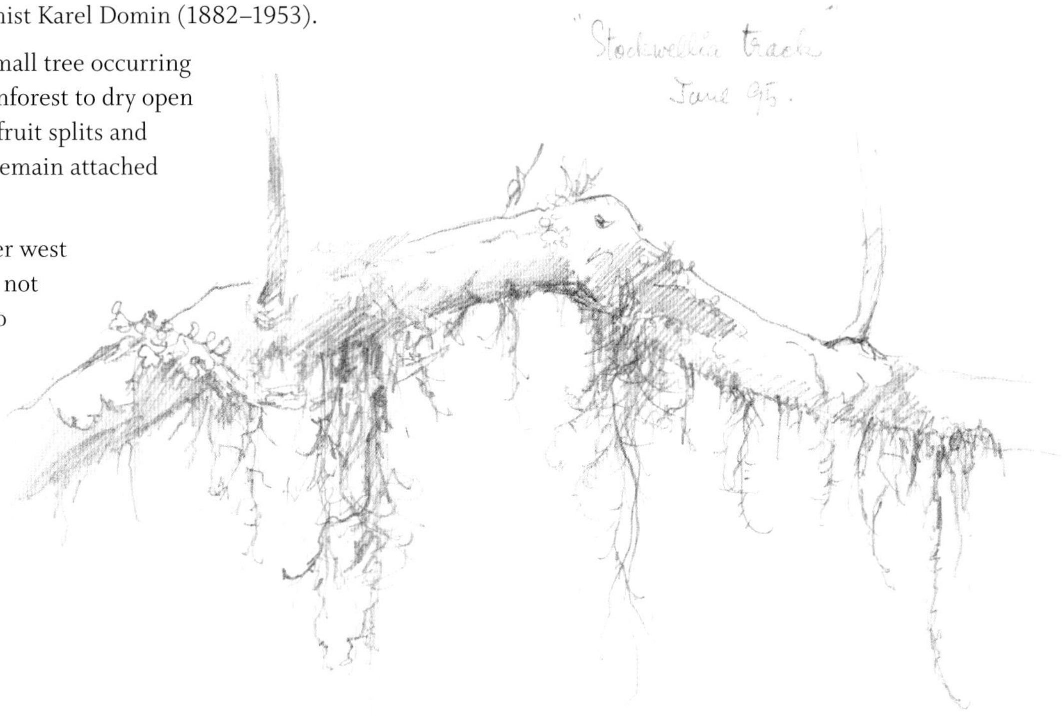

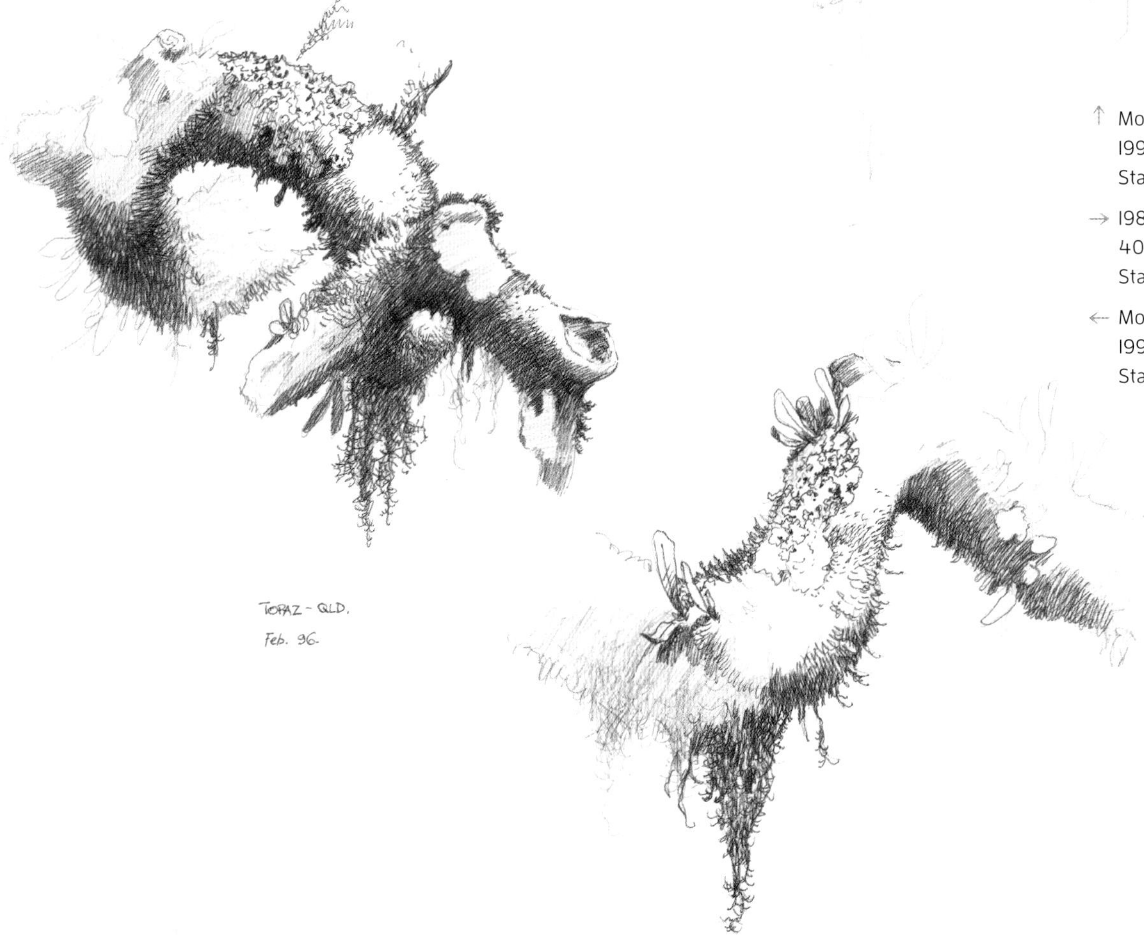

↑ Mossy Branches on the Stockwellia Track
1995; pencil; 210 x 280mm
State Library of New South Wales

→ 1983; pen, pencil and watercolour
400 x 300mm
State Library of New South Wales

← Mossy Branches at Topaz
1996; pen and pencil; 480 x 605mm
State Library of New South Wales

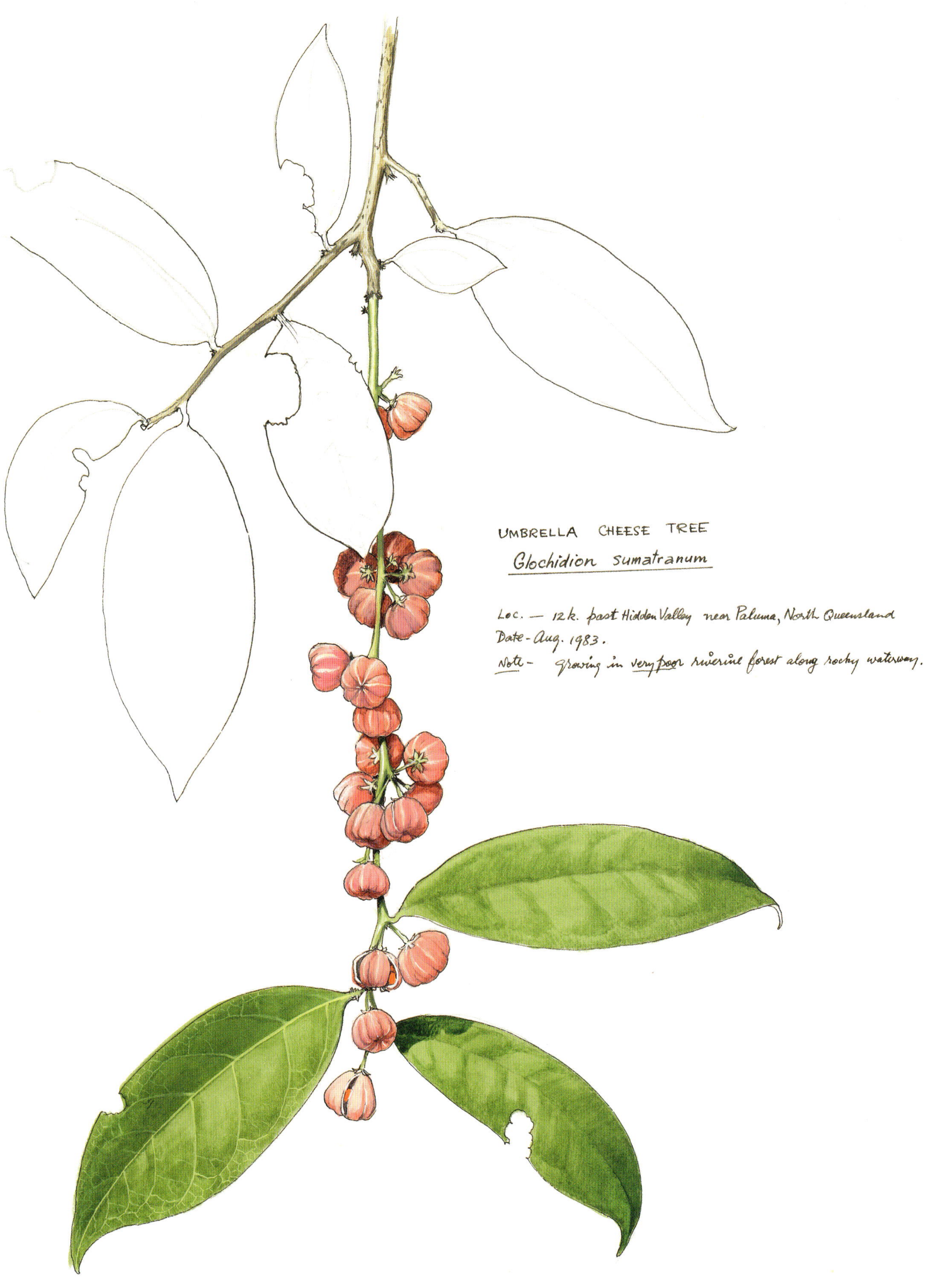
UMBRELLA CHEESE TREE
Glochidion sumatranum
Loc. — 12k. past Hidden Valley near Paluma, North Queensland
Date - Aug. 1983.
Note - growing in very poor riverine forest along rocky waterway.

Halfordia scleroxyla

Jitta or Ghittoe

RUTACEAE

Halfordia is named in honour of George Britton Halford (1824–1910), professor of anatomy, physiology and pathology, and founder of Australia's first medical school at the University of Melbourne in 1876.

scleroxyla is from *scleros* (hard) and *xylon* (wood).

Jitta is a tree of wet tropical rainforest, locally famous as an oily wood that will burn when wet. For that reason, the plant is also known as Kerosene-wood.

This particular tree fruits in abundance near our house almost every year. The somewhat nomadic Topknot Pigeons seem to particularly favour it, visiting in big flocks. The tree was an obvious choice for inclusion in the plate of this bird in *Pigeons and Doves in Australia*.

→ 2003; pen and watercolour; 390 x 570mm
State Library of New South Wales

↓ Topknot Pigeons
2005; watercolour; c.510 x 410mm
State Library of New South Wales

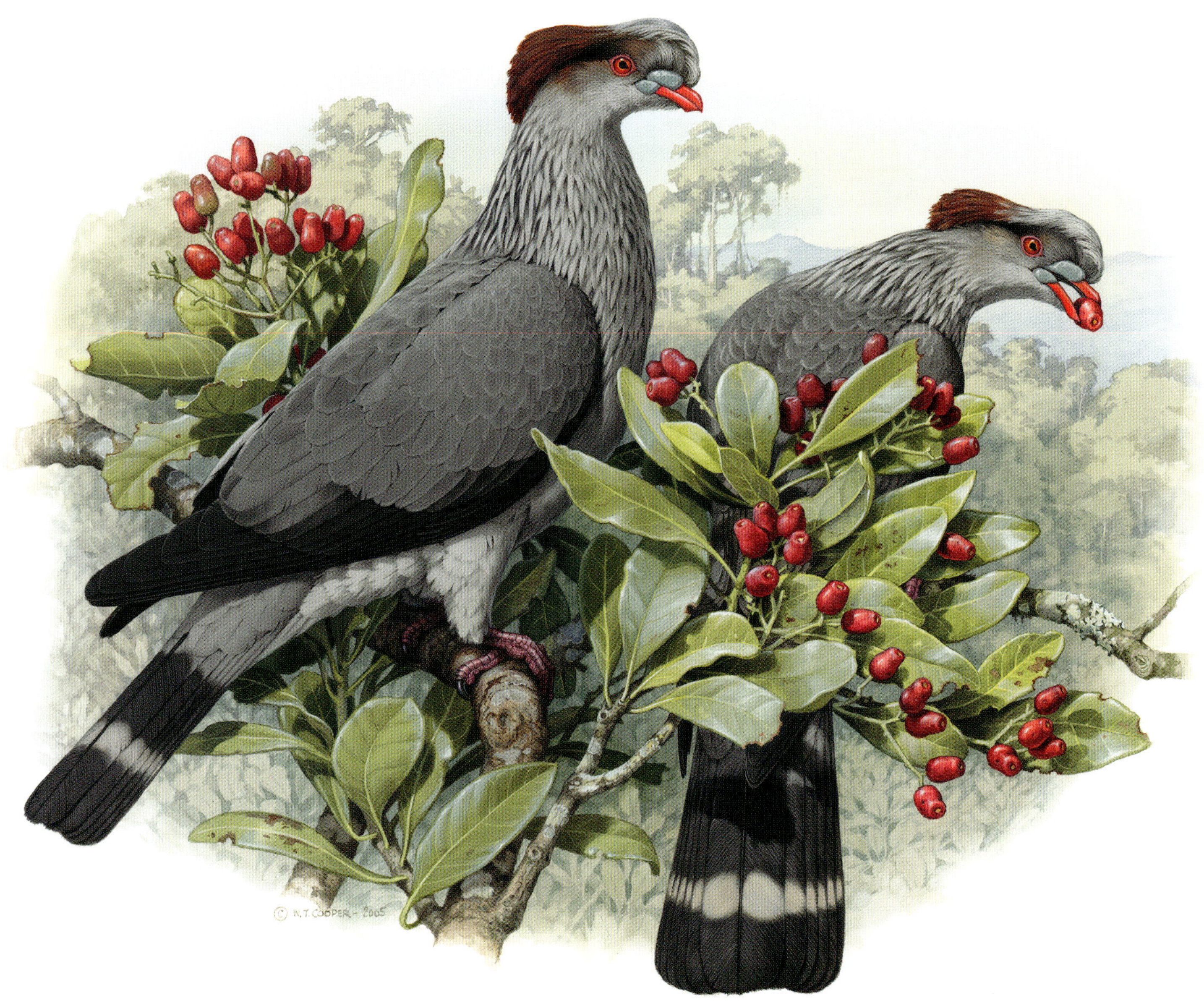

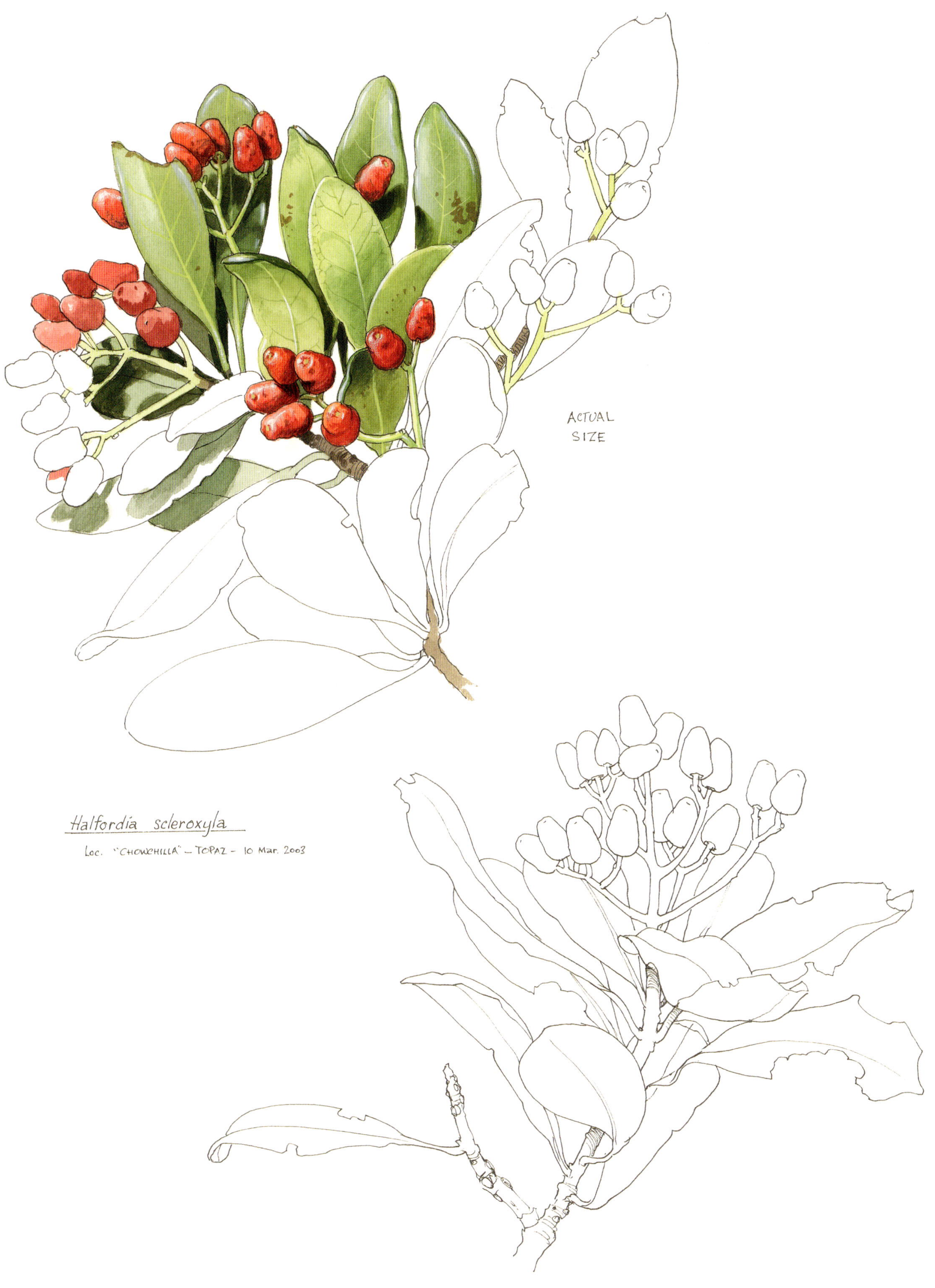
ACTUAL
SIZE
Halfordia scleroxyla
Loc. "CHOWCHILLA" – TOPAZ – 10 Mar. 2003

Homalanthus novo-guineensis

Tropical Bleeding Heart

EUPHORBIACEAE

Homalanthus is from *homalos* (even or level) and *-anthus* (flowered), referring to the flat flowers on the type species, *Homalanthus leschenaultianus*.

novo-guineensis means from New Guinea, where the type specimen was collected in what is now West Papua.

Tropical Bleeding Hearts are rainforest regrowth trees conspicuous for their old leaves, which turn bright red before falling. The tree occurs in northern Australia, throughout New Guinea, Malesia, Myanmar and Thailand. The fruit is eaten by birds, and the senescent leaves added colour to the Wreathed Hornbill plate in *Kingfishers and Related Birds*.

→ Wreathed Hornbill, colour plate from *Kingfishers and Related Birds*, 1993
National Library of Australia, 2494045

↓ 2011; pen, pencil and watercolour; 290 x 415mm
State Library of New South Wales

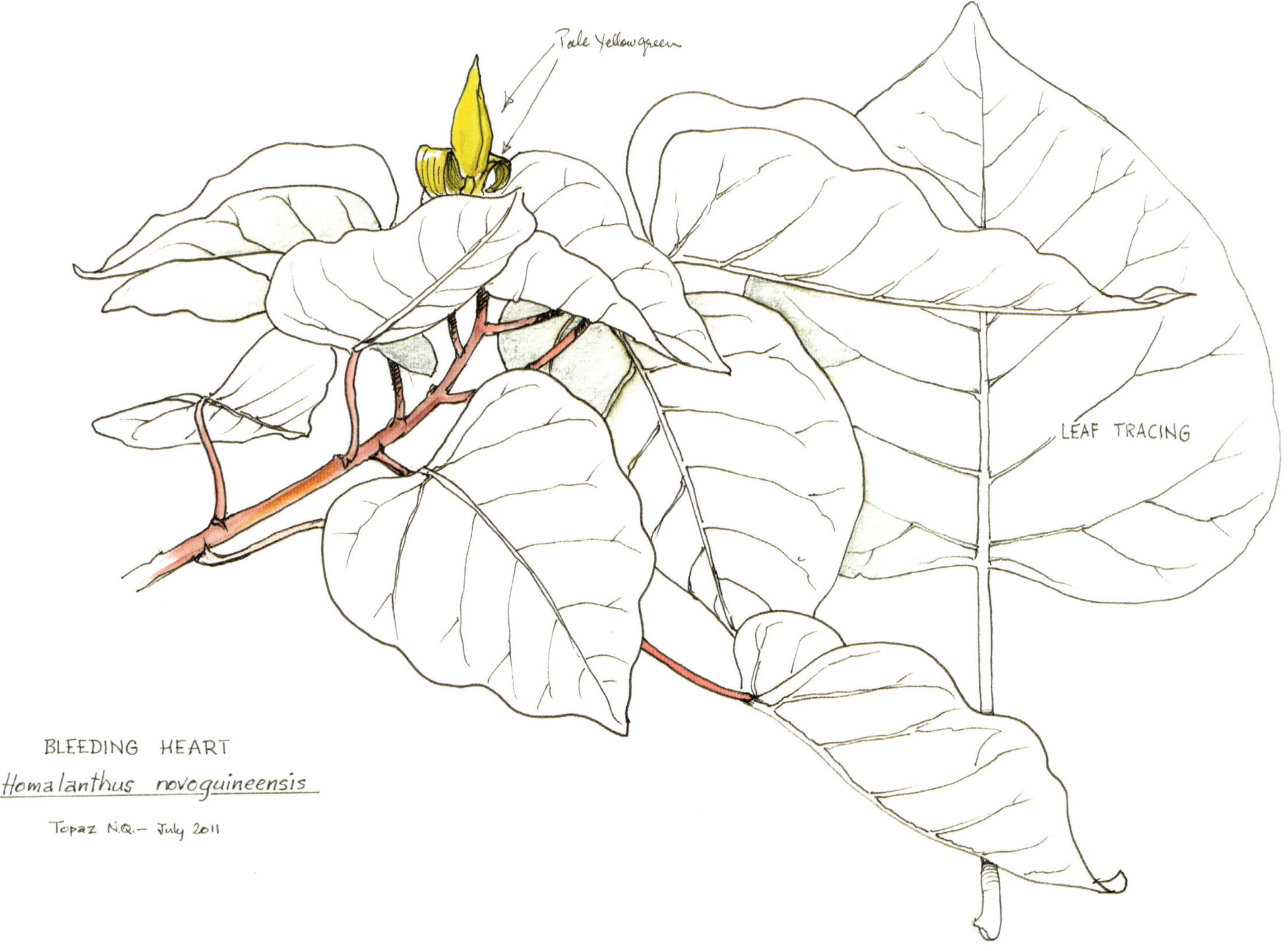

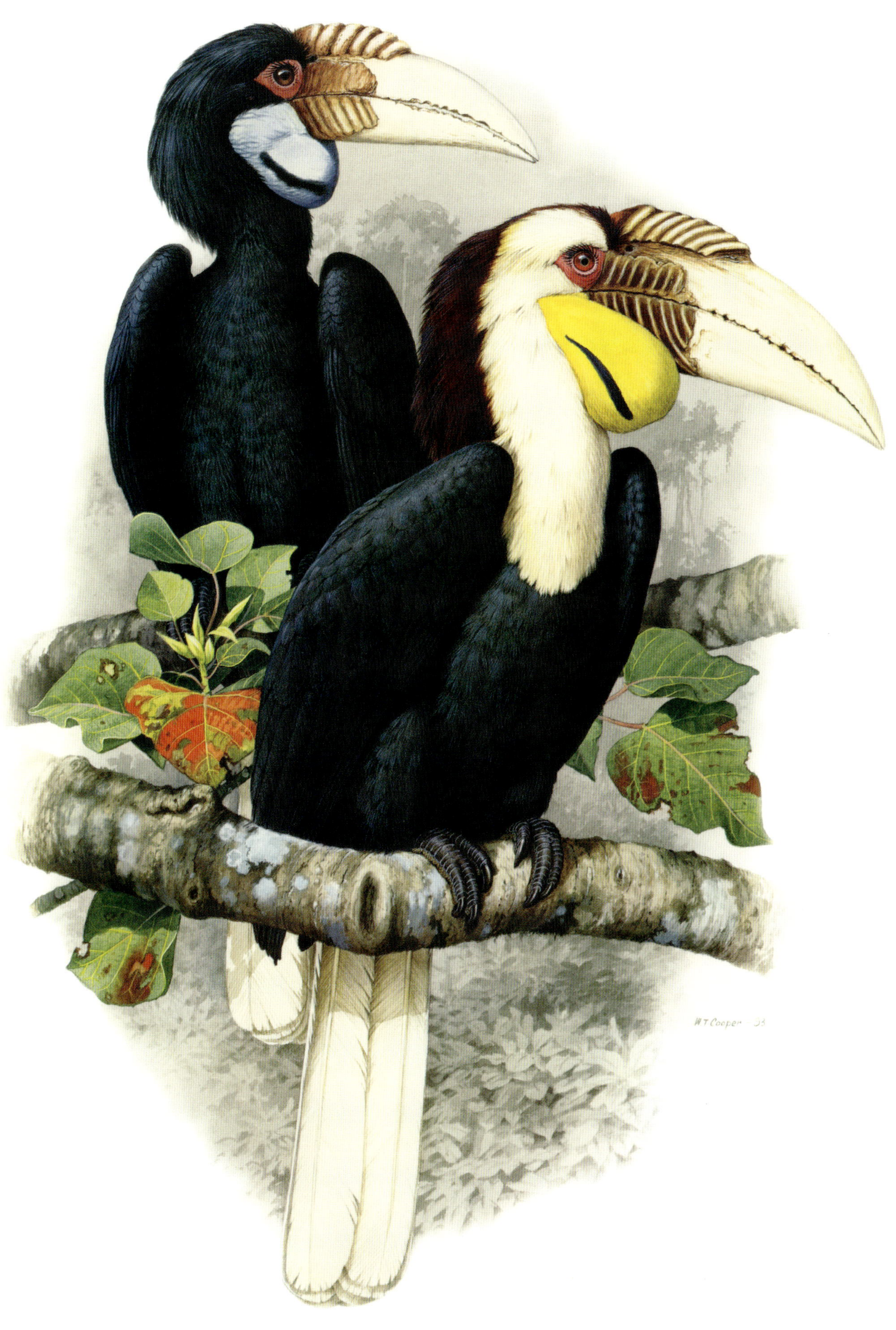
W T Cooper

Lepidozamia hopei

Hope's Cycad or Zamia Palm

ZAMIACEAE

Lepidozamia is from *lepis* (a scale) and after the genus *Zamia*, referring to the scale bases on the stems.

hopei is named in honour of Captain Louis Hope (1817–1894), grazier and sugar pioneer at Ormiston and Kilcoy, Queensland.

This stately cycad from the Wet Tropics rainforest is tall and palm-like but it belongs to an ancient lineage that existed long before palms evolved.

The cone depicted was nestled in the crown of the plant. It was so ripe that if we had cut it at the cone base it would have fallen and scattered into many pieces. Although this was in the Daintree rainforest, we were on private property. We noticed a pile of builder's essentials nearby, including some scaffolding, which we erected to about 4 metres. Bill then climbed to the top of this structure, where he sat to draw the fruit in situ, before we cut the sample for preserving as a herbarium specimen.

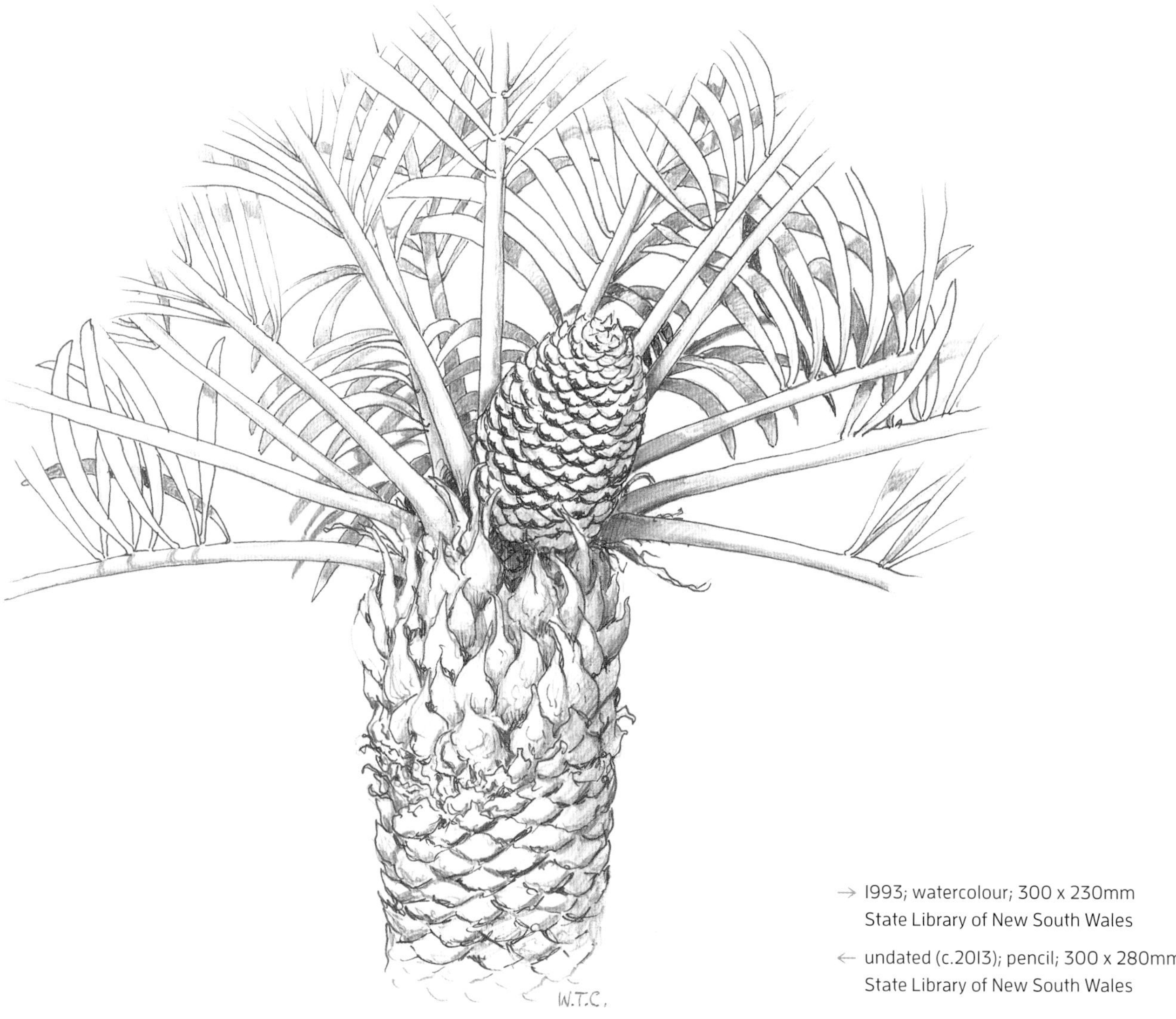

→ 1993; watercolour; 300 x 230mm
State Library of New South Wales

← undated (c.2013); pencil; 300 x 280mm
State Library of New South Wales

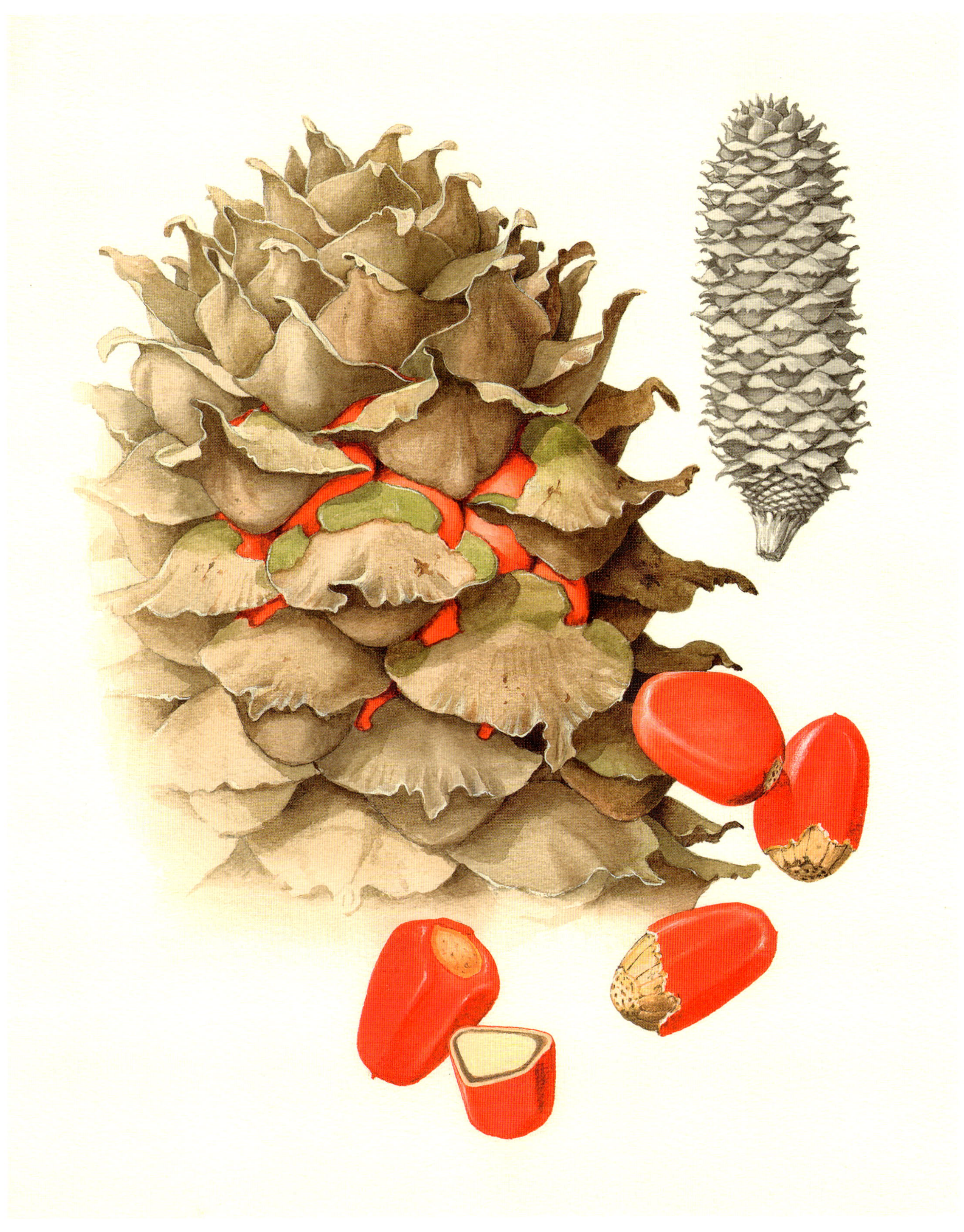

Levieria acuminata

Straw Beech

MONIMIACEAE

Levieria is named in honour of Émile Levier (1839–1911), French botanist and medical doctor who lived in Italy for most of his life, concentrating on Italian plants.

acuminata is from *acuminatus* (tapering to a narrow point), referring to the leaf apex.

Straw Beech is named for its superficial similarity to the beeches of Europe. Otherwise, it bears no resemblance to its namesake. This species is a slender tree, which grows to 20 metres in the Wet Tropics rainforest of north Queensland.

Many times we watched these small drupes being taken by fruit-doves in the forest around home. On 11 May 2001, Bill wrote in his diary, 'Wompoo Pigeons calling and some feeding on *Levieria* fruit'. The fruit is black when ripe but, because of the birds, they were disappearing quicker than Bill could get a specimen, so only unripe red fruit are shown here.

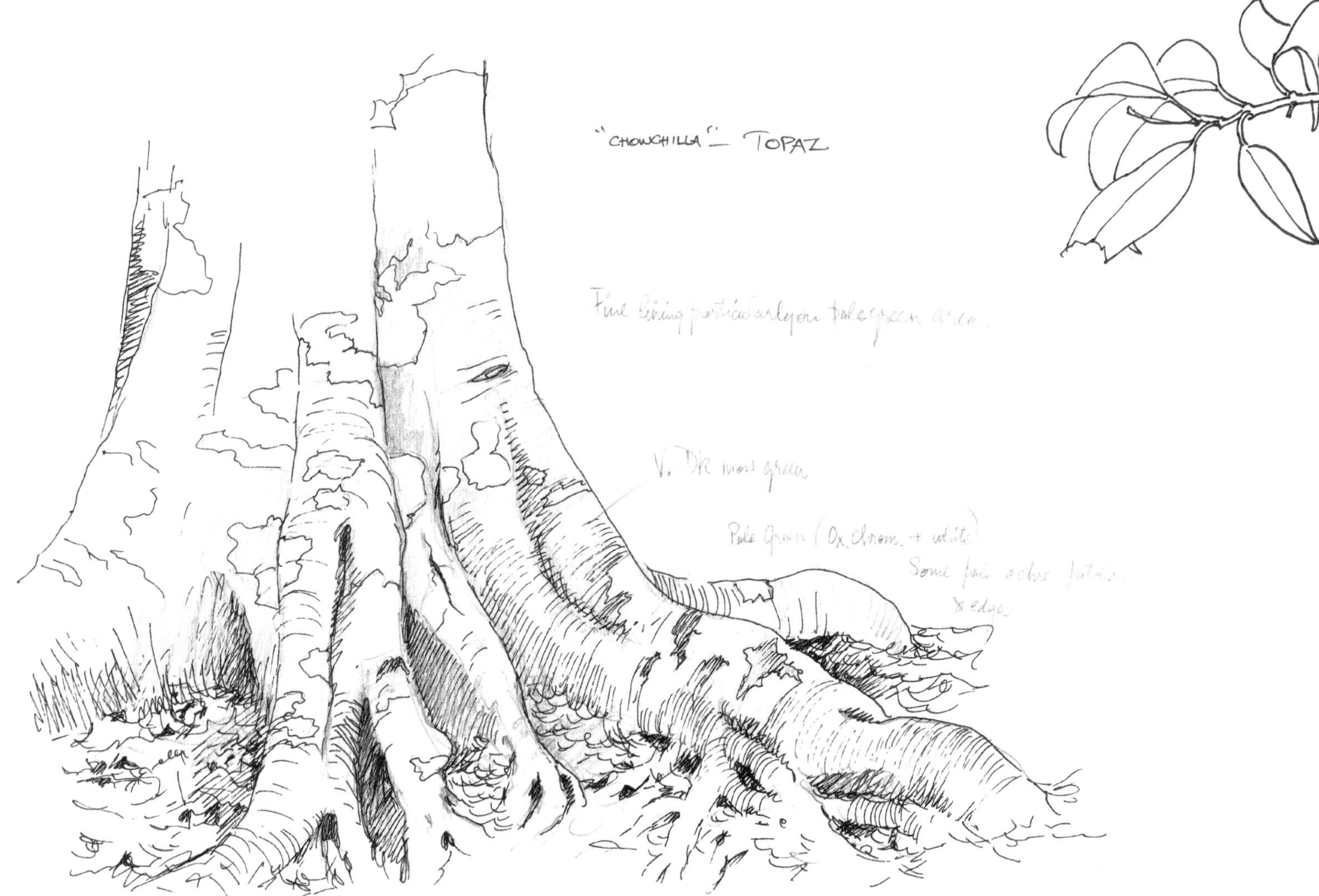

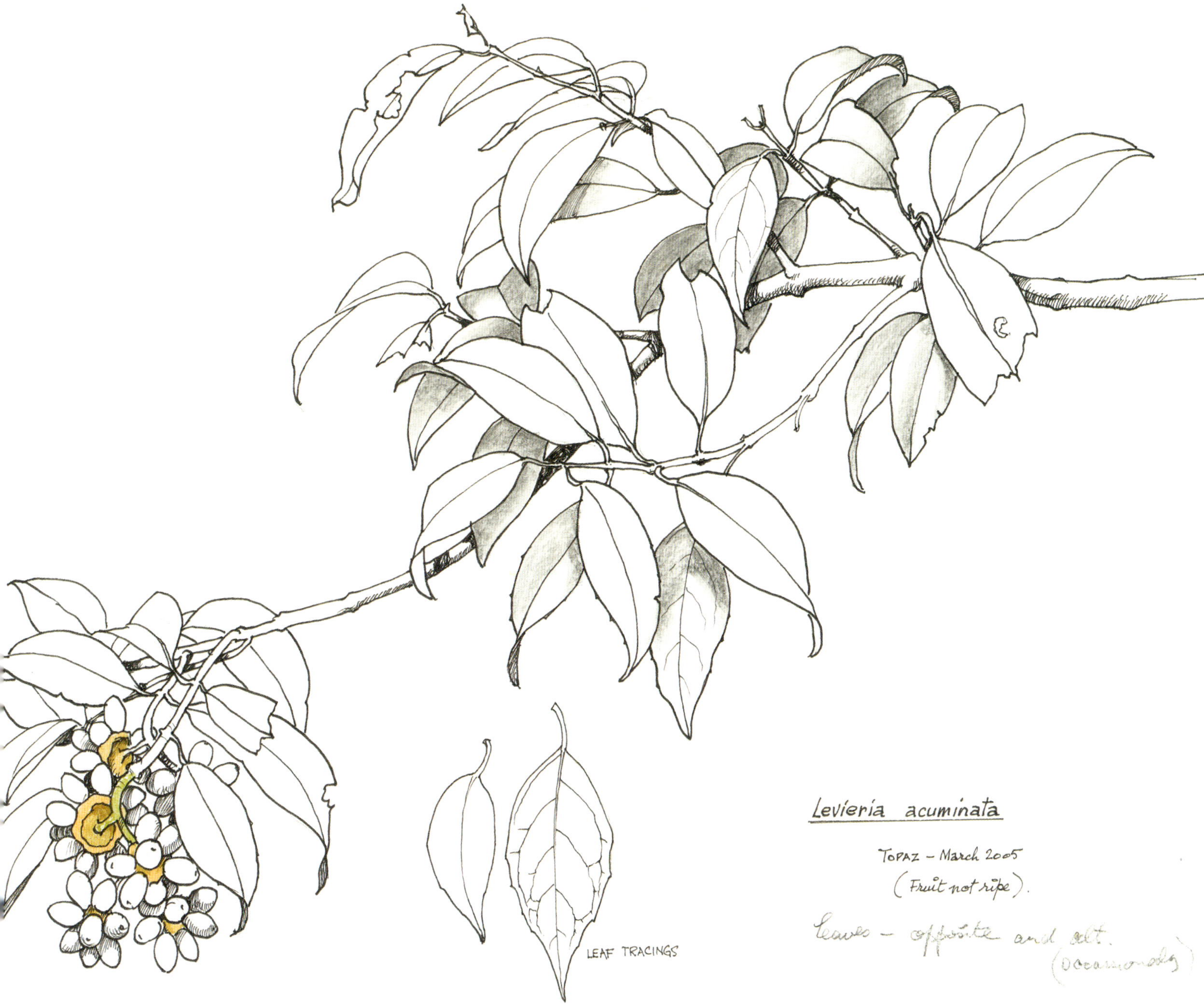

LEAF TRACINGS

Levieria acuminata

TOPAZ – March 2005
(Fruit not ripe).

Leaves – opposite and alt.
(occasionally)

↑ 2005; pen, pencil and watercolour; 370 x 480mm
State Library of New South Wales

← Unidentified Rainforest Tree Base
undated; pen and pencil; 210 x 280mm
State Library of New South Wales

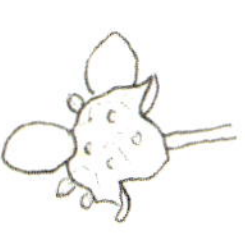

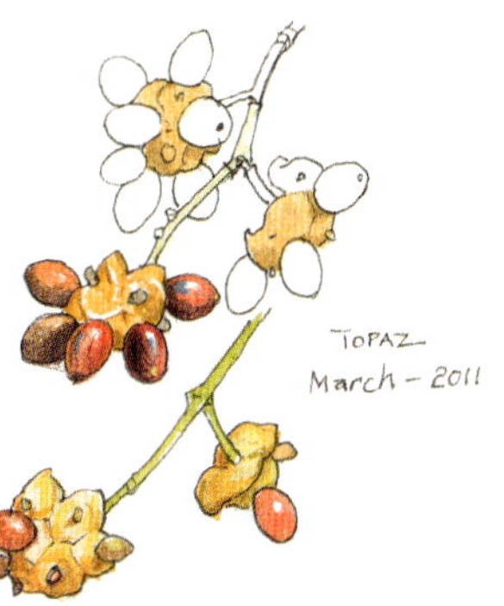

Linospadix microcaryus

Walking-stick Palm

ARECACEAE

Linospadix is from *linon* (net) and *spadix* (palm branch or palm inflorescence).

microcaryus is from *micro-* (small) and *caryon* (nut).

These slender multi-stemmed palms from the Wet Tropics bear distinctive broad bands on their trunks, left by the fallen leaves.

The plants are rather elegant among the forest undergrowth, with long pendant stems that carry the flowers and then the fruit. We once felt privileged and excited to watch at close range a Cassowary delicately pluck out the pea-sized red fruit.

The pen drawing opposite was made on Hinchinbrook Island. It represents the trunk bases and roots of Alexandra Palms (*Archontophoenix alexandrae*) and the leaves of the Fan Palm (*Licuala ramsayi*).

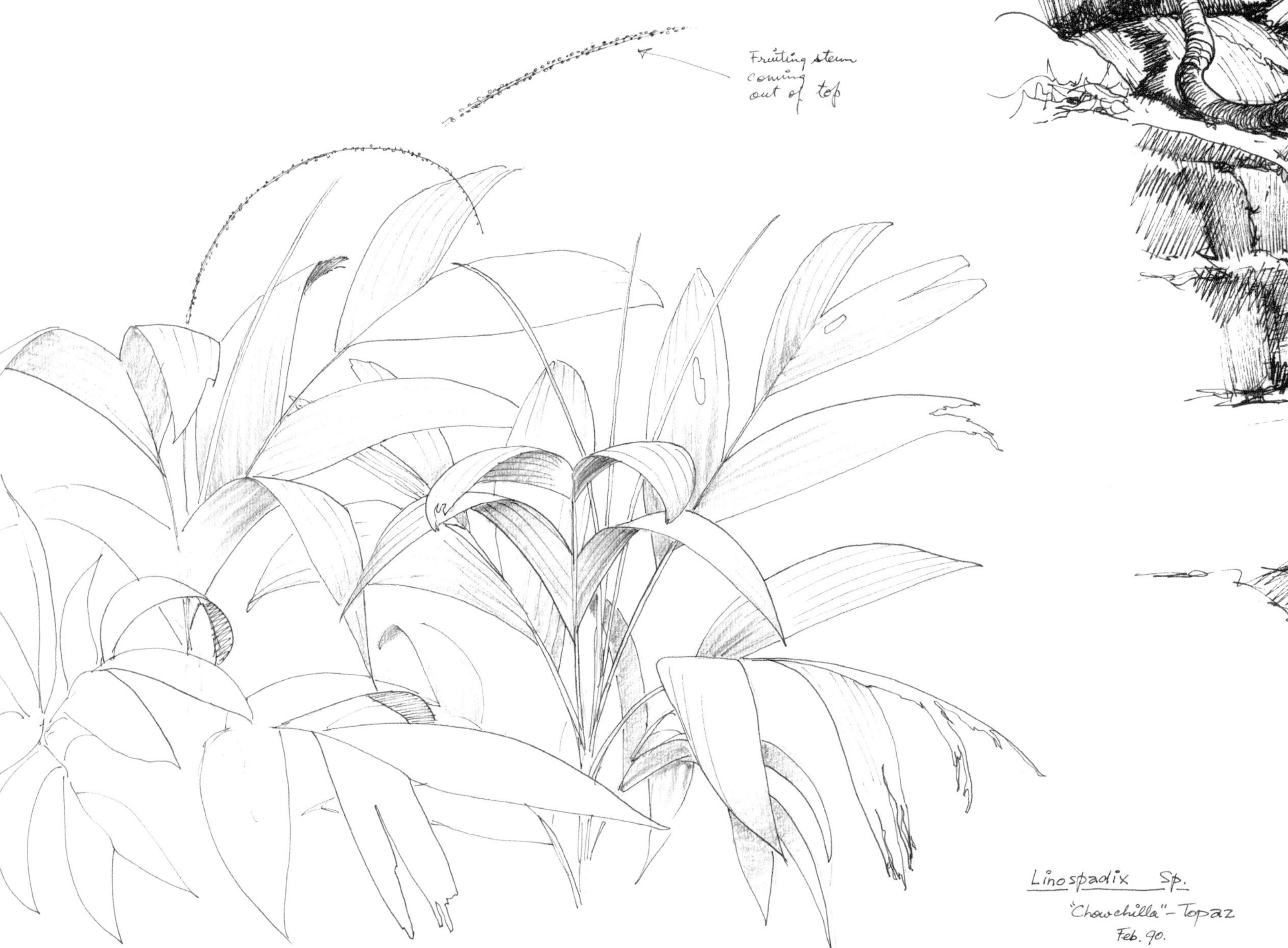

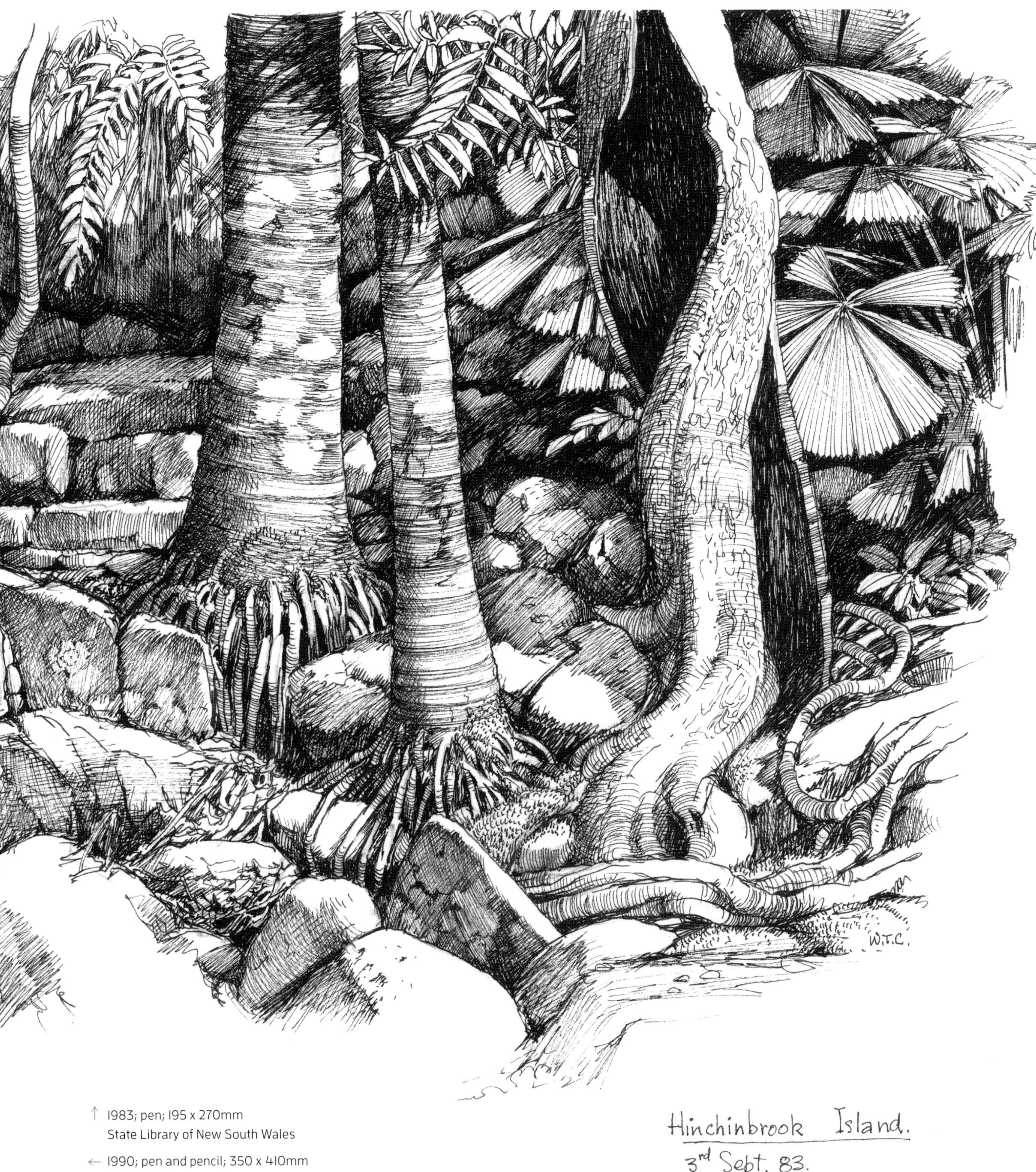

↑ 1983; pen; 195 x 270mm
State Library of New South Wales

← 1990; pen and pencil; 350 x 410mm
State Library of New South Wales

Mallotus philippensis

Red Kamala

EUPHORBIACEAE

Mallotus is from *mallotos* (woolly or fleecy), referring to the hairy leaves on some plants within this genus.

philippensis is from the Philippines.

Red Kamala is a shrub or tall tree that grows in rainforest from the Sydney area north to the Torres Strait Islands and through to South-East Asia and India. In parts of Asia, the gritty surface is removed from the red fruit for use as a fabric dye.

The widespread distribution of this plant and the fact that the fruits are eaten by many birds meant that it could be used for the plate of Purple-headed Bee-eaters in *Kingfishers and Related Birds*. Bee-eaters are not fruit eaters but the plant and the birds do occur together. We collected this specimen on the roadside near Gloucester in New South Wales.

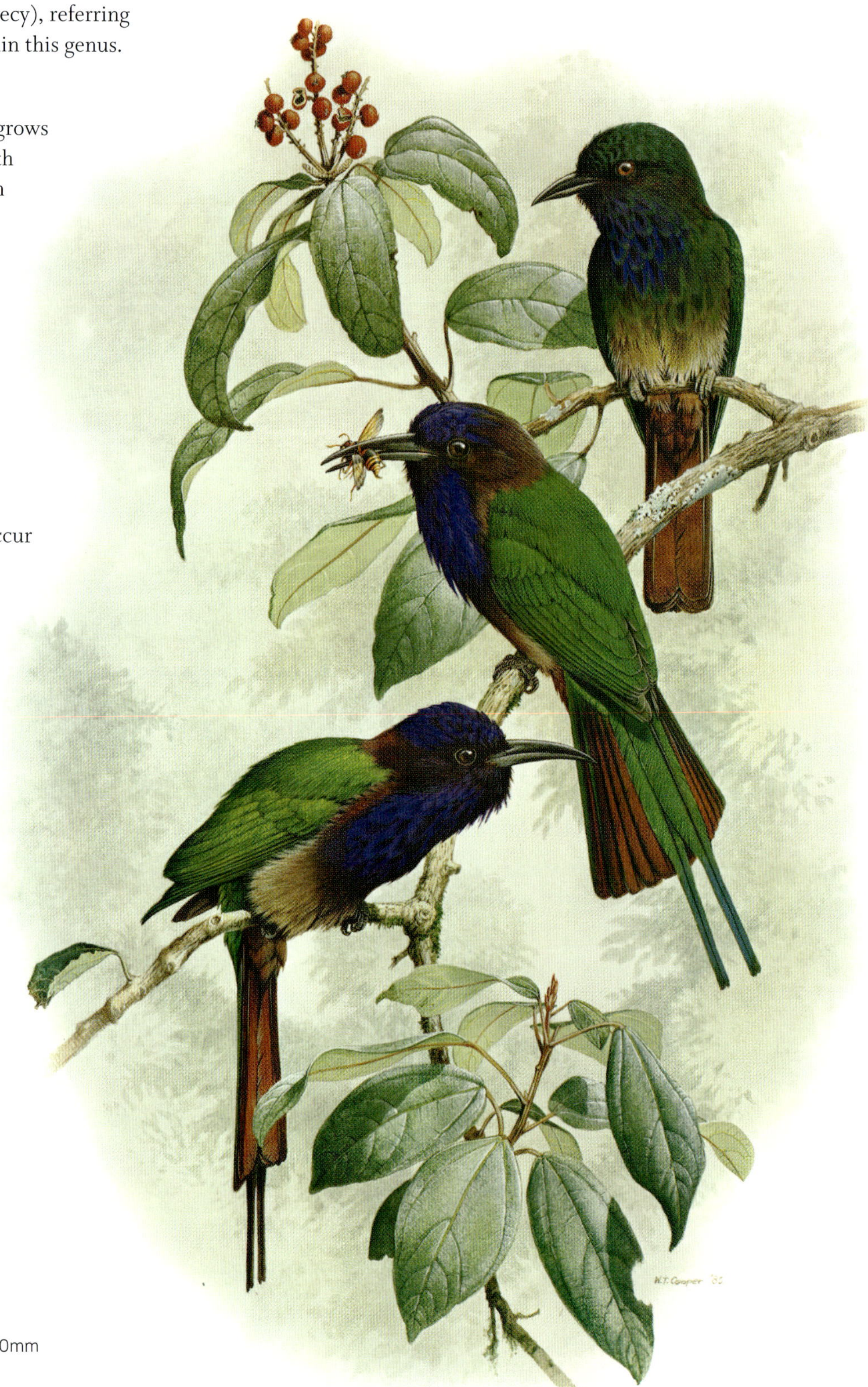

↗ Purple-headed Bee-eater, colour plate from *Kingfishers and Related Birds*, 1985
National Library of Australia, 2494045

→ 1985; pen, pencil and watercolour; 500 x 380mm
State Library of New South Wales

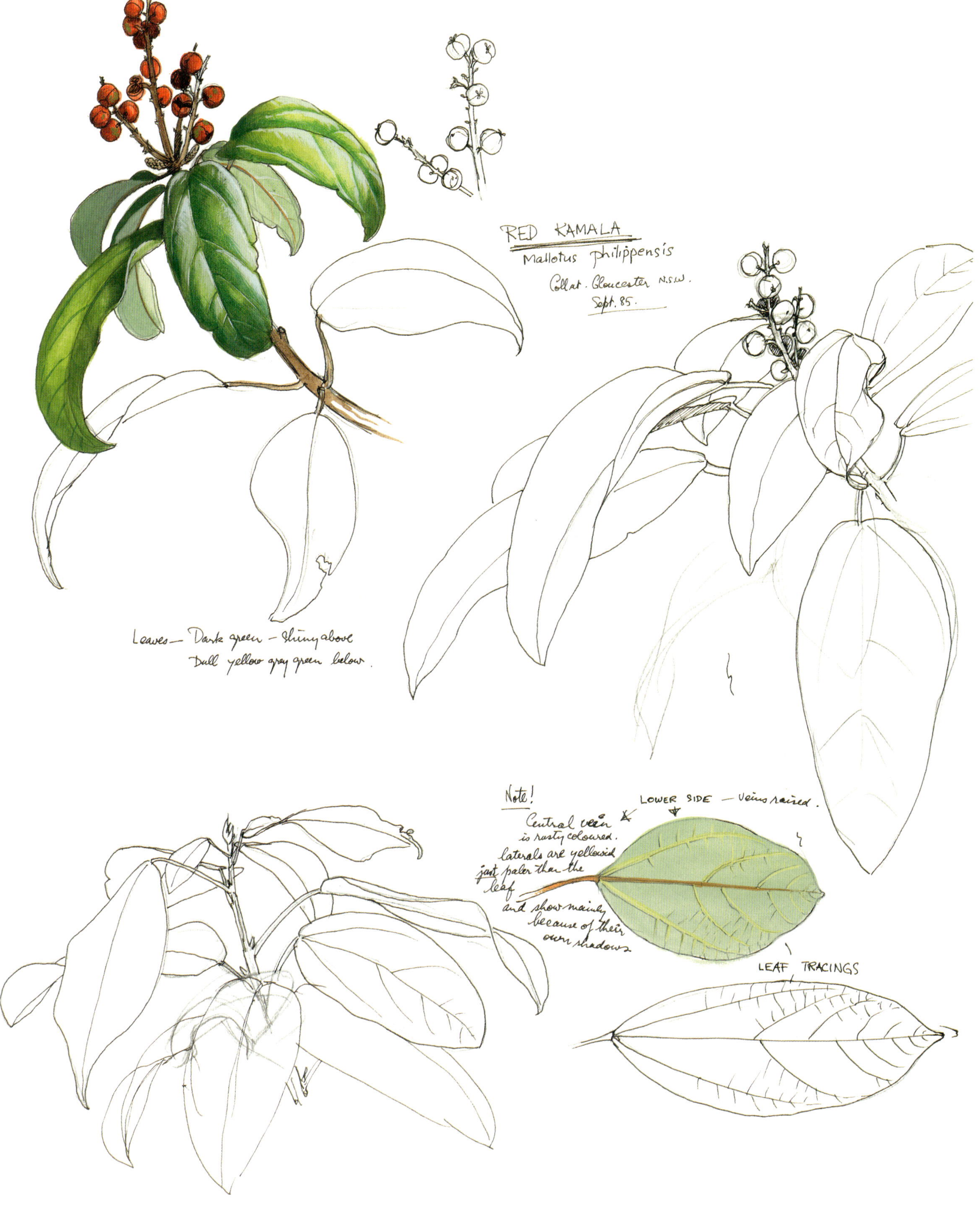
RED KAMALA
Mallotus philippensis
Collat. Gloucester N.S.W.
Sept. 85.
Leaves — Dark green — Shiny above
Dull yellow gray green below.
Note!
Central vein is rusty coloured.
laterals are yellowish just paler than the leaf
and show mainly because of their own shadows.
LOWER SIDE — Veins raised.
LEAF TRACINGS

Melicope broadbentiana

False Euodia

RUTACEAE

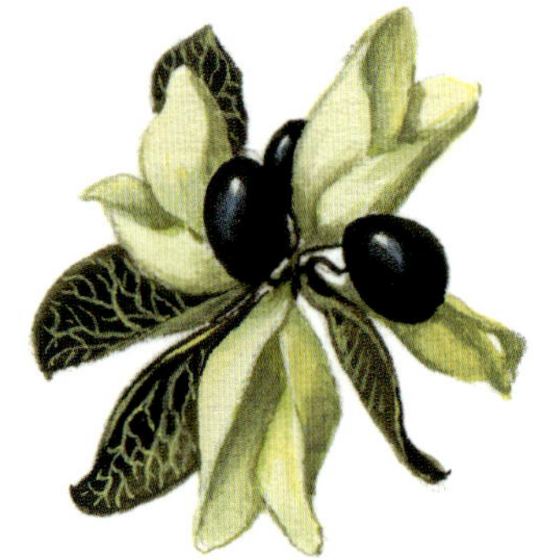

Melicope is from *meli* (honey) and *cope* (division), referring to the notched flower glands.

broadbentiana is named after Kendall Broadbent (1837–1911), English–Australian zoological collector and naturalist.

False Euodia is a spindly shrub or small tree growing in disturbed areas in tropical rainforest. The fully opened ripe fruit expose shiny blue-black seeds within a white papery inner section. Golden Bowerbirds use these open capsules as decoration on their bowers, as Bill described in his diary on 2 September 1992:

> *Beautiful day again … Walked out to look at the bowers of the Golden Bowerbirds. No. 1 was in good shape with plenty of lichen but hardly any Melicope pods … No. 2 bower was really well decorated with lichen and plenty of Melicope. The bird was in attendance when we arrived but when it knew we were there it sat about scolding. The colour of the back and wings is really very glowing and a spot of sunlight on it made the bird appear to have an extra yellow marking below the hindneck patch.*

Bill drew this False Euodia on a trip to Paluma, where he also sketched the rainforest canopy (below).

↑ undated; watercolour
State Library of New South Wales

→ 1983; pen and watercolour; 400 x 290mm
State Library of New South Wales

↓ 1983; pen; 215 x 270mm
State Library of New South Wales

RAINFOREST
Paluma – North Queensland. Aug. 83.

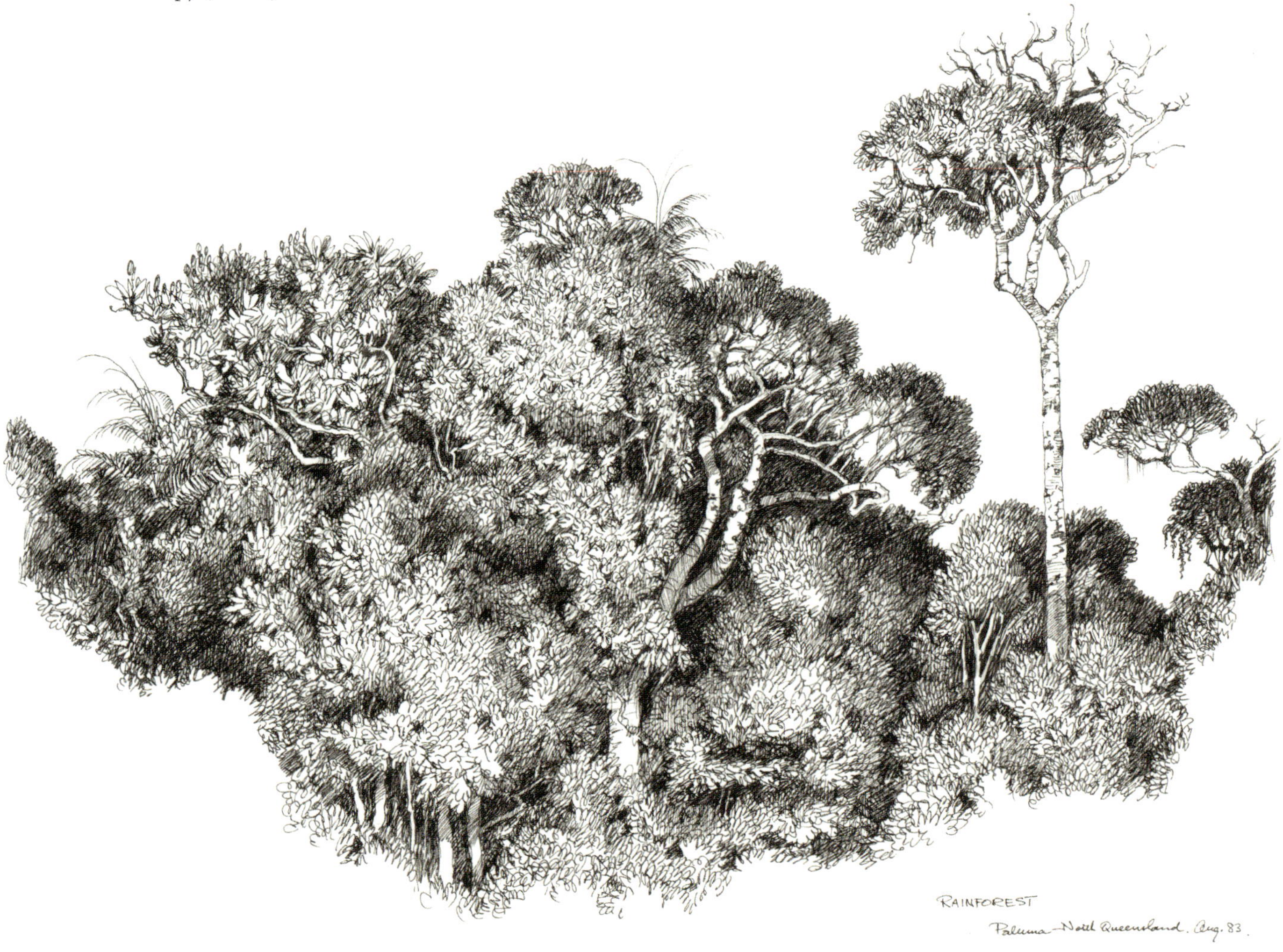

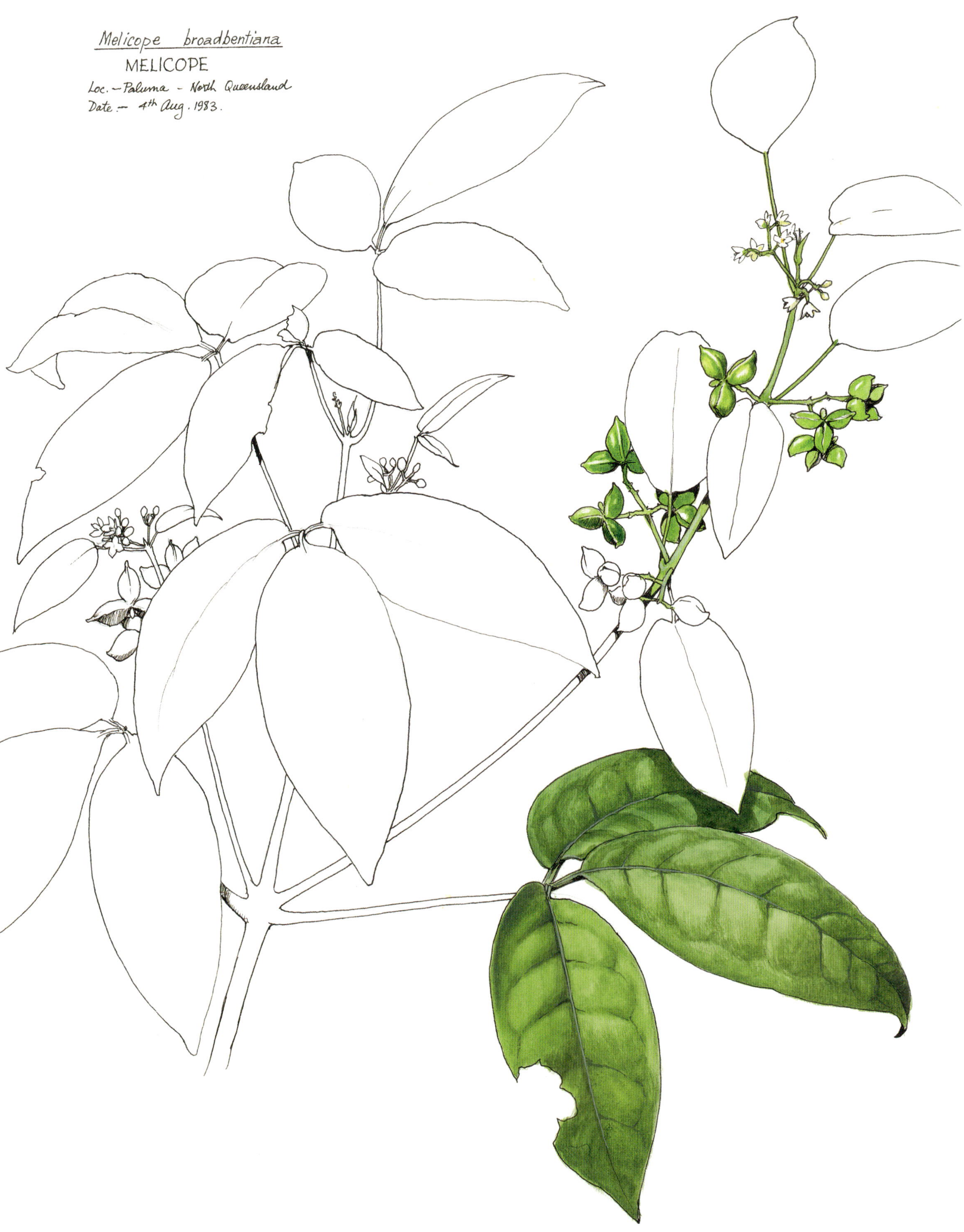
Melicope broadbentiana
MELICOPE
Loc. – Paluma – North Queensland
Date: – 4th Aug. 1983.

Musa banksii

Native Banana

MUSACEAE

→ 1993; pen and watercolour; 275 x 345mm
Private collection

↓ undated (c.1993); pencil; 350 x 335mm
State Library of New South Wales

Musa is from *Muz*, *Mauz* or *Mouz*, ancient Arabic names for banana.

banksii is named after Joseph Banks (1743–1820), the botanist who famously collected plants during James Cook's *Endeavour* voyage around the world.

Native Banana is a diminutive plant compared with the more commonly cultivated banana. These bananas are edible, but the numerous stony seeds make them difficult to eat.

The sketch opposite was done in the field at Cape Tribulation; the colour wash would have been added back at home. We often went bush to this botanically diverse area because it was what we so much enjoyed doing together: wandering, sharing delights in the forest while making small discoveries.

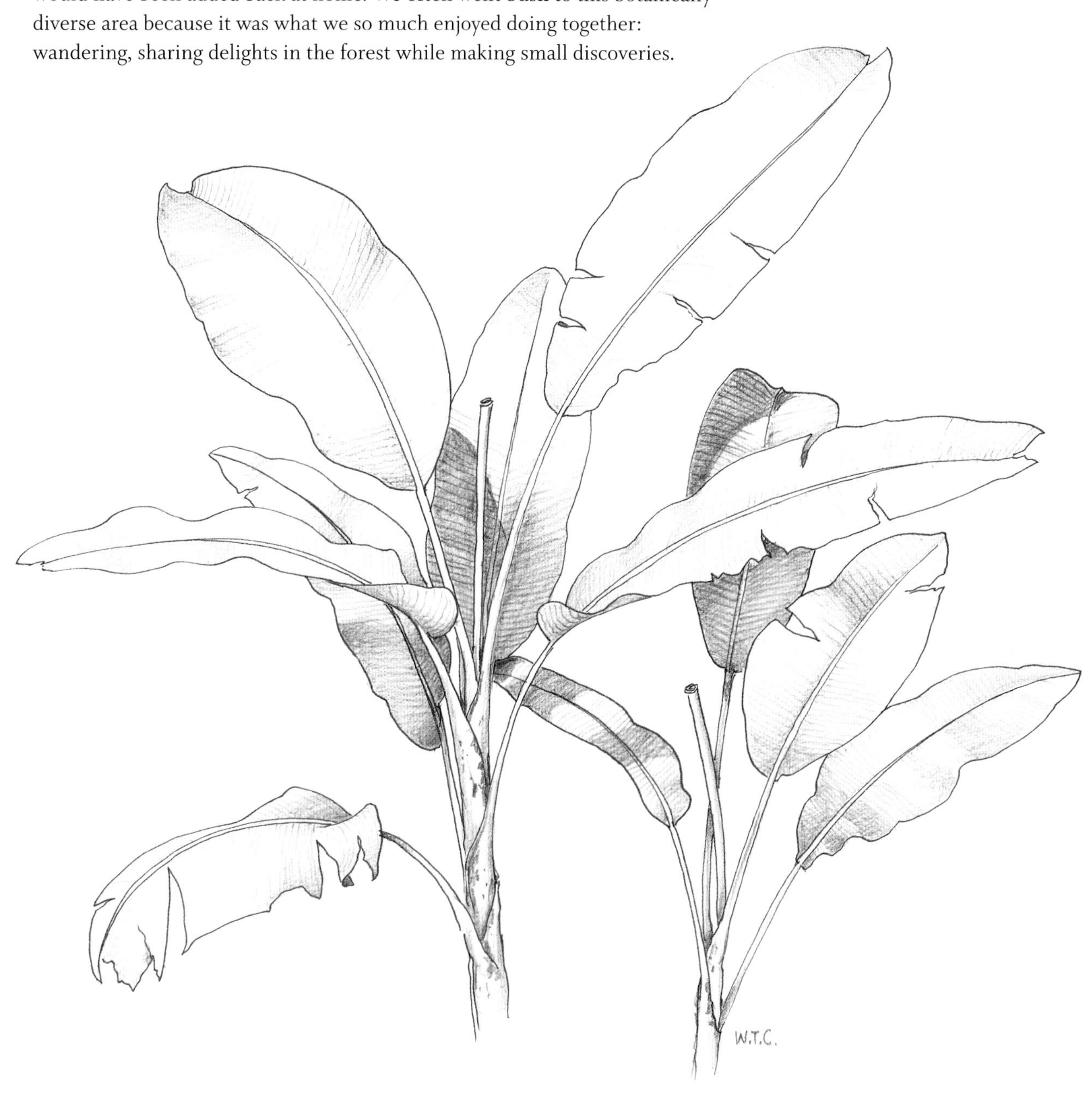

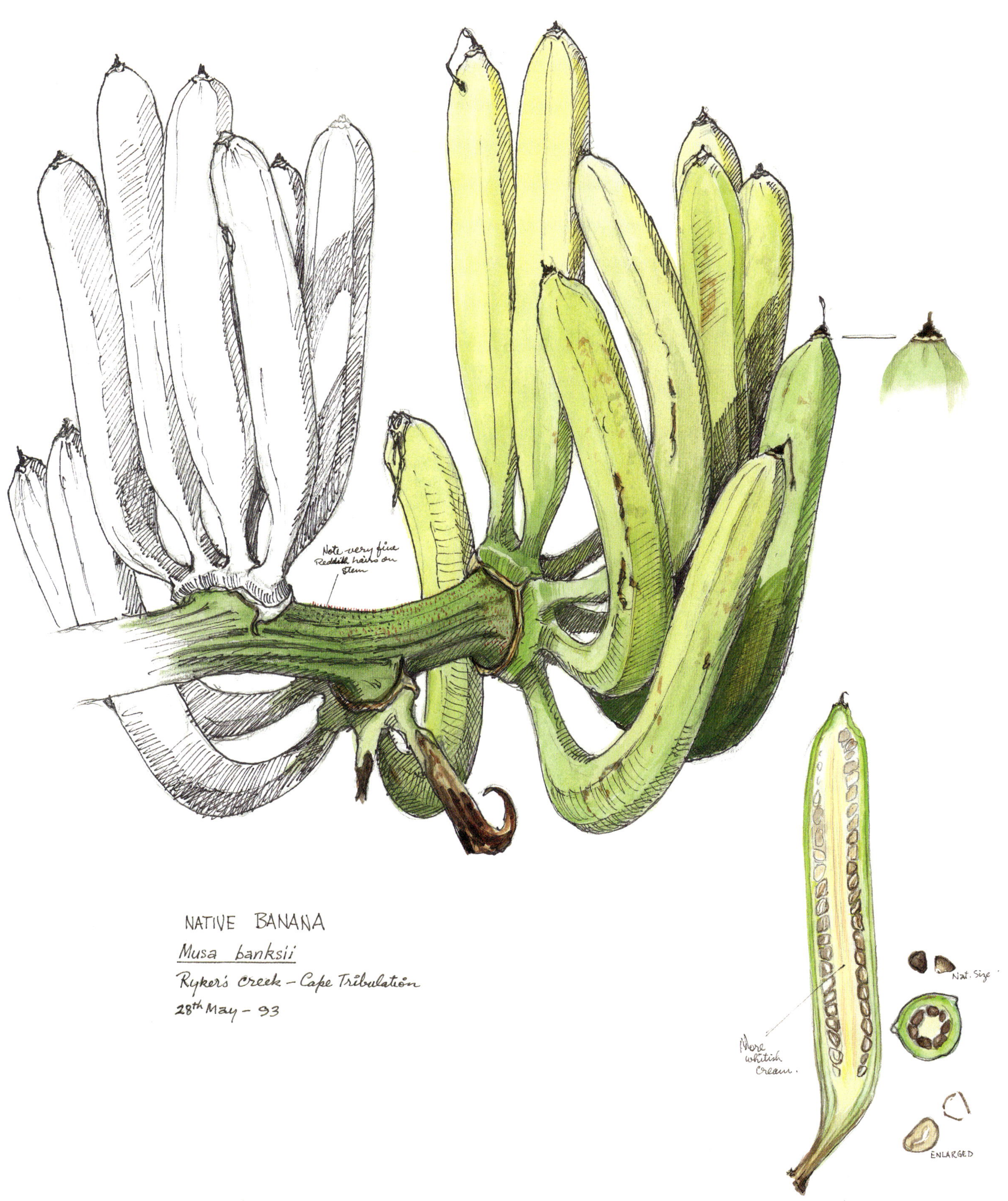
Note very fine Reddish hairs on stem
NATIVE BANANA
Musa banksii
Ryker's Creek – Cape Tribulation
28th May – 93
More whitish Cream.
Nat. Size
ENLARGED

Neololeba atra

Cape Bamboo

POACEAE

Neololeba is from *neo* (new) and *loleba*, a vernacular name.

atra is from *-atro* (black), possibly referring to the bamboo's black hairs.

This clumping Cape Bamboo grows to a height of about 7 metres in lowland tropical Queensland.

Both of these sketches were drawn during our first exciting trip, in 1981, to Iron Range National Park, where we camped among several of these bamboo stands near the idyllic West Claudie River. The drawing of the bamboo clump was done under the shelter of canvas. It was raining almost constantly for weeks, and the bamboo was in view from the camp table.

Bill noted in his diary on Saturday 11 July: 'Still raining! Built a lean-to of canvas so that we could get a fire going and dry some towels and clothes'.

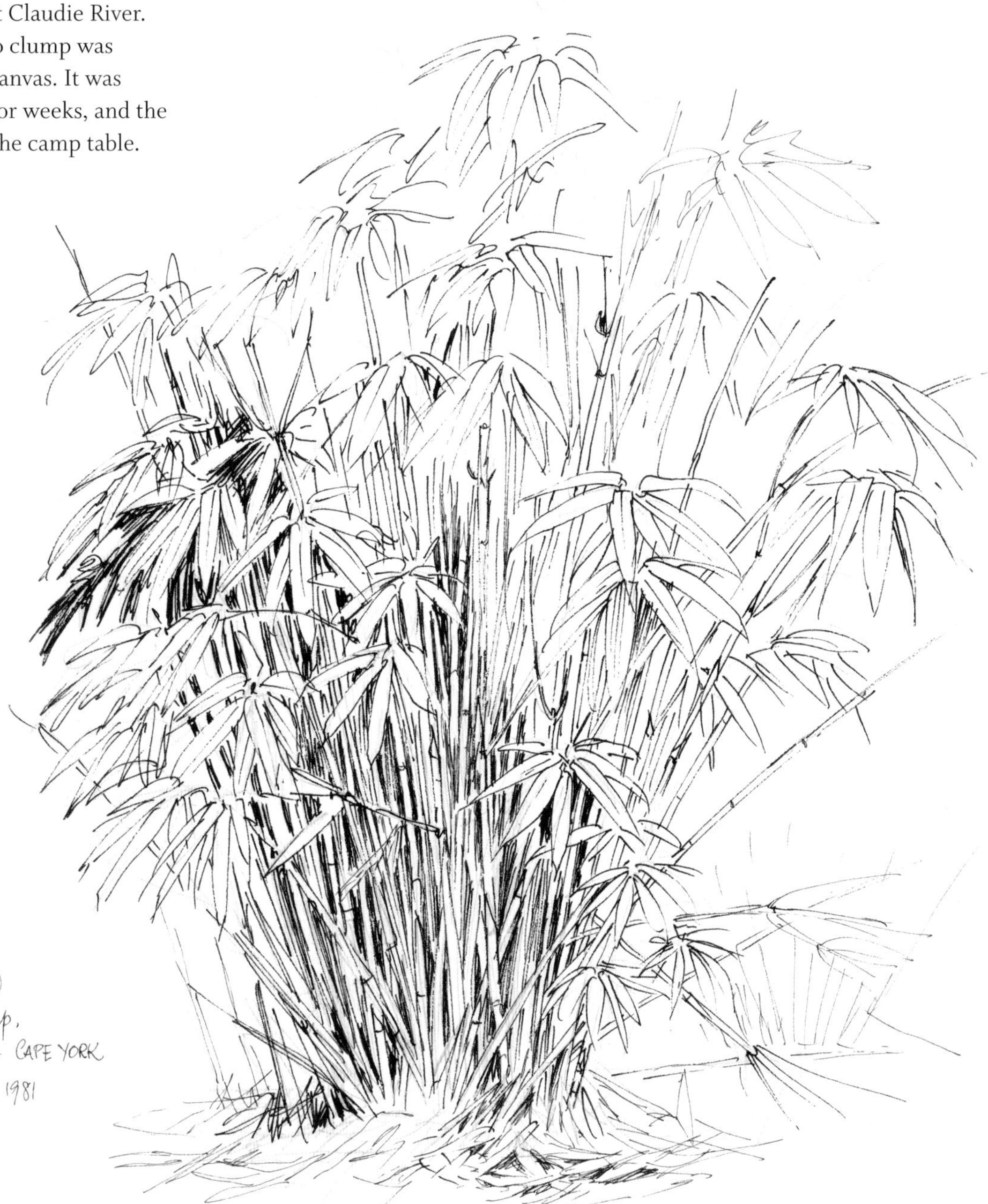

→ 1981; pen and pencil; 550 x 450mm
State Library of New South Wales

↓ 1981; pen and pencil; 370 x 210mm
State Library of New South Wales

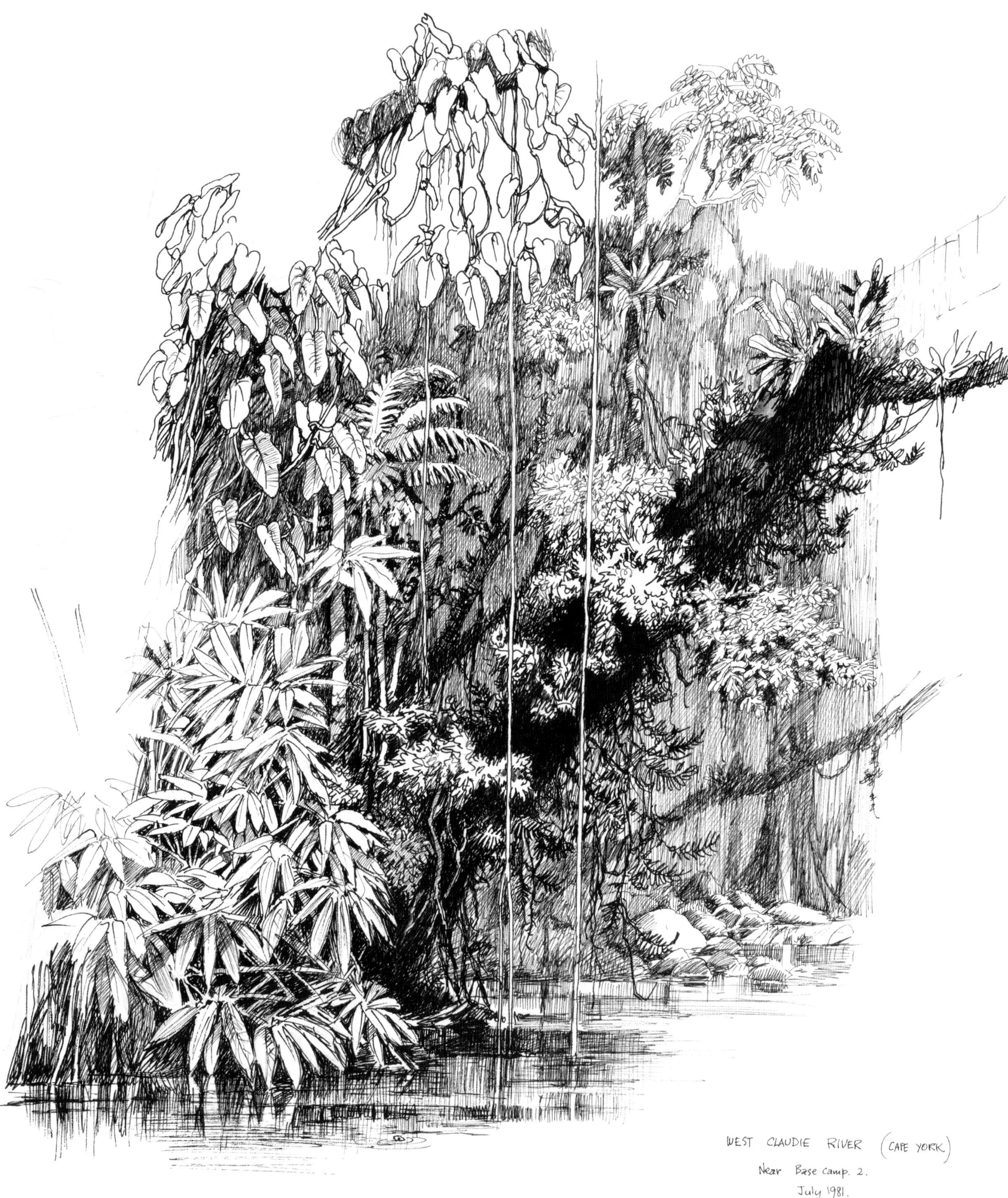
WEST CLAUDIE RIVER (CAPE YORK)
Near Base camp. 2.
July 1981.

Pararchidendron pruinosum

Snowwood or Monkey's Earrings

FABACEAE

Pararchidendron is from *para* (similar) and after the related genus *Archidendron*, referring to their similarities.

pruinosum is from *pruinosus* (a whitish powdery coating), probably referring to the pale timber.

Snowwood is a leguminous tree growing from southern New South Wales to Cooktown in north Queensland. These fruit paintings were done long before we concentrated on rainforest fruit but maybe the idea was already germinating. When we moved to tropical Queensland we didn't know many plants at all. Bill had the idea that he would paint any fruit we found and one day someone would give us names for them. From this emerged the fruit books.

→ undated; pen, pencil and watercolour
350 x 245mm
State Library of New South Wales

↓ 1983; pen and pencil; 163 x 270mm
National Library of Australia, 8070107

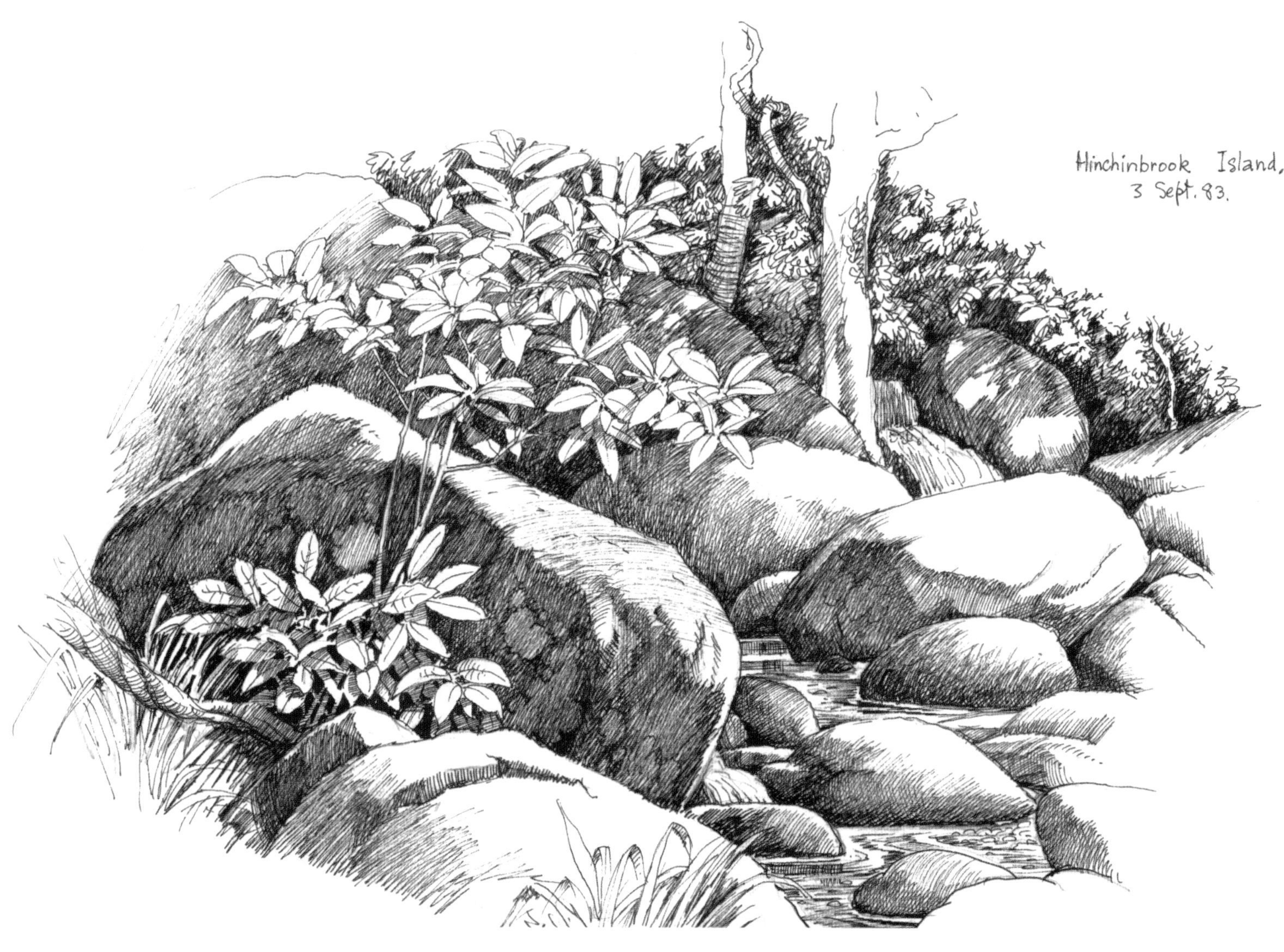

Parachidendron pruinosum
(Abarema sapindoides)
BUNGWAHL N.S.W.

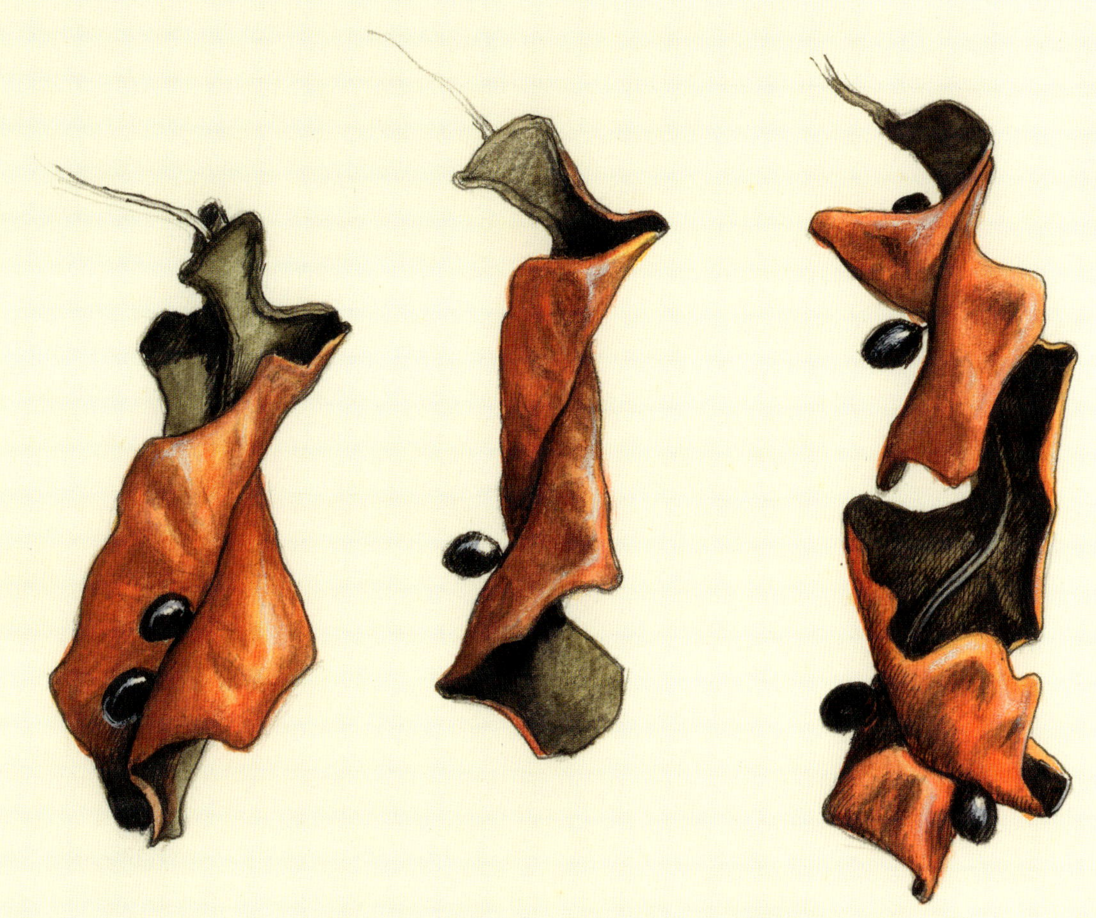

Seed Pods found on the
ground beneath the tree.
These were a little more brown than
the pods still hanging in the tree.
Some of the seeds that had come away
from the pods and lodged amongst
the leaf litter had already began
to sprout.

Pittosporum undulatum

→ 1985; pen and pencil; 370 x 300mm
State Library of New South Wales

↓ 1977; pen, pencil and watercolour
295 x 355mm
State Library of New South Wales

Sweet Pittosporum

PITTOSPORACEAE

Pittosporum is from *pitys* (pine or resin) and *spora* (seed), referring to the sticky seeds.

undulatum is from *undulatus* (undulating), referring to the wavy leaf margins.

Sweet Pittosporum is most commonly seen as a shrub but also grows to a tree 25 metres tall. It occurs naturally in various forms of rainforest and wet sclerophyll forest from East Gippsland in Victoria to the Gladstone area in Queensland. The sticky red seeds are dispersed mostly by birds, and Bill could have used this reference drawing with confidence for many bird paintings.

This plant was collected in the Ellenborough Falls area, not far from Gloucester Tops, where the drawings opposite were done.

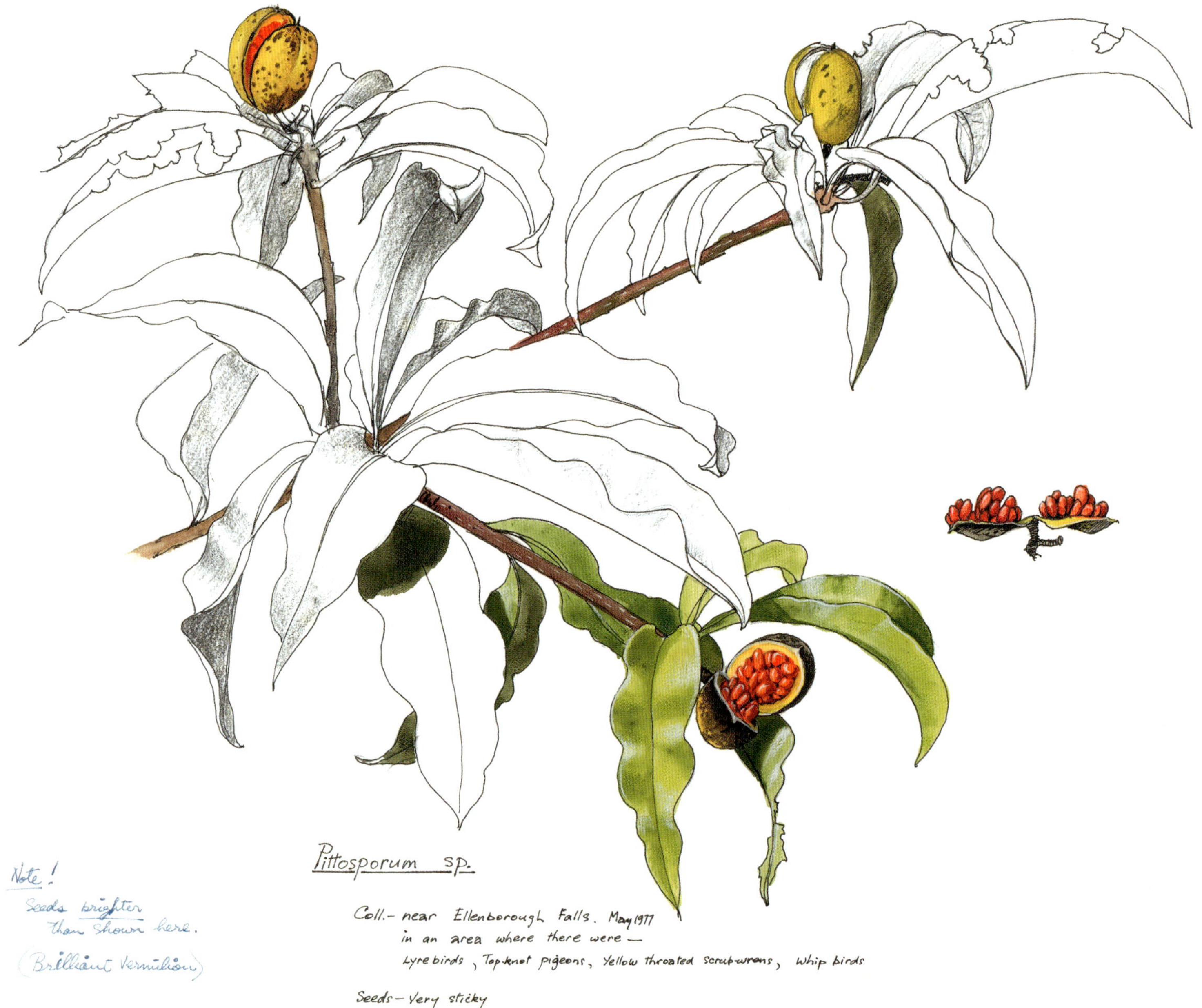

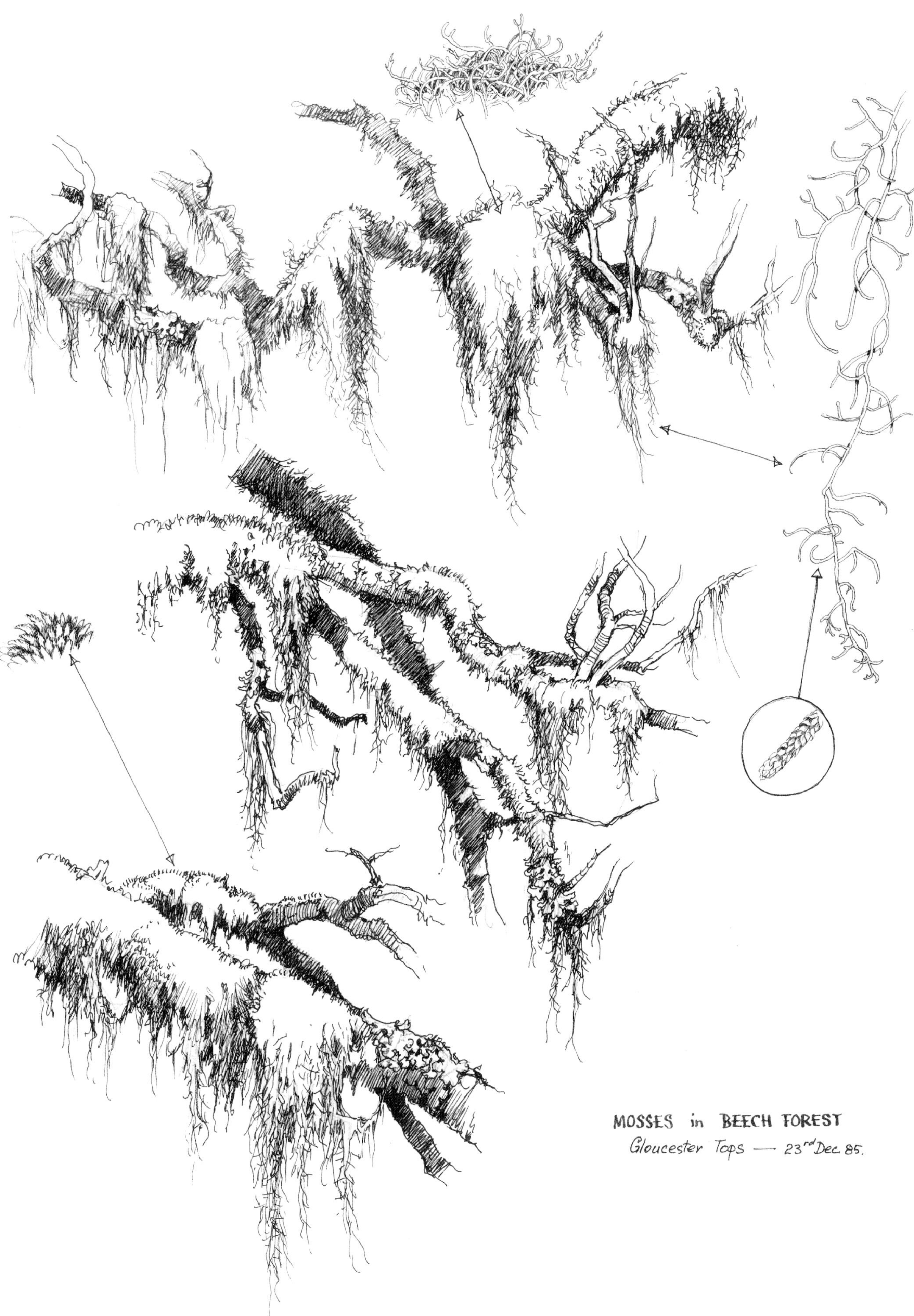
MOSSES in BEECH FOREST
Gloucester Tops — 23rd Dec 85.

Pleiogynium timorense

Burdekin Plum

ANACARDIACEAE

Pleiogynium is from *pleio* (more) and *gyne* (female), referring to the several carpels within the flower.

timorense is from Timor, where the type specimen was collected.

Burdekin Plum is a familiar tree in drier rainforest north from Brisbane. The fruit is edible, although quality is variable from tree to tree. Its wide distribution across the Pacific, through tropical Australia, New Guinea and South-East Asia, made it an ideal subject for use with the Umbrella Cockatoo or White Cockatoo, as both bird and plant occur together in Indonesia. The painting was published in *Cockatoos: A Portfolio of All Species*.

This specimen came from Forty Mile Scrub, an isolated deciduous vine thicket a couple of hours west of the Atherton Tablelands. It is an area of dry rainforest with many deciduous species. In this particular tropical forest, the plants are typically leafless for a month or two in spring, when it is usually dry.

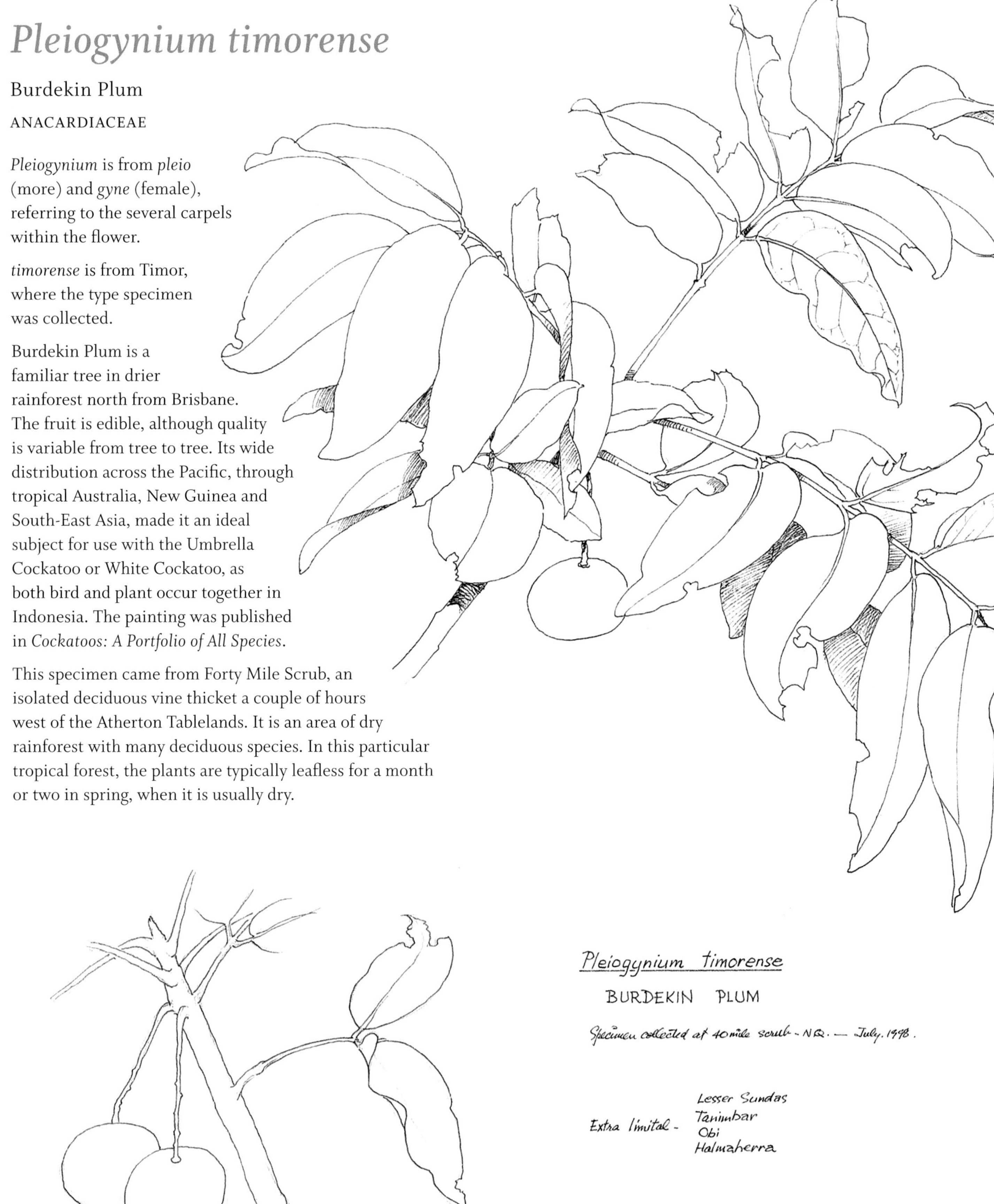

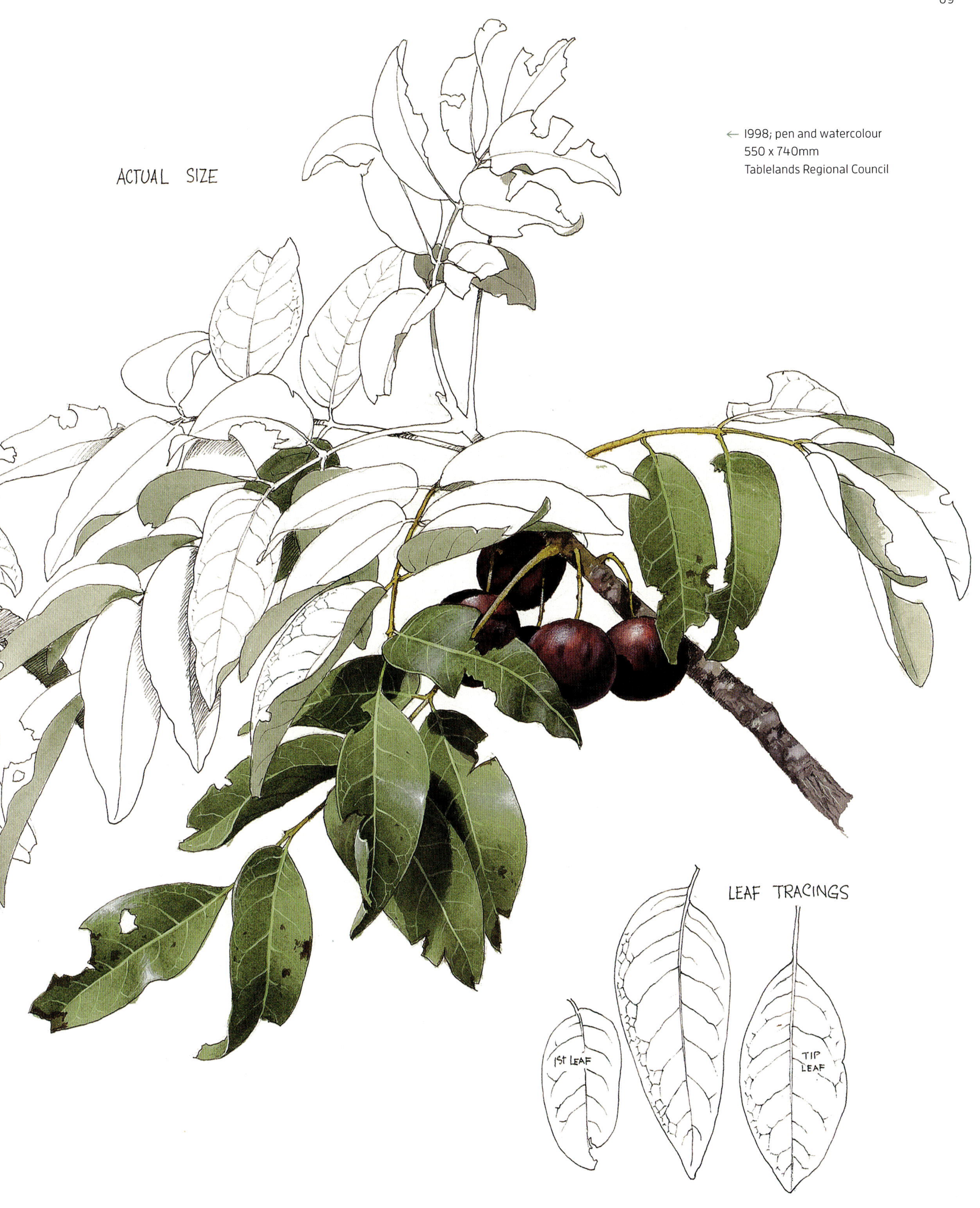

← 1998; pen and watercolour
550 x 740mm
Tablelands Regional Council

Pleomele angustifolia

Native Dracaena

ASPARAGACEAE

→ 1981; pen and watercolour; 380 x 380mm
State Library of New South Wales

↓ Tree Base, West Claudie River
1999; pen and pencil; 130 x 215mm
State Library of New South Wales

Pleomele is from *pleio* (more than usual) and *meli* (honey), referring to the copious amounts of nectar produced by the first species described as a *Pleomele* (*Pleomele fragrans*).

angustifolia is from *angustus* (narrow) and *folius* (leaved).

Native Dracaena is an understorey shrub occurring north from Mission Beach up to Cape York and the Northern Territory, and in New Guinea and Asia.

This sketch was made when Iron Range was much more remote than it is today. It is a small rainforested area on the east coast of Cape York Peninsula that contains elements of New Guinea among its plants, birds and mammals.

From Bill's diary:

> *Wed. 24th June 1981. Drew plants in camp. Saw Blue-breasted Pitta (Red Bellied) on the creek near camp, also Scrubfowl. After lunch we drove to the mission to get petrol. We had a 44 gal. drum put aside for us before we came so we filled both trucks and containers. Went down to Quintell Beach—beautiful—took some photographs. At night in camp saw Cus-Cus (2) in trees beside camp.*

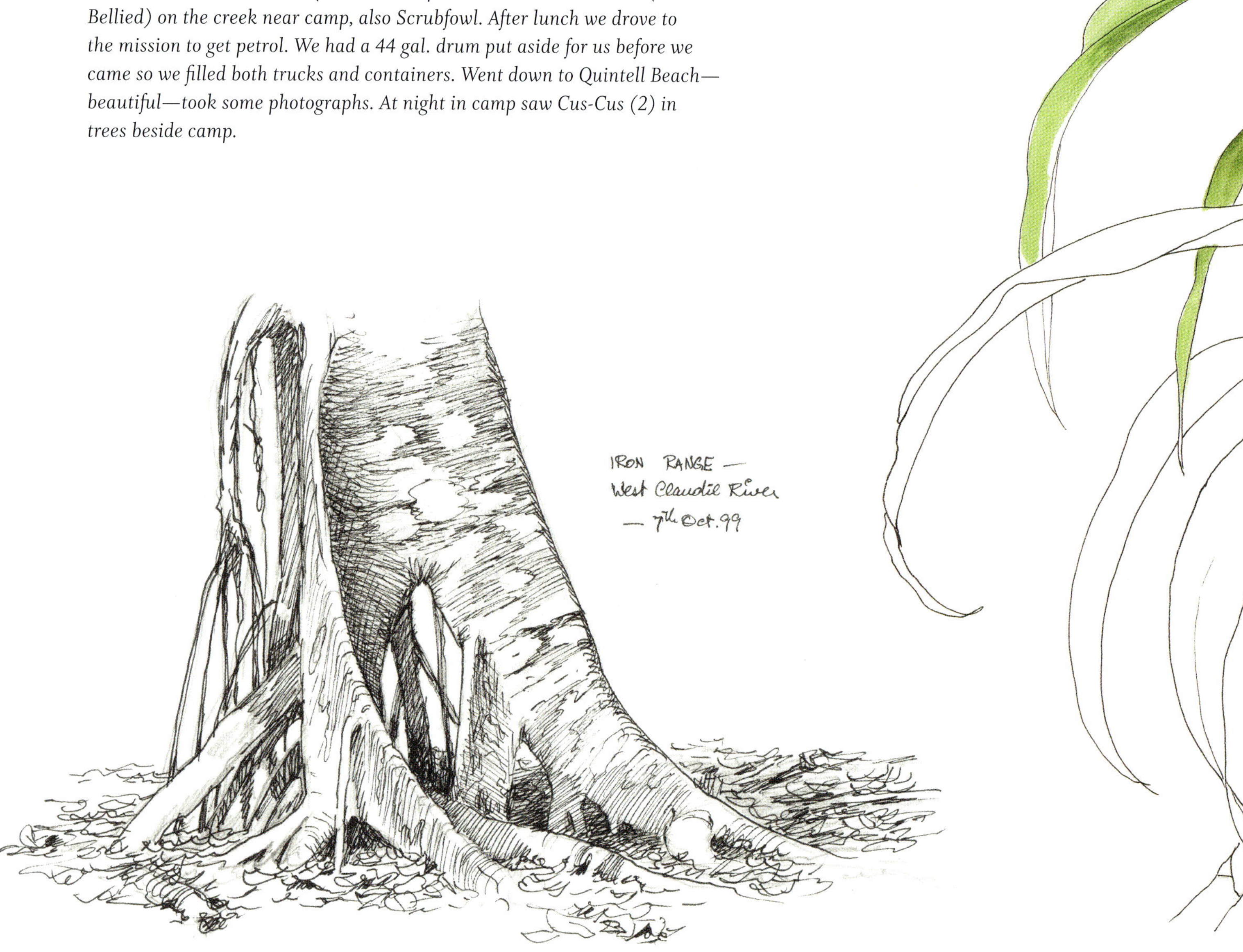

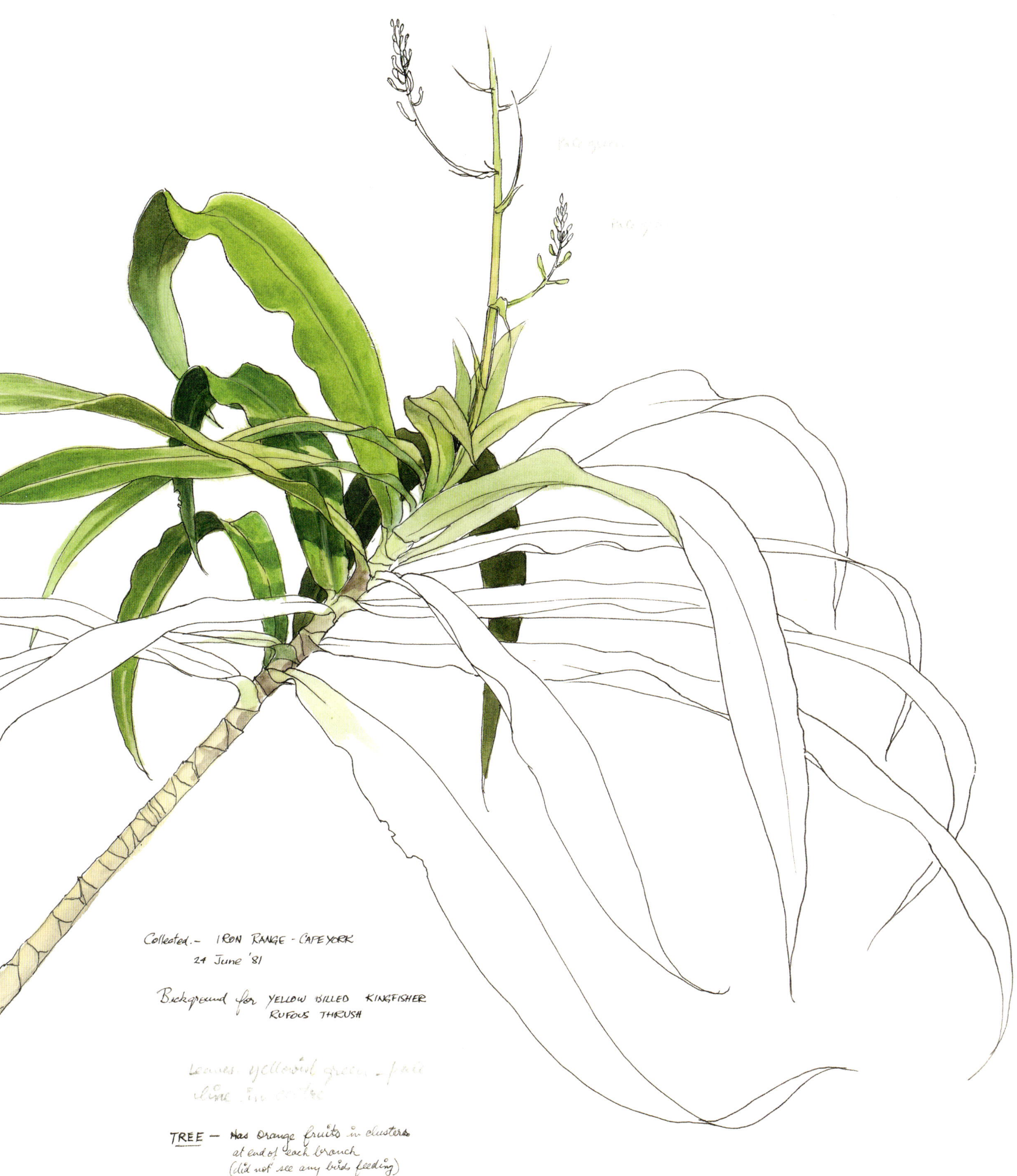
Collected.– IRON RANGE - CAPE YORK
24 June '81
Background for YELLOW BILLED KINGFISHER
RUFOUS THRUSH
TREE – Has orange fruits in clusters
at end of each branch
(did not see any birds feeding)

Pullea stutzeri

Hard Alder

CUNONIACEAE

Pullea is named in honour of August Adriaan Pulle (1878–1955), Dutch botanist and professor.

stutzeri is named after John Julius Stutzer (1827–1867), lecturer in English literature and modern history.

Hard Alder is a tree often conspicuous along rainforest edges because of the frequently present, bright red new leaves. The tree grows to more than 40 metres in the Wet Tropics of north Queensland.

Bill was always on the lookout for leaves with colour or deformities to add interest to paintings. This dead leaf was typical of such subjects.

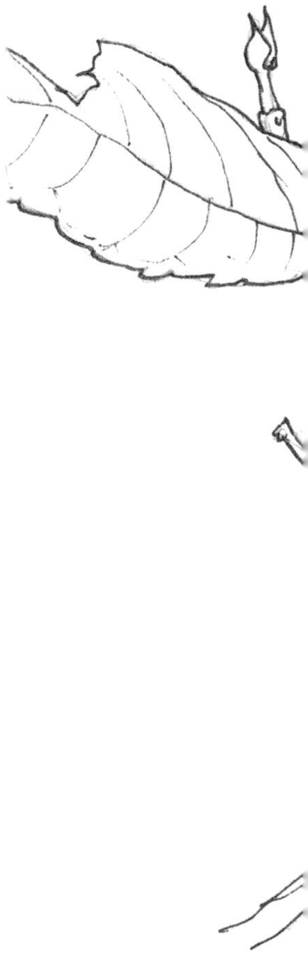

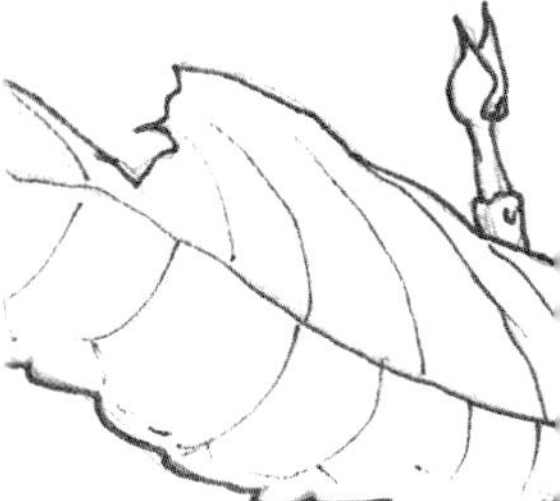

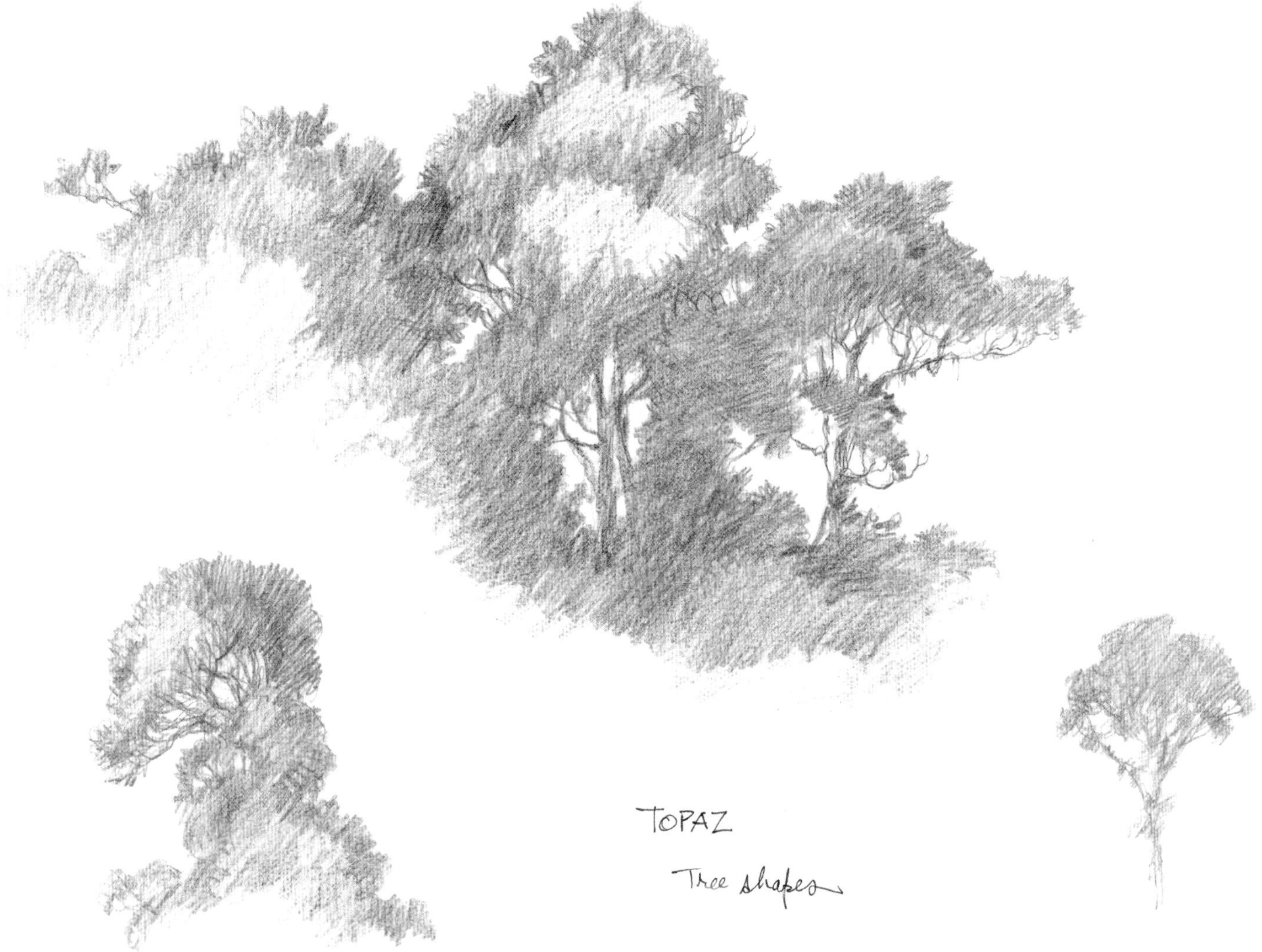

TOPAZ

Tree shapes

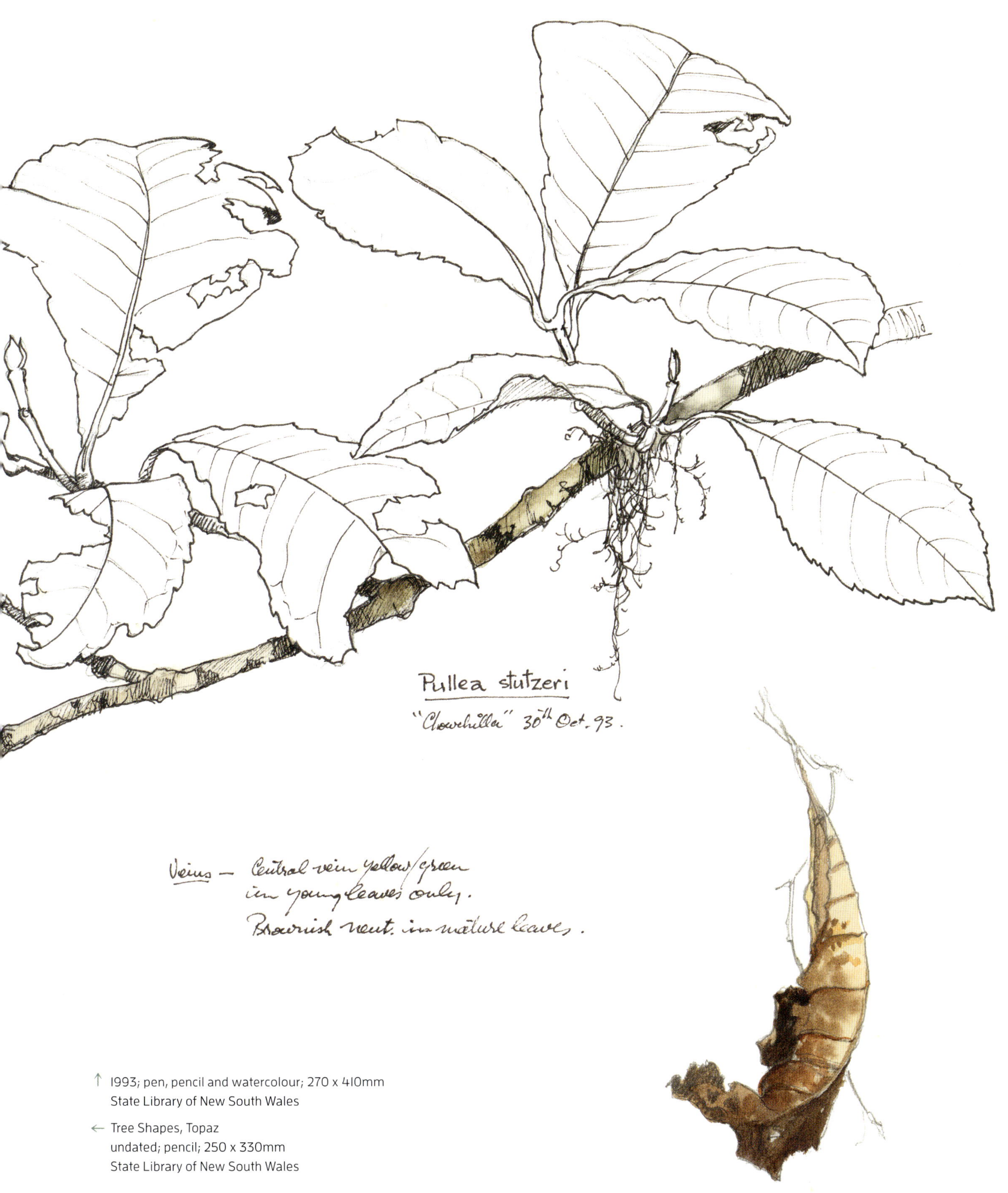

↑ 1993; pen, pencil and watercolour; 270 x 410mm
State Library of New South Wales

← Tree Shapes, Topaz
undated; pencil; 250 x 330mm
State Library of New South Wales

Rhus taitensis

Sumac

ANACARDIACEAE

Rhus is the Latin transcription of the Greek name *rhous* for *Rhus coriaria* or Tanner's Sumac, which is the Middle Eastern spice sumac, derived from the dried fruit. The leaves are also dried and used for dyeing.

taitensis means 'from Tahiti', where the type specimen was collected.

Sumac is a tree that grows to about 30 metres, mostly at lower altitudes. Because Sumac has clustered leaves and reddish new growth, Bill chose it for the plate of the Lesser Sulphur-crested Cockatoo and the Citron-crested Cockatoo in *Cockatoos: A Portfolio of All Species*. The plant and the birds occur on the Indonesian island of Sumba, where habitat loss is significant. Sumac probably persists there because it seems to be less affected by habitat disturbance than many other plants.

→ 1998; pen, pencil and watercolour; 440 x 560mm
State Library of New South Wales

ACTUAL SIZE

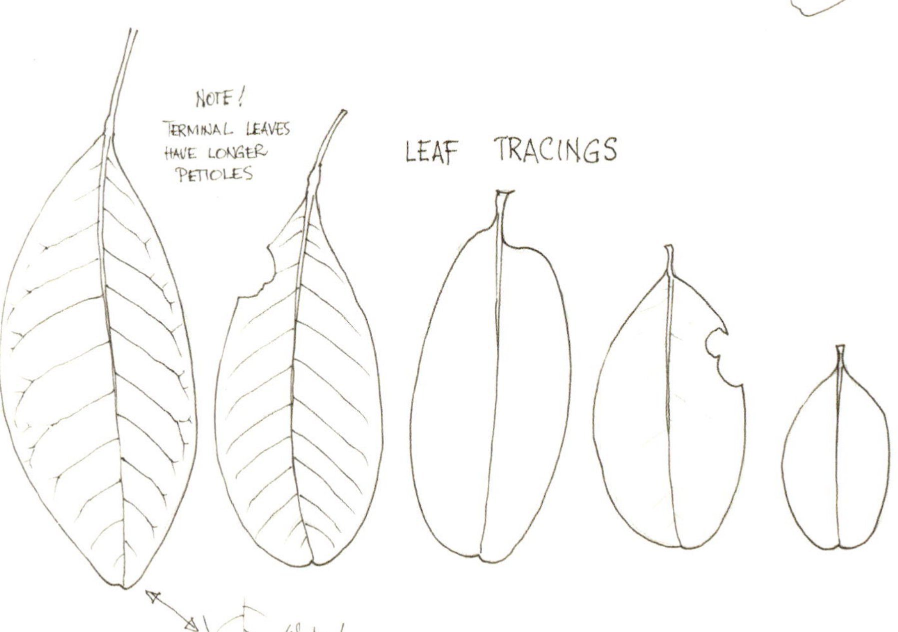

- SUMAC -

Rhus taitensis

Loc. – Daintree river — 10th June 98

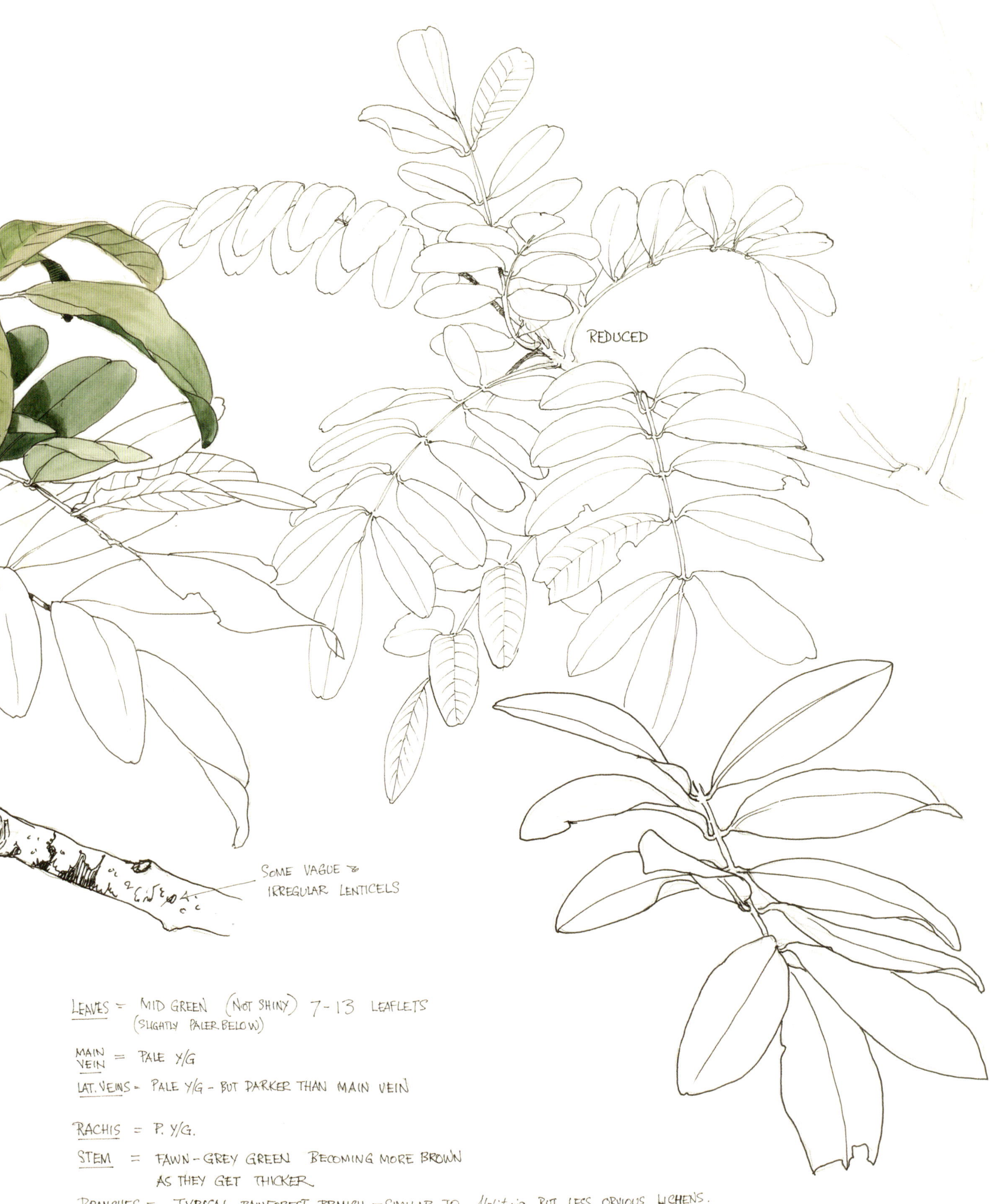
REDUCED
SOME VAGUE & IRREGULAR LENTICELS
LEAVES = MID GREEN (NOT SHINY) 7-13 LEAFLETS
(SLIGHTLY PALER BELOW)
MAIN VEIN = PALE Y/G
LAT. VEINS = PALE Y/G - BUT DARKER THAN MAIN VEIN
RACHIS = P. Y/G.
STEM = FAWN-GREY GREEN BECOMING MORE BROWN
AS THEY GET THICKER
BRANCHES = TYPICAL RAINFOREST BRANCH - SIMILAR TO *Alphitonia* BUT LESS OBVIOUS LICHENS.

Ryparosa kurrangii

ACHARIACEAE

→ Lowland Rainforest
1983; pencil and pen; 360 x 270mm
Private collection

↓ 1992; pen and pencil; 210 x 145mm
State Library of New South Wales

Ryparosa is from *rhyparos* (dirty or sordid), referring to the plant's hair colour.

kurrangii is after the Kuku Yalangi name *kurrangi* (Cassowary), referring in part to the fruit being taken by Cassowaries.

Ryparosa kurrangii is a sub-canopy rainforest tree restricted to the Cape Tribulation area within Daintree National Park. It is cauliflorous, meaning that the flowers and fruit emerge from the trunk. The fruit hang on pendulous stems and are eaten by Cassowaries with impunity, even though they are poisonous to humans.

This species was previously known as *Ryparosa javanica*, as annotated on the drawing.

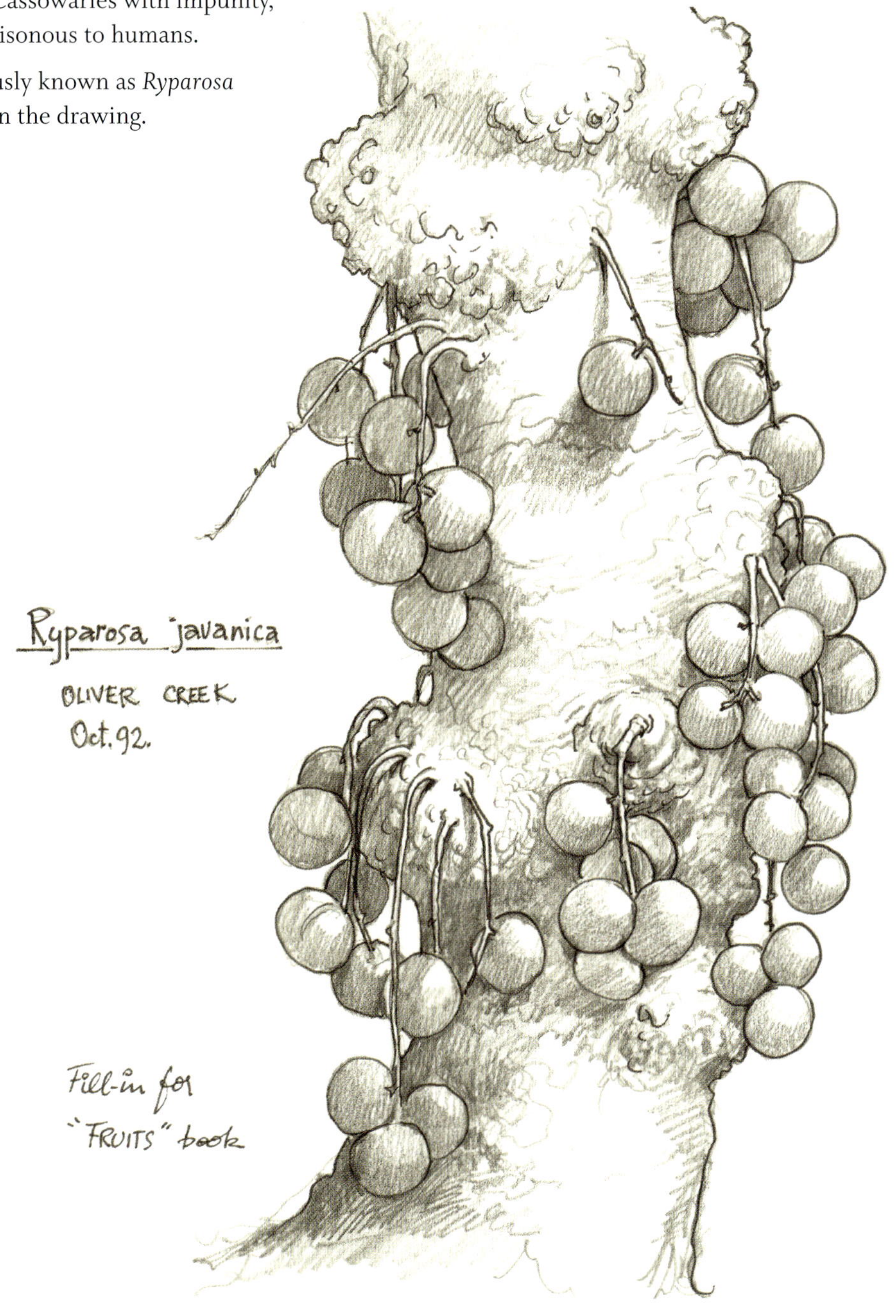

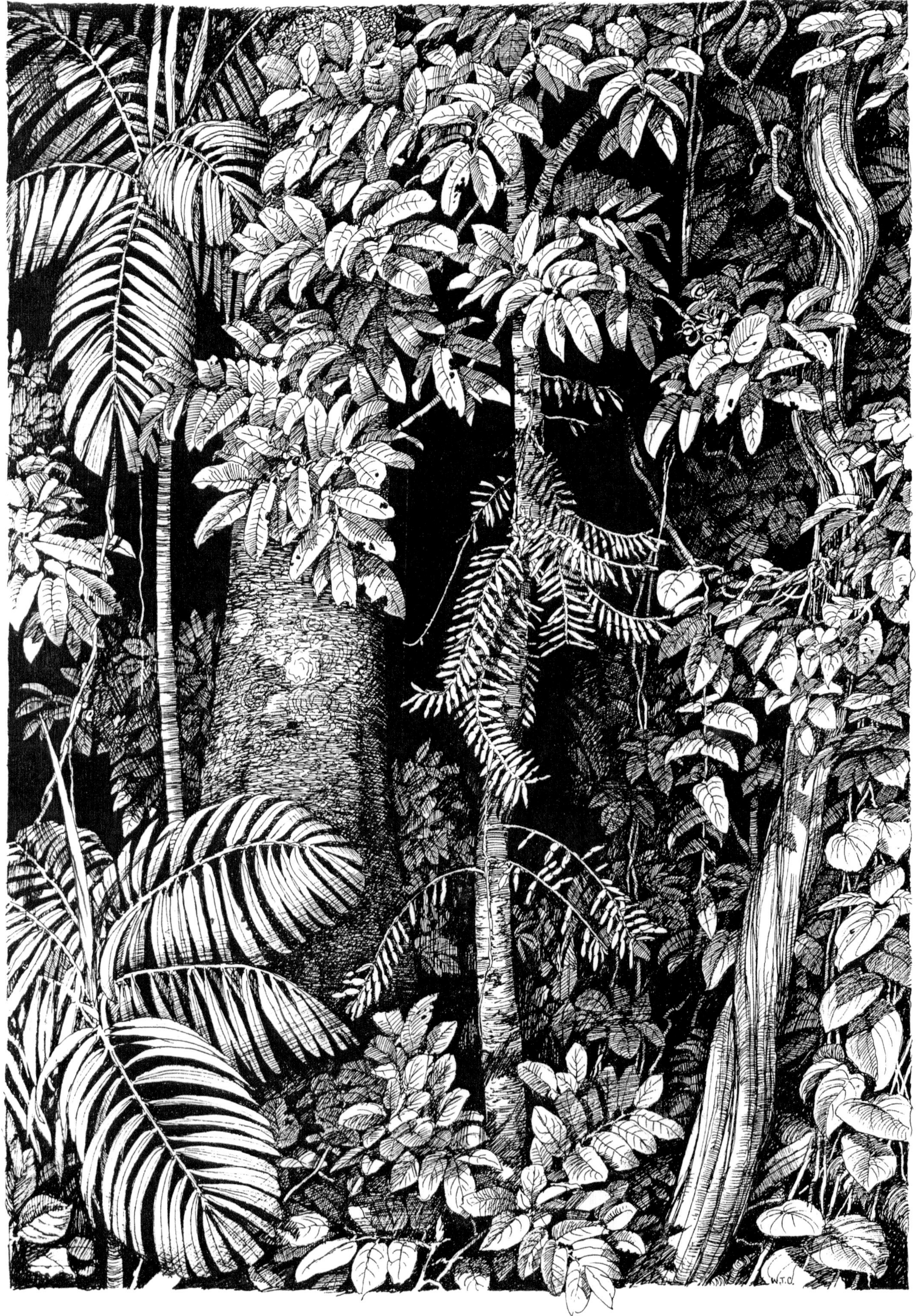

Sarcotoechia protracta

SAPINDACEAE

Sarcotoechia is from *sarco-* (fleshy) and *toechos* (wall or enclosure), referring to the fleshy-walled fruit.

protracta is from *protractus* (drawn out or lengthened), reference unknown.

Sarcotoechia protracta is a sub-canopy tree growing to about 13 metres within the Wet Tropics rainforests of north Queensland. The fruits are typical of the Sapindaceae family in having a black seed within a colourful aril held inside a capsule of a contrasting colour. Bill was commissioned to paint a large acrylic painting of Australian King Parrots in a lush jungle setting, with Bird's Nest Ferns and colourful fruit.

He wrote in his diary in November 1988:

> *Saw 2 male and at least 2 female riflebirds in the Sarcotoechia protracta in front of the walkway. They were extracting the seeds from the almost open fruits. I saw one male holding a fruit under his left foot while he prised at it with his bill in an attempt to open it. He finally flew off with the fruit in his bill so I didn't see whether he was successful. Another was holding a fruit under his right foot.*

He observed birds feeding on the fruit again on 12 May 1990: 'The *Sarcotoechia protracta* in front of the house is in fruit (coloured) … King Parrots feeding in this tree (eating fruits)'.

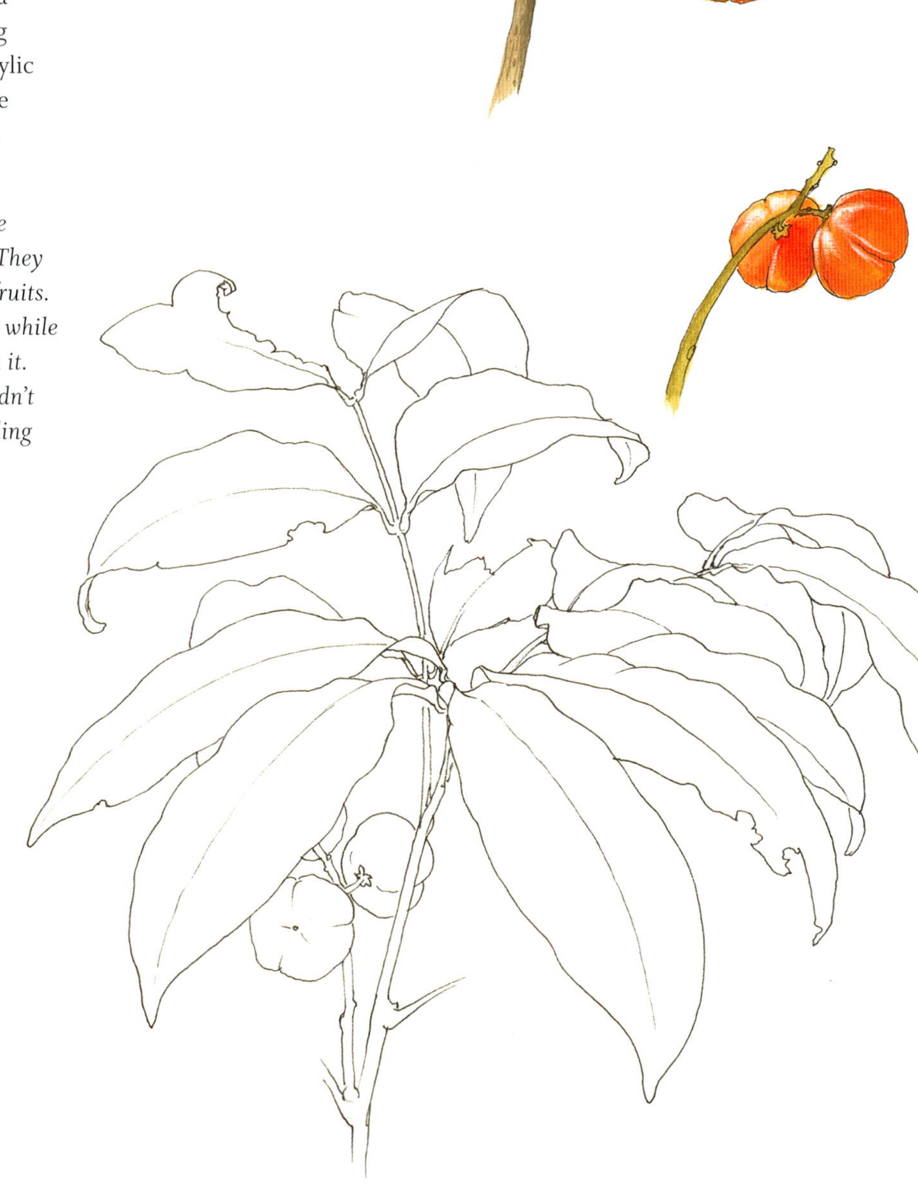

→ 1988; pen and watercolour; 540 x 780mm
State Library of New South Wales

SAPINDACEAE

Sarcotoechia protracta

"CHOWCHILLA"
TOPAZ — 17th Nov. 88

King Parrots feeding on fruits
(extracting seeds)

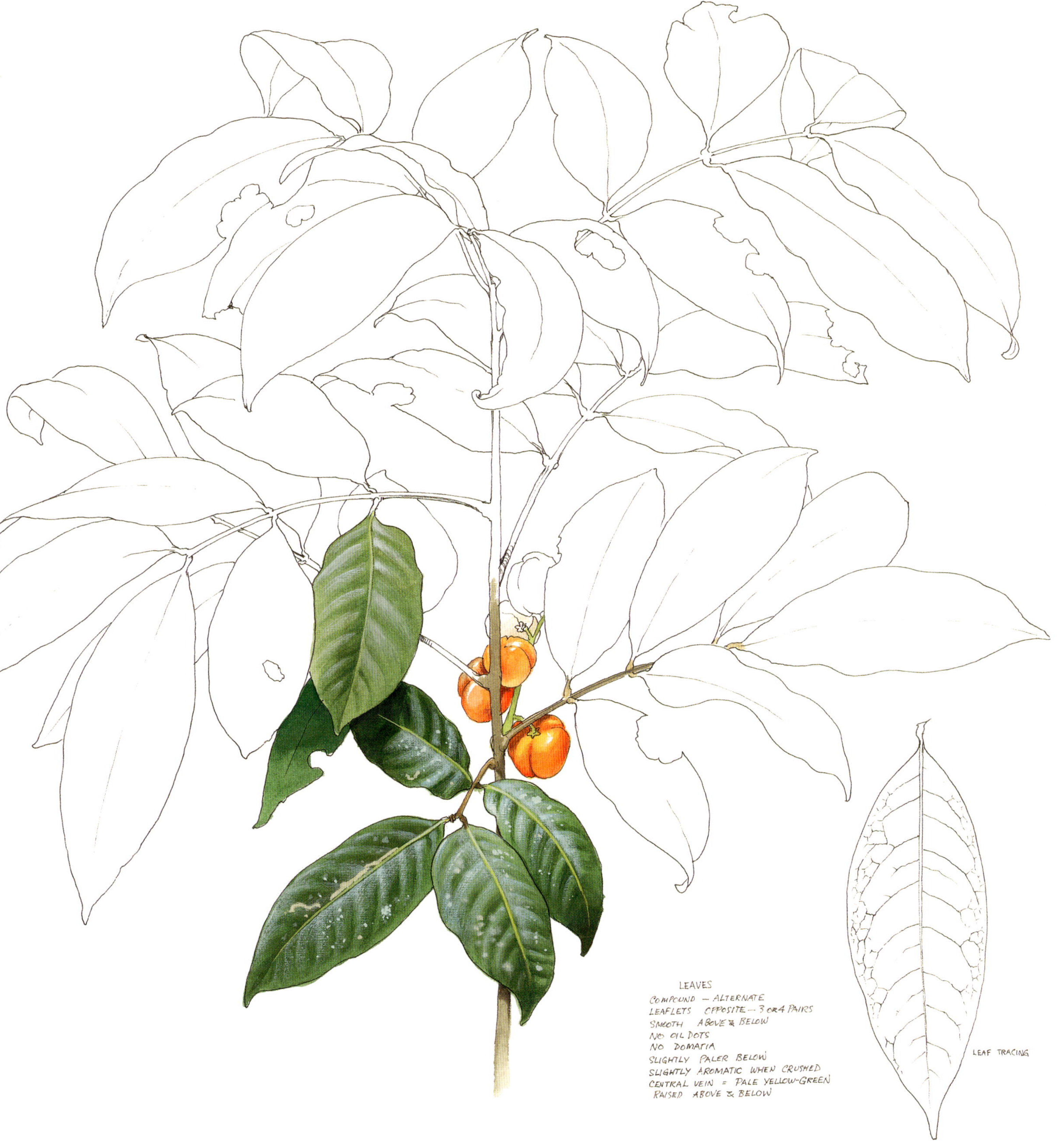

Schefflera actinophylla

Umbrella Tree

ARALIACEAE

Schefflera is named in honour of German physician and botanist Johann Peter Ernst von Scheffler (1739–1809), which is contrary to several sources erroneously bestowing that honour on writer and physician Jacob Christoph Scheffler.

actinophylla is from *actinos* (ray) and *-phyllus* (leaved), referring to the digitate leaves.

Umbrella Trees, with their especially large shiny leaves, are conspicuous plants often growing as epiphytes in tropical Australia north from Rockhampton. The plant also occurs in New South Wales, where it is now a naturalised weed. Bill especially liked to use Umbrella Tree leaves in forest backgrounds of paintings. The painting at right shows the common epiphytic habit in which seeds have sprouted on a branch and have sent roots towards the ground and branches upwards.

Bill made several mentions of bird behaviour around the Umbrella Tree in his diary:

> *26th Feb. 1993. A couple of days ago I saw 16 Figbirds feeding on the fruits of an Umbrella Tree over the driveway—later saw Catbird and a female riflebird feeding on the same bunch of fruit and in the company of 5 or 6 Figbirds.*
>
> ...
>
> *8th Feb. 1998. Yesterday saw a Pademelon on the driveway at about 17.30 hrs. It was feeding on fallen fruits of Schefflera actinophylla. These fruits are at the stage of being almost black.*
>
> ...
>
> *9th Dec. 1999 ... Watched a female coloured riflebird feeding on the flowers of a Schefflera on the driveway. I'm convinced it was taking nectar as they do on the Thunbergia in the garden. It went from open flower to open flower. A male riflebird was in the nearby tree. It went over and worked its way up to where the female coloured bird was feeding in the crown of the flowering head. When within 30cm of the 'female' it lowered its head (it was just above the female) and gaped to show the yellow mouth. It did this a few times and then went into full display—wings clapping and head from side to side. The brown bird took no notice whatsoever.*

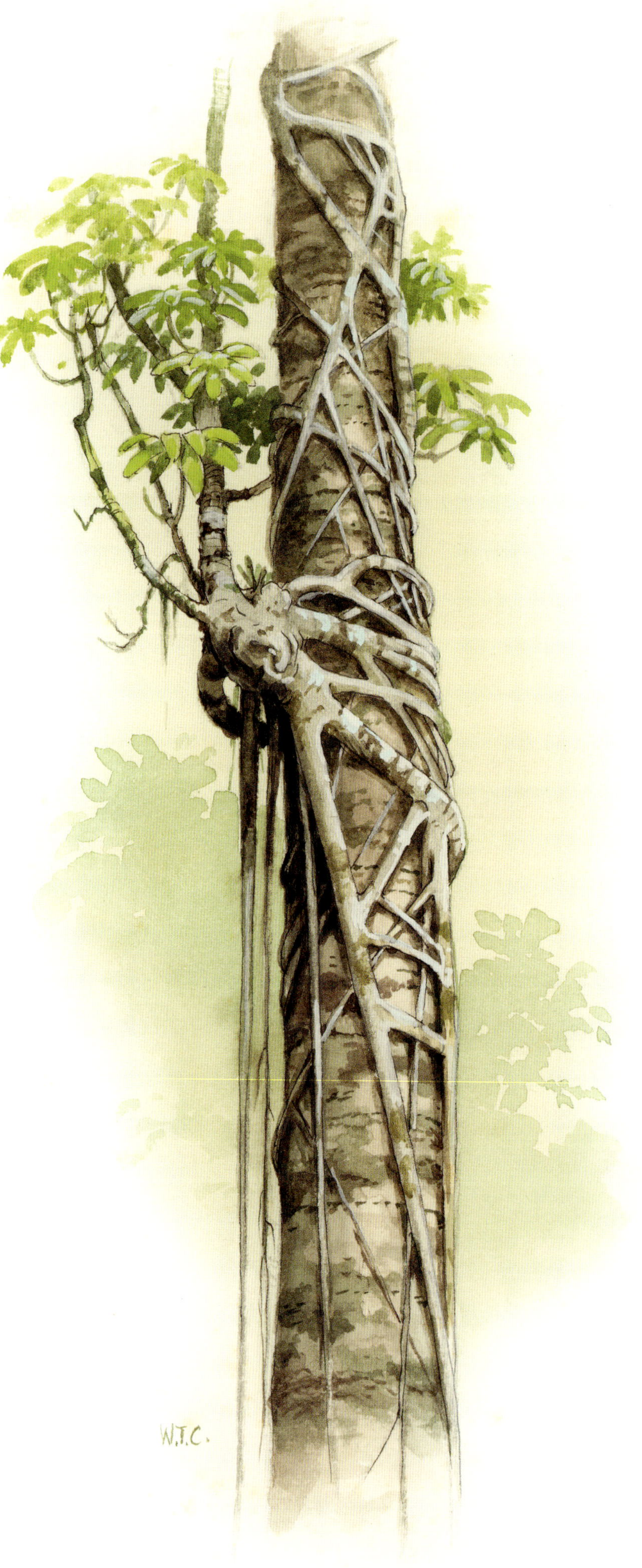

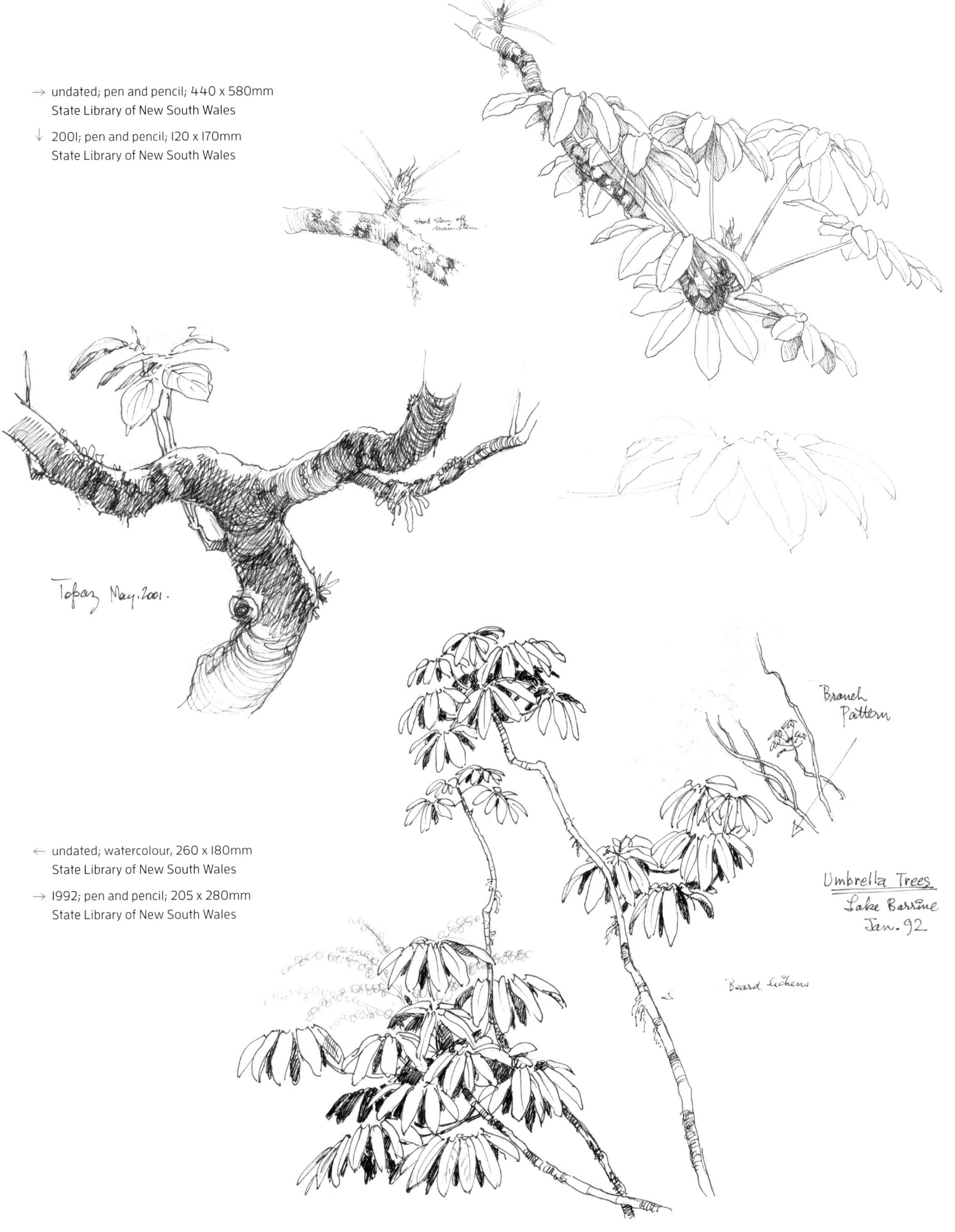

→ undated; pen and pencil; 440 x 580mm
State Library of New South Wales

↓ 2001; pen and pencil; 120 x 170mm
State Library of New South Wales

← undated; watercolour, 260 x 180mm
State Library of New South Wales

→ 1992; pen and pencil; 205 x 280mm
State Library of New South Wales

Semecarpus australiensis

Native Cashew or Tar Tree

ANACARDIACEAE

Semecarpus is from *semi-* (half) and *carpos* (-fruited), referring to what appears to be the fruit but is actually half fruit and half receptacle.

australiensis is from Australia.

Native Cashew is a tree of tropical lowlands north from Townsville, as well as in the Northern Territory and New Guinea. The fleshy orange fruit-like parts are actually fruit stalks and the dry nut section is the real fruit. Some people have severe allergic reactions to the irritant sap and fruit stalks.

Native Cashew is easy to find in north Queensland and also occurs on New Britain in Papua New Guinea along with the Blue-eyed Cockatoo, so it was chosen for the plate of that bird in *Cockatoos: A Portfolio of All Species*.

→ 1999; pen and watercolour; 550 x 750mm
Private collection

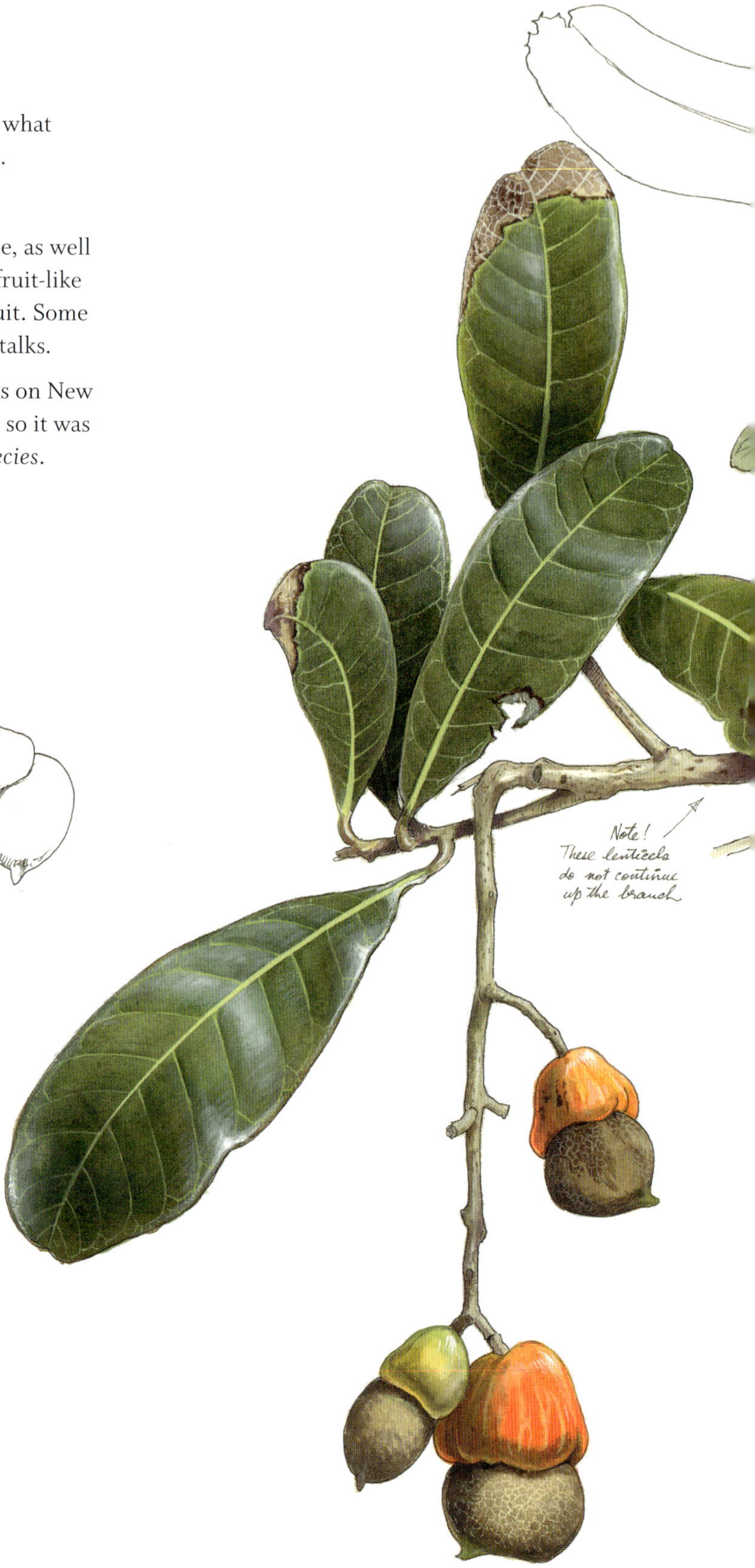

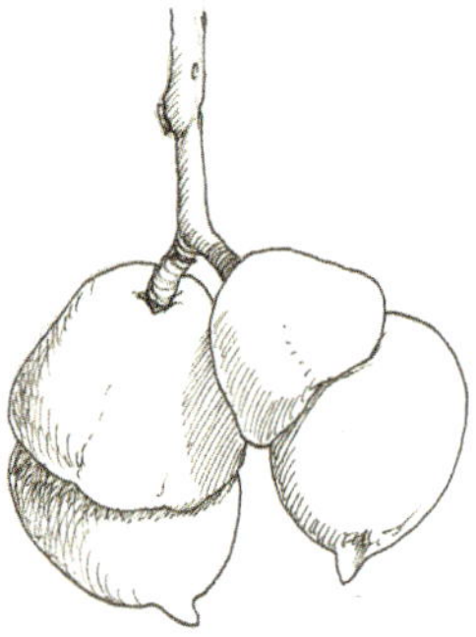

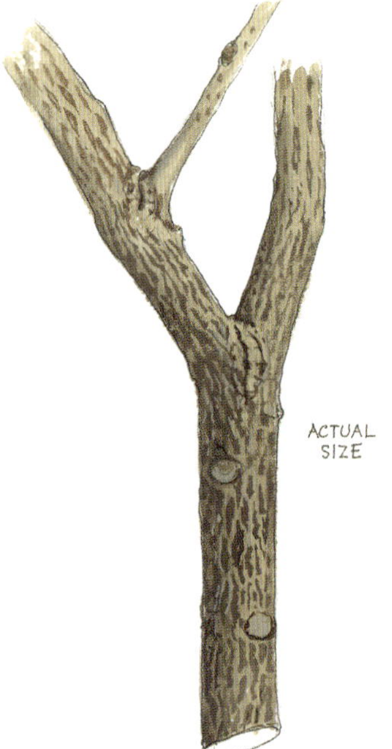

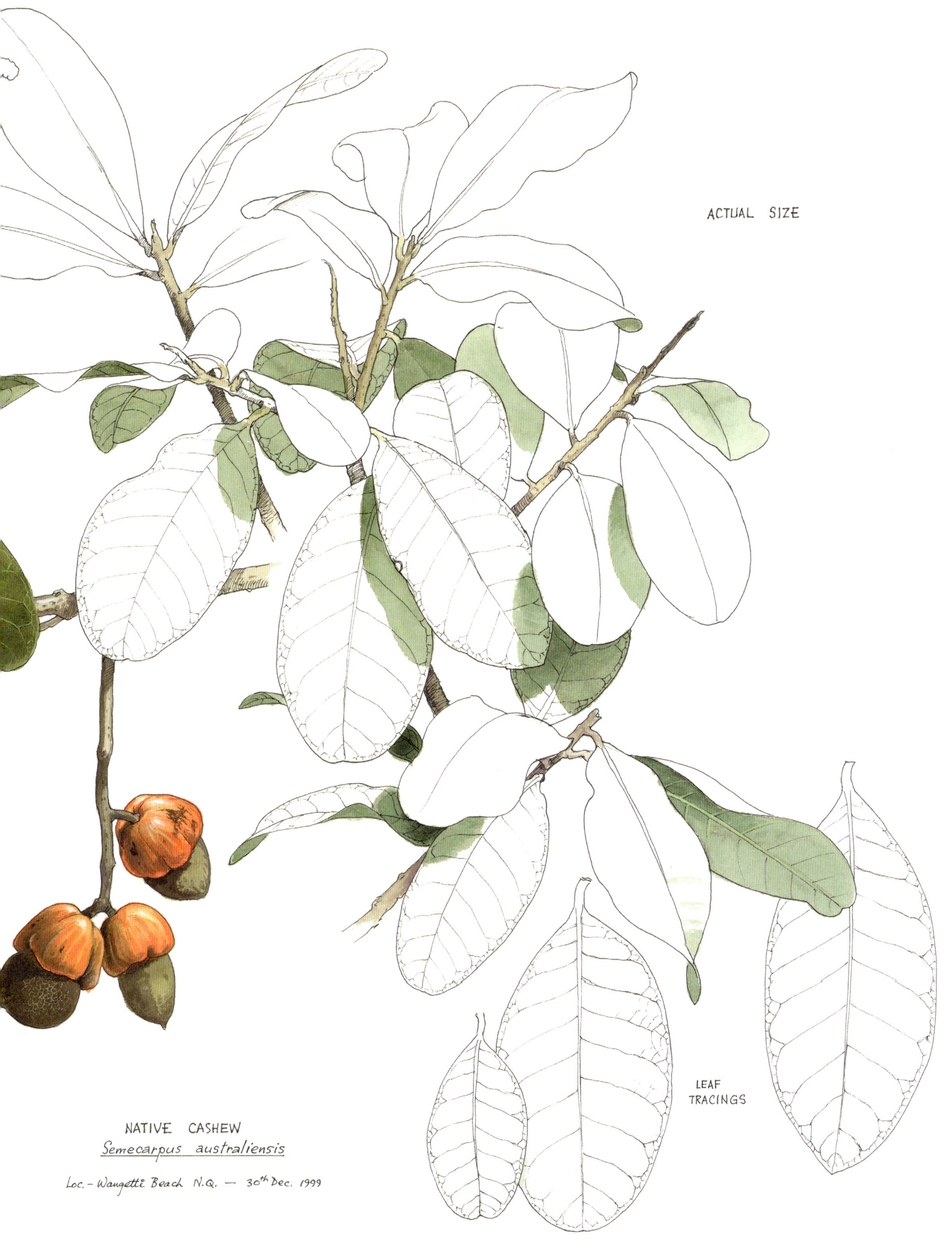
ACTUAL SIZE
LEAF TRACINGS
NATIVE CASHEW
Semecarpus australiensis
Loc. - Wangetti Beach N.Q. — 30th Dec. 1999

Sloanea australis subsp. *parviflora*

Maiden's Blush or Blush Alder

ELAEOCARPACEAE

Sloanea is named in honour of Hans Sloane (1660–1753), physician, naturalist and collector, whose private collection became the founding collection of the British Museum at his bequest.

australis is from southern.

parviflora is from *parvus* (small) and *-florus* (-flowered).

Maiden's Blush trees are forest giants with wonderful high-arching plank buttresses; this subspecies is restricted to the Wet Tropics. The large leaves for the painting of Regent Bowerbirds were easy enough to collect from low branches but Bill resorted to a .22 rifle to shoot down a small stem with fruit that was visible high up in the forest canopy.

He recorded animal activity around the tree in his diary:

> *27th Dec. 1995. Walked in our forest—saw a really colourful and well-marked Carpet Python between 7 and 8 ft. It was very 'stroppy' and struck at me quite a few times without any provocation. It was morning and the snake looked really nice as it crawled over the buttresses of a Sloanea tree.*
>
> ...
>
> *23rd Feb. 1996 ... Sulphur-Crested Cockatoos ... there was a group (10 birds approximately) feeding in a Sloanea australis (on the fruits). Beautiful views as they clambered about in the attractive foliage with dappled sun on the white of the birds. There was a short shower and then the sun came out again as this group of birds were excitedly leaf bathing in the top of another tree—flying out and back to plummet into or onto the topmost foliage—Lovely to watch.*

The sketch below shows buttresses developing on a young *Sloanea* tree.

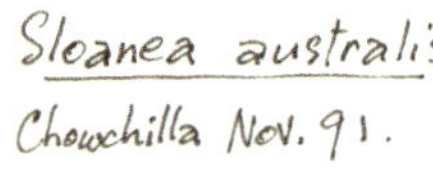

→ 2013; acrylic; 550 x 440mm
Private collection

← Blush Alder Buttresses
c.1991
State Library of New South Wales

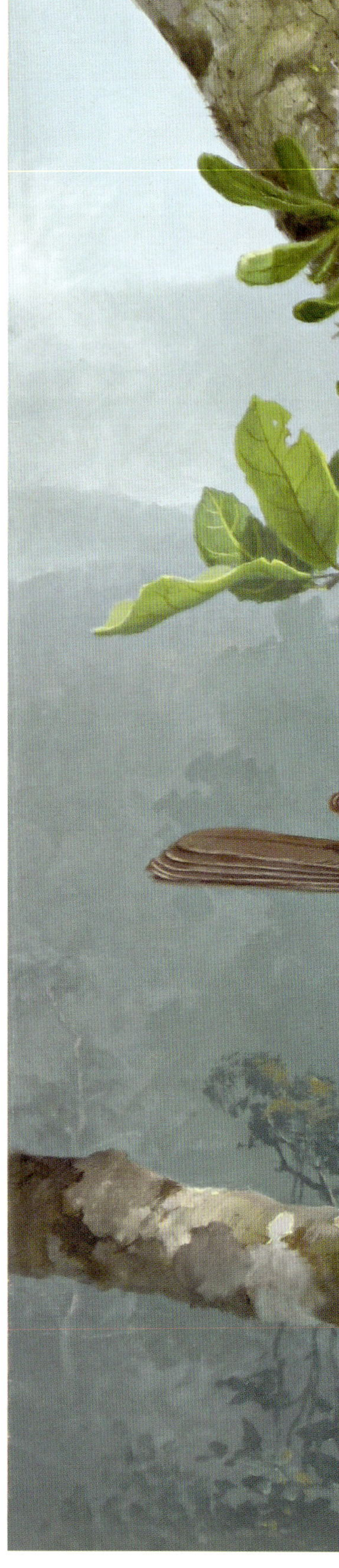

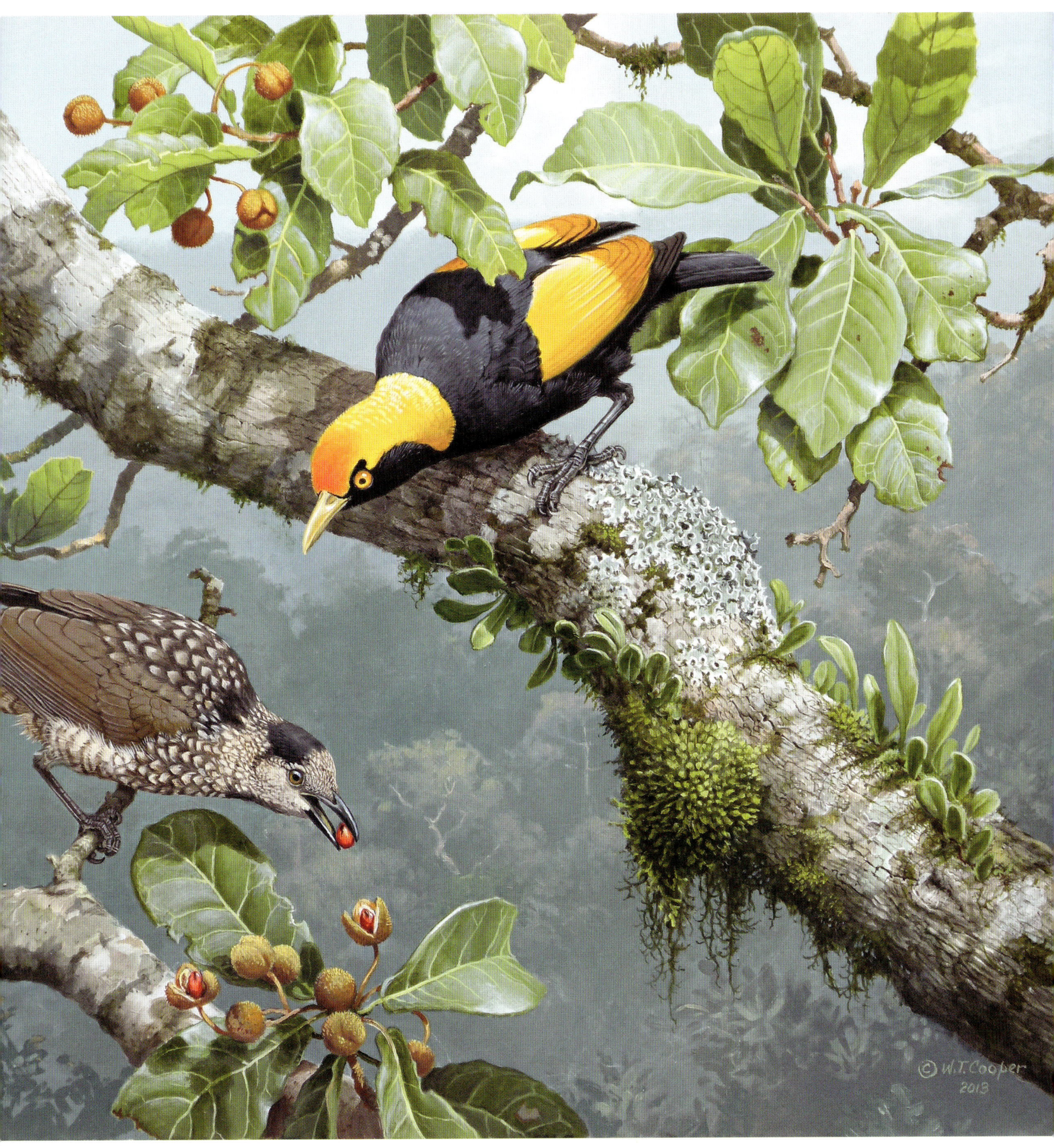
© W.T.Cooper
2013

Steganthera laxiflora

Tetra Beech

MONIMIACEAE

Steganthera is from *steg-* (shelter or roof) and *anthera* (anther), referring to the floral receptacle.

laxiflora is from *laxus* (loose) and *-florus* (-flowered), referring to the open panicles of flowers.

Tetra Beech is a sub-canopy tree growing to about 20 metres in tropical Queensland. The flowers of Tetra Beech form on the inside wall of a ball-like receptacle. That receptacle has a tiny opening at the apex allowing insects to enter and pollinate the flowers. As the fertilised flowers and immature fruit expand and ripen, the receptacle turns inside out with the fruit protruding outwards, resembling a cow's udder. Many species of birds feed on these fruits at our rainforested property. As it is not a canopy tree, it was an easy specimen to collect with a long-handled picker pole and was an ideal plant for a painting with Superb Fruit-Doves.

As annotated by Bill, this plant was previously known as *Tetrasynandra laxiflora*.

→ 1988; pen and watercolour; 470 x 580mm
Private collection

↙ Tree Shapes, Topaz
pen and pencil
State Library of New South Wales

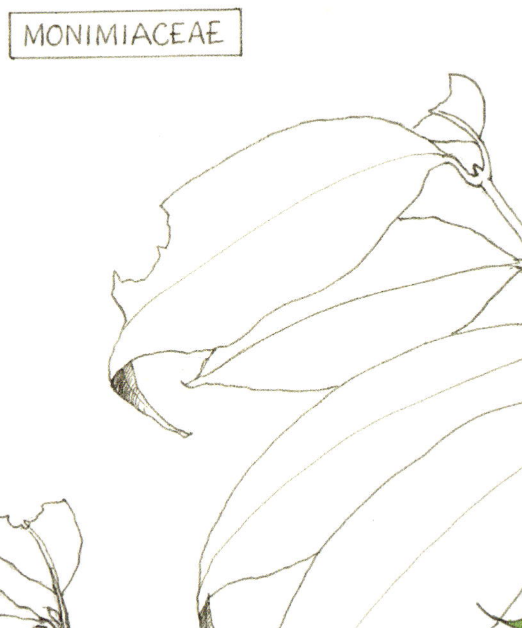

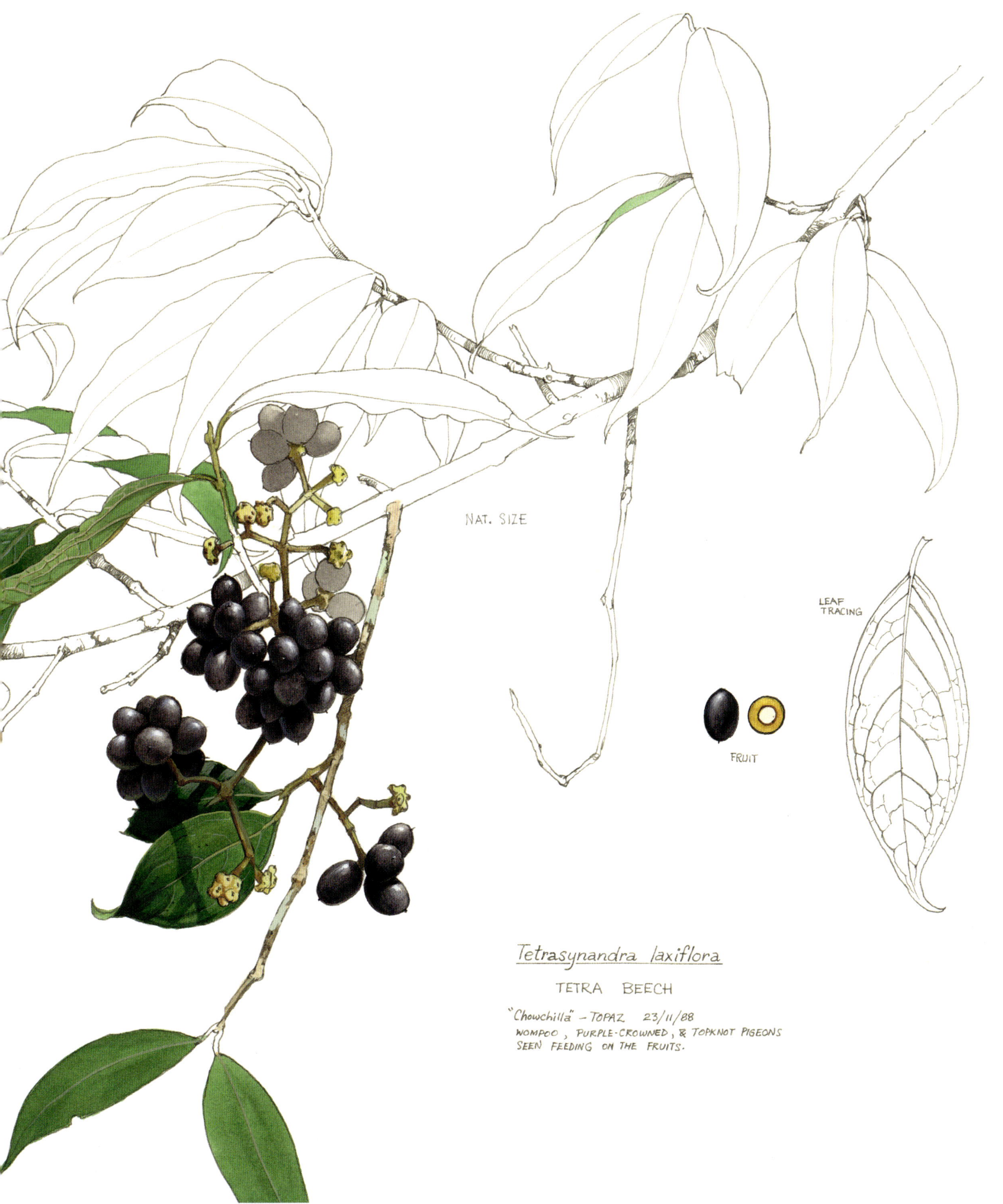
NAT. SIZE
LEAF TRACING
FRUIT
Tetrasynandra laxiflora
TETRA BEECH
"Chowchilla" – TOPAZ 23/11/88
WOMPOO, PURPLE-CROWNED, & TOPKNOT PIGEONS
SEEN FEEDING ON THE FRUITS.

Synima macrophylla

Topaz Tamarind

SAPINDACEAE

Synima is from *syn-* (with or together) and *ima* (cloak), referring to the basal aril.

macrophylla is from *macro-* (large) and *-phyllus* (-leaved).

Topaz Tamarind is a slender, often single-stemmed tree that grows to 11 metres but is usually much shorter than that. Its distribution is confined to the Wet Tropics bioregion and it is most common at Topaz.

The hot pink fruit have a fleshy yellow aril at the base of each seed. Victoria's Riflebird is dexterous with its slender long bill, which it uses to remove the fruit's shiny black seed intact along with the aril. This is all swallowed whole and the aril digested before the naked seed is later regurgitated. Often at our back door! Some of these birds were frequently hand-fed by Bill and in order to be able to fit in some mealworms they needed to first regurgitate useless seeds.

→ Victoria's Riflebirds at Topaz 1993; acrylic on canvas on board 750 x 500mm Collection of Robyn Lewis

↓ 2001; pen; 300 x 410mm State Library of New South Wales

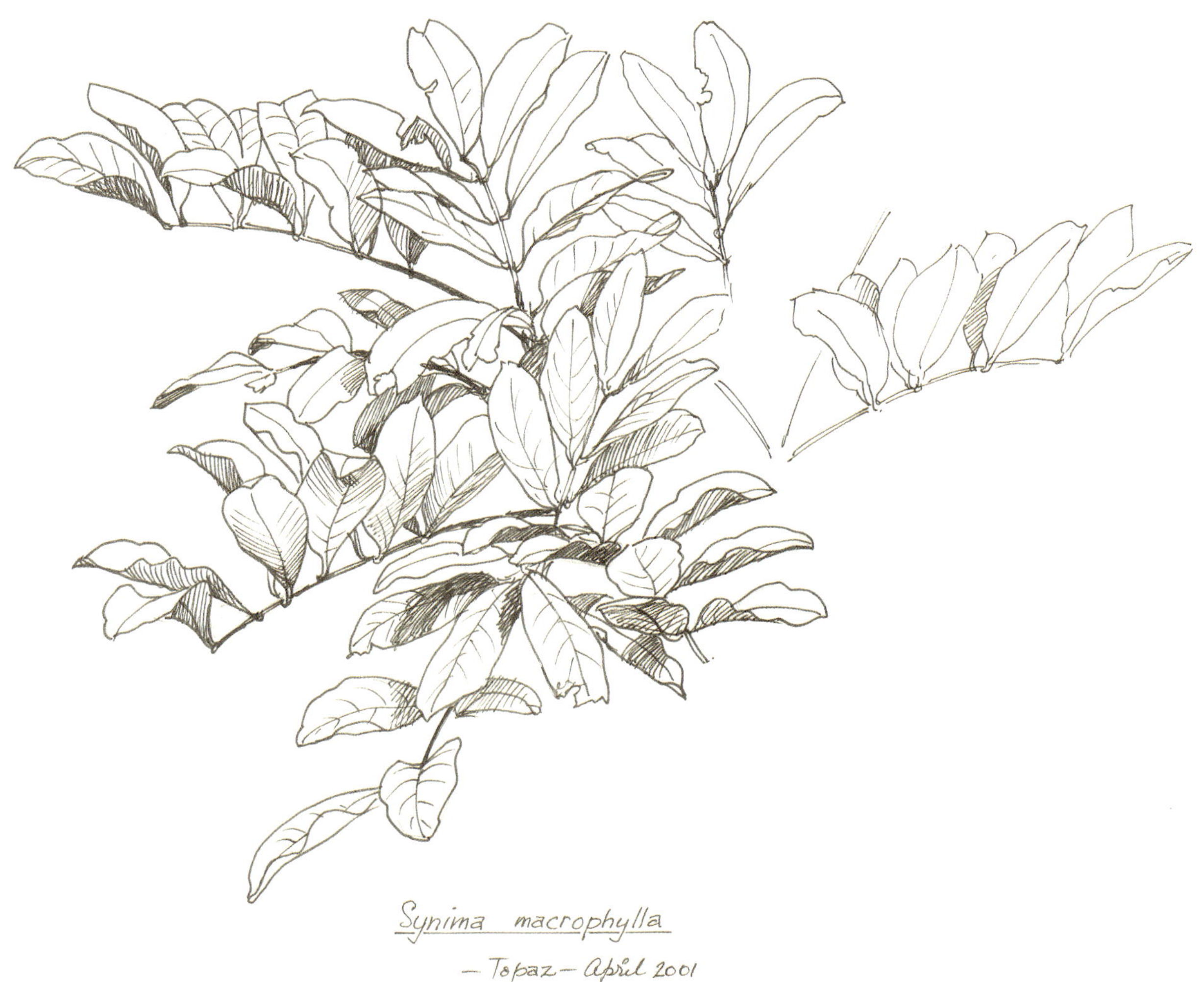

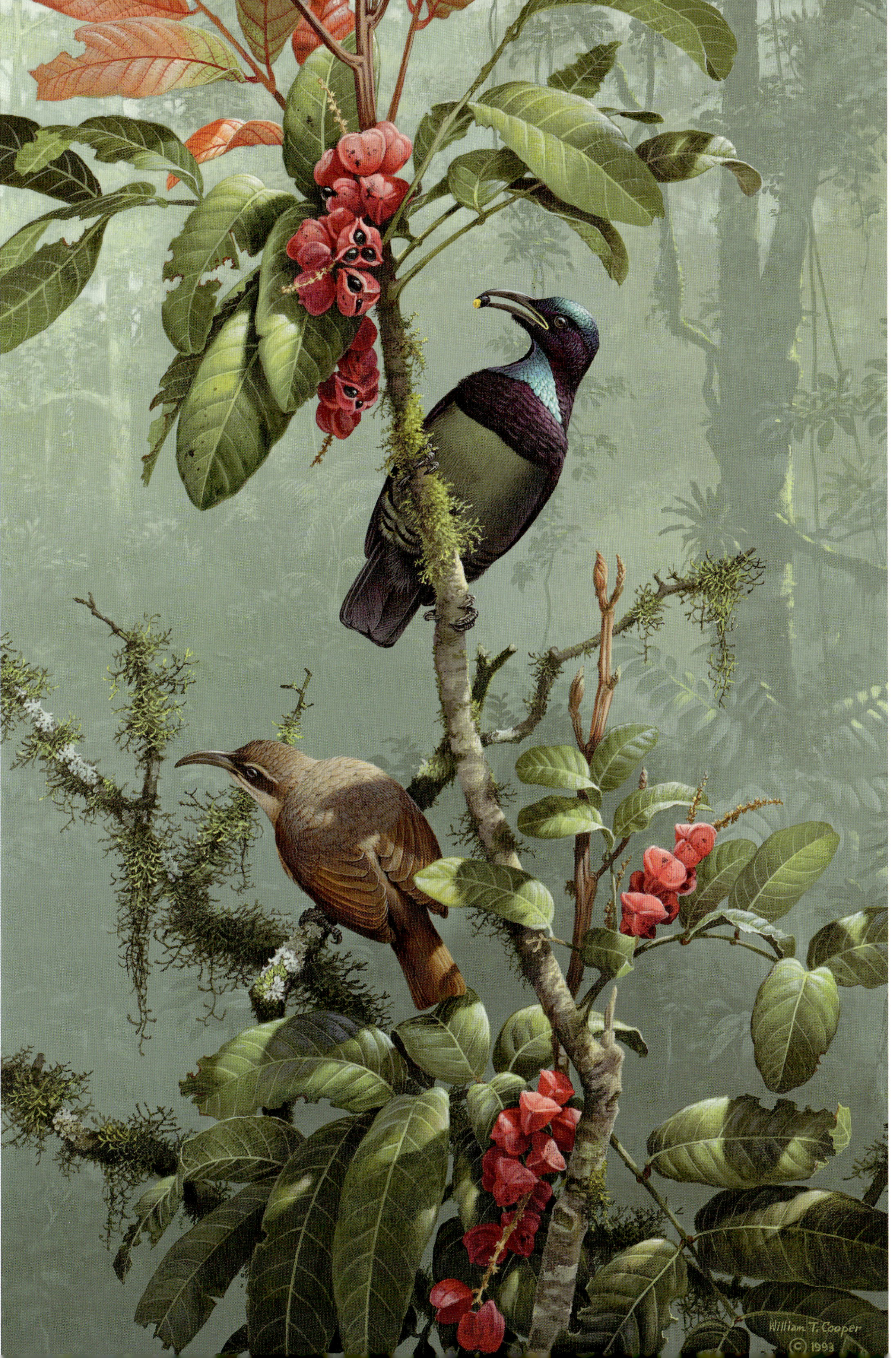
William T. Cooper
© 1993

Synoum glandulosum

Scentless Rosewood

MELIACEAE

Synoum from *syn-* (together) and *oon* (ovum), referring to the two ovaries within each valve.

glandulosum from *glans* (gland) and *-osus* (well-developed or abundant), referring to the large domatia on the leaflets.

Scentless Rosewood, with its dark green leaves, is a lush rainforest shrub or sub-canopy tree in the Myall Lakes area in New South Wales (it also occurs from southern New South Wales to north-east Queensland). Bill had always been attracted to the romance of rainforest, which took him to Papua New Guinea several times and eventually induced him to move from the Myall Lakes area to live in north Queensland. As Bill's favourite bird, the Regent Bowerbird, feeds on Scentless Rosewood, both bird and fruit provided a colourful combination in a painting for *The Birds of Paradise and Bower Birds*.

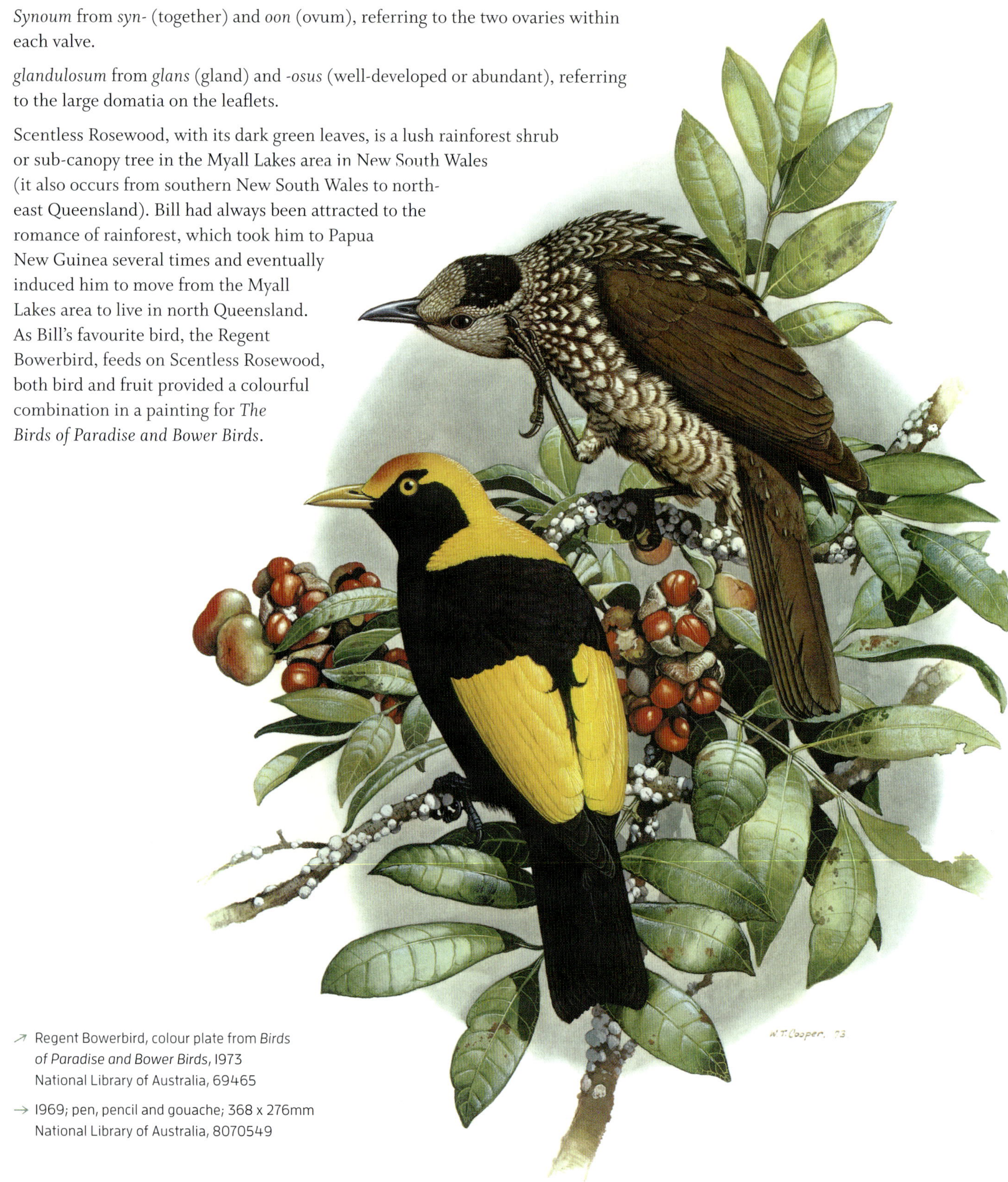

↗ Regent Bowerbird, colour plate from *Birds of Paradise and Bower Birds*, 1973
National Library of Australia, 69465

→ 1969; pen, pencil and gouache; 368 x 276mm
National Library of Australia, 8070549

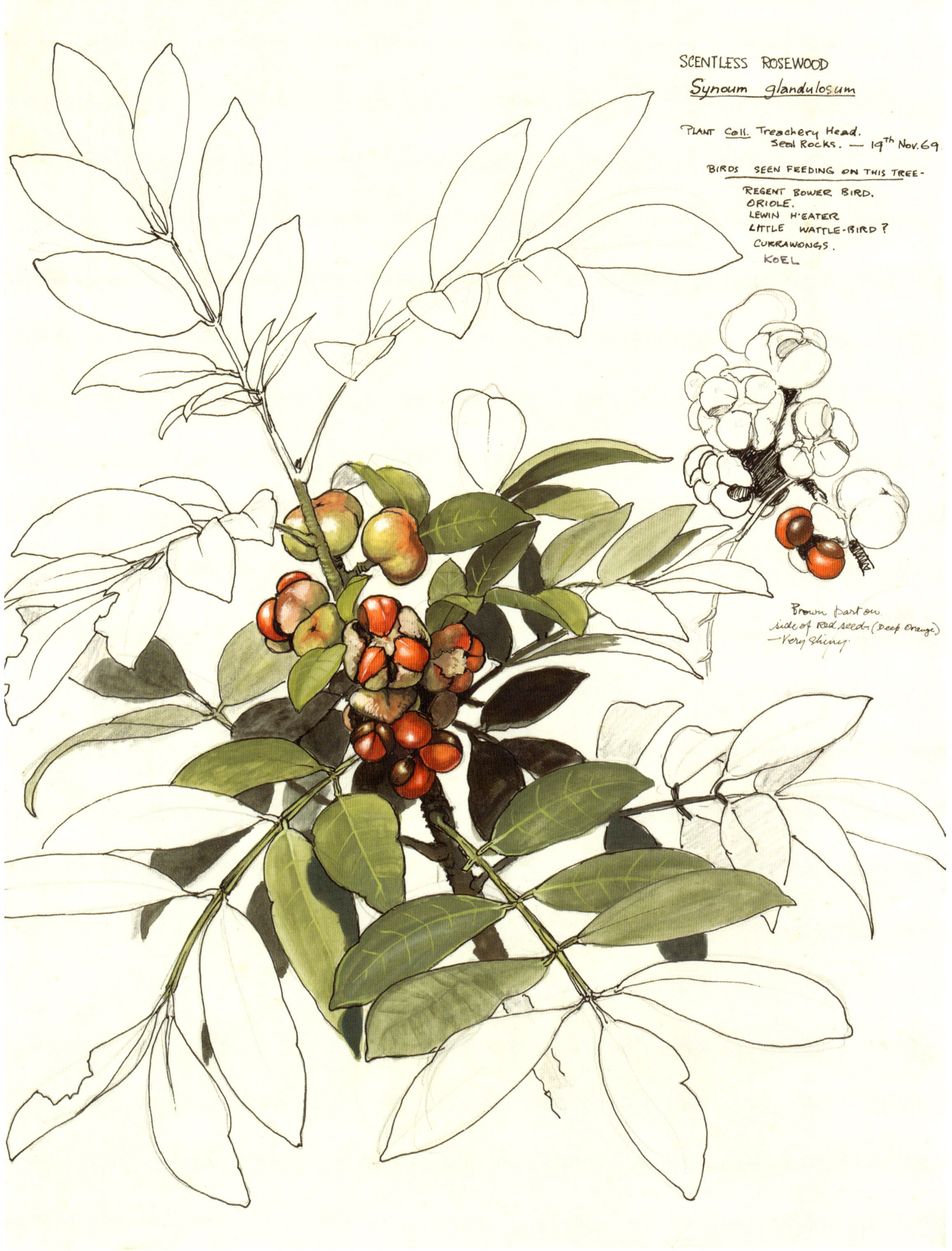
SCENTLESS ROSEWOOD
Synoum glandulosum
PLANT Coll. Treachery Head. Seal Rocks. — 19th Nov. 69
BIRDS SEEN FEEDING ON THIS TREE-
REGENT BOWER BIRD.
ORIOLE.
LEWIN H'EATER
LITTLE WATTLE-BIRD ?
CURRAWONGS.
KOEL
Brown part on side of Red seed- (Deep Orange) —Very shiny.

Syzygium cormiflorum

Bumpy Satinash

MYRTACEAE

Syzygium is from *syzgos* (joined together), referring to the flower segments, which are joined together forming a cap that is shed as the flowers open.

cormiflorum is from *cormus* (trunk) and *-florus* (flowered), referring to the flowers and fruit that emerge from the tree trunk and major branches.

Bumpy Satinash is a tall tree growing in Queensland's tropical rainforest. The presence of flowers and fruit studded on the trunk (cauliflory) gives the plant an exotic jungle look. Cauliflory is restricted to tropical forests.

Bill wrote about a visit to our property from a film crew in June 2002 in his diary:

> *ABC/BBC film crew working at Syzygium cormiflorum on the edge of the road (100m from house). They were filming hand reared animals within an enclosure on the tree – generator running – bright lights and 4 or 5 people. On more than one occasion a Striped Possum came and fed (2 animals) and 2 Herbert River Ringtails on separate nights. During the day there were riflebirds taking nectar and Lewins and McLeays Honeyeaters. A very productive tree!*

→ undated; pen and watercolour
690 x 505mm
National Library of Australia, 8069877

↓ 2009; pen, pencil and watercolour
850 x 580mm
State Library of New South Wales

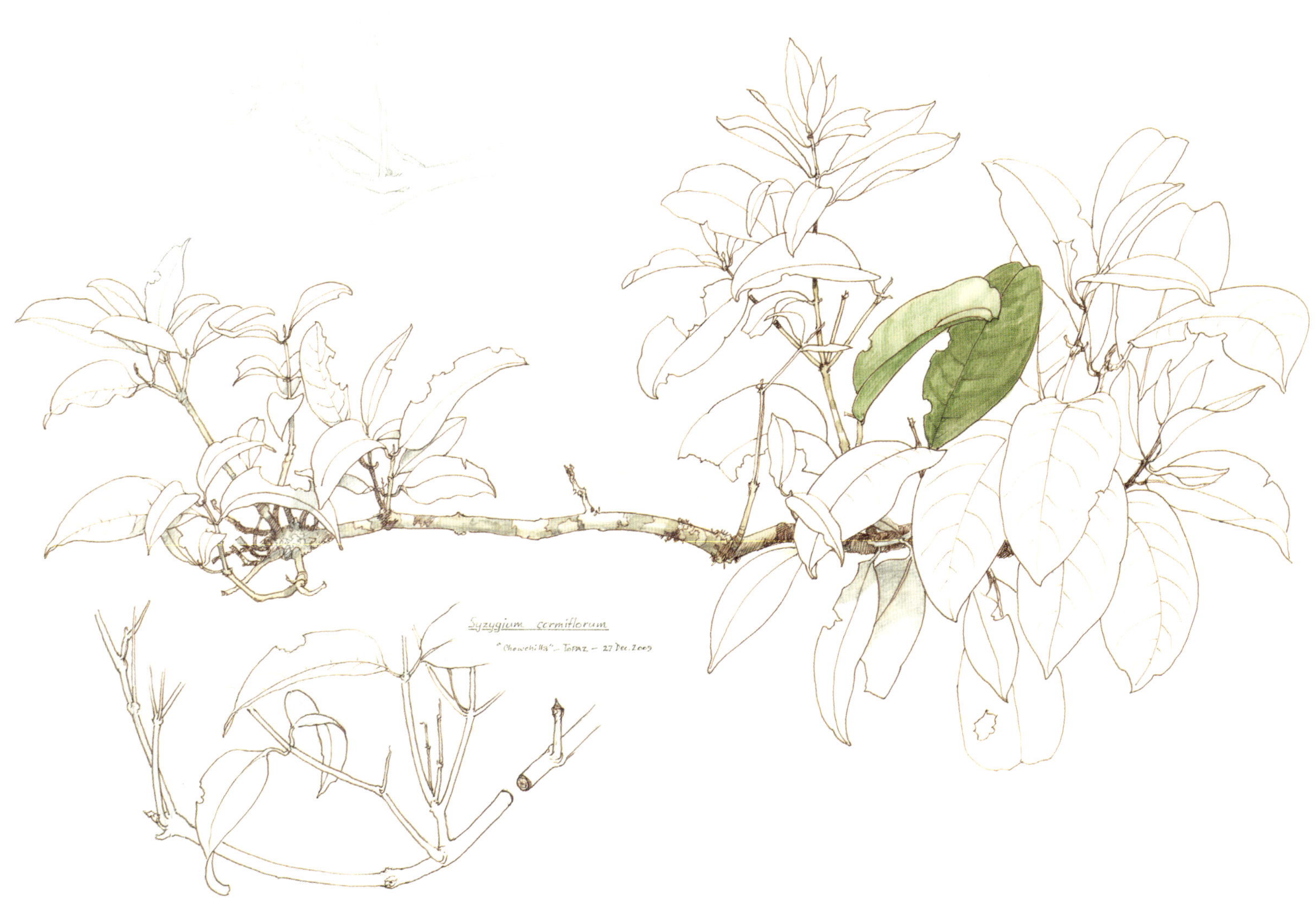

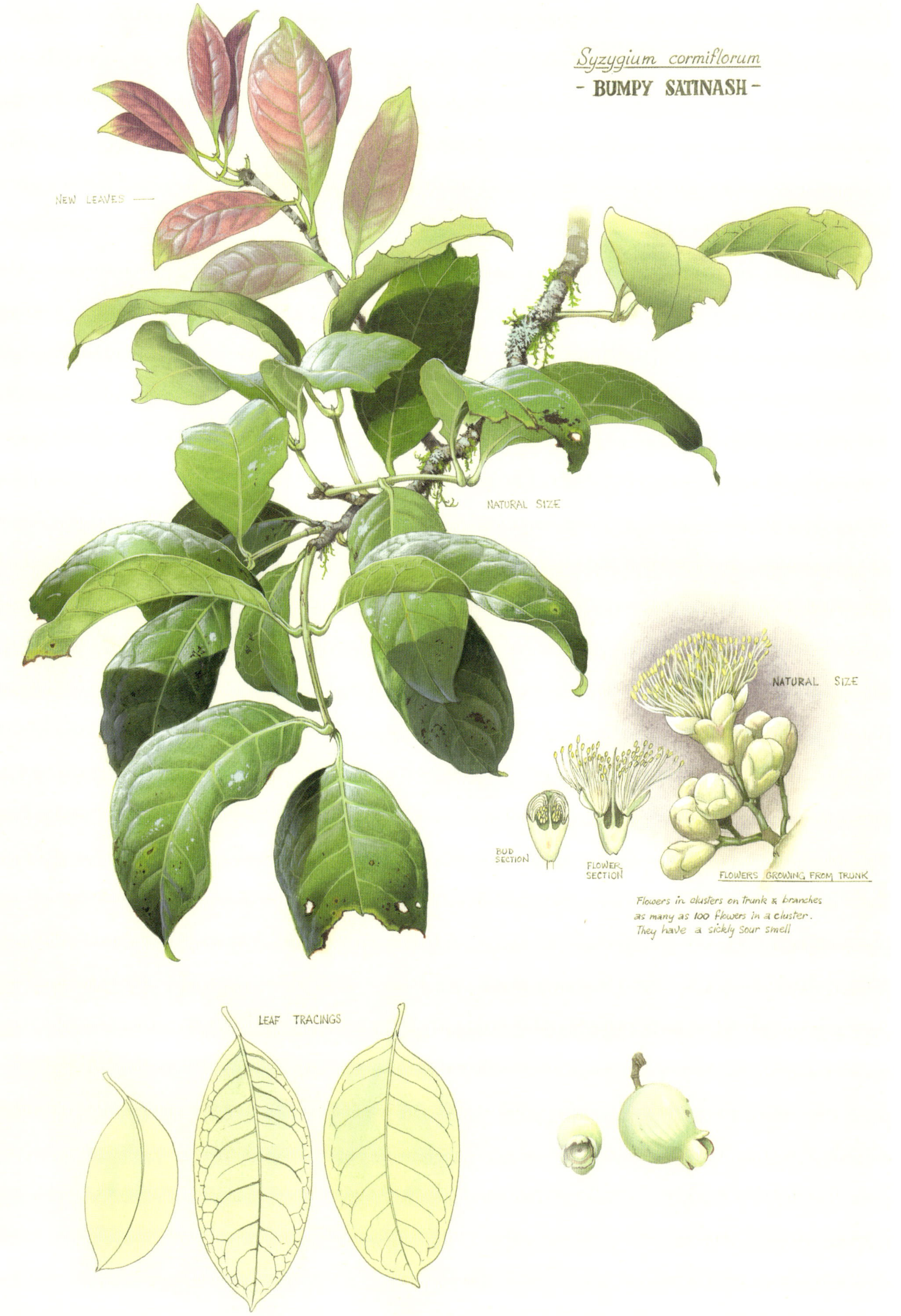
Syzygium cormiflorum
- BUMPY SATINASH -
NEW LEAVES
NATURAL SIZE
NATURAL SIZE
BUD SECTION
FLOWER SECTION
FLOWERS GROWING FROM TRUNK
Flowers in clusters on trunk & branches as many as 100 flowers in a cluster. They have a sickly sour smell
LEAF TRACINGS

Telopea truncata

Tasmanian Waratah

PROTEACEAE

Telopea is from *telopus* (seen from afar), referring to the bright red flowers evident from a distance.

truncata is from *truncatus* (abruptly cut), reference unknown but possibly connected to the flat-topped flower head or the truncated wing on the seeds.

Tasmanian Waratah is mostly a shrub growing to 3 metres, endemic to high country in Tasmania.

Bill had been obsessed by the extinct Tasmanian Thylacines for some years. He read all that he could on them and bought a short length of old film of an animal in captivity. He played and replayed the footage numerous times analysing and getting to know the animal so that he could draw it in the way that he wanted to show the animal in the final painting.

For the painting's background, we spent time at locations known to have been Thylacine habitat. Bill wrote of one such trip in his diary:

> *Friday 3rd Dec. 2004, Cradle Mt. Most wonderful sight. The lake and mountain spectacular in the low morning light. Everywhere you look are wonderful little pictures, mosses, plants, lichens and twisted trees. It was worth coming to Tasmania just for this morning's experience. After lunch I drew details of the waratah with a view to doing a painting of the Green Rosella feeding on it as we saw. Wendy said she twice saw a bird with an individual flower in its bill after pulling it from the flower head.*

The waratah flowers were used in the Thylacine painting, as well as one of a Green Rosella.

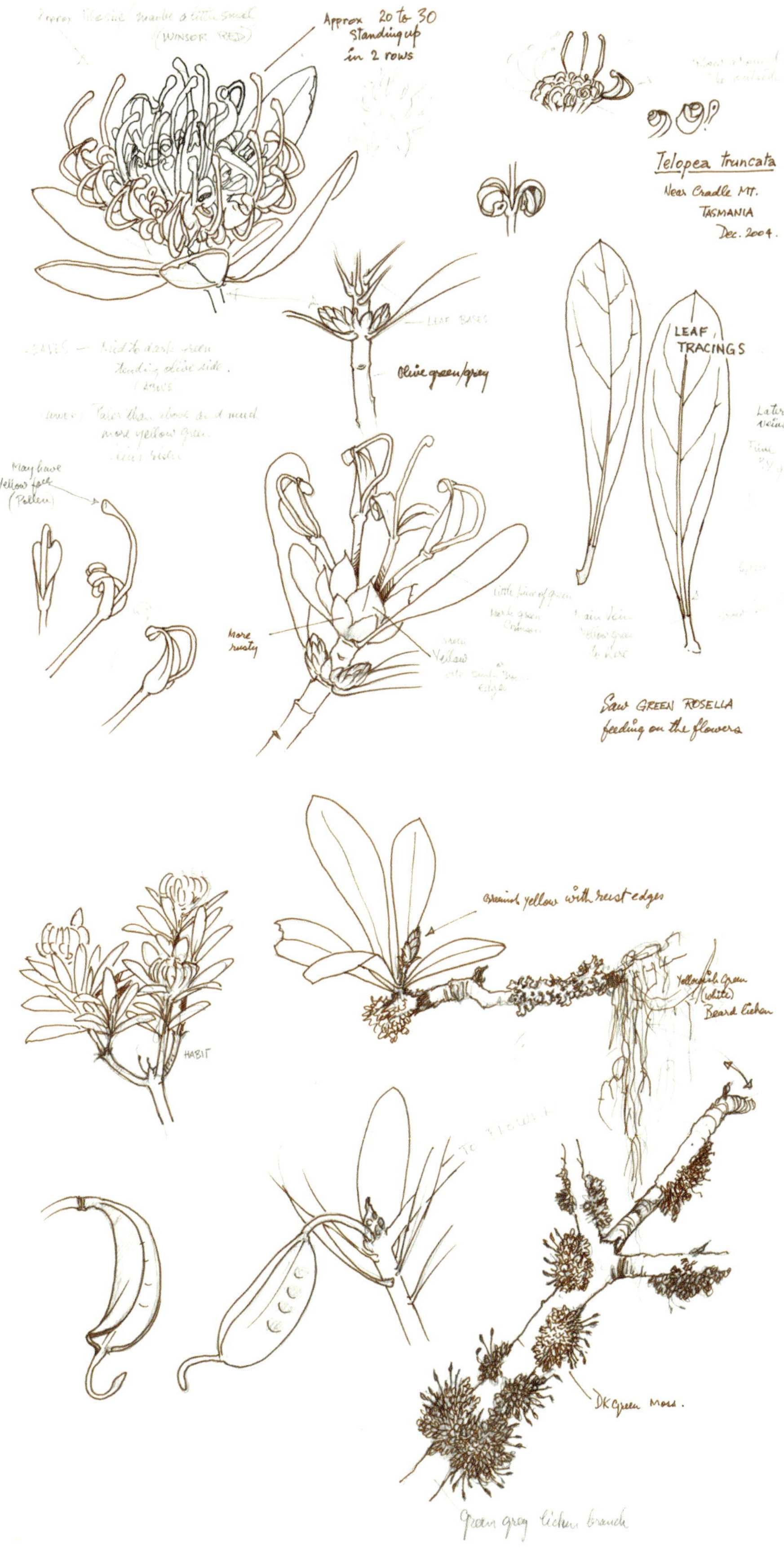

↗ 2004; pen and pencil; 245 x 260mm
State Library of New South Wales

↗ 2004; pen and pencil; 245 x 260mm
State Library of New South Wales

→ Green Rosellas
2005; watercolour
Private collection

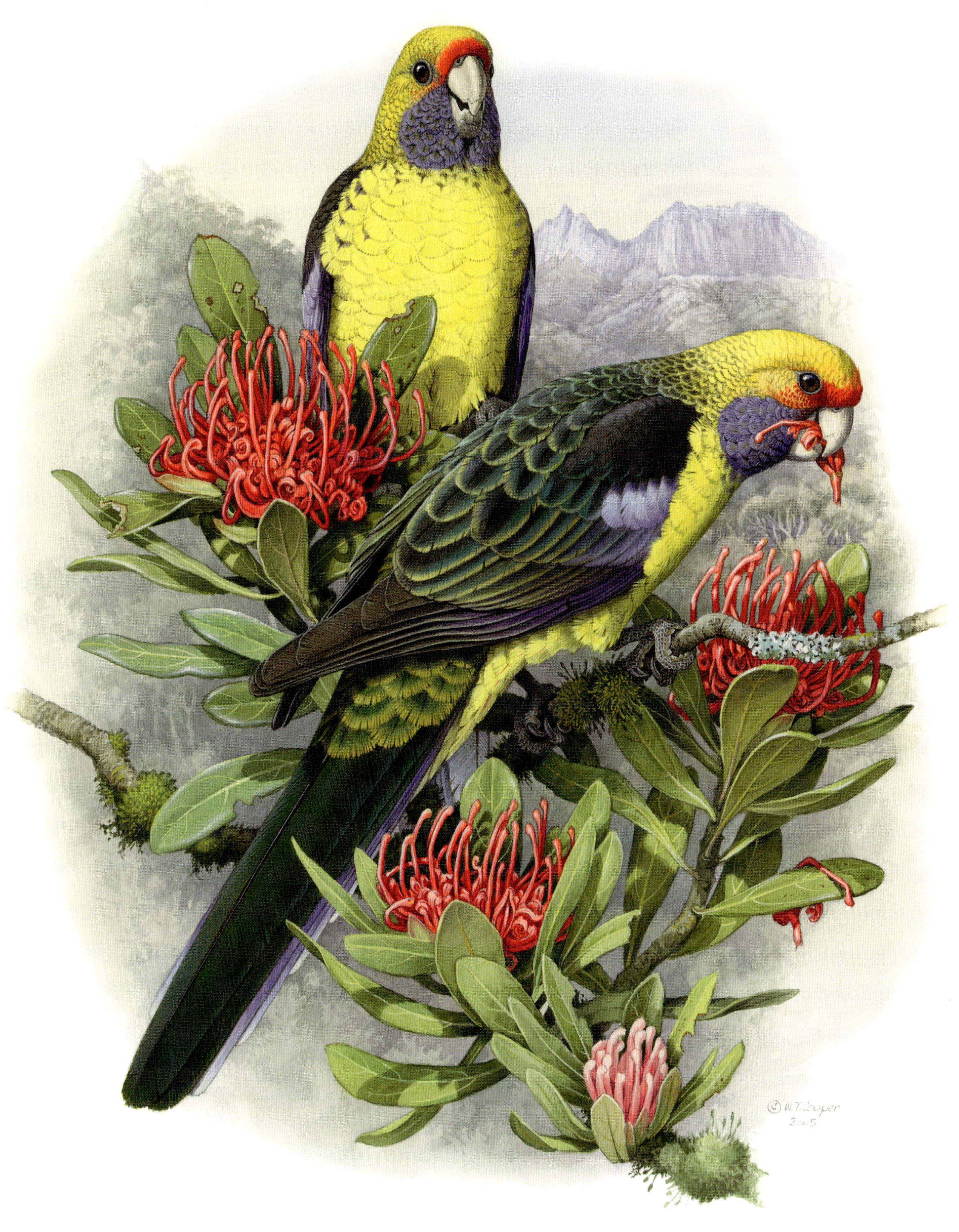
© W.T.Cooper
2005

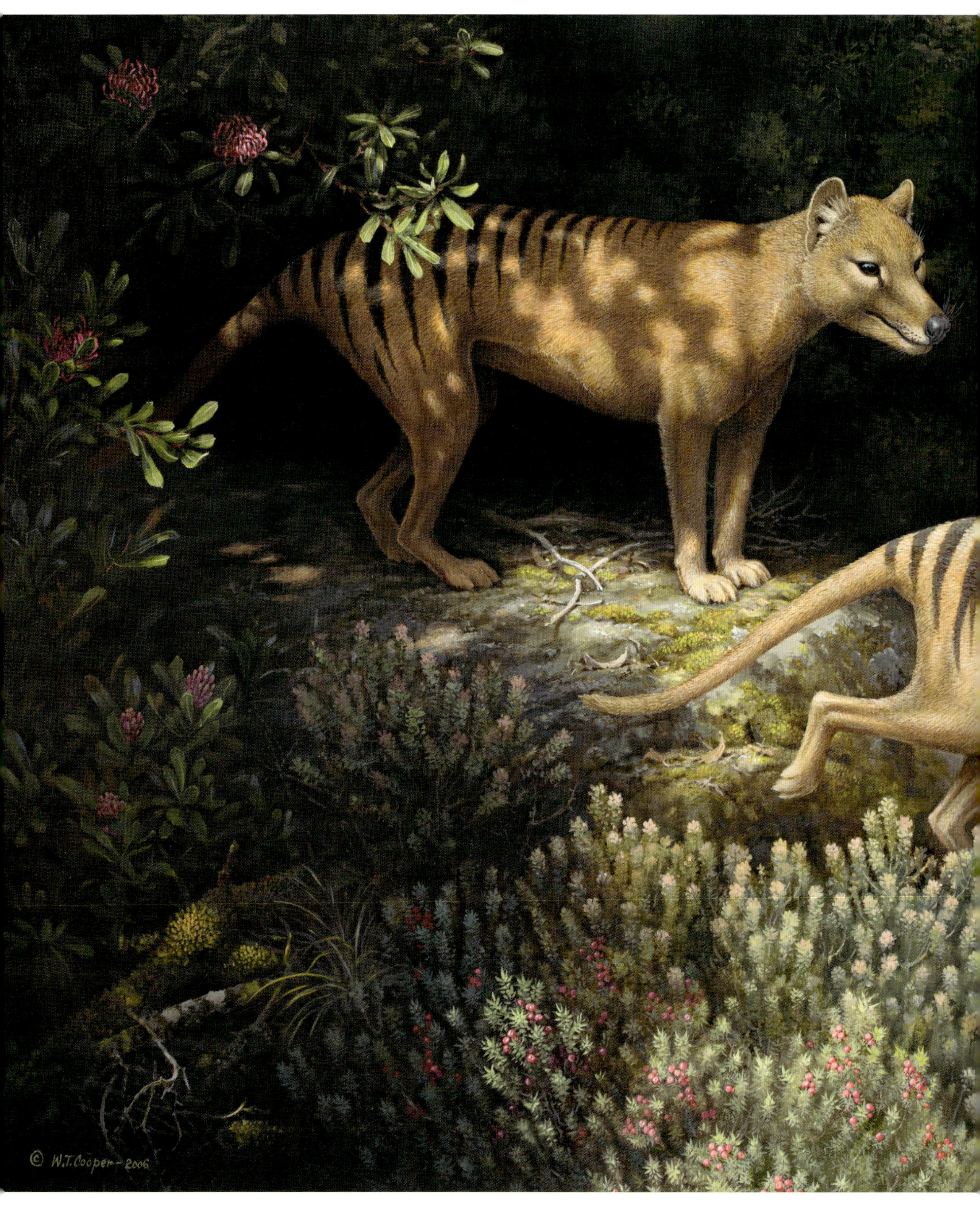
© W.T.Cooper – 2006

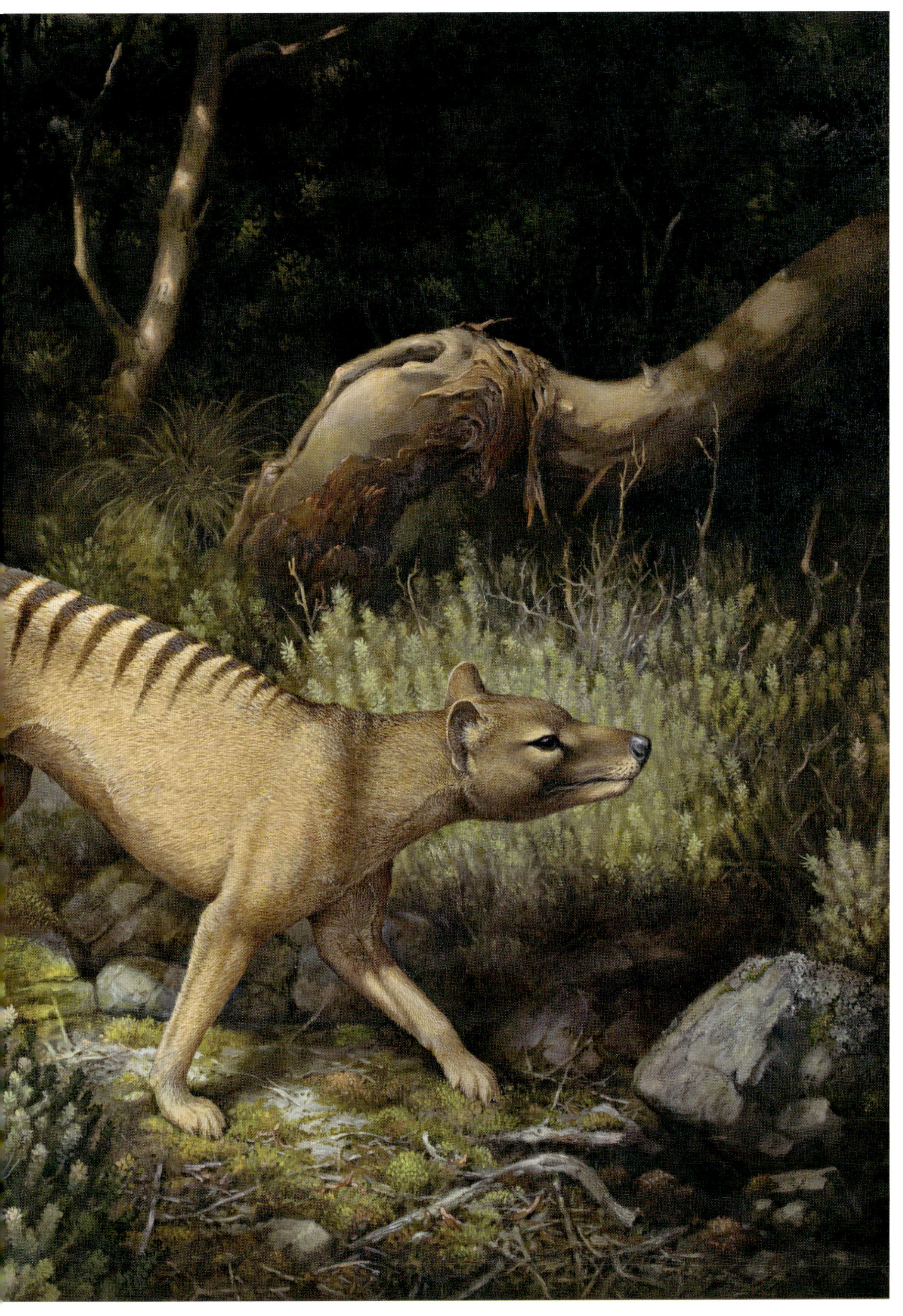

← Thylacine
2006; oil on canvas
146 x 223cm
Tasmanian Museum
and Art Gallery

Terminalia sericocarpa

Sovereignwood or Damson

COMBRETACEAE

Terminalia is from *terminalis* (terminal), referring to the clustered leaves at the tips of the branches in many species.

sericocarpa is from *sericeus* (clothed in silky hairs) and *-carpos* (-fruited).

Damson is a widespread very tall tree in northern Australia's drier rainforests. The bluish-purple fruit are keenly sought by many birds. The specimen painted here was from a plant in a dear friend's garden. Rigel, with a slingshot, fired a line over a selected branch and was then able to snap it off, providing an excess of leafy material for Bill to choose from. The plant was used in a painting of Palm Cockatoos (overleaf).

↓→ 2014; pen and watercolour
390 x 585mm
State Library of New South Wales

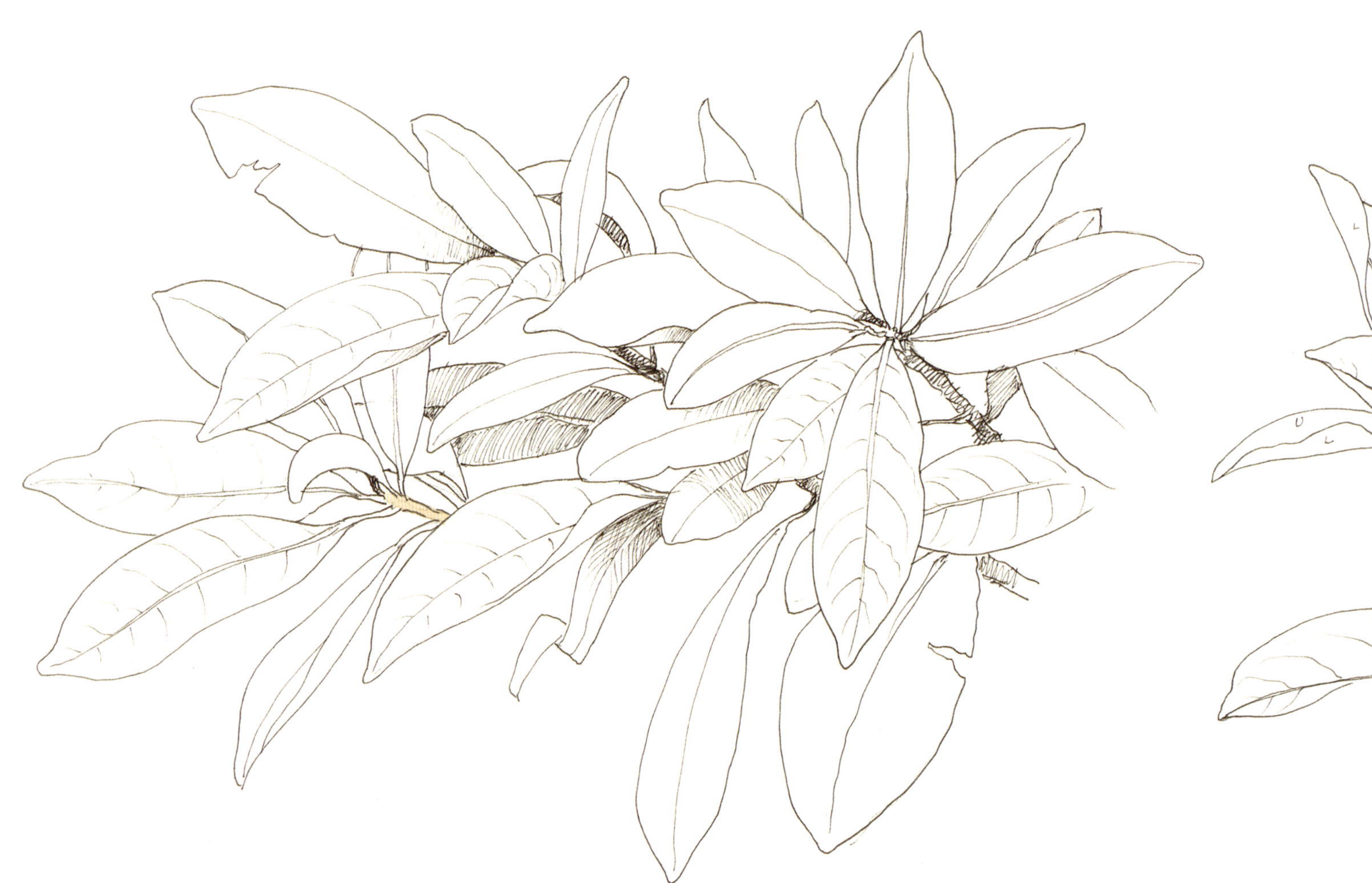

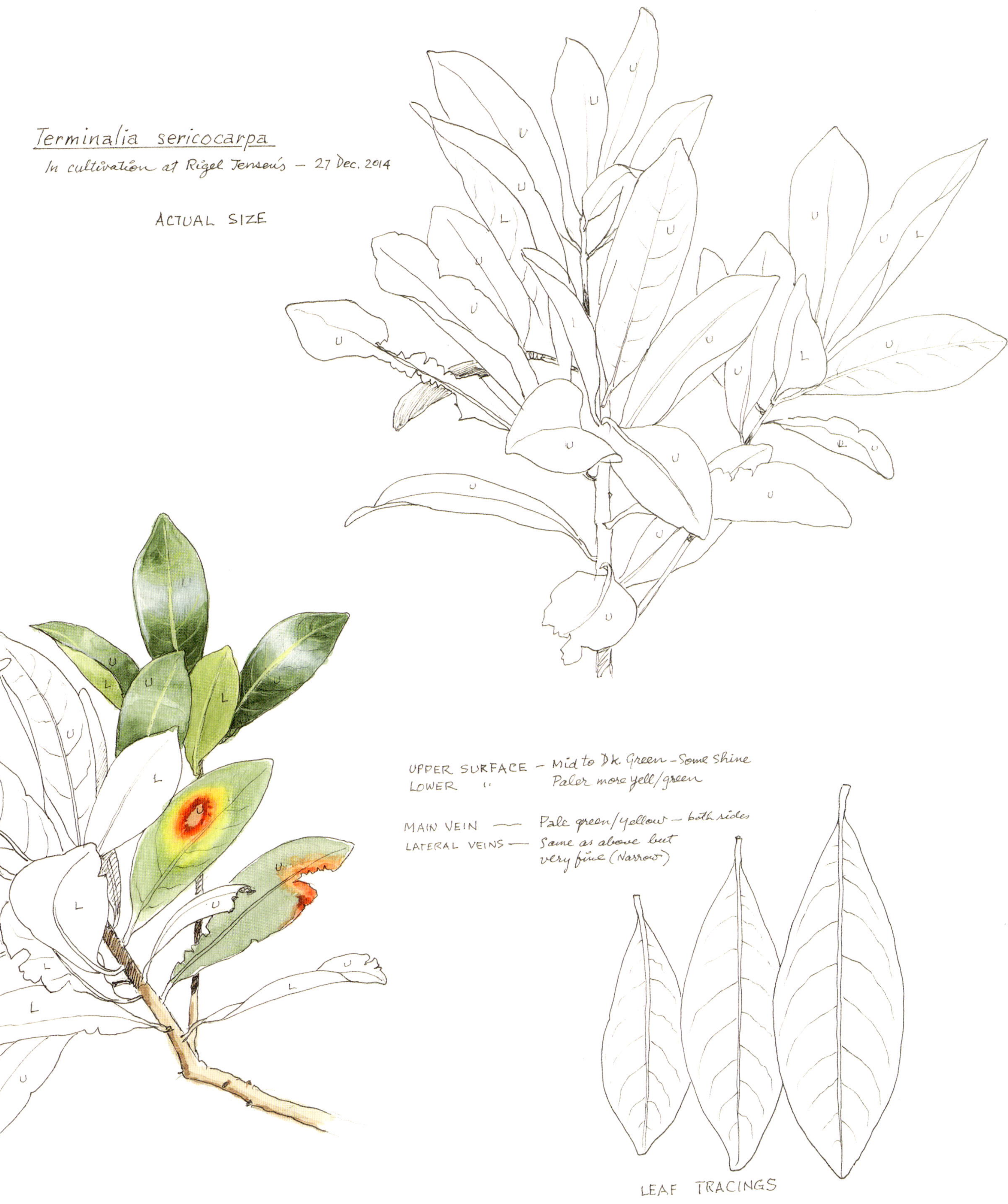
Terminalia sericocarpa
In cultivation at Rigel Jensen's – 27 Dec. 2014
ACTUAL SIZE
UPPER SURFACE – Mid to Dk. Green – Some Shine
LOWER " Paler more yell/green
MAIN VEIN — Pale green/yellow – both sides
LATERAL VEINS — Same as above but very fine (Narrow)
LEAF TRACINGS

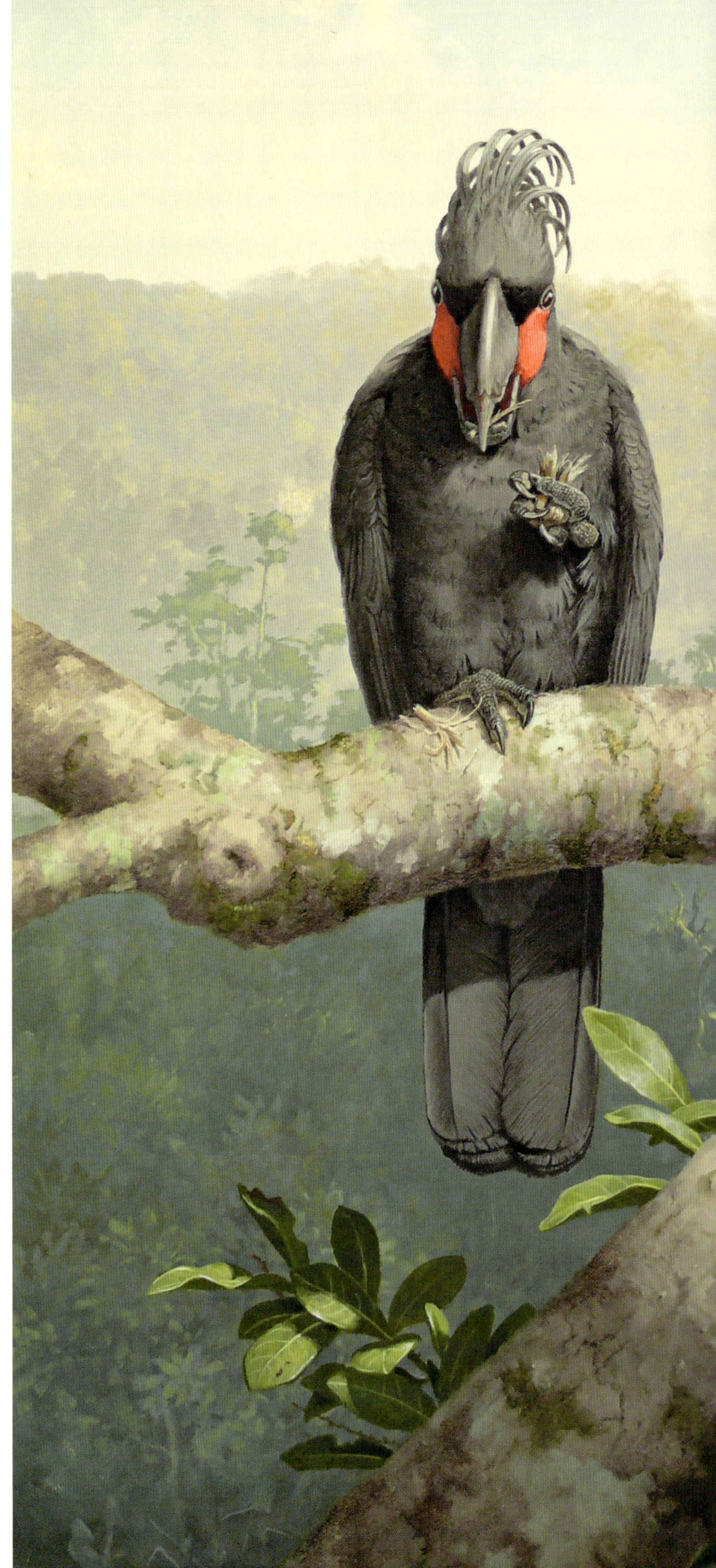

→ Palm Cockatoos
2015; oil on board; 685 x 763mm
Private collection

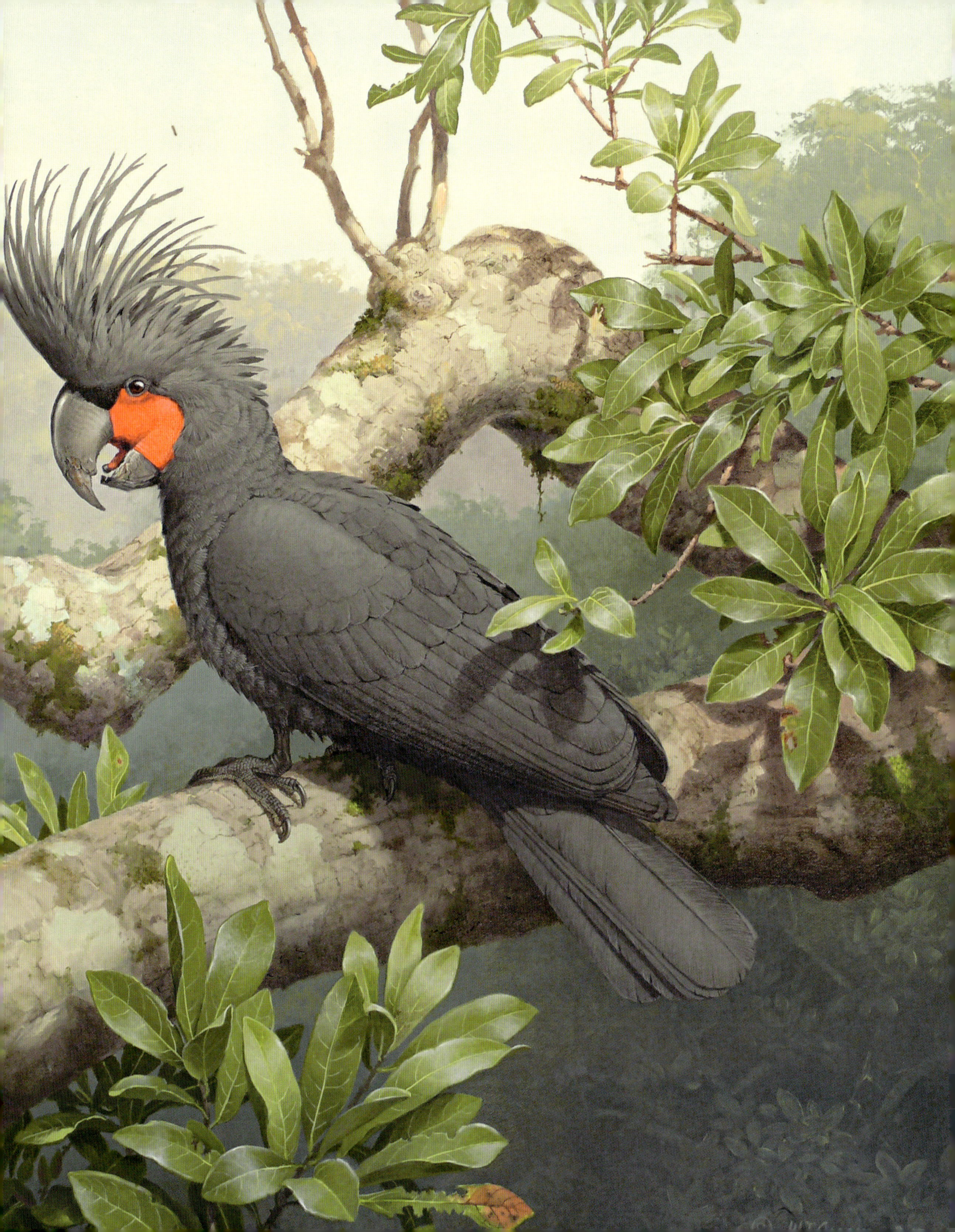

Triunia erythrocarpa

Spice Bush

PROTEACEAE

Triunia is from *tri-* (three) and *unus* (one), referring to the flowers, which consist of three united tepals and one that is free.

erythrocarpa is from *erythros* (red) and *-carpos* (fruited).

Spice Bush is a sub-canopy tree that grows to 20 metres in wetter rainforests of the Wet Tropics bioregion. The leaves are shiny and mostly clustered towards branch tips.

Spice Bush is so named for the spicily fragrant flowers. These are followed by luscious-looking bright-red fruit. Even though they are related to edible Macadamias, they are extremely poisonous to people. Cassowaries and Musky Rat-kangaroos take them with impunity.

→ Mossy Trunks
2005; pen, pencil and watercolour; 285 x 340mm
State Library of New South Wales

↓ undated (c.1991); watercolour; 200 x 260mm
Private collection

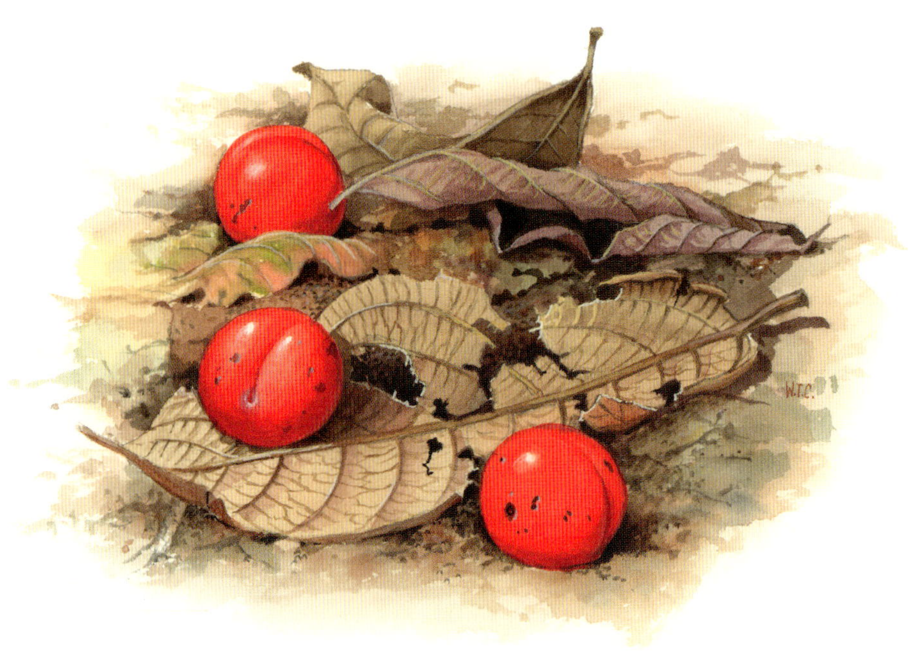

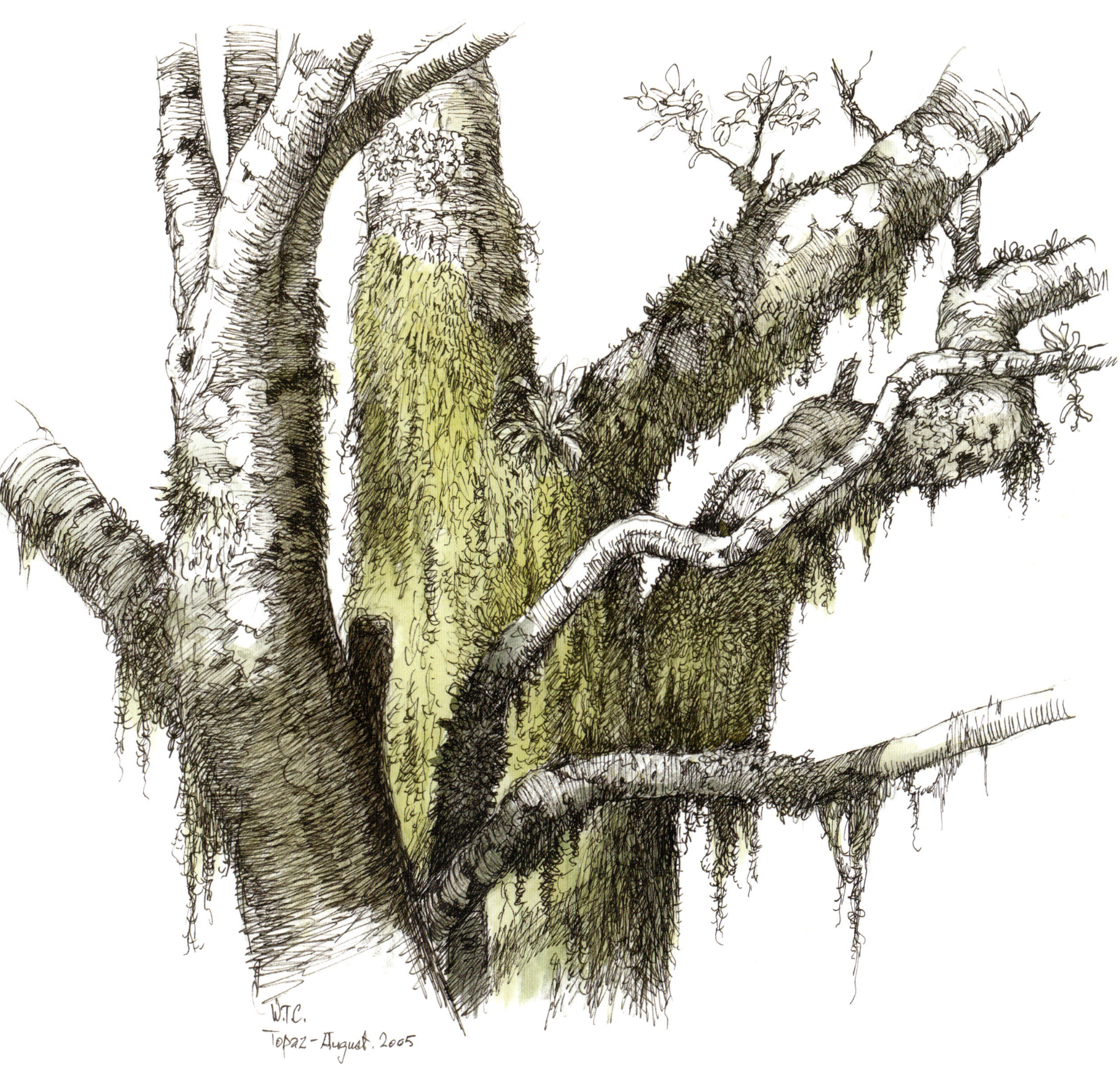
W.T.C.
Topaz - August. 2005

Xanthophyllum octandrum

Yellow Boxwood or Soft Ghittoe

POLYGALACEAE

Xanthophyllum is from *xanthos* (yellow) and *-phyllus* (-leaved), referring to the often yellowish leaves.

octandrum is from *oct-* (eight) and *-andrus* (male), referring to the flowers with eight anthers.

Yellow Boxwood is a rainforest tree, growing to 35 metres, but it frequently fruits as a small tree just a few metres tall. It occurs in Queensland, north from the Mackay area.

Fruit burst open and tear in a ragged pattern to reveal a shiny textured aril on the seeds. The seeds are attached at the side of the capsule rather than at the base near the fruit stalk, which is an unusual place for them to emerge. Yellow Boxwood seeds are taken by many birds, as noted on Bill's painting. This piece was done while on holiday in north Queensland:

> *8th June 1992 … For the past 2 weeks there has been a flock of White Cockatoos about. I would guess at 60 birds in the flock. They seem to be feeding in the forest and I suspect they are feeding on Xanthophyllum octandrum by the mess we are seeing under the trees.*

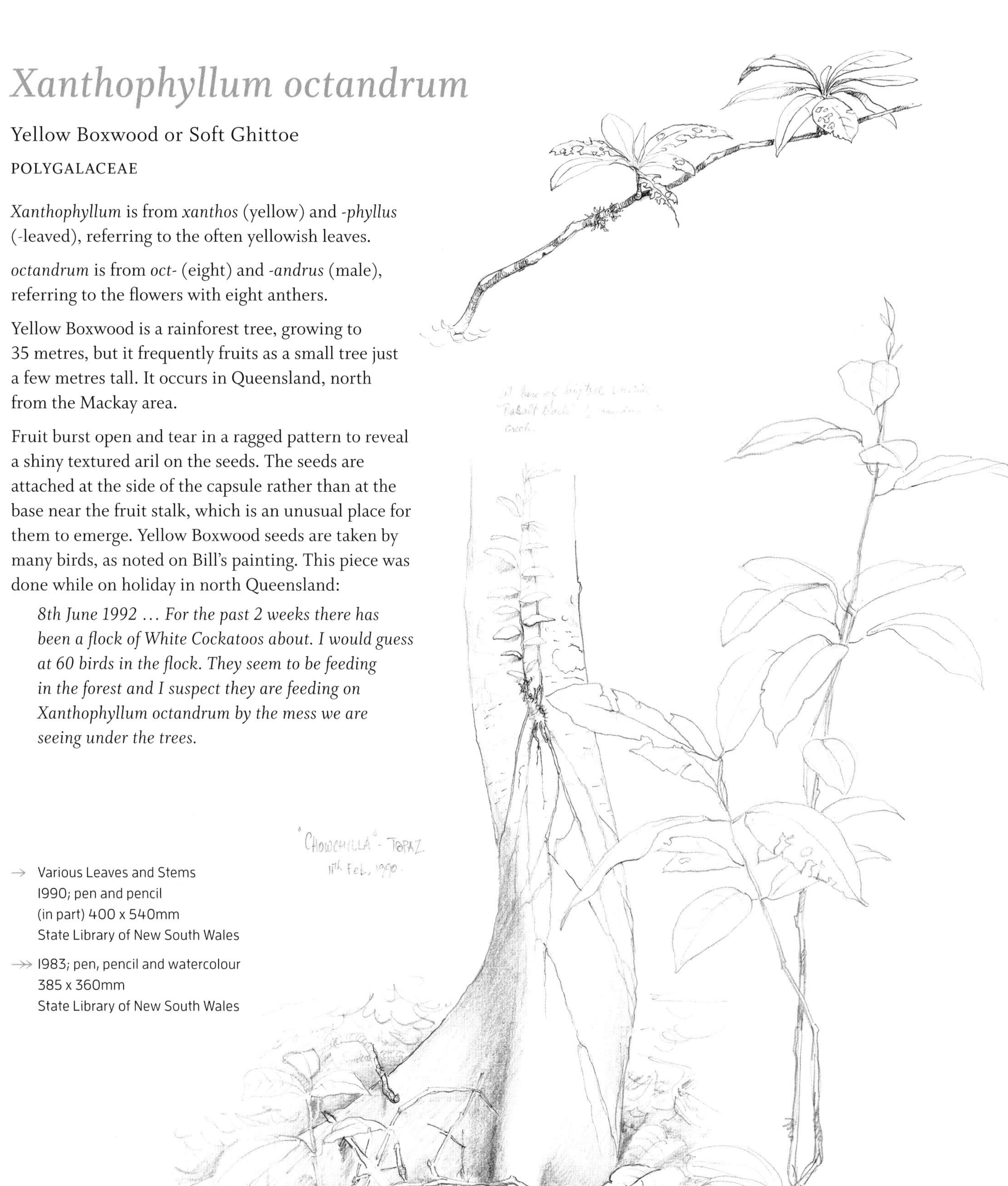

→ Various Leaves and Stems
1990; pen and pencil
(in part) 400 x 540mm
State Library of New South Wales

→→ 1983; pen, pencil and watercolour
385 x 360mm
State Library of New South Wales

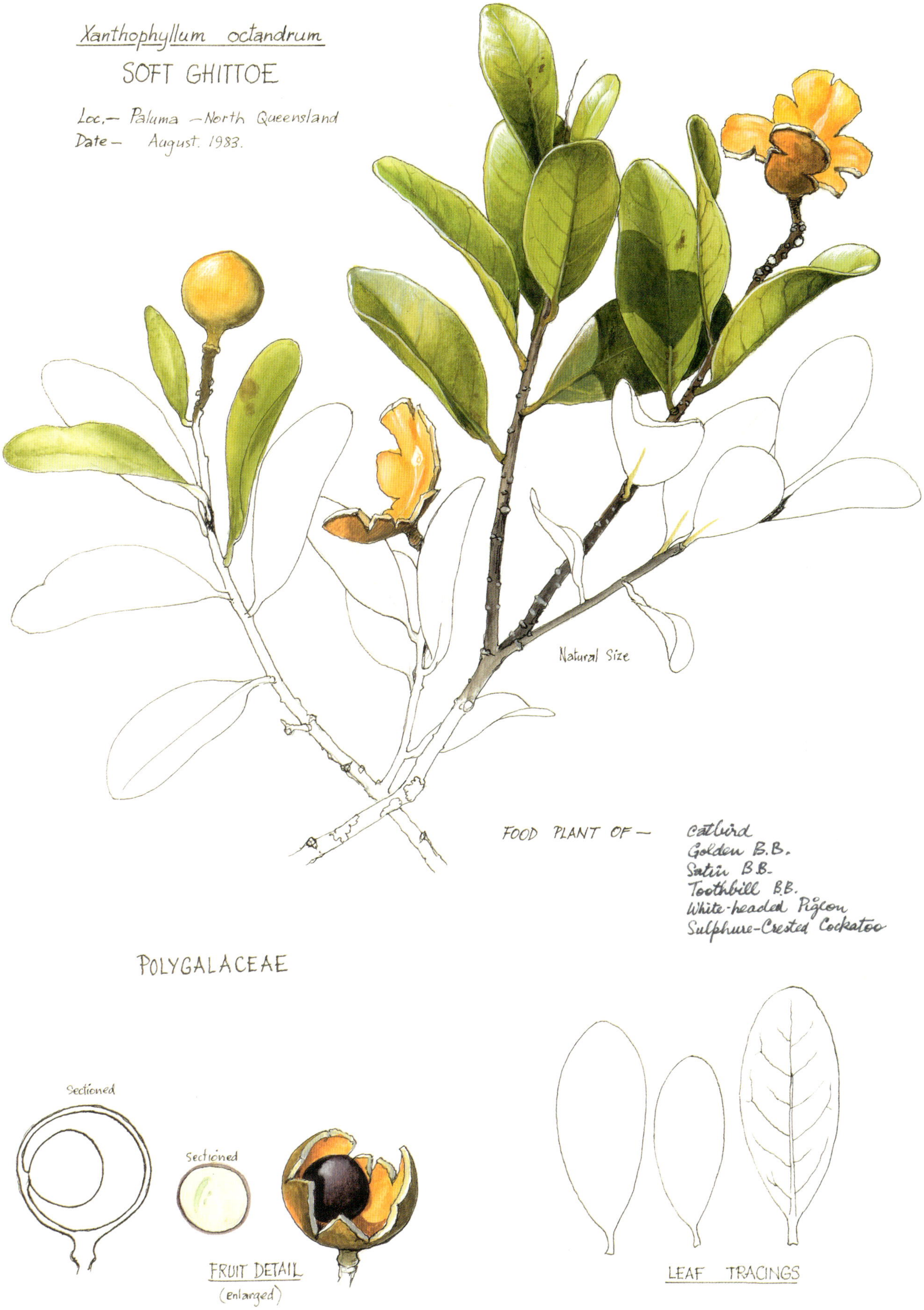

Xanthophyllum octandrum

SOFT GHITTOE

Loc.— Paluma —North Queensland
Date— August. 1983.

Natural Size

FOOD PLANT OF — Catbird
Golden B.B.
Satin B.B.
Toothbill B.B.
White-headed Pigeon
Sulphure-Crested Cockatoo

POLYGALACEAE

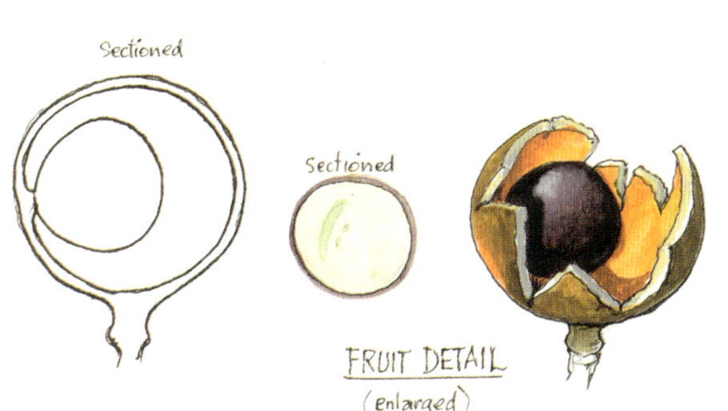

FRUIT DETAIL
(enlarged)

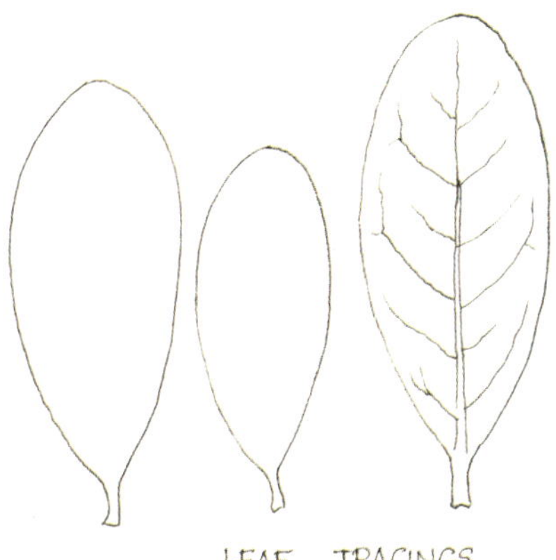

LEAF TRACINGS

Chapter 2

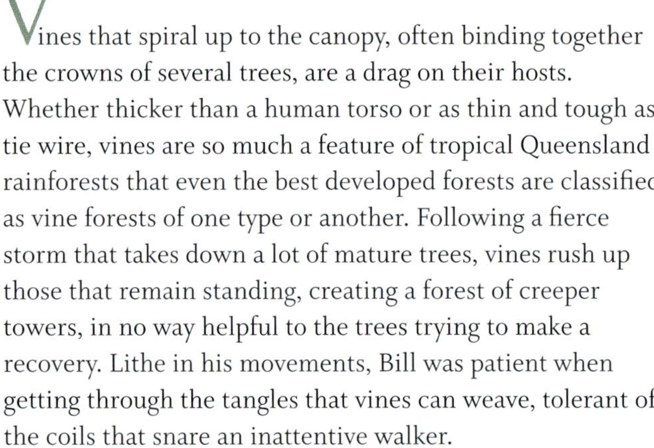

Vines, Ferns & Epiphytes

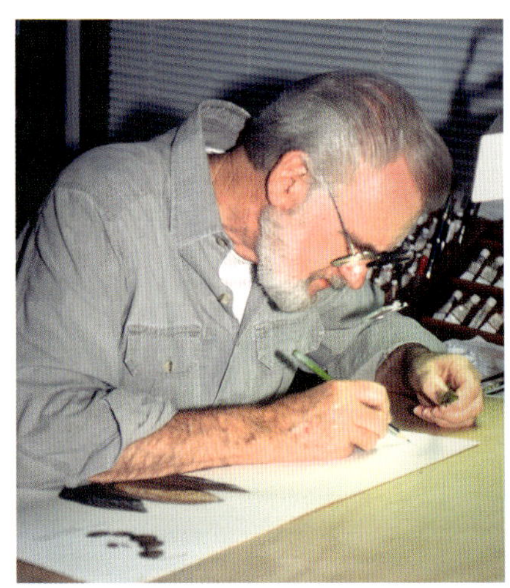

Vines that spiral up to the canopy, often binding together the crowns of several trees, are a drag on their hosts. Whether thicker than a human torso or as thin and tough as tie wire, vines are so much a feature of tropical Queensland rainforests that even the best developed forests are classified as vine forests of one type or another. Following a fierce storm that takes down a lot of mature trees, vines rush up those that remain standing, creating a forest of creeper towers, in no way helpful to the trees trying to make a recovery. Lithe in his movements, Bill was patient when getting through the tangles that vines can weave, tolerant of the coils that snare an inattentive walker.

Although they are insistent hangers-on, many vines do provide fruit that entice birds and mammals. The leaves of some are eaten by Green Possums; others have leaves eaten by the caterpillars of magnificent butterflies. And big vines, lianas, take on magical shapes as they spiral their way up the trunk of a tree. For Bill, lianas were part of the complexity and beauty in a forest. In his paintings, he celebrated the eccentric twists and turns of thick lianas, giving them a role when parts of a composition might be best linked by a loop or coils of a vine.

Because they do not need rigidity, vines can be unruly and mystifying, heading up a tree for six or ten metres, then inexplicably sending out a branch that plunges back to earth, where it takes root before veering off in another direction, starting up another hapless tree. Sometimes a vine that has begun an ascent but failed to properly bind its host will have had several coils slip to the base of the tree, from where a fresh climb has then been made. Because nearly every vine in the forests of north Queensland spirals up in an anticlockwise direction, Bill enjoyed keeping a look out for those that wind clockwise up a tree, which does happen on occasion. Most vines in the Northern Hemisphere also climb in an anticlockwise direction, and some research suggests this spiralling might result from asymmetry in the structure of proteins comprising the stem.

There are several vines that do not depend on spiralling around a support to get off the ground. Some take a direct route up a tree, sending roots out to either side at close intervals as they make their upward scramble. Among these 'root climbers', there are delicate species like *Pothos* and small-leaved *Rhaphidophora*, which go up using fine, short roots that adhere firmly to the bark of the host. For at least some species, a secretion from the roots fixes them to the tree. More muscular root climbers like the large-leaved species of *Rhaphidophora* send out a number of roots at every node, holding limpet-like to the trunk they have ascended. These big climbers, which kink at every node, can be so numerous on the host tree that the main trunk is quite obscured.

Some vines use thorns as a climbing aid; for other species thorns are a defence mechanism. *Maclura* has hooks for climbing but also very fierce straight spikes on a vine that yields deliciously acid fruit. Another climber with palatable fruit is *Elaeagnus*, which also protects itself with spikes. Most notorious of the climbers using thorns are several species of palm known as Lawyer Vines.

Lawyer Vines begin as an ordinary rosette of palm fronds, which may change little for decades, until some impulse—perhaps an increase in the amount of light—prompts the palm to grow a cane that heads up to the canopy. Each leaf along the cane starts from a spiny leaf-sheath more than 30 centimetres in length that clasps the stem very tightly. The spines on this sheath hook onto any support the ascending palm contacts but a long 'fishing line' also grows out of the sheath. These fishing lines, known as flagella, each several metres in length, are equipped with strongly curved, closely spaced hooks. Whipping back and forth in strong winds, the hooks on a long flagellum snare a tree, giving the cane a sure grip as it climbs higher. It is the needle-sharp thorns

← Bill Painting the Fruit Collected from the Vine *Gouania exilis*, July 2001

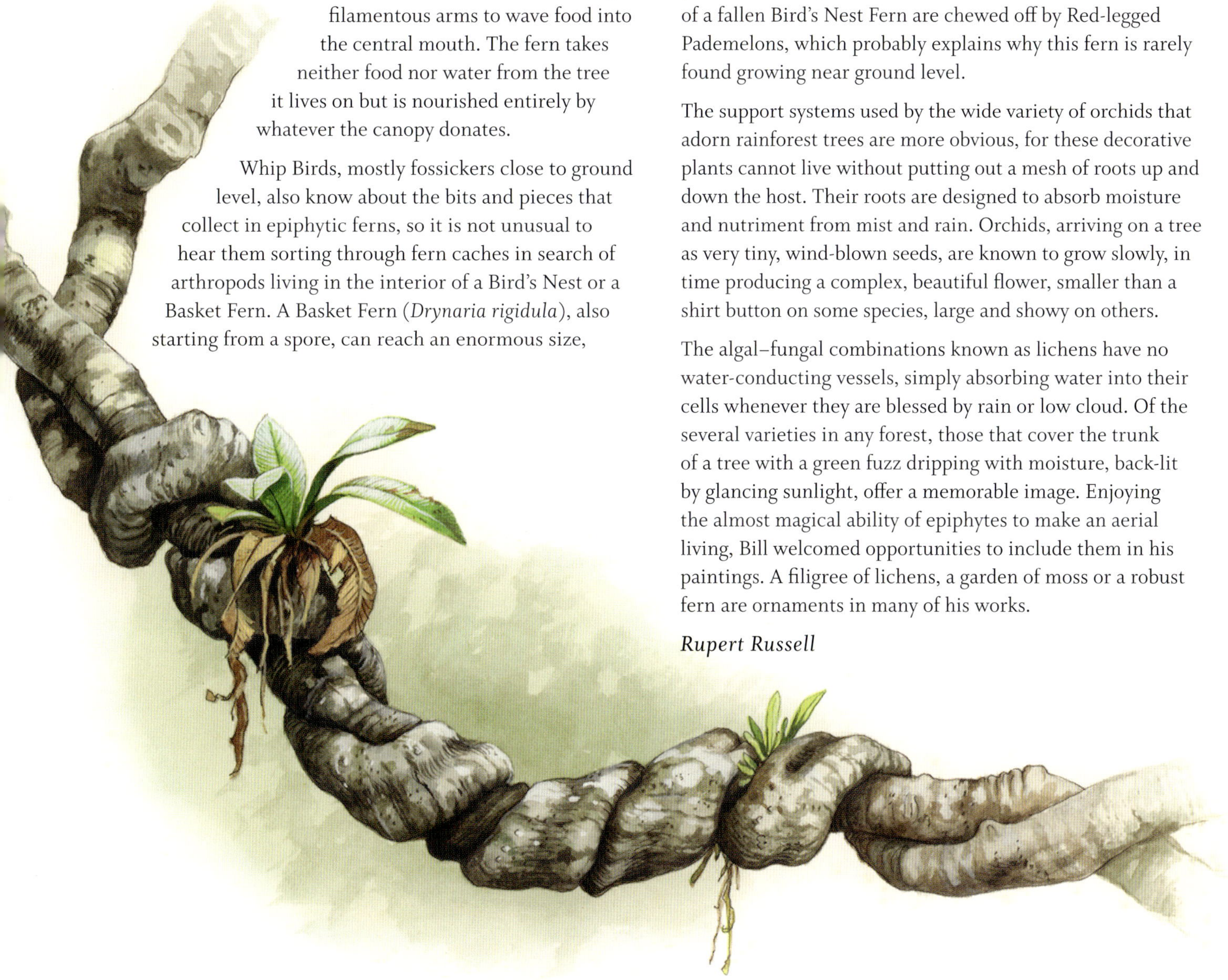

↓ Bird's Nest Ferns
undated (c.1993); pen and watercolour; 265 x 390mm
State Library of New South Wales

on lengths of these fine lines that dangle unnoticed until they snag a visitor, giving full meaning to the 'wait-a-while' description for this group of palms.

On occasion, a stout liana will itself be set upon by an epiphyte, most often a Bird's Nest Fern (*Asplenium*). Starting as one of the myriad spores released from a mature fern, a single spore in a raindrop can, following a complicated sperm and egg conjunction, grow into a tiny fern. If settled on a vine, Bird's Nest Ferns do not usually grow very large, but when they start on a big tree these ferns can send up fronds more than a metre in length, radiating from all around the base. Debris from the canopy is caught in the centre, helped by the fronds funnelling leaf litter inward, reminiscent of a sea anemone in the ocean using filamentous arms to wave food into the central mouth. The fern takes neither food nor water from the tree it lives on but is nourished entirely by whatever the canopy donates.

Whip Birds, mostly fossickers close to ground level, also know about the bits and pieces that collect in epiphytic ferns, so it is not unusual to hear them sorting through fern caches in search of arthropods living in the interior of a Bird's Nest or a Basket Fern. A Basket Fern (*Drynaria rigidula*), also starting from a spore, can reach an enormous size, spreading for several metres along the length of a supporting branch, easily reaching more than a metre across. Contemplating these great structures prompted Bill into thinking about the huge weight of water they would hold during persistent rain—a tremendous additional weight for the put-upon branch to support—and provoked a smiling alacrity when he sidestepped one, narrowly avoiding a drenching. The ability to reach a great size—despite subsisting on the meagre quantities of leaves, flowers and fruit available to a fern living on a lofty branch close to the upper canopy—suggests that Basket Ferns may be slow growing but very long lived.

It is unknown whether there is chemical bonding between a fern and its support, but if not it is mystifying that Crow's Nest and Basket Ferns can cling to a branch that is swayed and showered by every storm, using roots that neither penetrate nor encircle the support. Huge agglomerations do fall occasionally, discomfiting any arboreal snake or pygmy possum that they are known to sometimes shelter. Leaves of a fallen Bird's Nest Fern are chewed off by Red-legged Pademelons, which probably explains why this fern is rarely found growing near ground level.

The support systems used by the wide variety of orchids that adorn rainforest trees are more obvious, for these decorative plants cannot live without putting out a mesh of roots up and down the host. Their roots are designed to absorb moisture and nutriment from mist and rain. Orchids, arriving on a tree as very tiny, wind-blown seeds, are known to grow slowly, in time producing a complex, beautiful flower, smaller than a shirt button on some species, large and showy on others.

The algal–fungal combinations known as lichens have no water-conducting vessels, simply absorbing water into their cells whenever they are blessed by rain or low cloud. Of the several varieties in any forest, those that cover the trunk of a tree with a green fuzz dripping with moisture, back-lit by glancing sunlight, offer a memorable image. Enjoying the almost magical ability of epiphytes to make an aerial living, Bill welcomed opportunities to include them in his paintings. A filigree of lichens, a garden of moss or a robust fern are ornaments in many of his works.

Rupert Russell

Adenia heterophylla

→ c.1993; watercolour; 315 x 415mm
Collection of Peter and Marilyn Chapman

↘ undated; pen, pencil and watercolour
335 x 460mm
State Library of New South Wales

↓ undated; pen and pencil; 330 x 280mm
State Library of New South Wales

Lacewing Vine

PASSIFLORACEAE

Adenia is from *aden* (gland), referring to the raised glands at the base of the leaves.

heterophylla is from *heteros* (different) and *-phyllus* (-leaved).

Lacewing Vine is a large canopy climber whose leaves are food for caterpillars of the Red Lacewing, Cruiser and Glasswing butterflies. It is found in northern Australian rainforests, New Guinea and Malesia.

The small white flowers develop into spectacular red fruits. Empty capsules are often found but the creature that takes the fleshy white aril enveloping the black seeds remains unknown. We found this plant dangling low enough for easy plucking beside the Mossman River in north Queensland. The completed painting (opposite) was used as a means of breaking up the page design in *Fruits of the Rain Forest*.

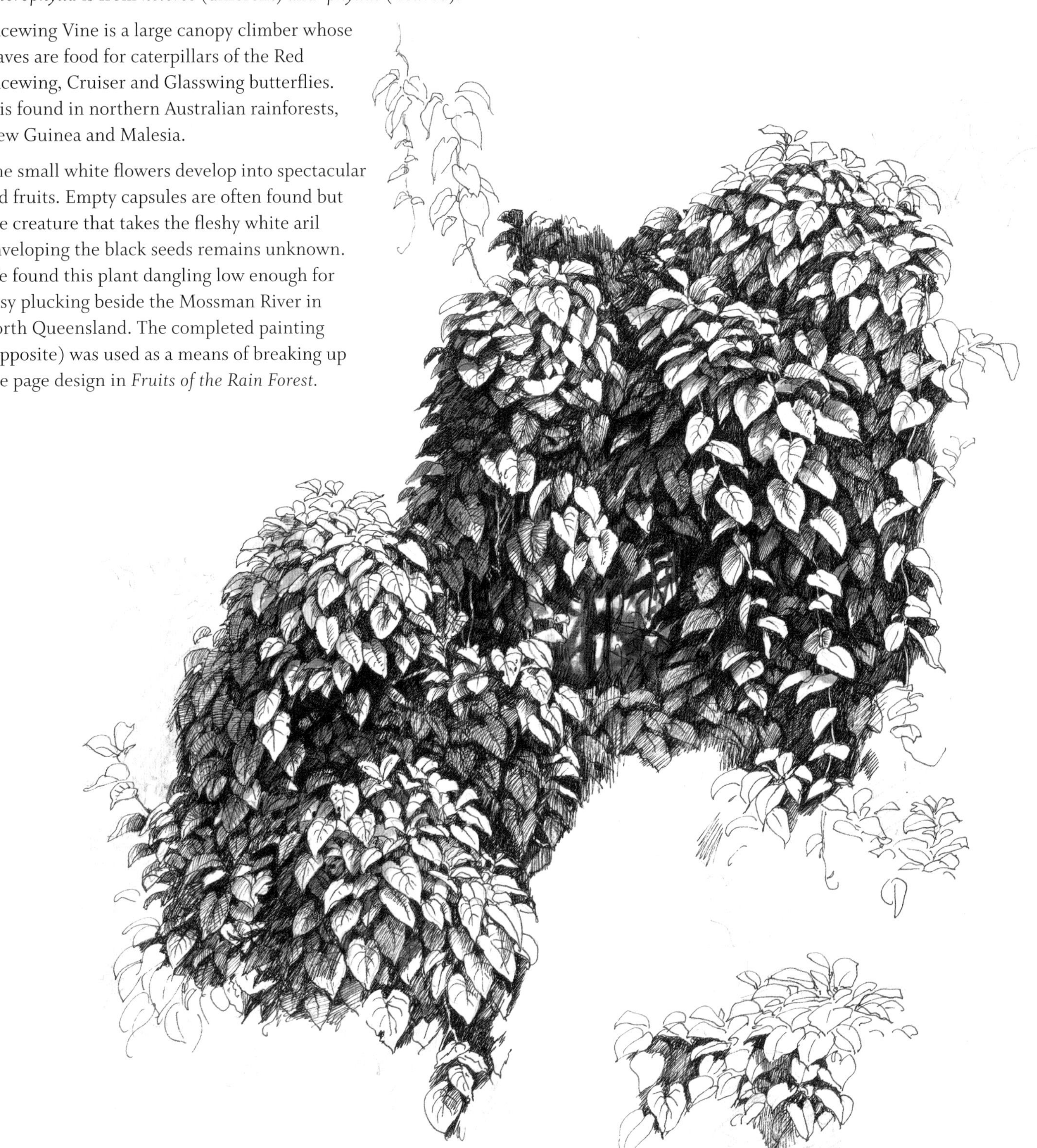

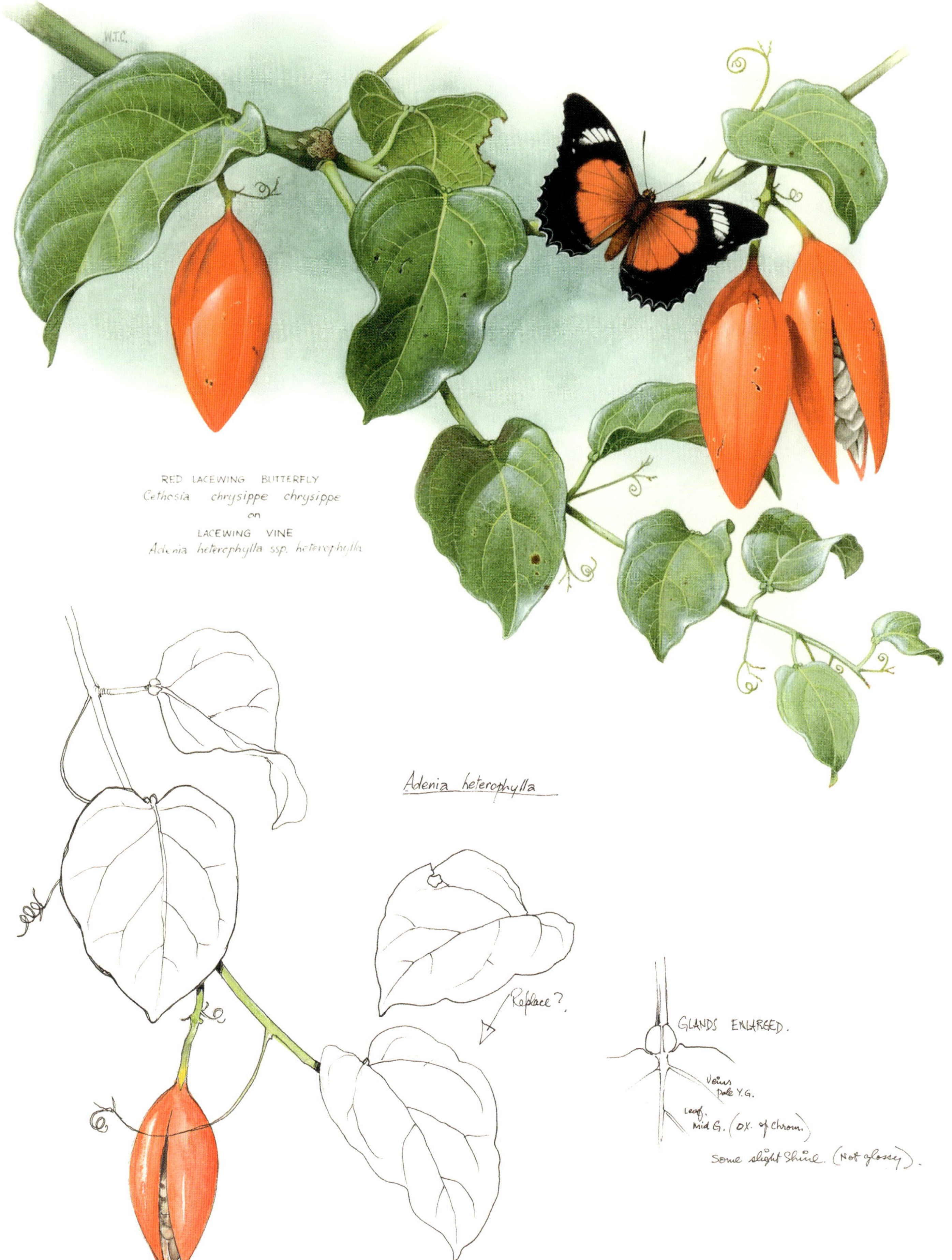
W.T.C.
RED LACEWING BUTTERFLY
Cethosia chrysippe chrysippe
on
LACEWING VINE
Adenia heterophylla ssp. heterophylla
Adenia heterophylla
Replace?
GLANDS ENLARGED.
Veins pale Y.G.
Leaf. Mid G. (OX. of Chrom.)
Some slight Shine. (Not glossy).

Amyema

LORANTHACEAE

Amyema is from *a-* (not) and *myeo* (I initiate), referring to the genus being separated from *Loranthus*.

Amyema are erect or pendulous mistletoes occurring in Australia as well as the Pacific, New Guinea and South-East Asia.

Amyema queenslandica

Mistletoe

queenslandica is from Queensland.

This mistletoe is an erect or spreading hemiparasitic plant that occurs on rainforest trees in Queensland, north from the Seaview mountain range and into New Guinea. The leaves are green and fleshy, with flowers that are yellow and red and hang in threes within a pendulous inflorescence. They are visited regularly by honeyeaters, who sip the nectar.

Amyema quandang var. *quandang*

Grey Mistletoe

quandang is after the Quandang (*Santalum acuminatum*), a plant on which this mistletoe occurs, as noted by the surveyor and explorer Thomas Mitchell.

Grey Mistletoe is a hemiparasitic pendulous plant that grows on *Acacia* trees including *Acacia aneura*. It occurs in every mainland state and mostly in arid areas. New leaves are clothed in short whitish hairs. The flowers grow in clusters of three, with the central one on a much shorter stalk.

→ Grey Mistletoe
1975; pen, pencil and watercolour; 410 x 310mm
State Library of New South Wales

↓ Mistletoe
undated; pen and pencil; 195 x 300mm
State Library of New South Wales

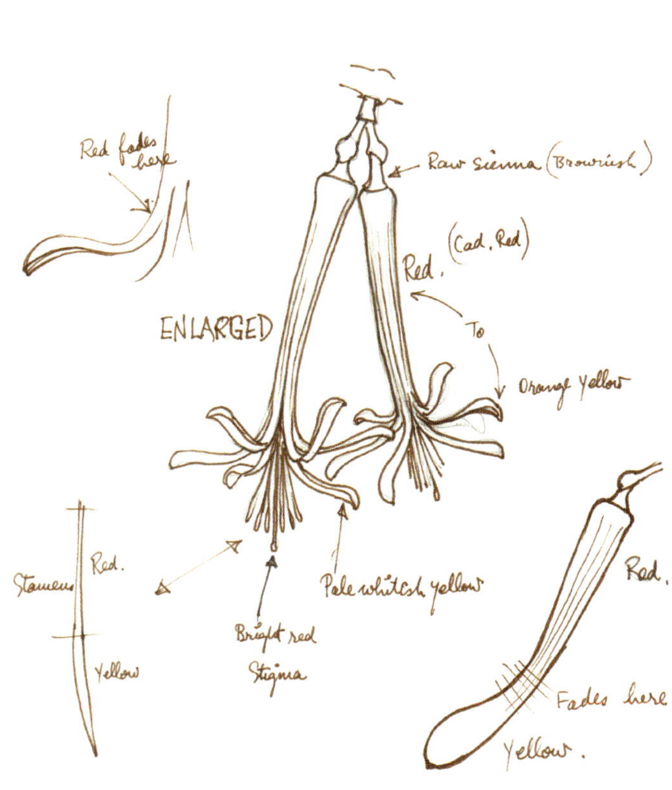

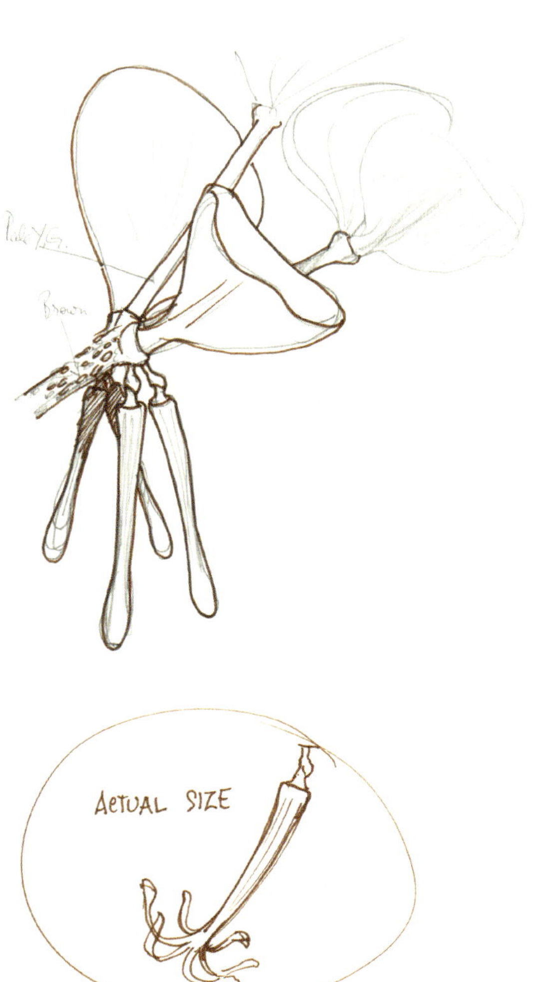

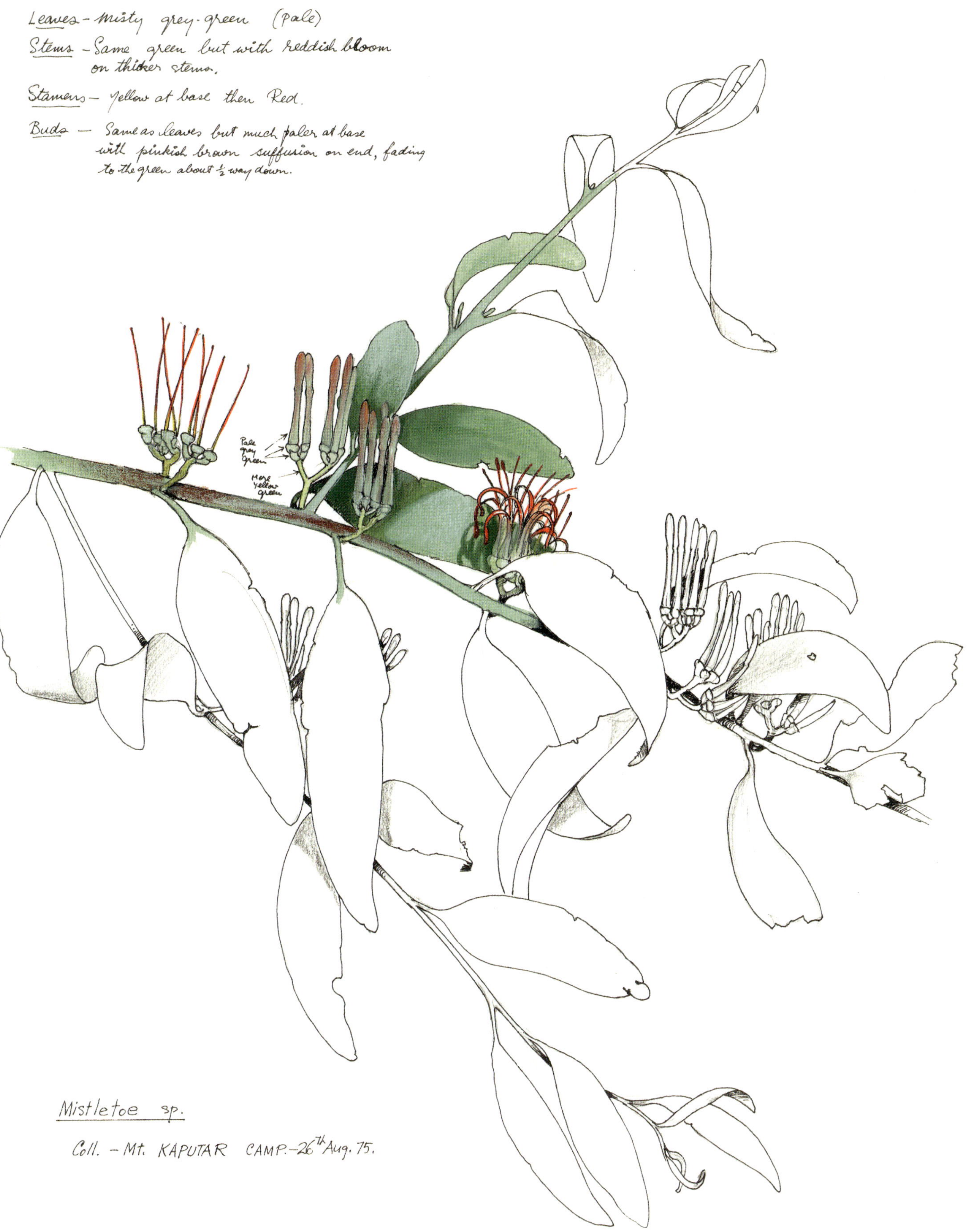
Leaves - misty grey-green (pale)
Stems - Same green but with reddish bloom on thicker stems.
Stamens - Yellow at base then Red.
Buds - Same as leaves but much paler at base with pinkish brown suffusion on end, fading to the green about ½ way down.
Pale grey green
More yellow green
Mistletoe sp.
Coll. - Mt. KAPUTAR CAMP. - 26th Aug. 75.

Asplenium

Bird's Nest Fern

ASPLENIACEAE

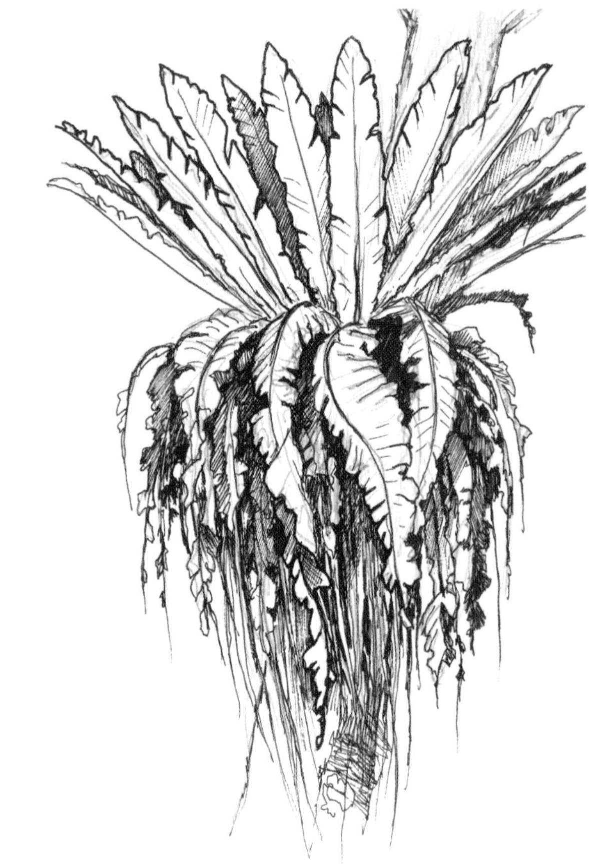

Asplenium is from *a* (without) and spleen, referring to the plant's traditional use as a cure for spleen disorders.

Bird's Nest Fern species are many and widespread in distribution. Their compact epiphytic growth habit and large glossy leaves made them excellent candidates for use in the foreground or background of Bill's bird paintings. There are many examples of them in Bill's archives from the forests of north and southern Australia to New Guinea, and on our own property at Topaz:

> *30th Oct. 1990. The Catbirds whose nest was predated at the end of the garage built again in an Asplenium in front of the studio, on about the 27th Oct. it had two eggs. On the late afternoon of the 29th the bird was sitting tight. On the morning of the 30th the nest was empty—predated again!*

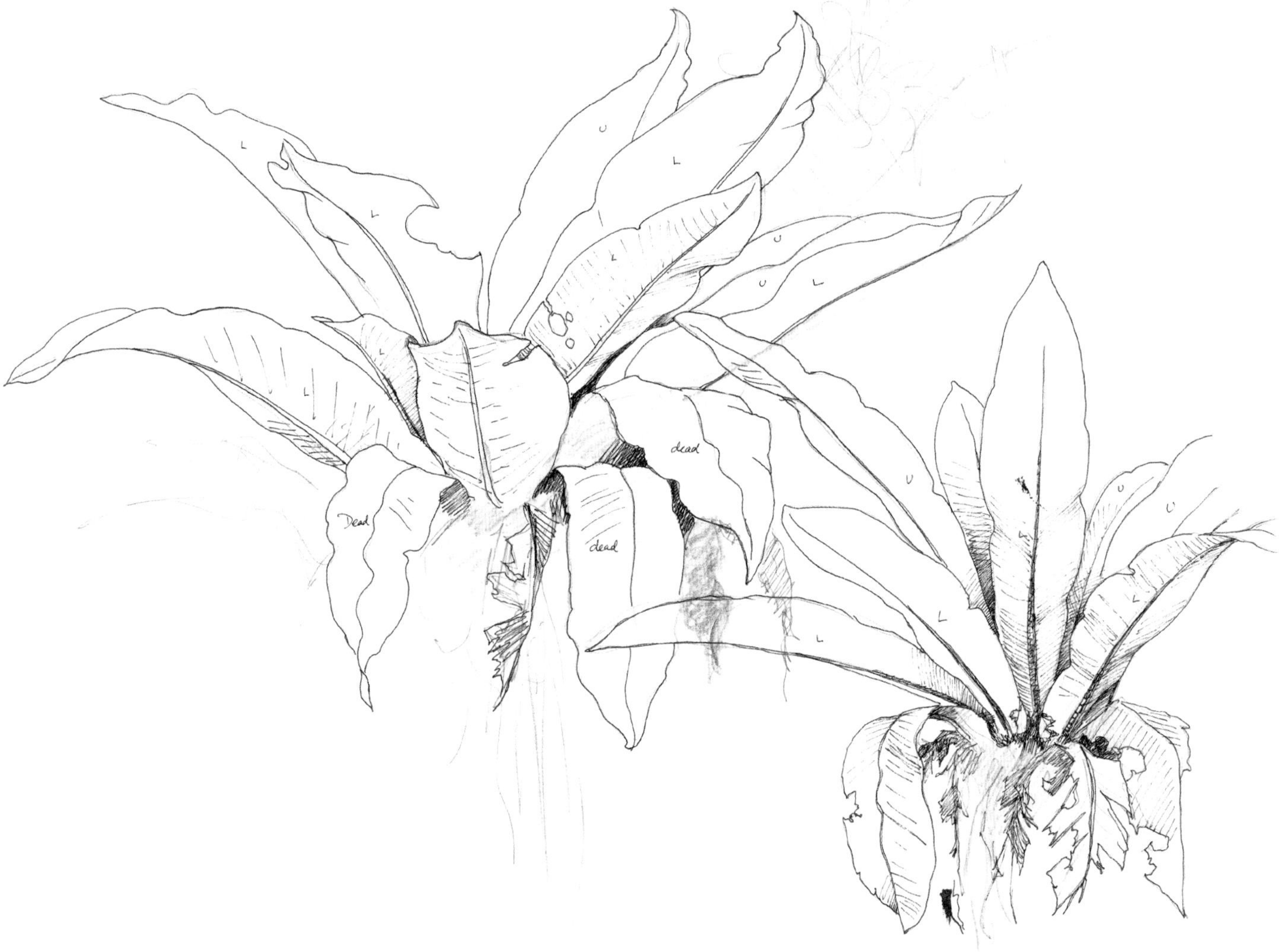

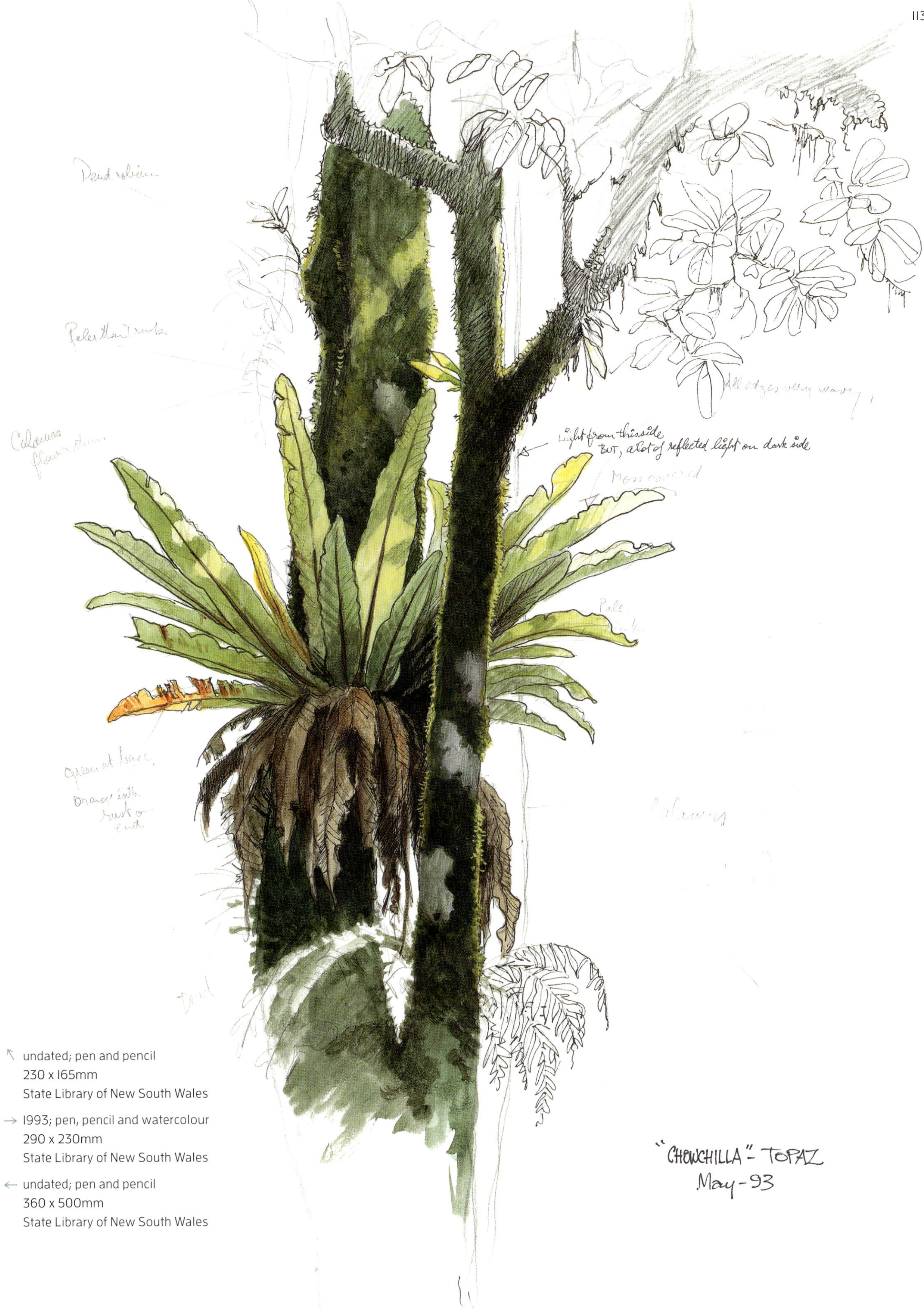

↖ undated; pen and pencil
230 x 165mm
State Library of New South Wales

→ 1993; pen, pencil and watercolour
290 x 230mm
State Library of New South Wales

← undated; pen and pencil
360 x 500mm
State Library of New South Wales

Blechnum cartilagineum

Gristle Fern

BLECHNACEAE

Blechnum is from the Greek name for a fern, *blechnon*.

cartilagineum is from *cartilagineous* (firm and tough), referring to the fronds.

As the name Gristle Fern implies, this species has rather stiff fronds when mature. This specimen was collected at Topaz but it is a widespread terrestrial species in eastern Australia, also occurring in the Philippines and Melanesia. The delicately coloured new growth could have been used by Bill to add colour to a painting of a ground-dweller like the Emerald Dove or Wonga Pigeon depicted on the otherwise brown leaf litter.

→ Ferny Branch, Iron Range
1981; pen and pencil; 280 x 265mm
State Library of New South Wales

↘ Sketch of Large Epiphytic Ferns Including Basket Fern (*Drynaria rigidula*) and a *Staghorn* (*Platycerium*) Species
1992; pen and pencil; 145 x 210mm
State Library of New South Wales

↓ 2010; pen, pencil and watercolour; 380 x 305mm
State Library of New South Wales

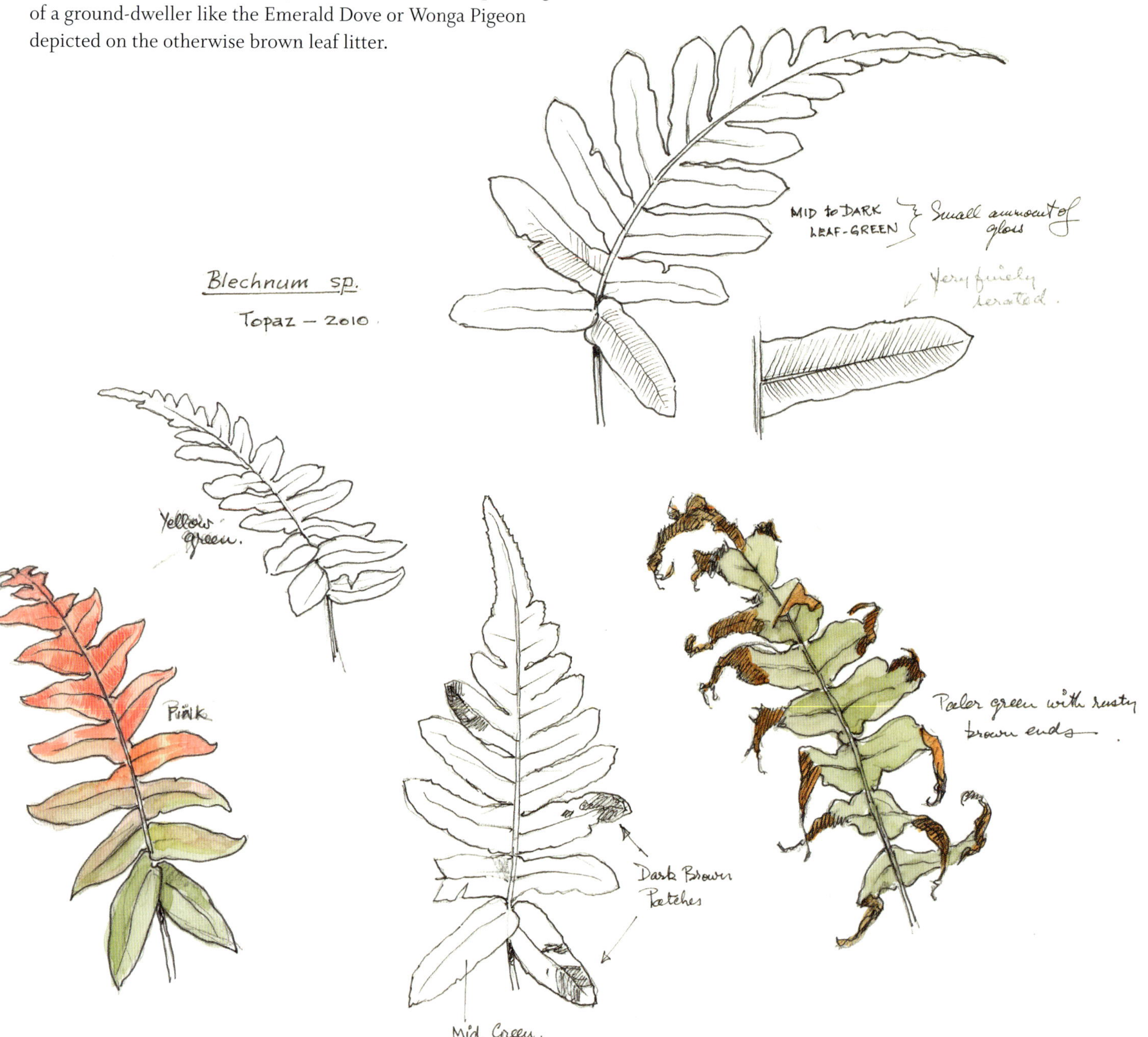

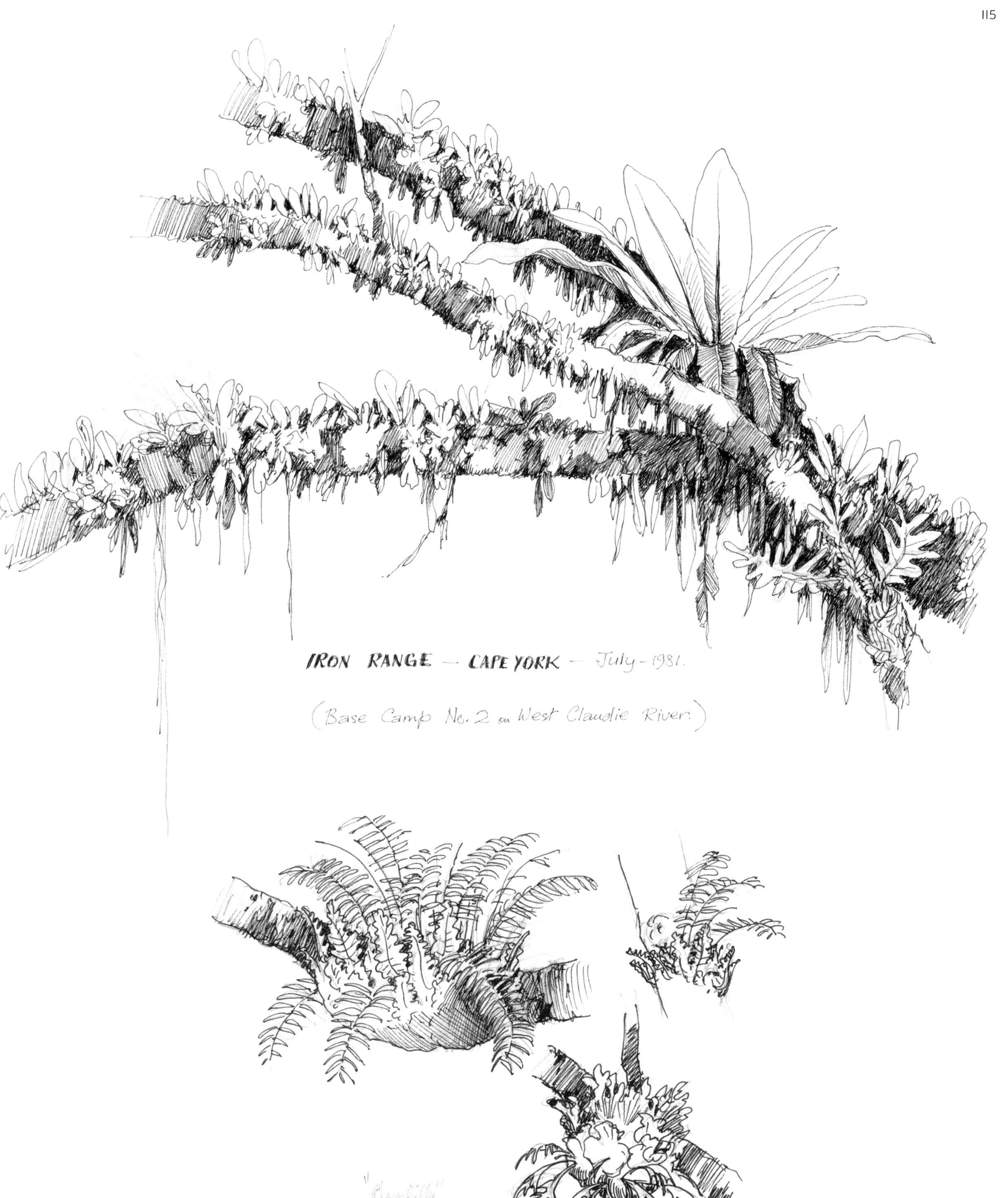
IRON RANGE – CAPE YORK – July – 1981.
(Base Camp No. 2 on West Claudie River.)

Bulbophyllum longiflorum

Pale Umbrella Orchid

ORCHIDACEAE

Bulbophyllum is from *bulbos* (bulb) and *-phyllus* (-leaved), referring to the bulb-like base to the leaves.

longiflorum is from *longus* (long) and *-florus* (-flowered).

Pale Umbrella Orchid is an epiphytic orchid that occurs in humid rainforests across the Pacific, Asia, Africa (including Uganda) and in the McIlwraith Range area on Cape York Peninsula, Australia.

Bill needed to paint the Black-casqued Wattled Hornbill from Uganda for *Kingfishers and Related Birds*. Not many Australian plants also occur in the rainforests of Africa but, luckily, a friend who cultivates native orchids loaned Bill this flowering plant to draw so that it could be used as an authentic plant in the hornbill plate.

→ undated; pen, pencil and watercolour; 470 x 355mm
National Library of Australia, 8070530

↓ Rainforest Tree Trunk with Epiphytic Vine
1991; pen and pencil
State Library of New South Wales

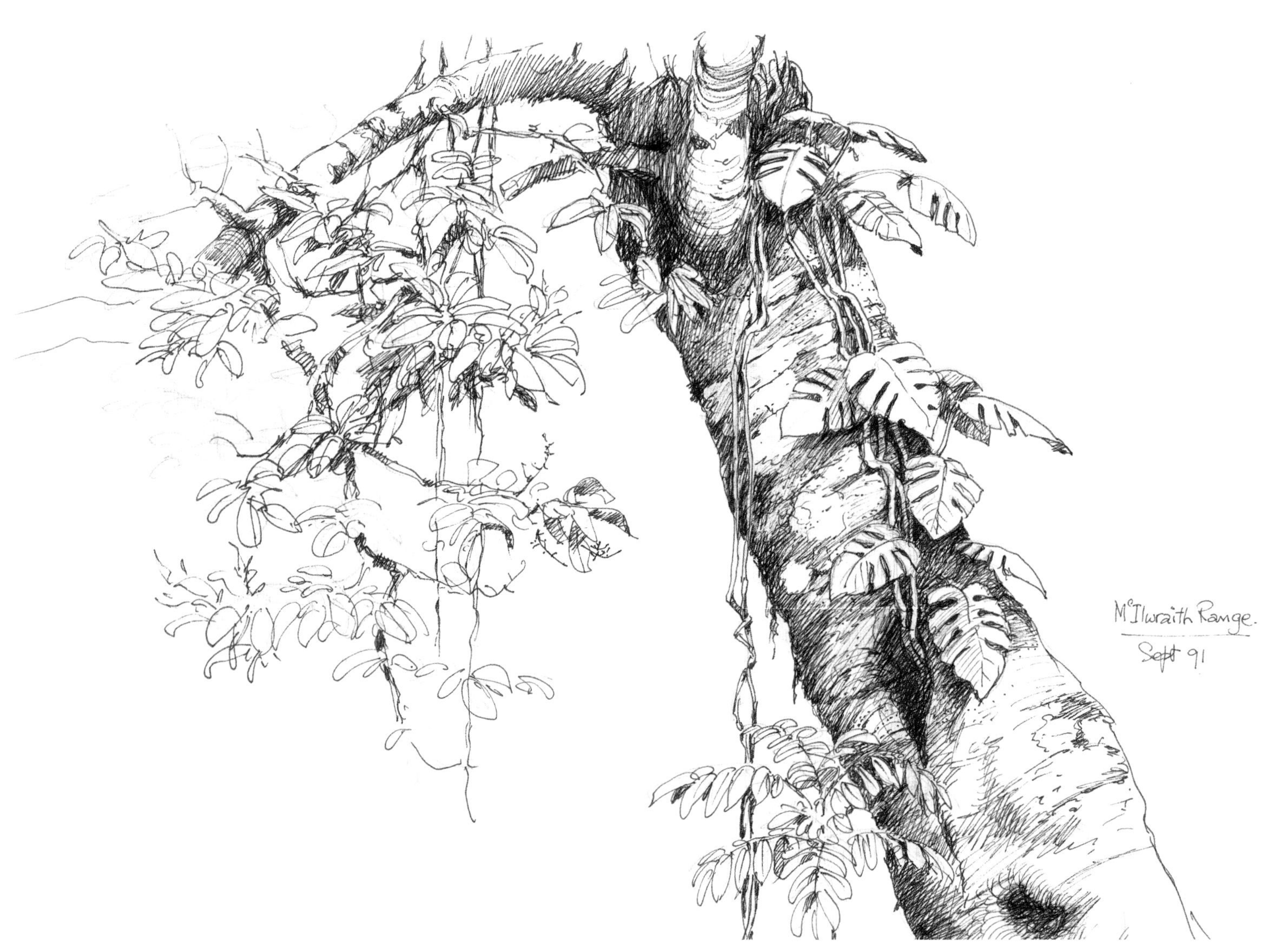

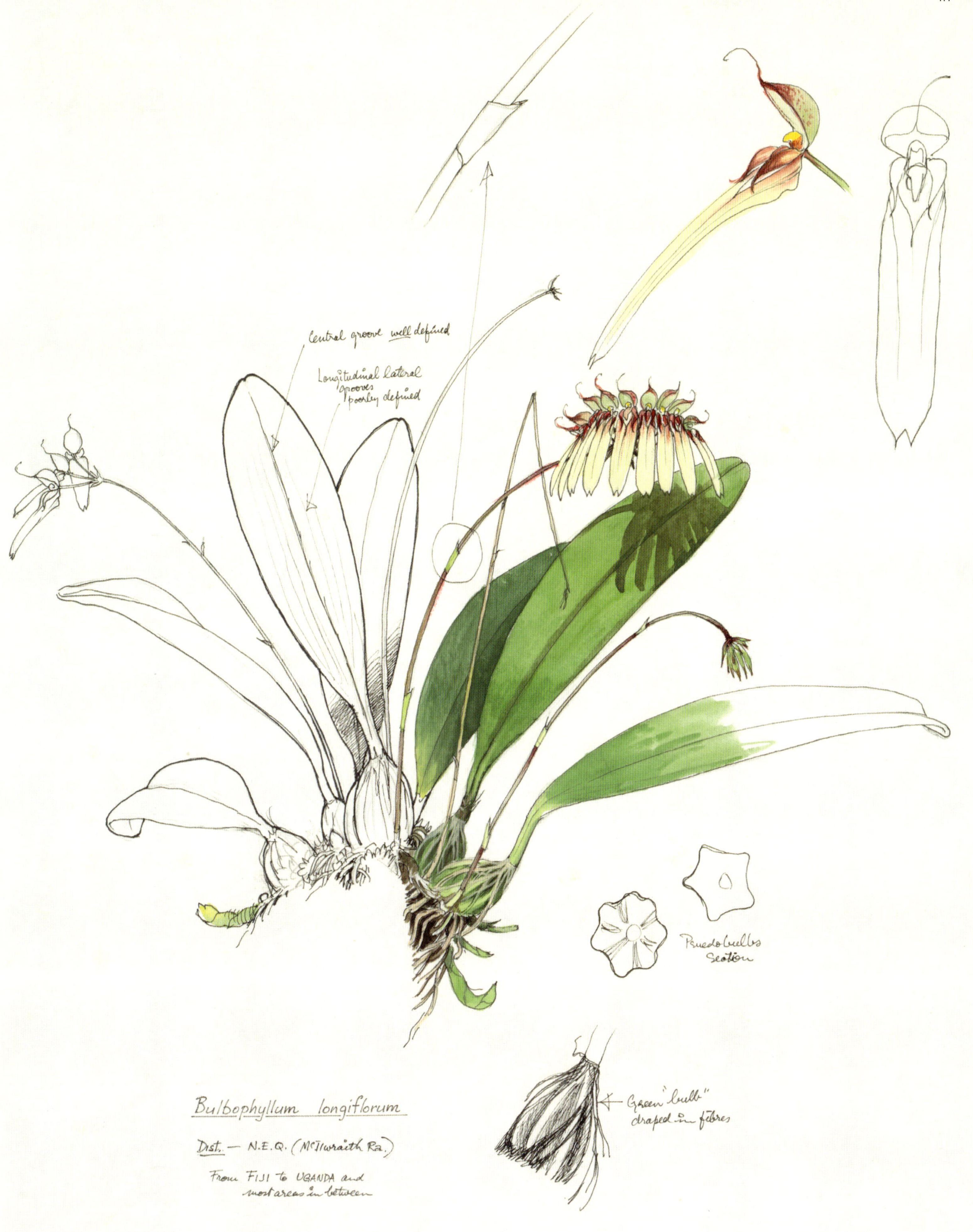
Central groove well defined
Longitudinal lateral grooves poorly defined
Psuedobulbs Section
Green "bulb" draped in fibres
Bulbophyllum longiflorum
Dist. — N.E.Q. (McIlwraith Ra.)
From FIJI to UGANDA and most areas in between

Cissus

VITACEAE

Cissus is from the Greek name for Ivy, referring to these plants being climbers.

Cissus vines or Native Grapes are a worldwide group of vines. They are closely related to European wine grapes (*Vitis*).

→ Superb Fruit-Dove
undated; watercolour
State Library of New South Wales

↘ 2002; pen, pencil and watercolour; 395 x 510mm
State Library of New South Wales

↓ 2003; pen, pencil and watercolour; 420 x 300mm
State Library of New South Wales

Cissus hypoglauca

Five-leaved Grape

hypoglauca is from *hypo-* (under or beneath) and *glaucus* (blue-grey), referring to the colour of the leaflet underside.

Five-leaved Grape is a tendril climber, often growing on the forest edge. It has an extensive distribution in rainforest and wet sclerophyll forest from Cooktown in north Queensland to Victoria.

The fruits of this plant are keenly sought by many bird species. This sketch was done for use in the plate of the Superb Fruit-Dove in *Pigeons and Doves in Australia*.

Cissus hypoglauca

loc. - 'CHOWCHILLA' - TOPAZ - 9th Mar. 2003

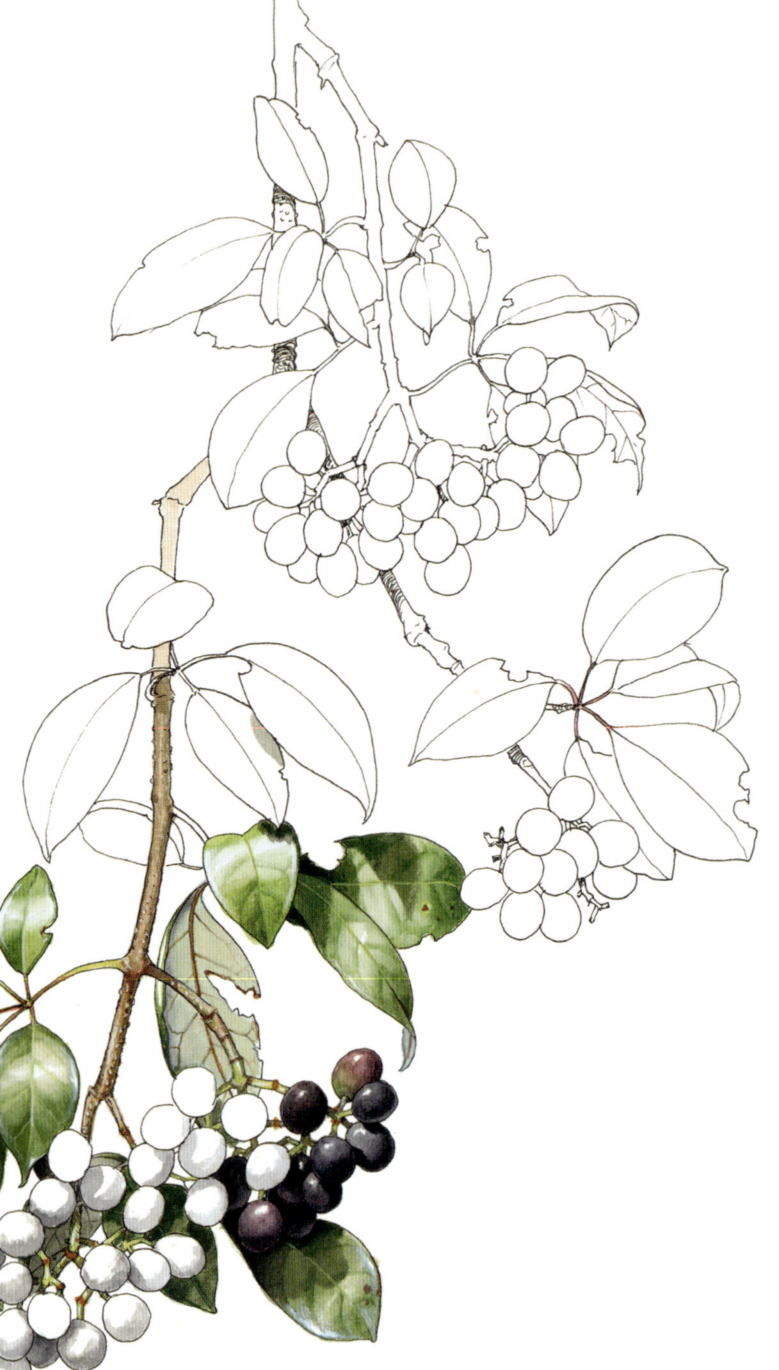

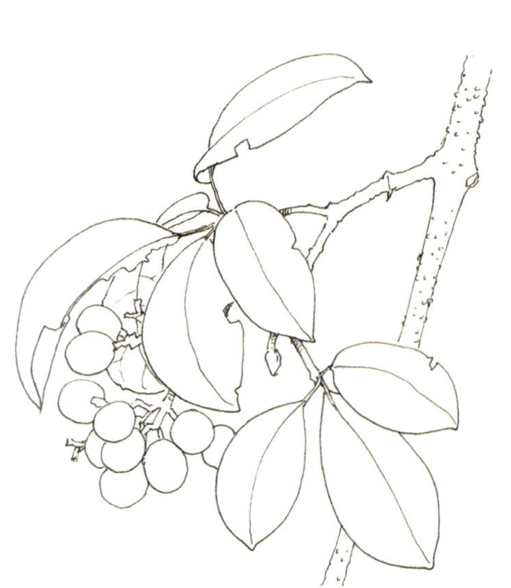

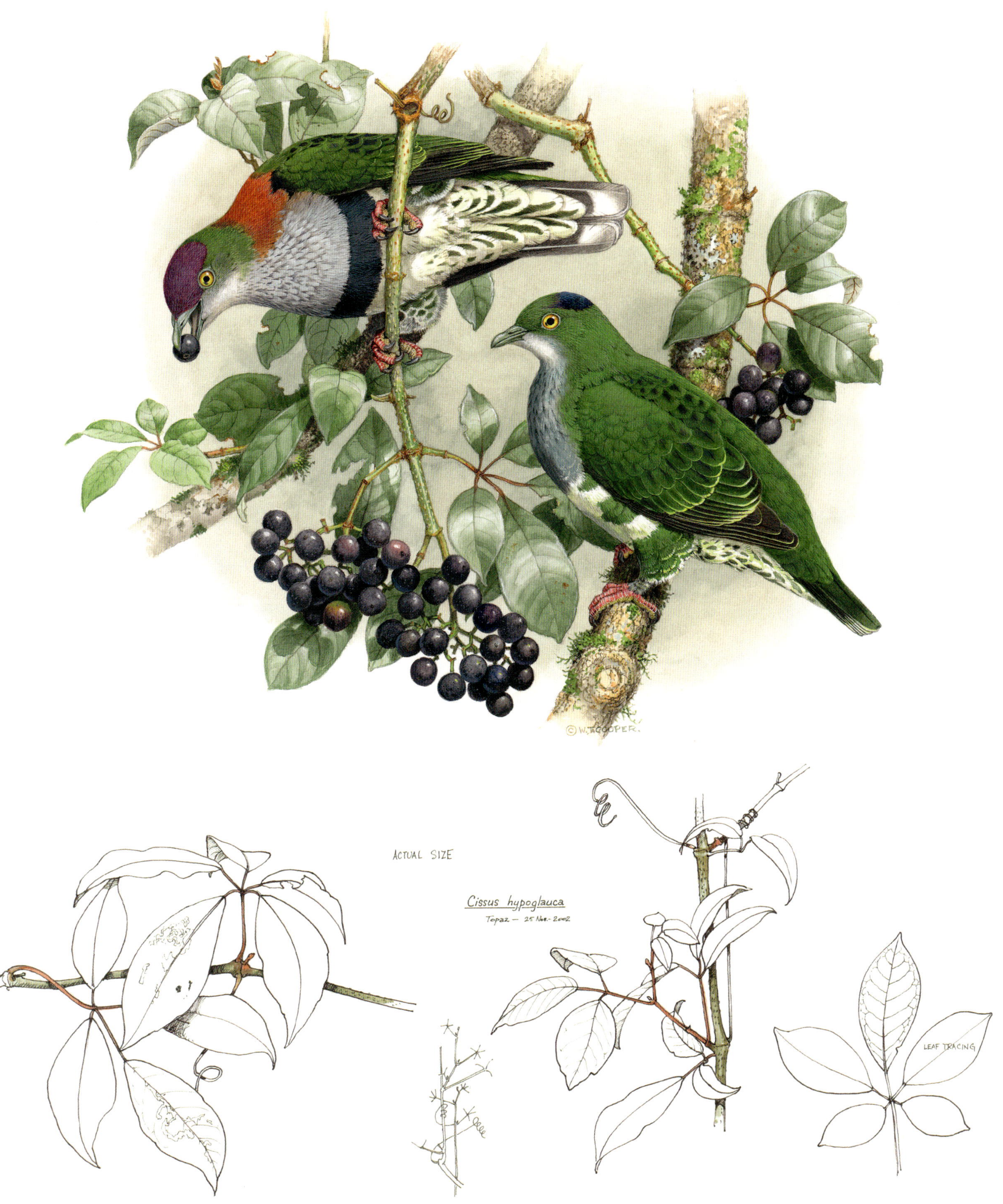
© W.T.COOPER.
ACTUAL SIZE
Cissus hypoglauca
Topaz — 25 Nov. 2002
LEAF TRACING

Cissus sterculiifolia

Long-leaved Grape

sterculiifolia is after the genus *Sterculia* and *-folius* (-leaved), referring to the similar leaves.

Long-leaved Grape is a tendril climber from the rainforests of eastern Australia and New Guinea.

Since Long-leaved Grape fruits are eaten by many birds, Bill made a reference drawing of this specimen growing near home at Topaz, knowing he would use it one day. The vine looked particularly attractive with its multi-coloured fruits, a feature that is shown to advantage in a painting of King Parrots.

→ 2005; pen, pencil and watercolour; 490 x 340mm
State Library of New South Wales

↓ Australian King Parrot
2005; watercolour; 500 x 350mm
Private collection

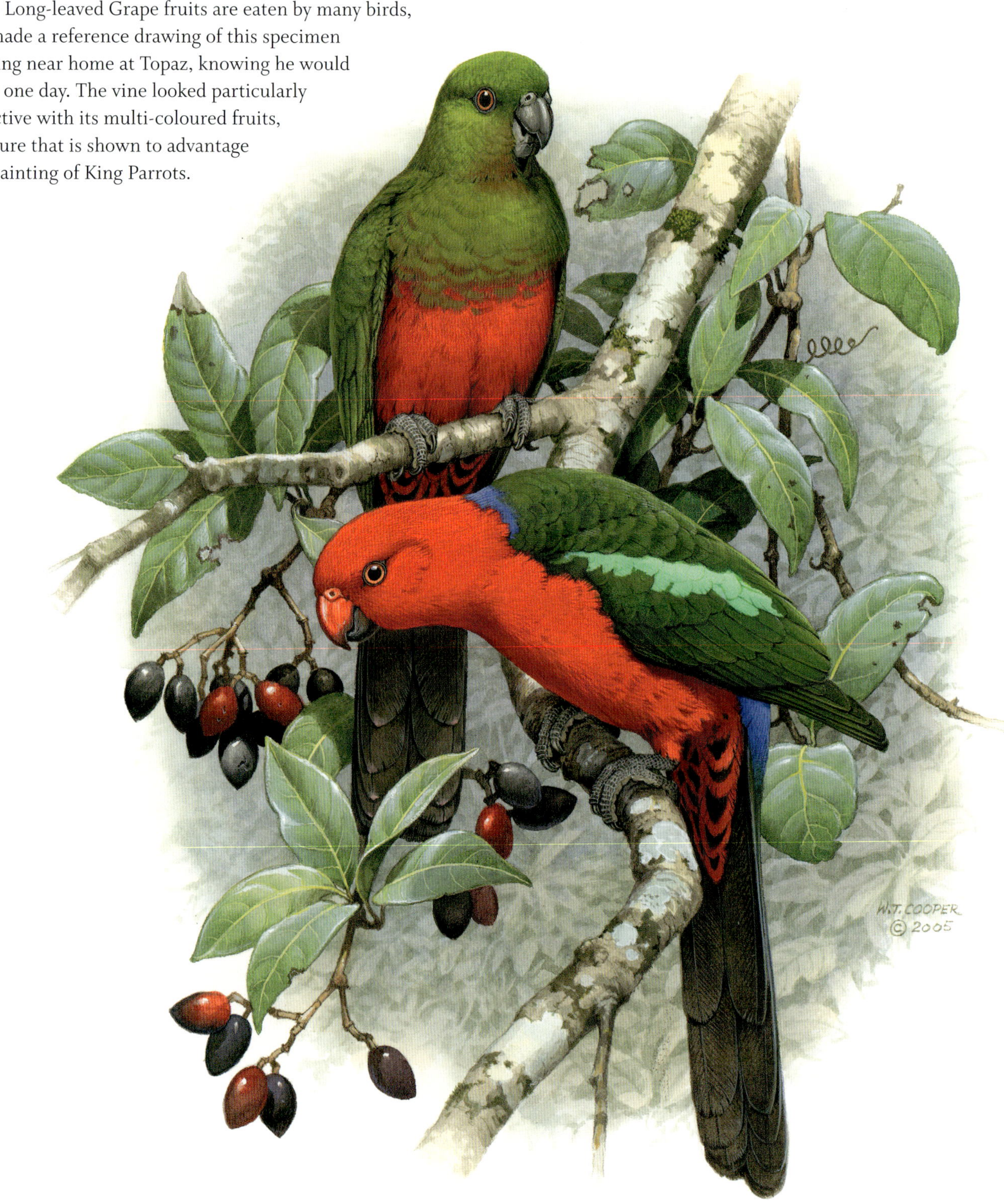

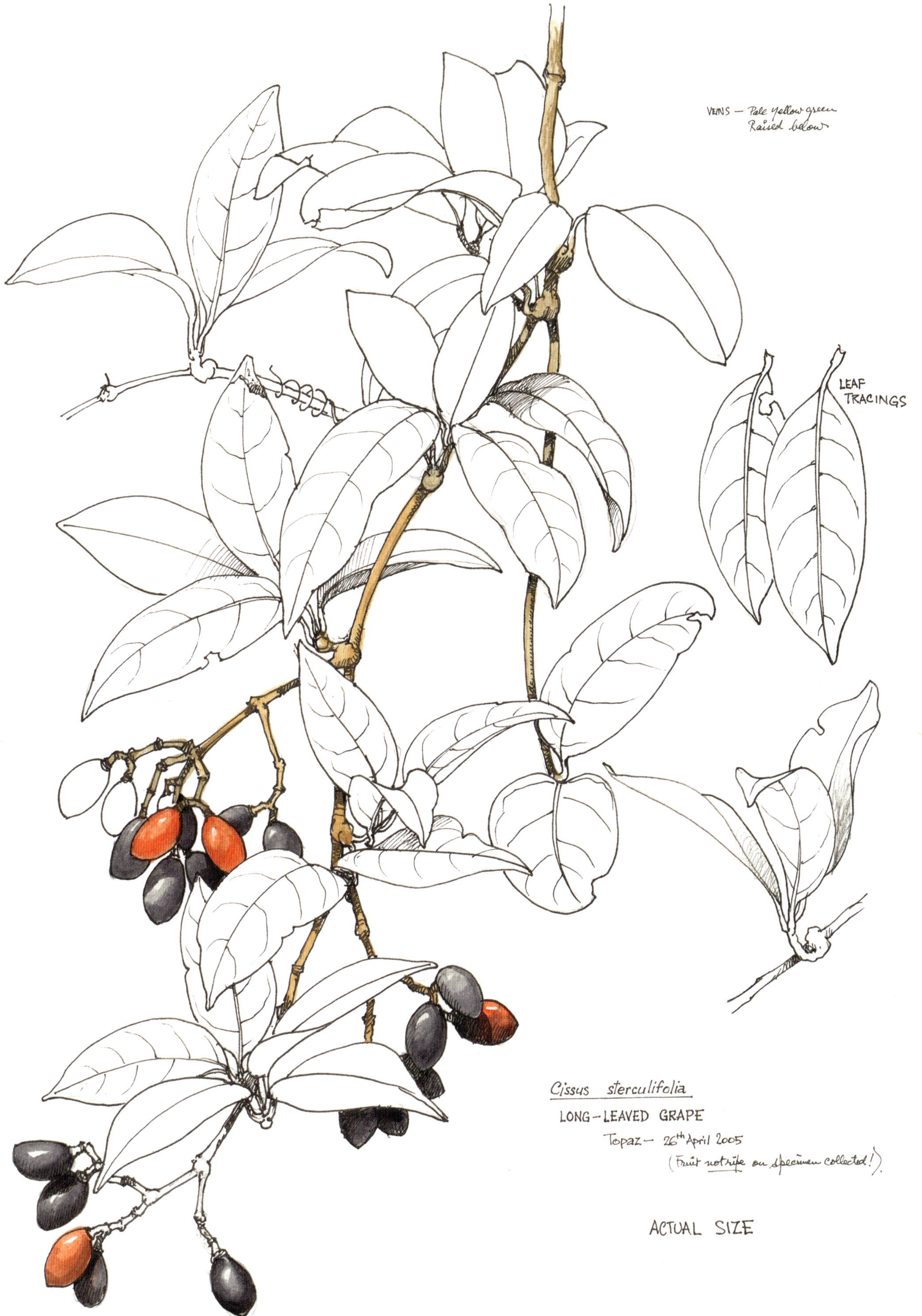
VEINS — Pale yellow green
Raised below
LEAF TRACINGS
Cissus sterculifolia
LONG-LEAVED GRAPE
Topaz — 26th April 2005
(Fruit not ripe on specimen collected!)
ACTUAL SIZE

Dischidia

APOCYNACEAE

Dischidia is from *dischides* (divided into two parts), referring to the flower structure.

Four species of *Dischidia*, which grow as epiphytes, occur in Australia on Cape York Peninsula and islands in the Torres Strait. There are 30 to 40 species worldwide, mostly in Asia.

Dischidia major

Rattle Skulls

major means greater, referring to the large leaves.

Rattle Skulls is an epiphytic vine that occurs in drier habitats throughout the Asian tropics and in northern Australia. It has two types of leaves: clusters of large balloon-like or bladdery leaves and strings of smaller, flat circular leaves. The plant's swollen leaves are inhabited by ants in a symbiotic relationship whereby ants gain shelter in exchange for elevated nutrition from nitrogen and carbon dioxide.

They occur naturally only as far south as Cape Tribulation, but some people grow them as a curiosity. The sketch at right was done in a Cairns garden and the colour was added back home in the studio. Bill disliked immensely painting in the field as he was too often discomforted by a disturbing breeze or missing some vital piece of equipment for plein-air painting.

The branch below has a tiny Prickly Ant-house Plant seedling just beginning life. The hanging plant is a *Dischidia*.

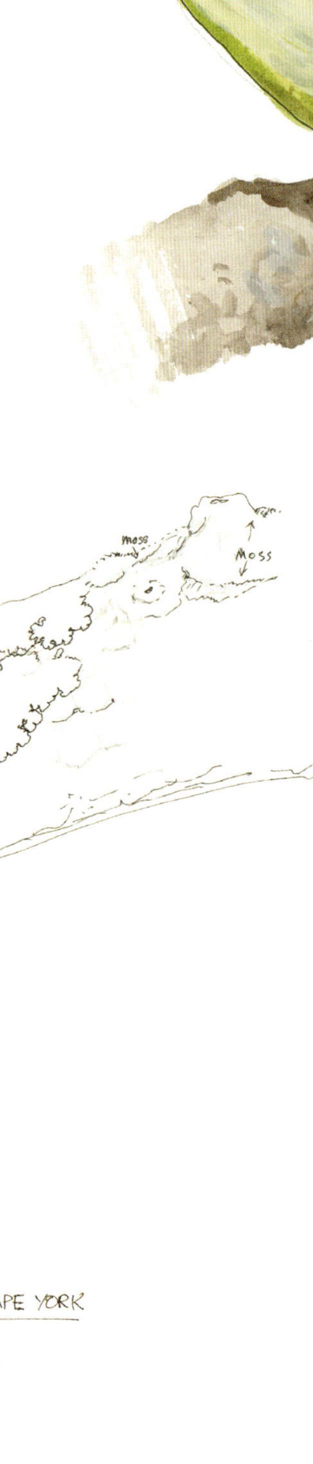

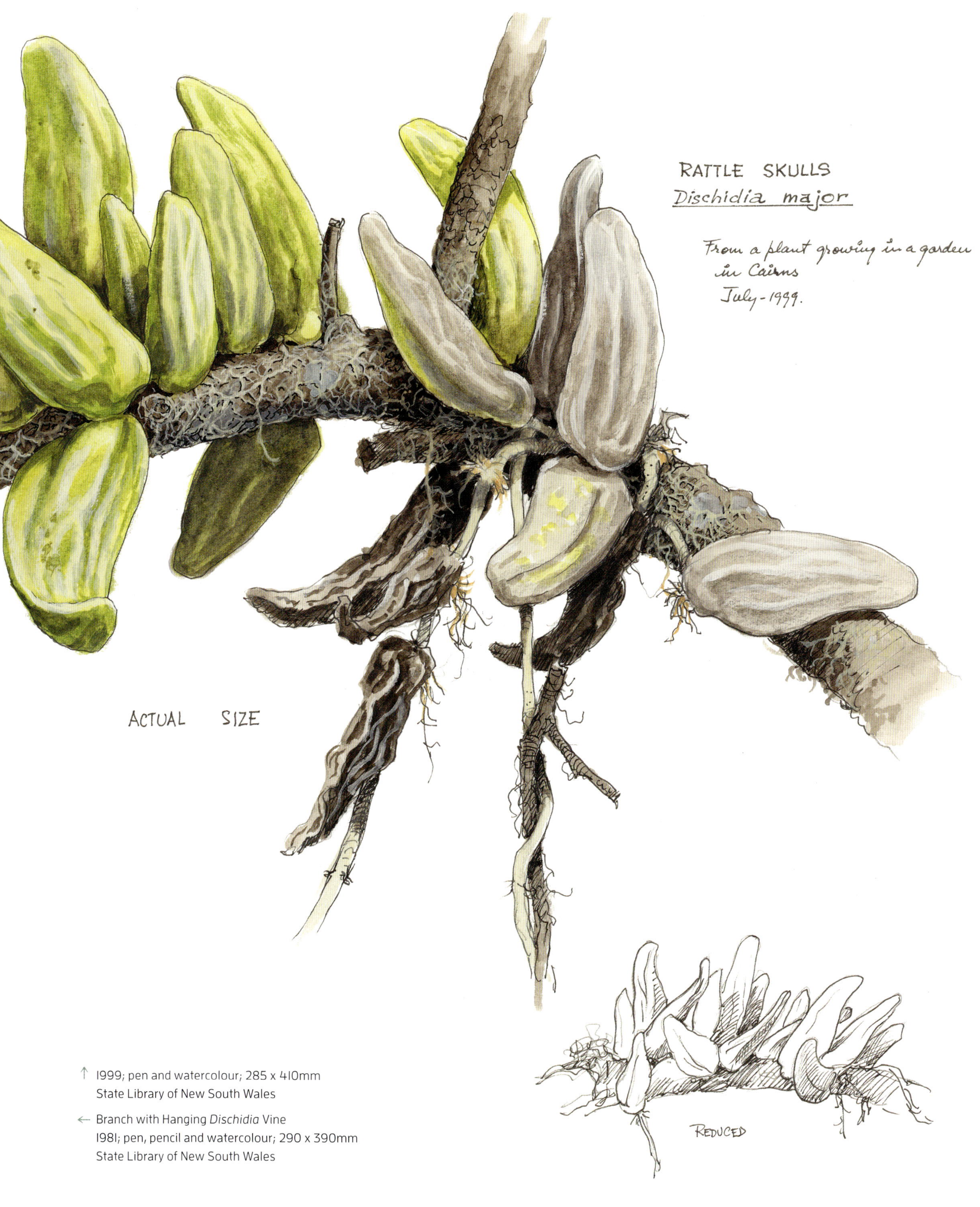

↑ 1999; pen and watercolour; 285 x 410mm
State Library of New South Wales

← Branch with Hanging *Dischidia* Vine
1981; pen, pencil and watercolour; 290 x 390mm
State Library of New South Wales

Dischidia nummularia

Button Orchid

nummularia is from *nummus* (coin), referring to the coin-shaped leaves.

The name Button Orchid is a misnomer. The leaves might be button-like but the plant is not an orchid. It is a vine that has adventitious roots that attach to branches. The leaves are succulent and covered in a white powdery film. Long strings of pendulous leafy stems hang like curtains from trees, most conspicuously on paperbarks around swampy areas in north Queensland, New Guinea and South-East Asia.

Dischidia nummularia.

Cooktown – July. 2003.

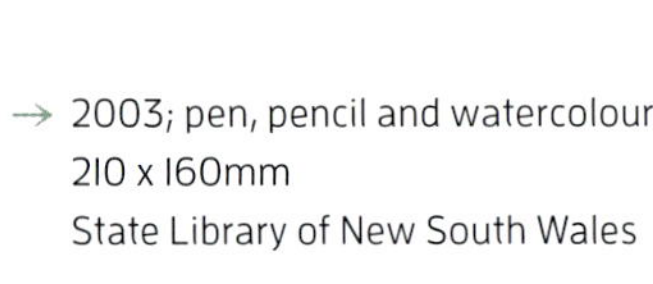

→ 2003; pen, pencil and watercolour
210 x 160mm
State Library of New South Wales

Dischidia ovata

Watermelon Dischidia

ovata is from *ovatus* (egg-shaped), referring to the leaf shape.

This plant is another epiphytic vine of tropical Australia, distinguished by the white veins on the fleshy, heart-shaped leaves. The common name, Watermelon Dischidia, is given for the distinctive markings on the leaves. It is found in rainforest, vine thickets, swamp forest and woodland on Cape York Peninsula and New Guinea.

This painting was one of the very few executed in the field. It was done at one of our favourite Cape York camps on the remote Rocky River, where the water was clear and the river was wide. We would often watch colourful Eclectus Parrots and Palm Cockatoos fly above us.

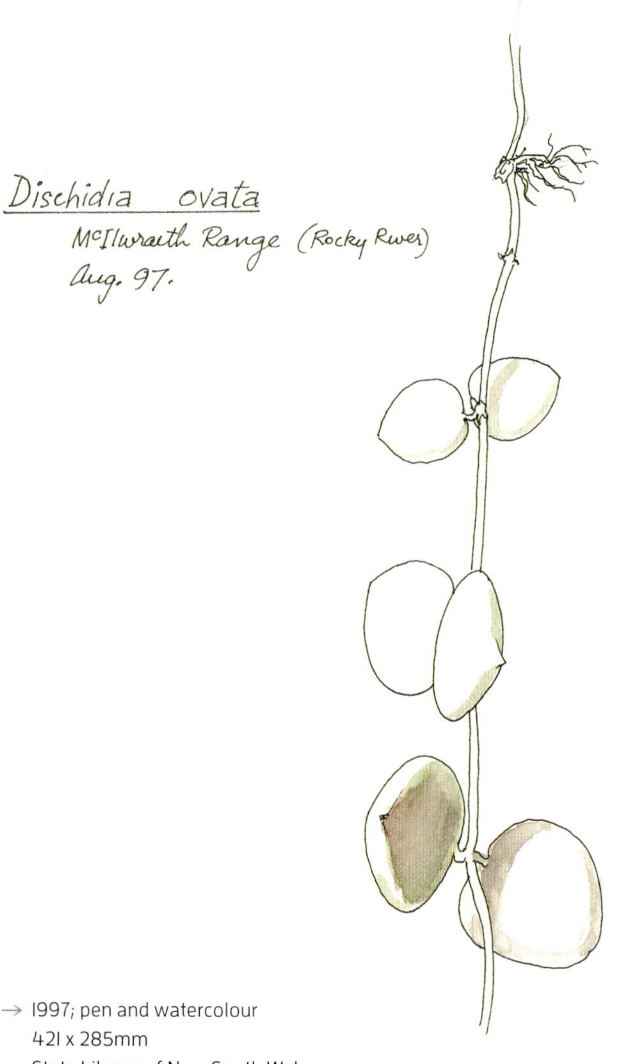

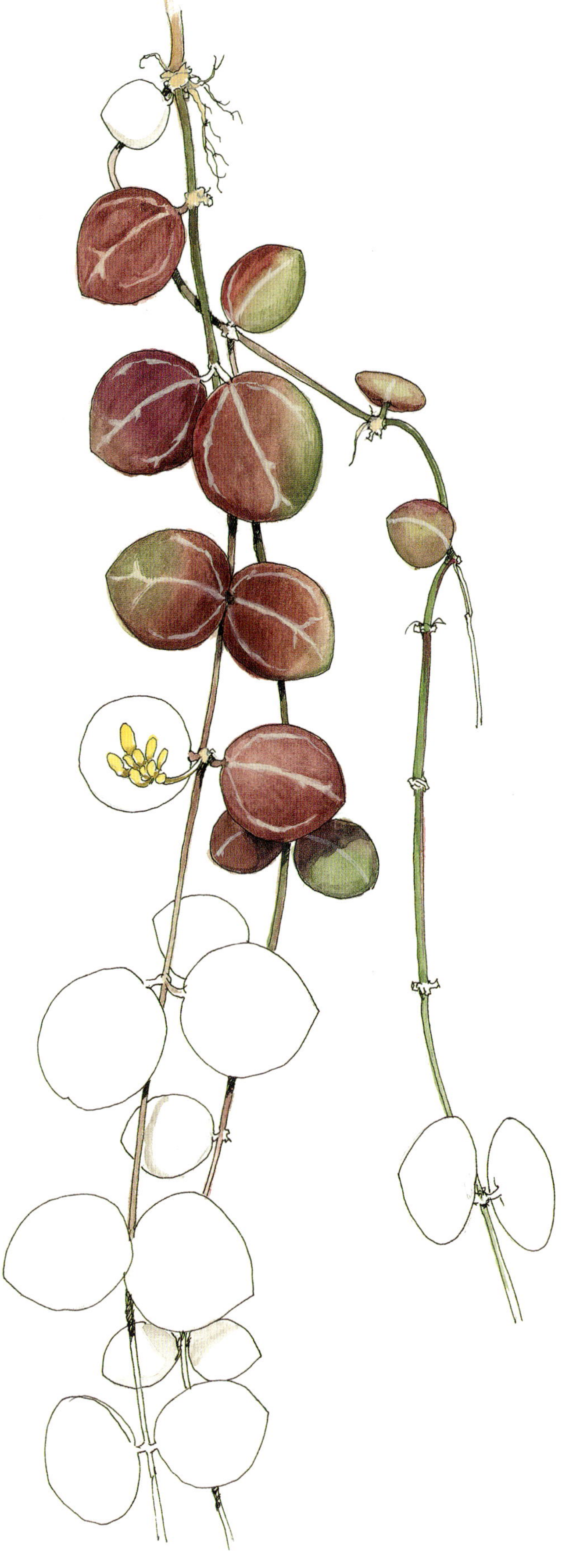

→ 1997; pen and watercolour
421 x 285mm
State Library of New South Wales

Entada phaseoloides

Matchbox Bean

FABACEAE

Entada is from a Malabar name.

phaseoloides is after a related genus of beans, *Phaseolus*, and *-oides* (resembling).

Matchbox Beans are canopy climbers, distinguished by their twisted, flattened stems, huge woody pods and hard, buoyant dark-red seeds. They are widespread in the coastal rainforests and mangroves of north Queensland, New Guinea, the Pacific and Asia.

The common name, Matchbox Bean, is derived from the use of the seeds as waterproof matchboxes in pioneering times. The seed was hollowed out and a lid attached with a brass clasp and hinge.

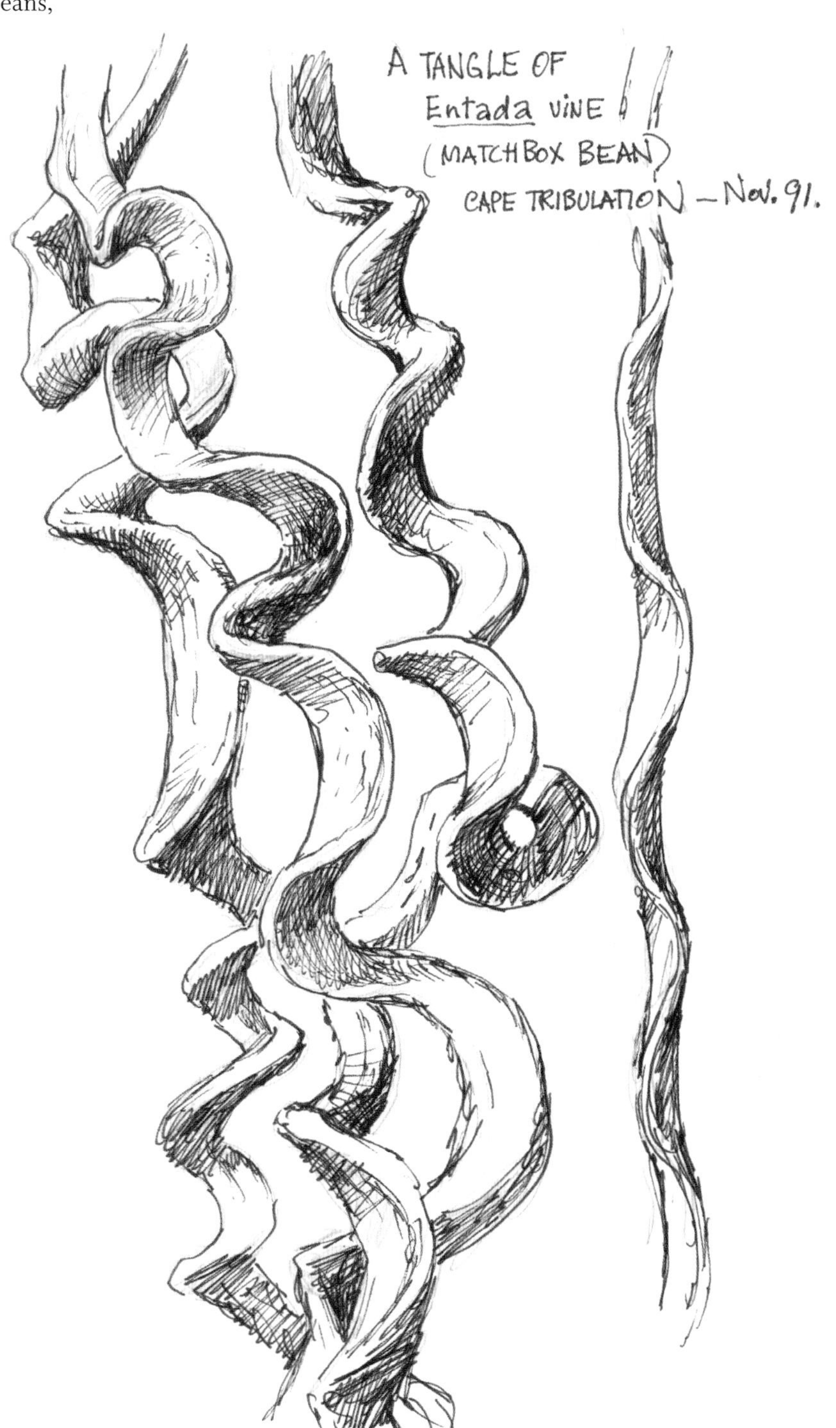

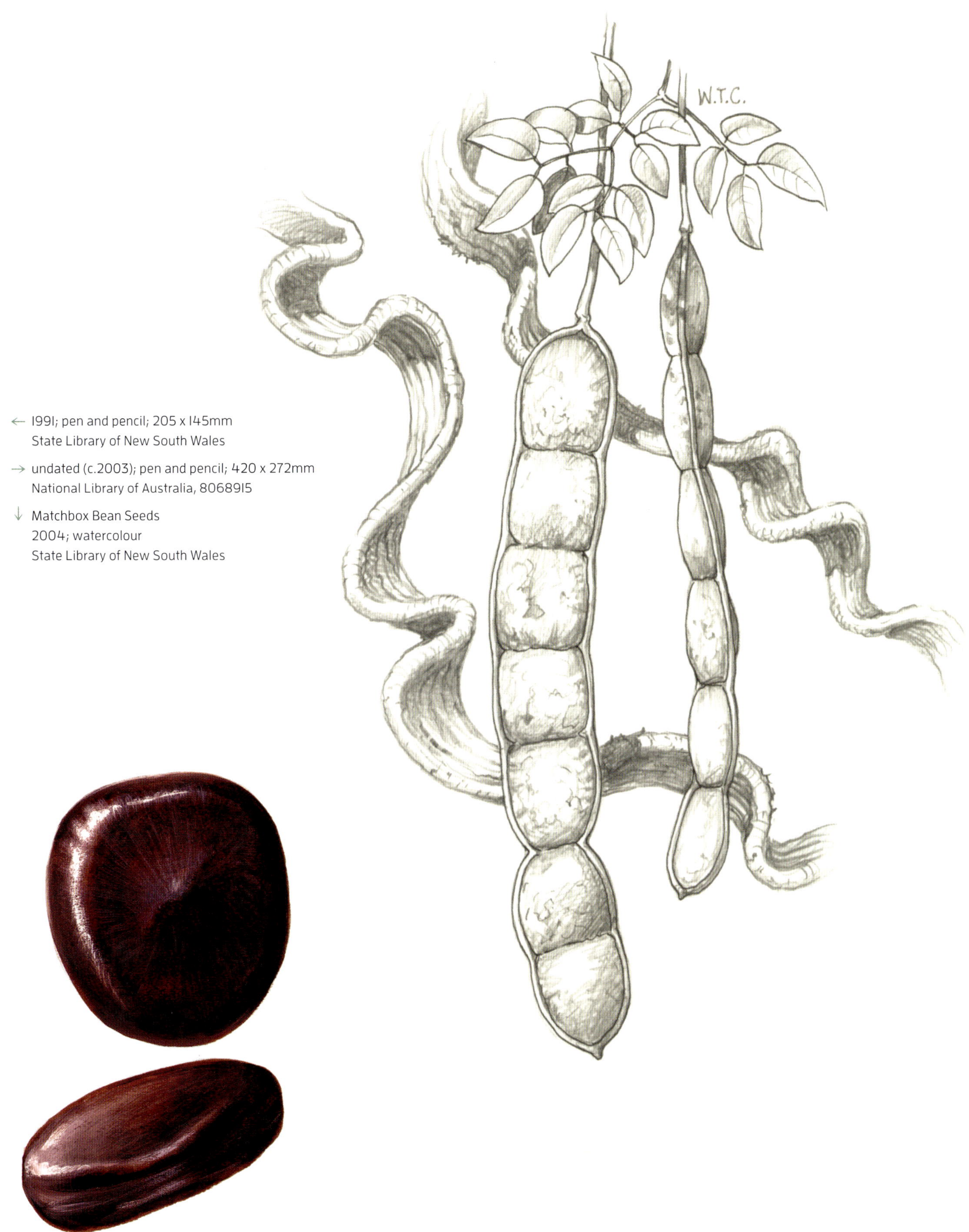

← 1991; pen and pencil; 205 x 145mm
State Library of New South Wales

→ undated (c.2003); pen and pencil; 420 x 272mm
National Library of Australia, 8068915

↓ Matchbox Bean Seeds
2004; watercolour
State Library of New South Wales

Freycinetia excelsa

Slender Climbing Pandan

PANDANACEAE

Freycinetia is named in honour of Louis Claude de Saulces de Freycinet (1779–1842), a French naval captain who, in 1811, published the first map of Australia with an entire coastline.

excelsa is from *excelsus* (high), referring to the plant's ability to climb into the forest canopy.

Slender Climbing Pandans are closely related to *Pandanus* trees and have similarly spiky leaves and segmented fruit. Adventitious roots that emerge from the stems and adhere to host trees enable them to ascend to great heights. These pandans occur in rainforest, from northern New South Wales to Cape York, as well as the Northern Territory and New Guinea.

The tight head of minute greenish-yellow flowers ripens to a large multiple fruit made up of small, fleshy red segments. These are relished by many birds, including Victoria's Riflebirds.

The sketch at right includes the climbing habit of a related species, *Freycinetia scandens* or Climbing Pandan.

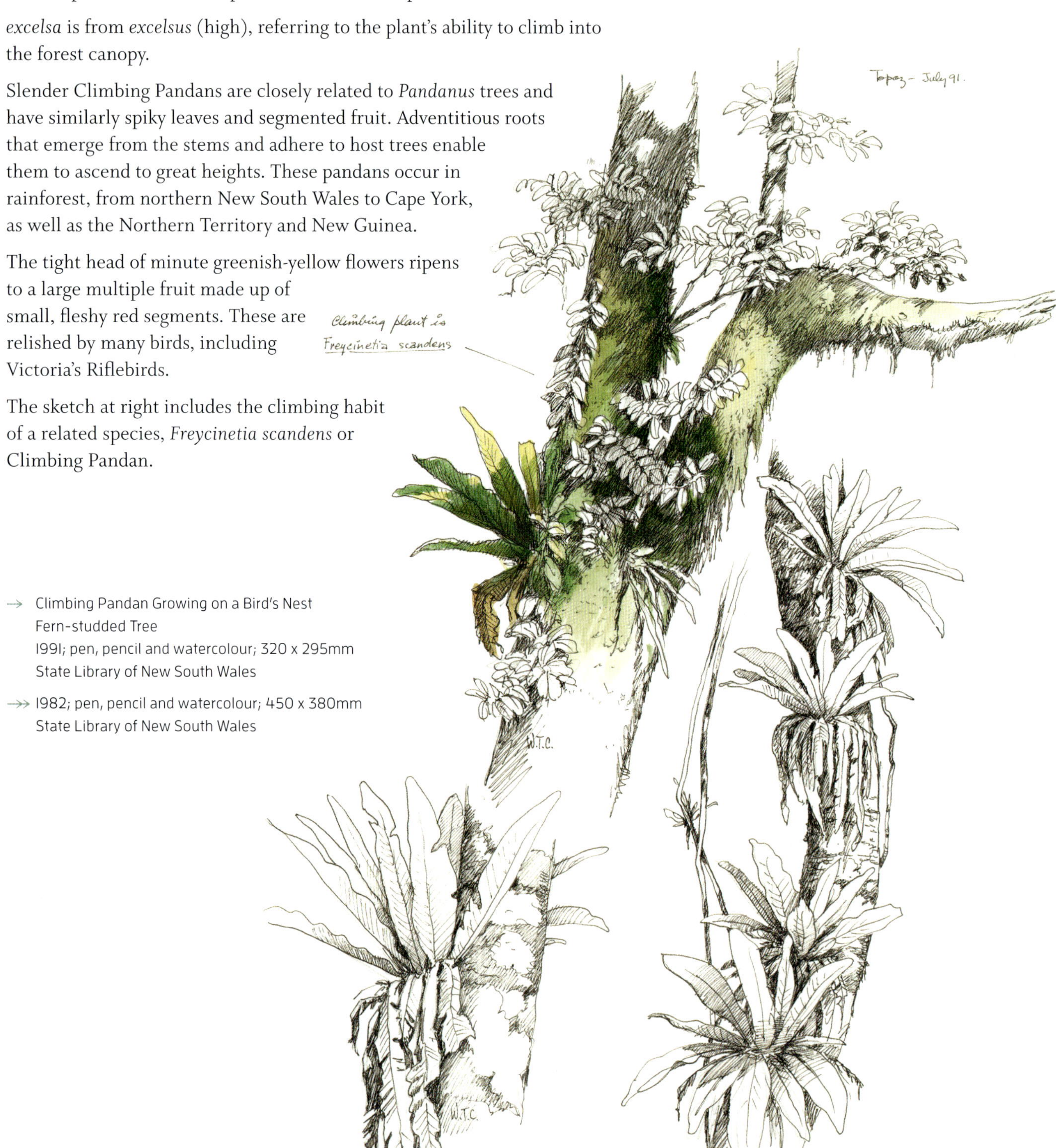

→ Climbing Pandan Growing on a Bird's Nest Fern-studded Tree
1991; pen, pencil and watercolour; 320 x 295mm
State Library of New South Wales

→→ 1982; pen, pencil and watercolour; 450 x 380mm
State Library of New South Wales

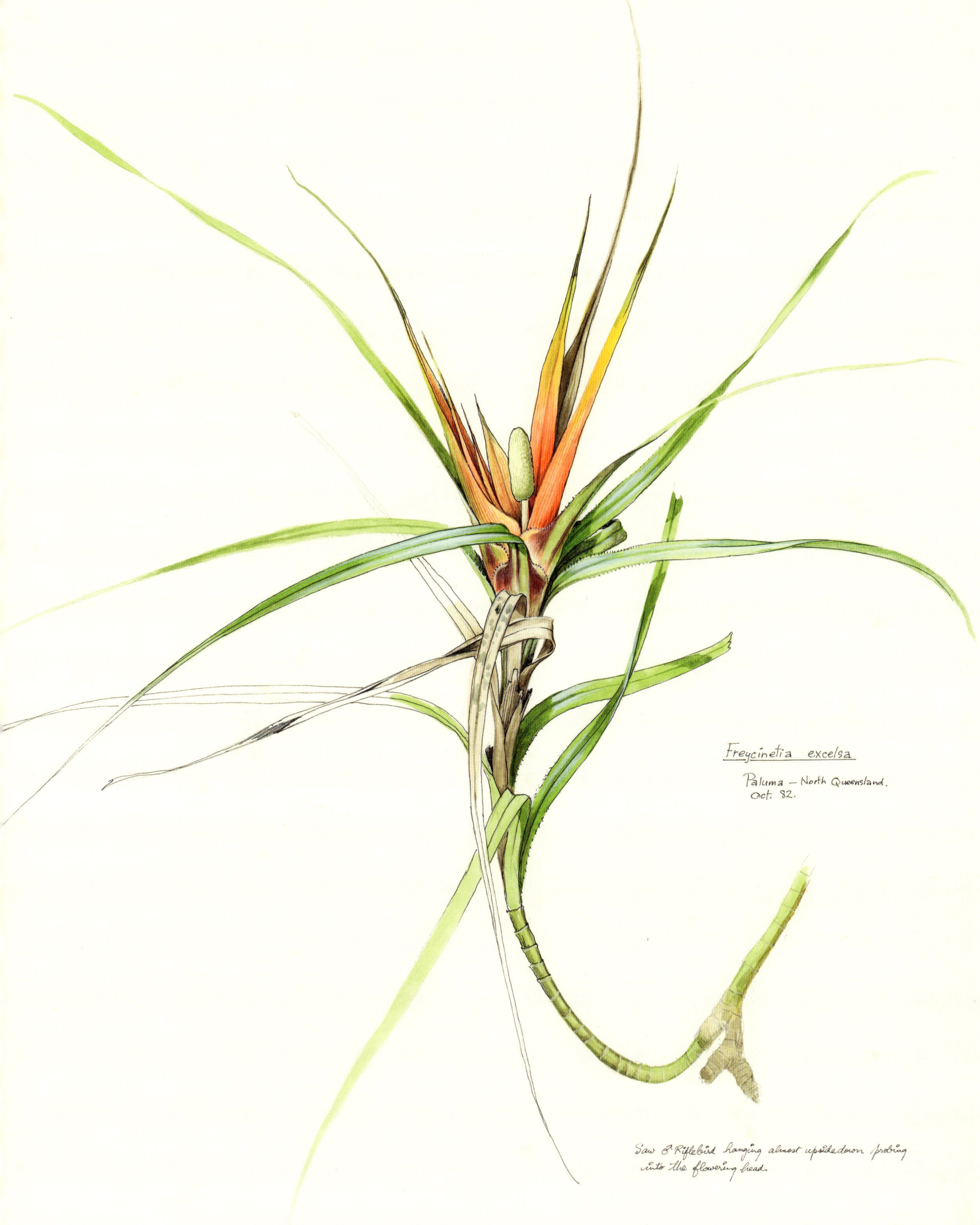

Freycinetia excelsa

Paluma – North Queensland.
Oct. 82.

Saw ♂ Riflebird hanging almost upsidedown probing into the flowering head.

Hibbertia scandens

Climbing Guinea Flower

DILLENIACEAE

Hibbertia is named in honour of George Hibbert (1757–1837), politician, opponent of the slave trade abolition bill, slave-trader and ship-owner, amateur botanist and owner of a private botanic garden at Chelsea.

scandens means climbing, referring to the plant's habit.

Most *Hibbertia* species are shrubs, but this is a twiner and scrambler, occurring mostly in the wet sclerophyll forests of eastern Australia.

Climbing Guinea Flower was a common species near our home at Bungwahl, New South Wales, and again at Topaz, Queensland, so material was always easily accessible. Because of the showy yellow flowers and fleshy orange-red fruit eaten by various birds, Bill used this species several times in paintings.

→ 1984; pen, pencil and watercolour; 410 x 290mm
State Library of New South Wales

↓ undated; pen, pencil and watercolour; 370 x 300mm
State Library of New South Wales

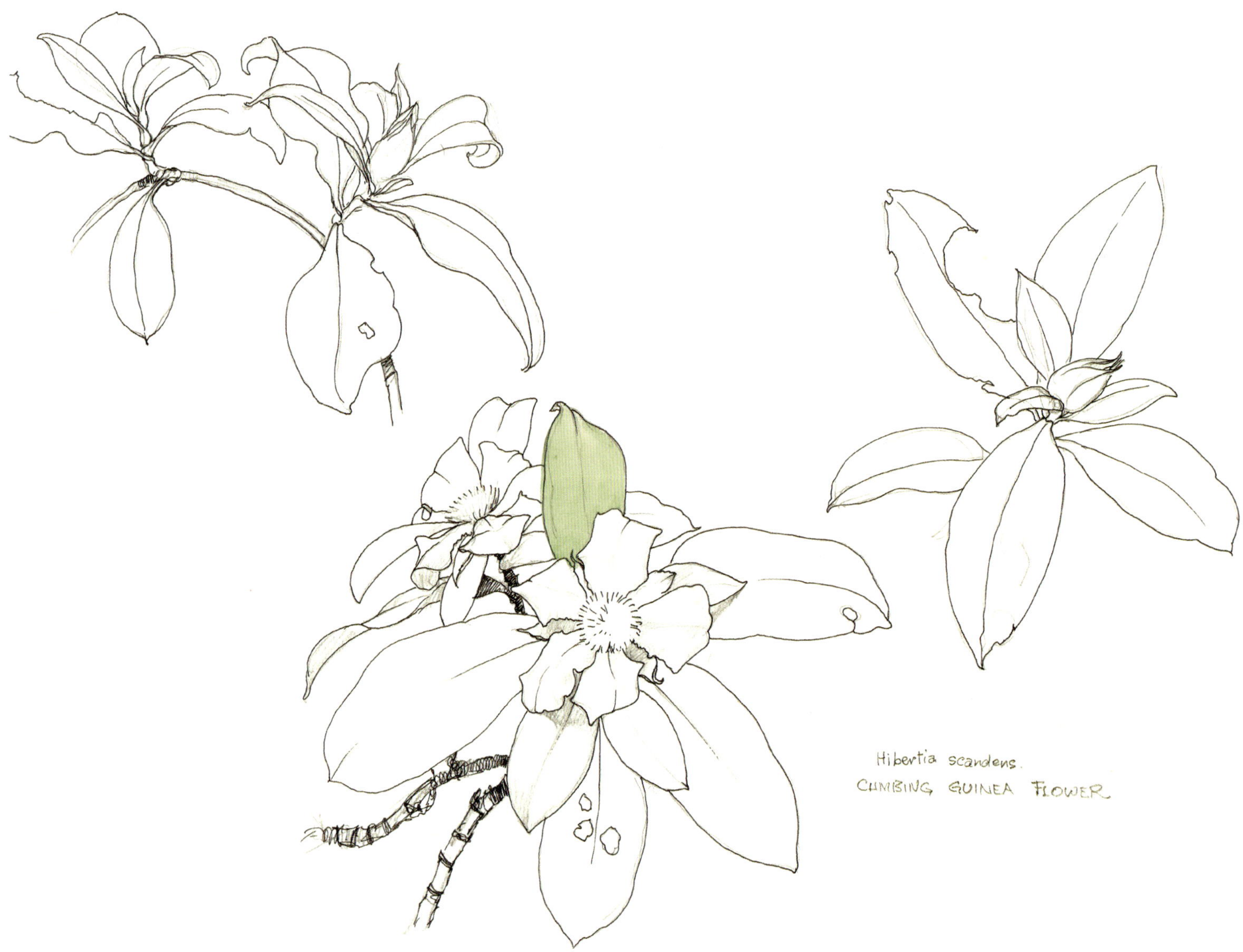

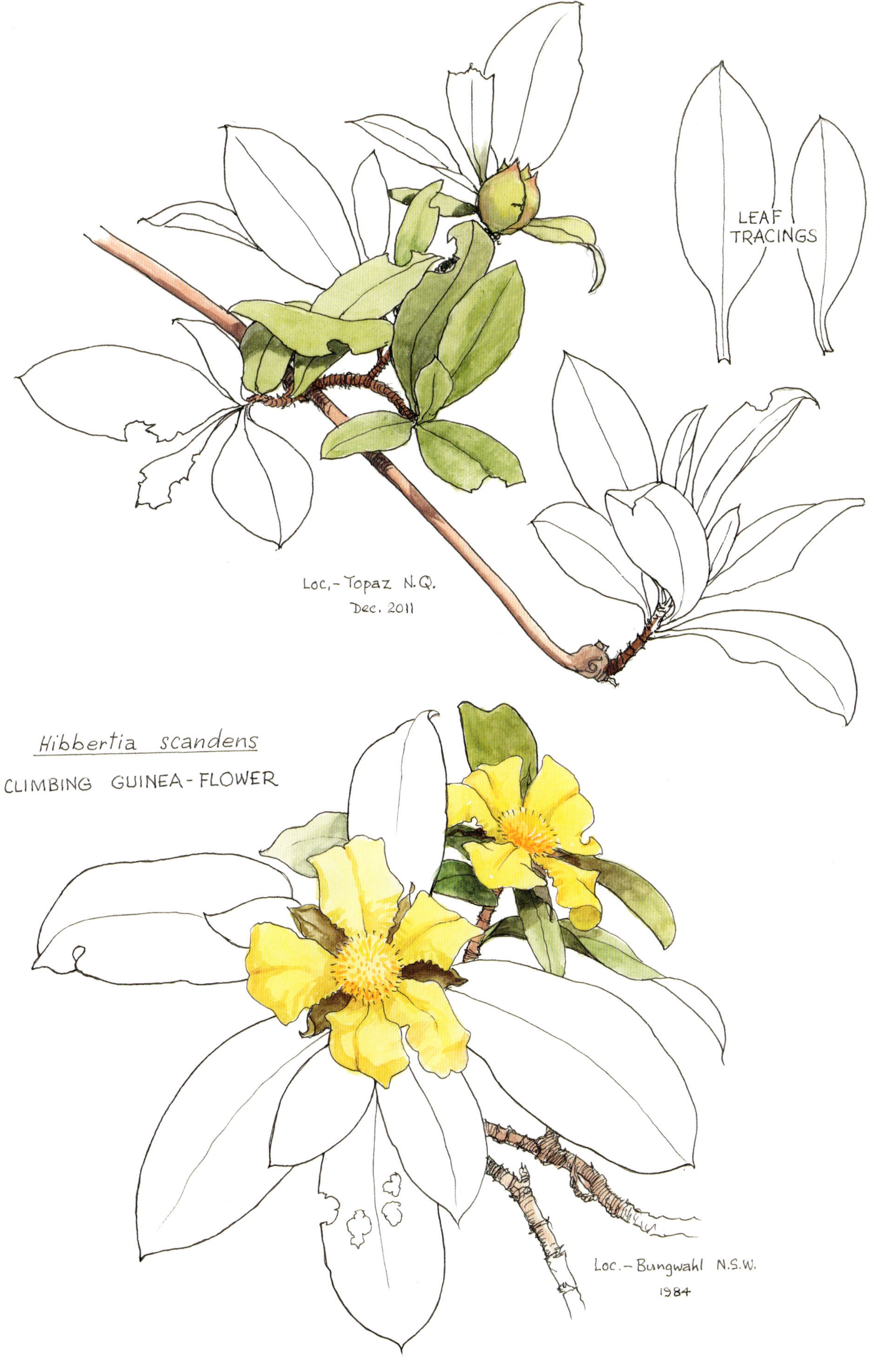
LEAF
TRACINGS
Loc.-Topaz N.Q.
Dec. 2011
Hibbertia scandens
CLIMBING GUINEA-FLOWER
Loc.-Bungwahl N.S.W.
1984

Ibatiria furfuracea

FABACEAE

→ undated (c.1993); watercolour; (in part) 550 x 750mm
State Library of New South Wales

Ibatiria: I named this plant in honour of Bill. Iba Tiri is the water spirit of the Huli tribe in Papua New Guinea, represented at ceremonies with the plumes of the Ribbon-tailed Bird of Paradise. The Ribbontail was a favourite bird of Bill's; a stained-glass specimen decorated his studio door. Iba Tiri is a joker and a trickster, and because of this Bill was likened to the spirit by long-time friend and ABC Natural History Unit filmmaker David Parer. *Ibatiria*, the plant, is also a trickster, with fruit that is atypical for a legume.

furfuracea is derived from *furfuraceus* (scurfy), referring to the fruit's surface.

Ibatiria is a rare vine known only from seven plants in the wild. It twines up to the rainforest canopy on the southern Atherton Tablelands. The fruit is large and somewhat woody. One can imagine that it might once have been dinosaur food, especially as nothing appears to eat and disperse the seed today. The leaflets have wavy margins and are shiny and silky on the underside. The pea-shaped flowers are dark red with lime-green nectar guides at the centre.

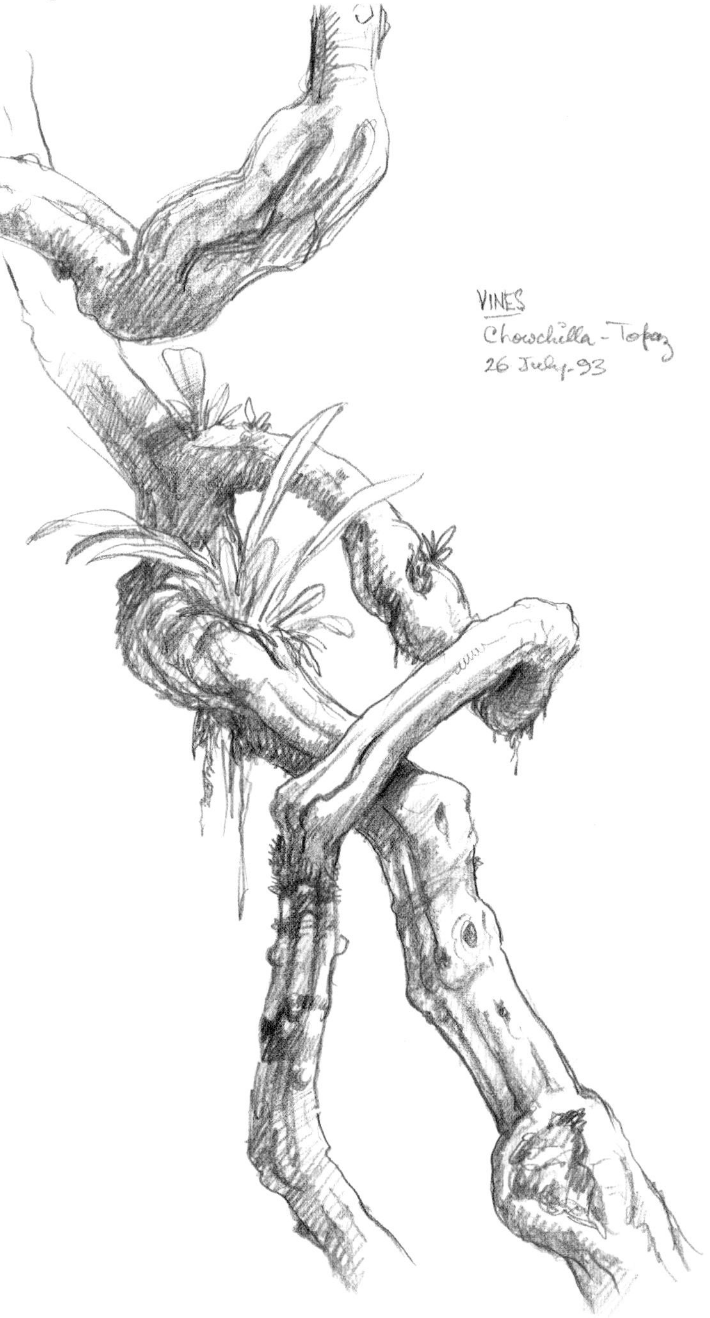

↓ 1993; pencil
State Library of New South Wales

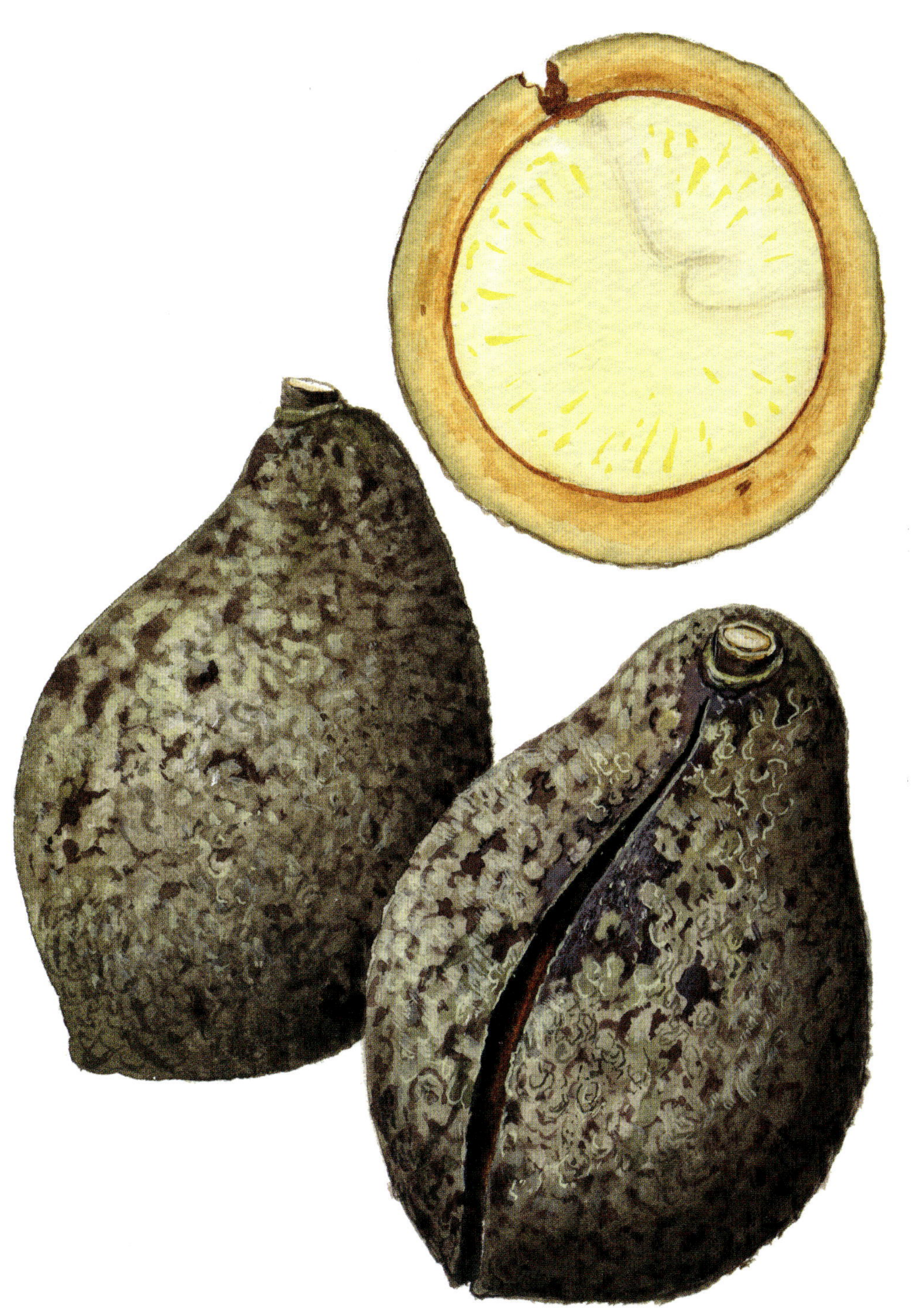

Marsdenia liisae

Large-flowered Milk Vine

APOCYNACEAE

→ 1982; pen, pencil and watercolour; 400 x 325mm
State Library of New South Wales

↓ 1985; pen, pencil and watercolour; 510 x 360mm
State Library of New South Wales

Marsdenia is named in honour of William Marsden (1754–1836), botanical collector and author.

liisae is named in honour of Liisa Atherton (nee Lapinpuro, b.1954), who studied Asclepiadaceae, which have since been absorbed into the family Apocynaceae.

Large-flowered Milk Vine is a twiner with milky latex. The windmill-like flowers are followed by slender pods packed tightly with silky-tufted seeds. When the pods open, the seeds disperse on the wind. This plant grew on our property at Bungwahl, which is not far from Bulahdelah, its southernmost location. Its northern limit is near Lismore in northern New South Wales.

The liana stems at right are from another Apocynaceae vine, *Parsonsia straminea* or Common Silkpod, and were drawn in littoral rainforest at nearby Sandbar.

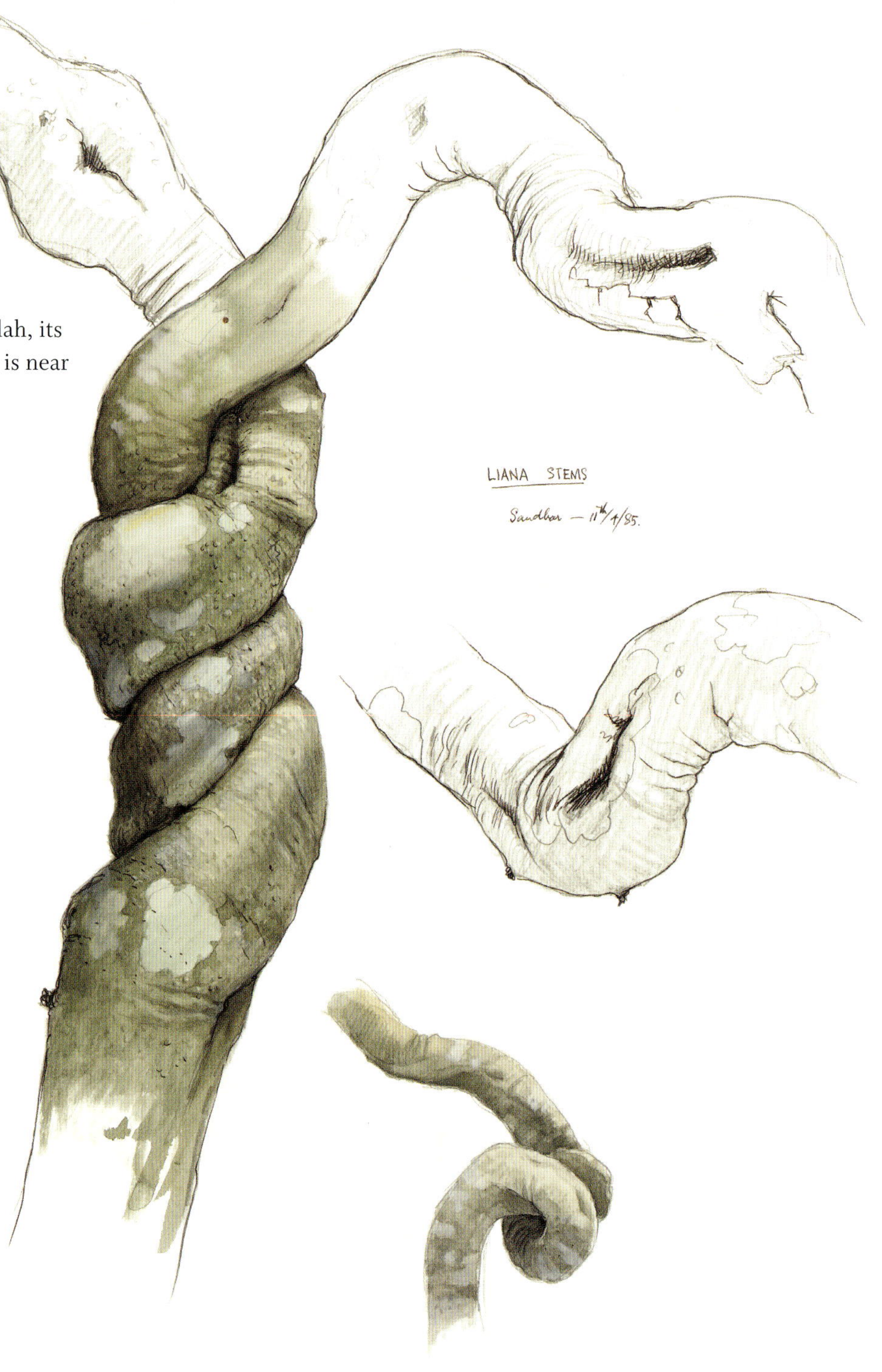

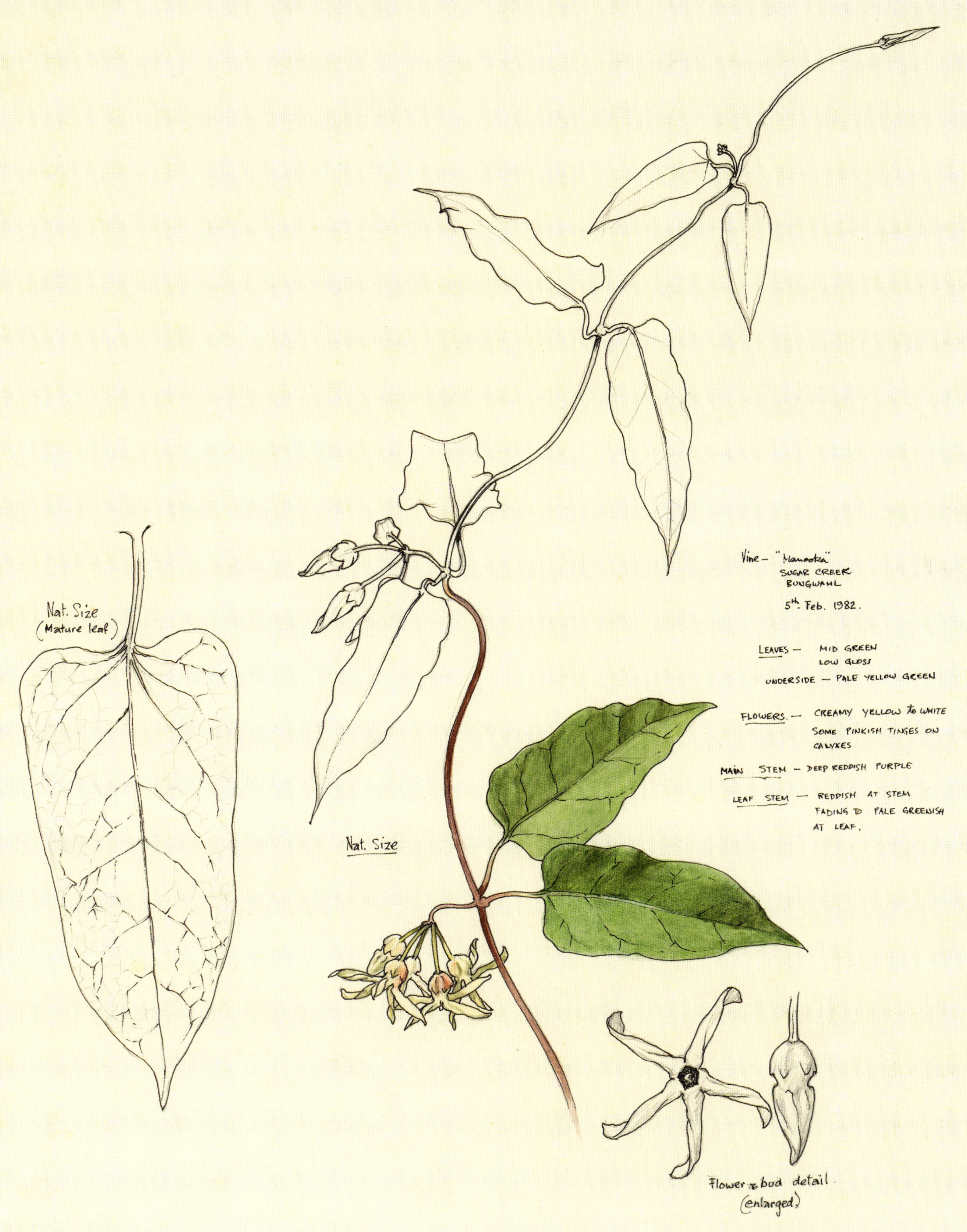

Nat. Size
(Mature leaf)
Nat. Size
Vine – "Manootka"
SUGAR CREEK
BUNGWAHL
5th. Feb. 1982.
LEAVES – MID GREEN
LOW GLOSS
UNDERSIDE – PALE YELLOW GREEN
FLOWERS. – CREAMY YELLOW to WHITE
SOME PINKISH TINGES ON
CALYXES
MAIN STEM – DEEP REDDISH PURPLE
LEAF STEM – REDDISH AT STEM
FADING TO PALE GREENISH
AT LEAF.
Flower & bud detail.
(enlarged)

Myrmecodia beccarii

Ant-house Plant

RUBIACEAE

→ Forest Kingfisher, colour plate from *Kingfishers and Related Birds*, 1983
National Library of Australia, 2494045

↓ 1981; pen and watercolour; 365 x 245mm
State Library of New South Wales

Myrmecodia is from *murmekodes* (full of ants), referring to the plant being a house for ants.

beccarii is named in honour of Odoardo Beccari (1843–1920), an Italian botanist who worked extensively in Indonesia.

These epiphytes are often called Ant-house Plants, as they have a fascinating symbiotic relationship with ants. Inside the tuberous base is a maze of chambers with numerous ants in residence. While living within the chambers, the ants supply nutrition to the plant.

This painting was done during our first trip to Iron Range in 1981, when it took three days to drive from Cairns. Today, because of improved roads, it takes only one day. We had never been so far north previously, so the trip was full of surprises, with many new discoveries for us. After numerous subsequent trips to the area, we still found it an exciting place, as its proximity to New Guinea ensures a variety of flora and fauna common to both areas.

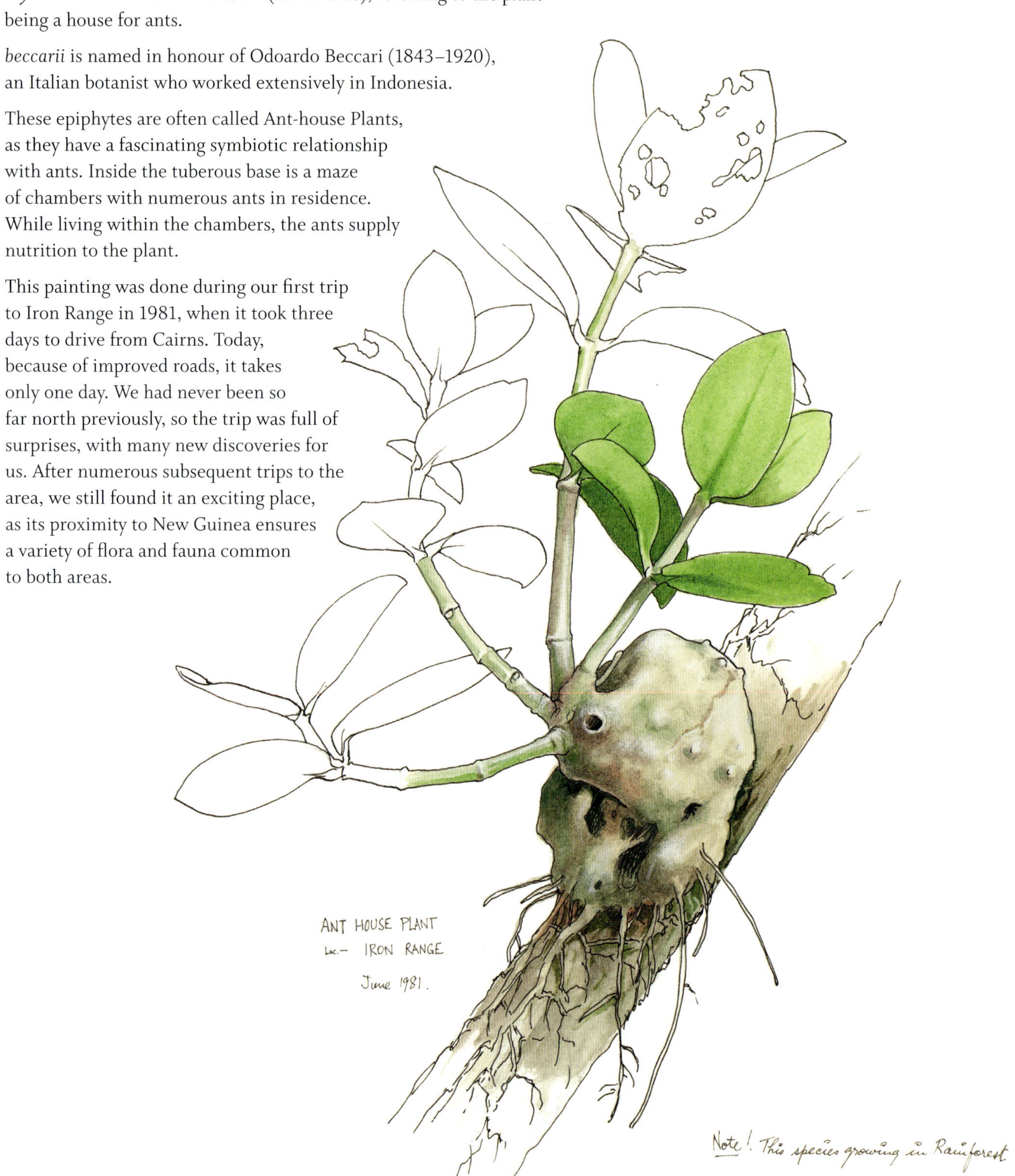

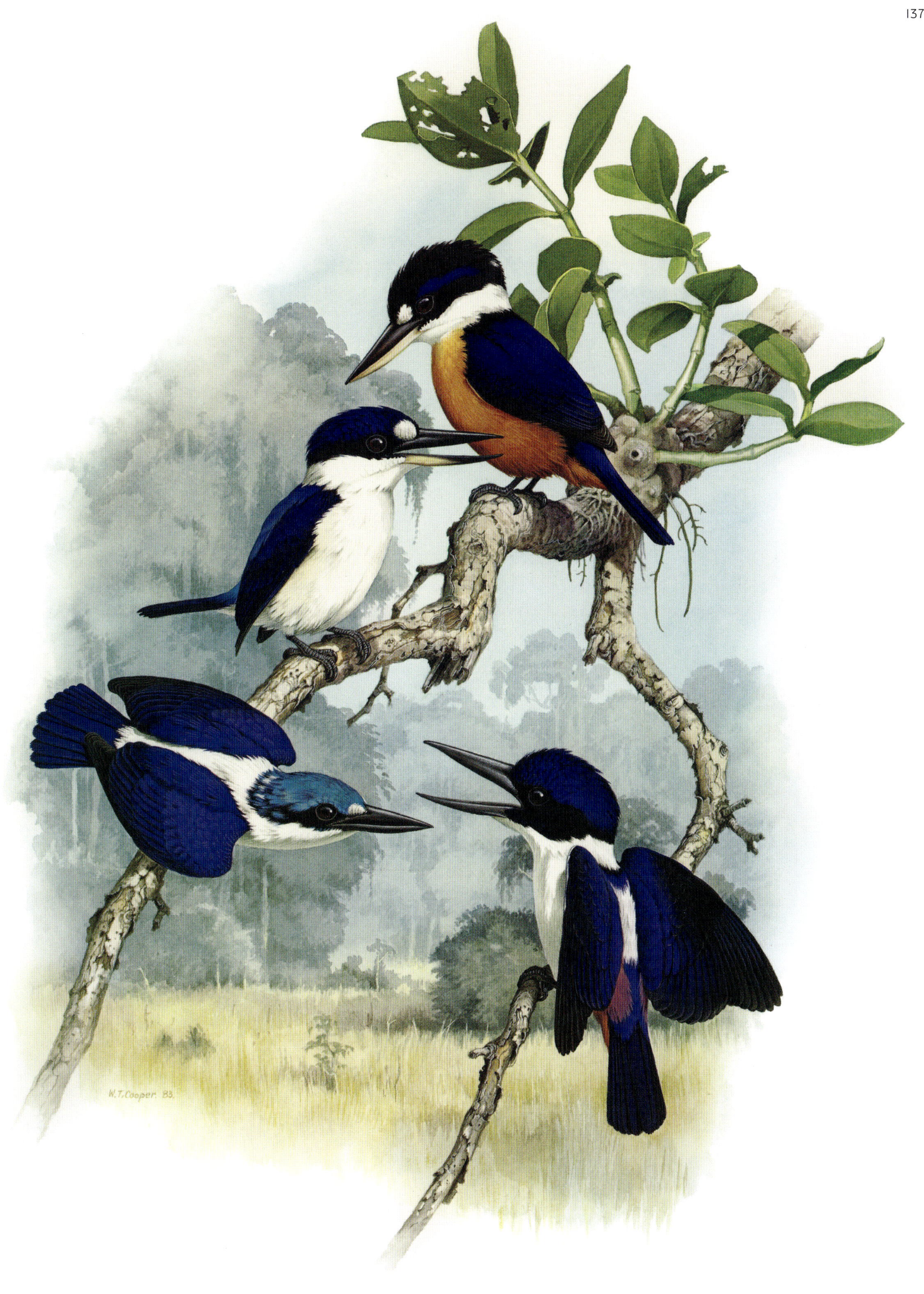
W.T.Cooper. 83.

Neosepicaea jucunda

Jungle Vine

BIGNONIACEAE

→ 2003; watercolour
Private collection

↓ 2003; pen, pencil and watercolour
395 x 495mm
State Library of New South Wales

Neosepicaea is from *neo* (new) and *sepicaea* (Sepik), referring to this new genus described from plants in the area of the Sepik River in Papua New Guinea.

Jungle Vine is a canopy-reaching twining vine of the Wet Tropics in north Queensland. It is mostly noticed only when the pink tubular flowers have fallen to the forest floor.

This sketch was done for the painting with the Ulysses Swallowtail. As was often the case, Bill studied the plant enough to be able to add or subtract parts to suit a composition without compromising the plant's natural features. This can be seen when comparing these two images.

From Bill's diary:

> *25th August 1998. Saw a male Pademelon on the driveway eating the fallen flowers of Neosepicaea jucunda. It would appear that it could not see the flowers among the leaf litter (even though they stood out being red) because he worked his way over the ground sniffing and only when a flower was quite close to his nose did he take it. Often he passed flowers just to the side of where he was smelling, even though to my eye they stood out.*

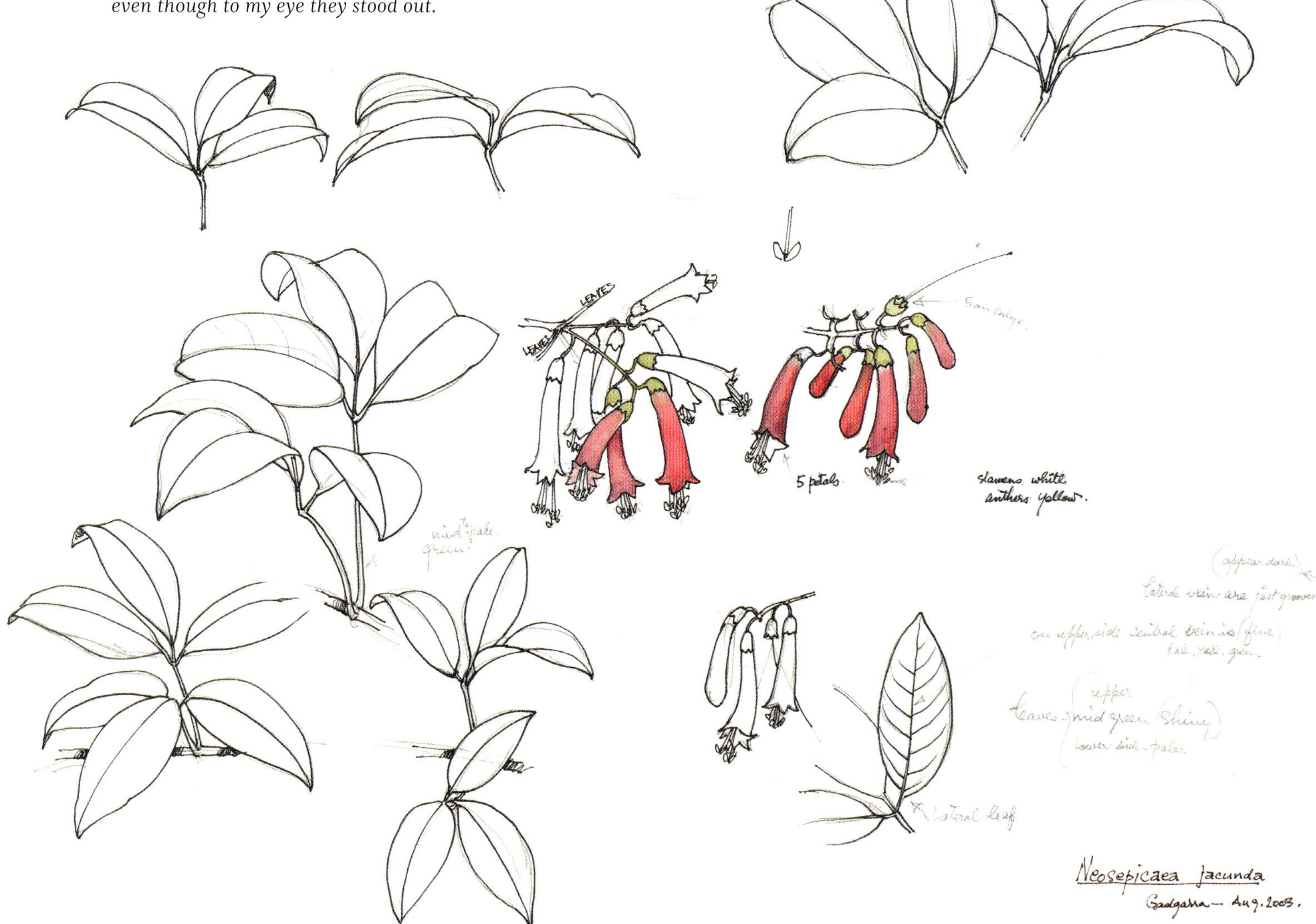

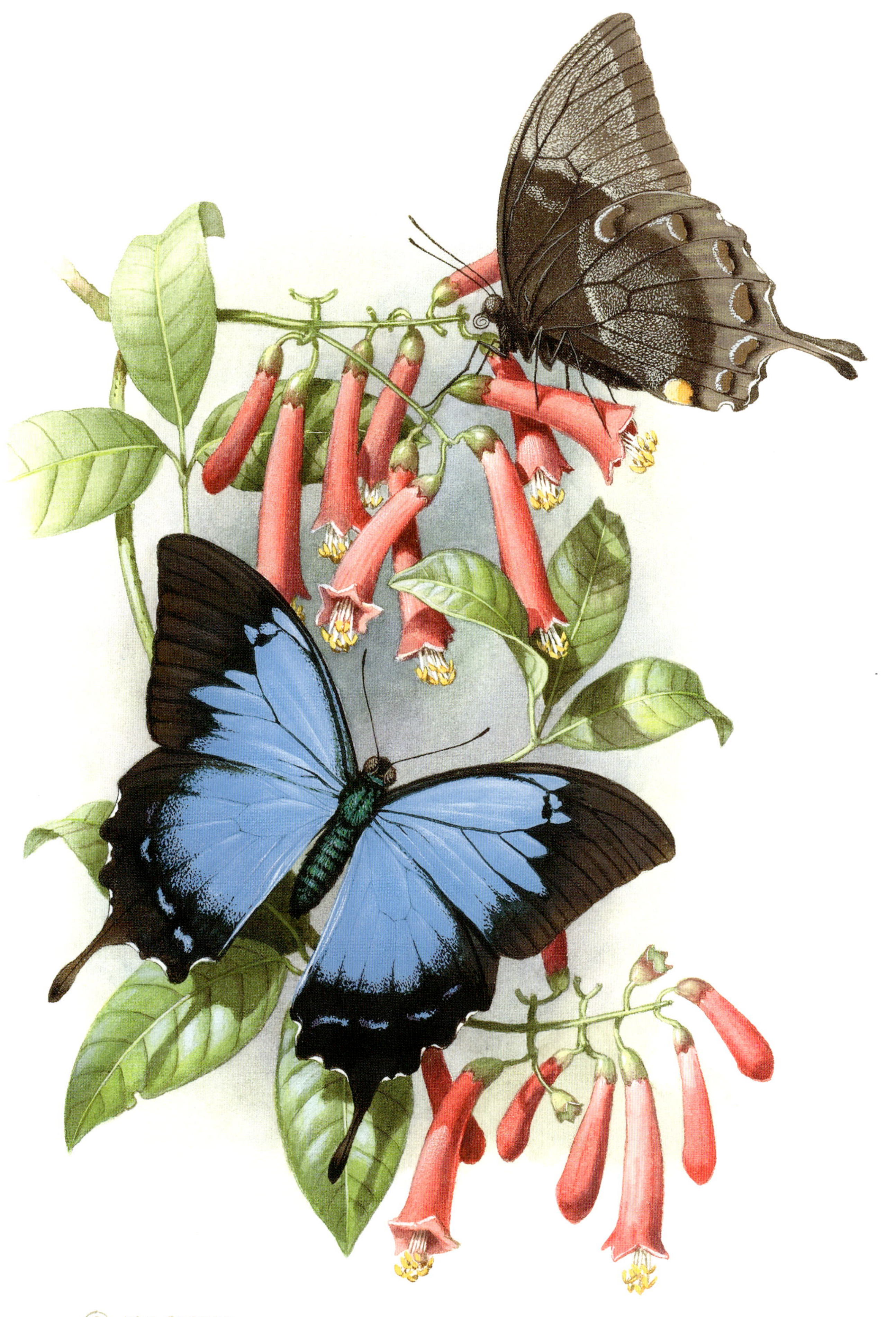

Omphalea queenslandiae

Omphalea

EUPHORBIACEAE

Omphalea is from *omphalos* (navel), referring to the flowers' navel-like anthers.

queenslandiae is from Queensland.

Omphalea is a large vine, climbing to the rainforest canopy in Queensland's Wet Tropics. Once the flowers fall from the inflorescences, the stems continue to grow, becoming the hooked tendrils by which the vine climbs. The large day-flying Zodiac Moth lays its eggs on this plant, and once the larvae emerge they feed on the leaves.

Bill especially painted this plate and two other larger pieces for our first botanical book, *Fruits of the Rain Forest*. As double-spreads, they were used to break up the otherwise uniform page design.

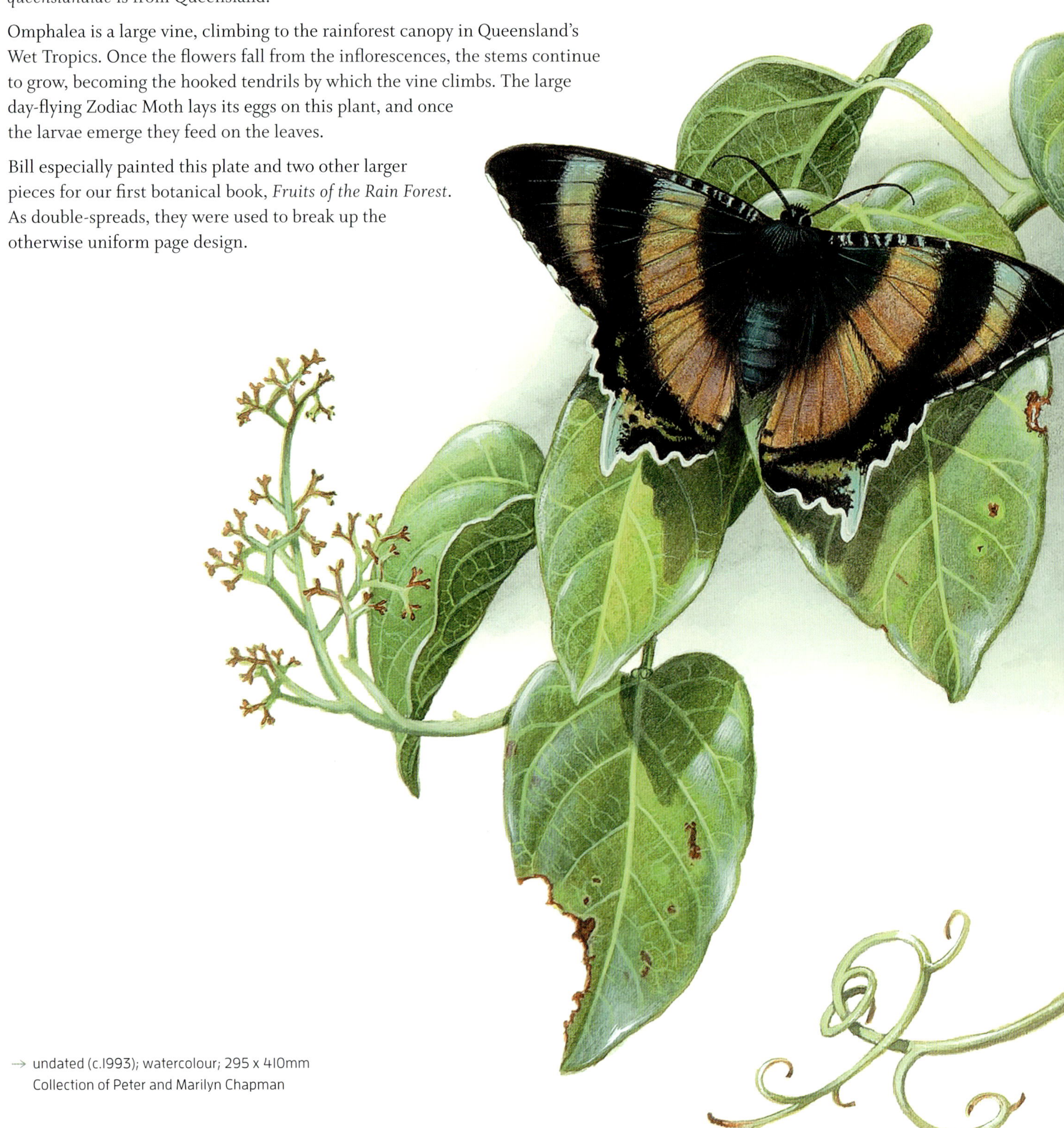

→ undated (c.1993); watercolour; 295 x 410mm
Collection of Peter and Marilyn Chapman

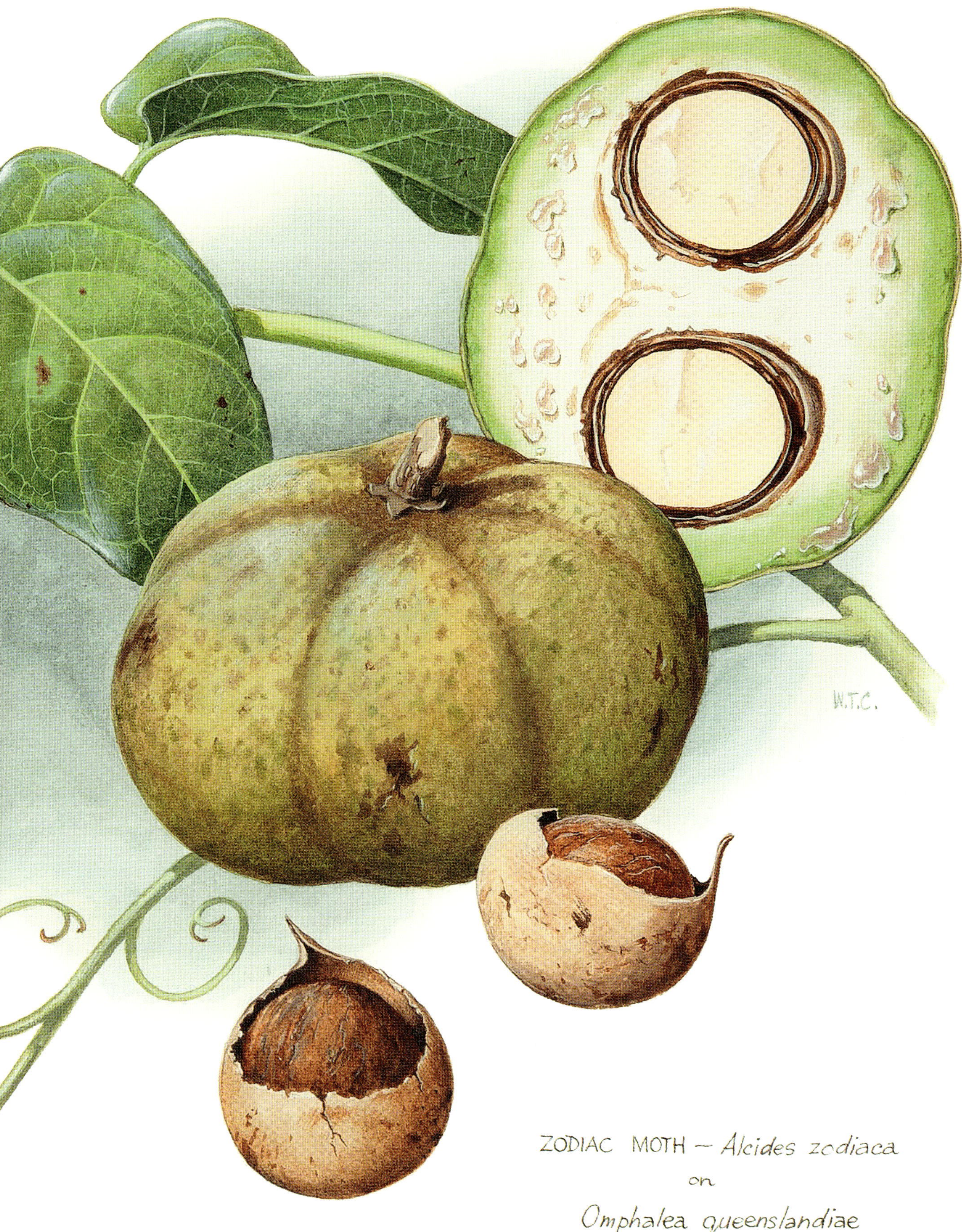

ZODIAC MOTH – *Alcides zodiaca*
on
Omphalea queenslandiae

Parsonsia latifolia

→ 1983; pen and watercolour; 520 x 350mm
State Library of New South Wales

↓ 1985; pen and pencil; 530 x 370mm
State Library of New South Wales

Milky Silkpod

APOCYNACEAE

Parsonsia is named in honour of James Parsons (1705–1770), English physician and author of *The Microscopical Theatre of Seeds*.

latifolia is from *lati-* (broad or wide) and *-folius* (leaved).

Milky Silkpod is a liana that twines to the rainforest canopy in north Queensland. When the long capsules split, they release numerous plumed seeds, which disperse in the breeze. The silky plumes are used by Lewin's Honeyeaters for nest lining.

These glossy dark-green leaves and fruits were hanging conspicuously on the edge of the upland rainforest at Paluma, north Queensland, just asking to be painted. The muscular liana stems sketched on this page belong to *Parsonsia straminea*, a closely related vine.

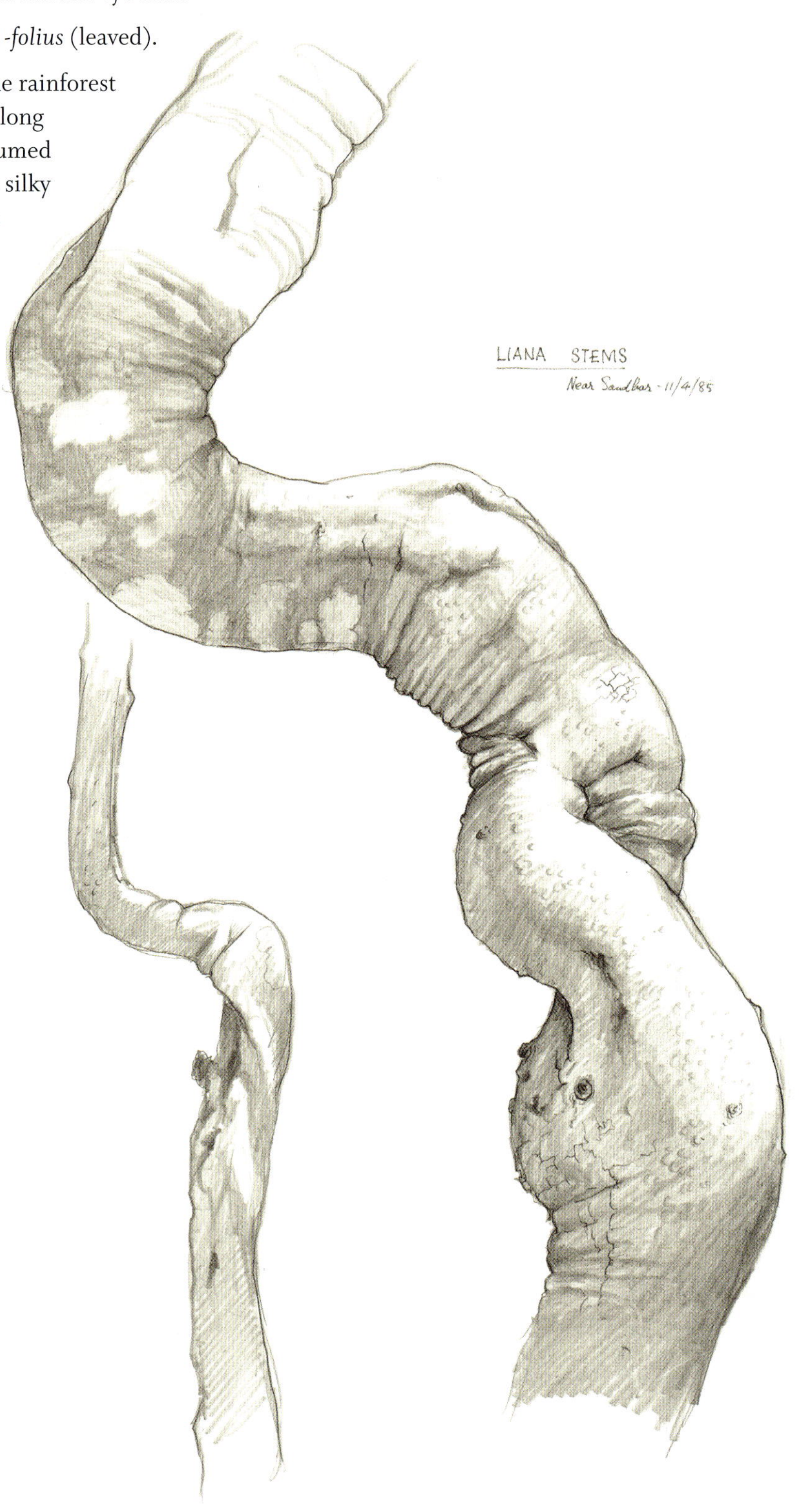

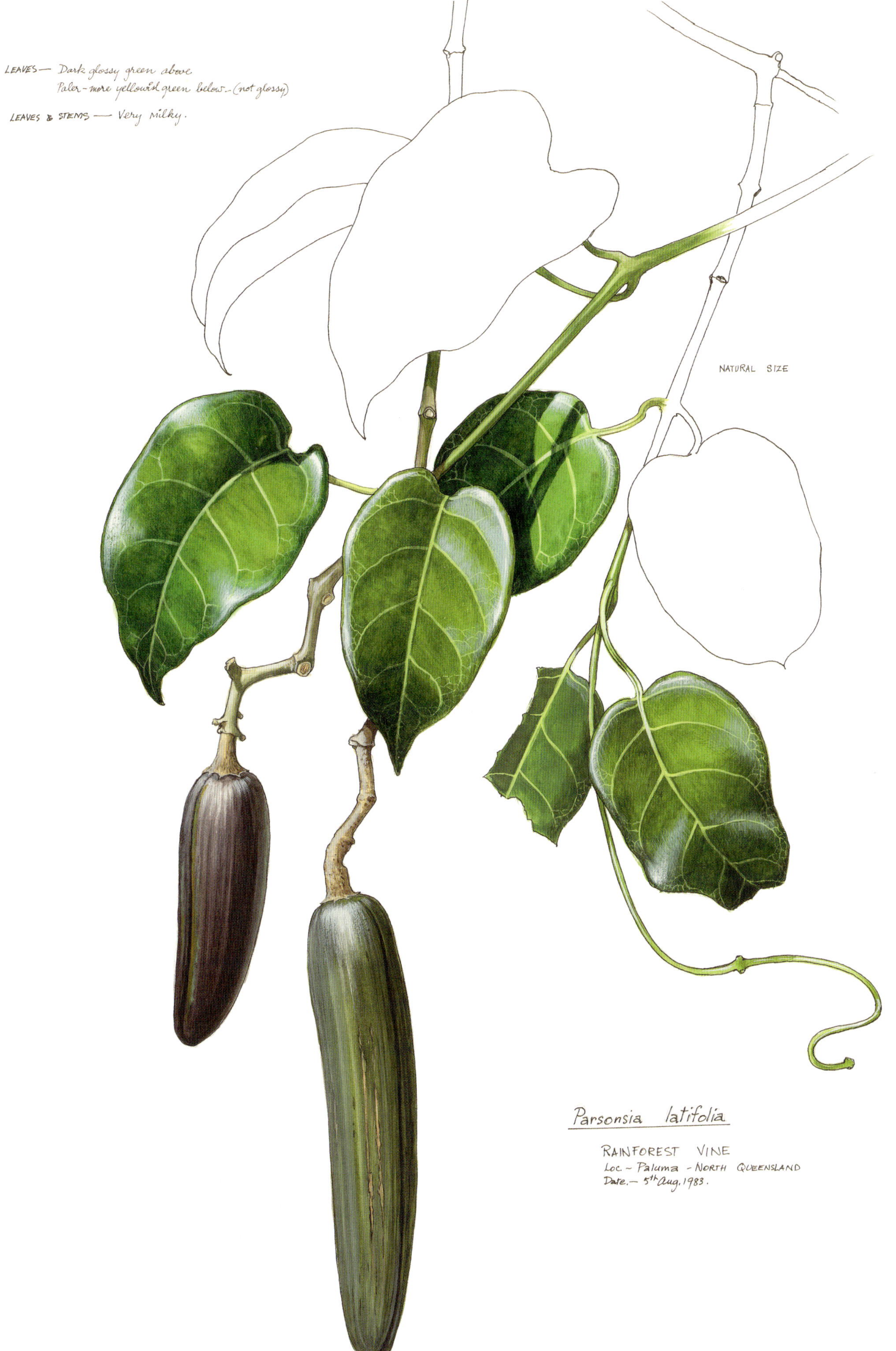
LEAVES — Dark glossy green above
Paler - more yellowish green below. - (not glossy)
LEAVES & STEMS — Very milky.
NATURAL SIZE
Parsonsia latifolia
RAINFOREST VINE
Loc. - Paluma - NORTH QUEENSLAND
Date. - 5th Aug. 1983.

Piper hederaceum

Giant Pepper Vine

PIPERACEAE

Piper is from *peperi* (pepper), a Bengali name for pepper vines.

hederaceum is after *Hedera* (Ivy) and *-aceus* (resembling).

Giant Pepper Vine climbs to the forest canopy, adhering to host tree trunks by adventitious roots. It has a broad distribution from north Queensland to southern New South Wales.

The fruit of this vine is eaten by birds and, because Bill found the leaf shape and knuckles at the junctions on the stems attractive, he used it a few times in major bird paintings, including those of White-headed Pigeons and Victoria's Riflebirds, which are always keen to seek out the peppery-scented fruit.

This species was previously known as *Piper novae-hollandiae*, as annotated on the sketch on page 147.

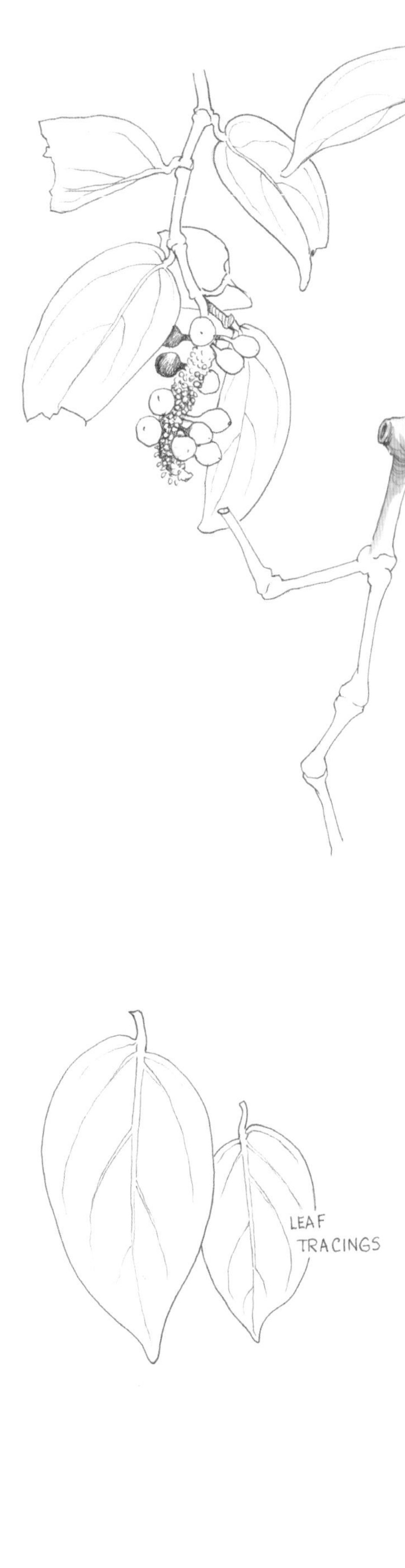

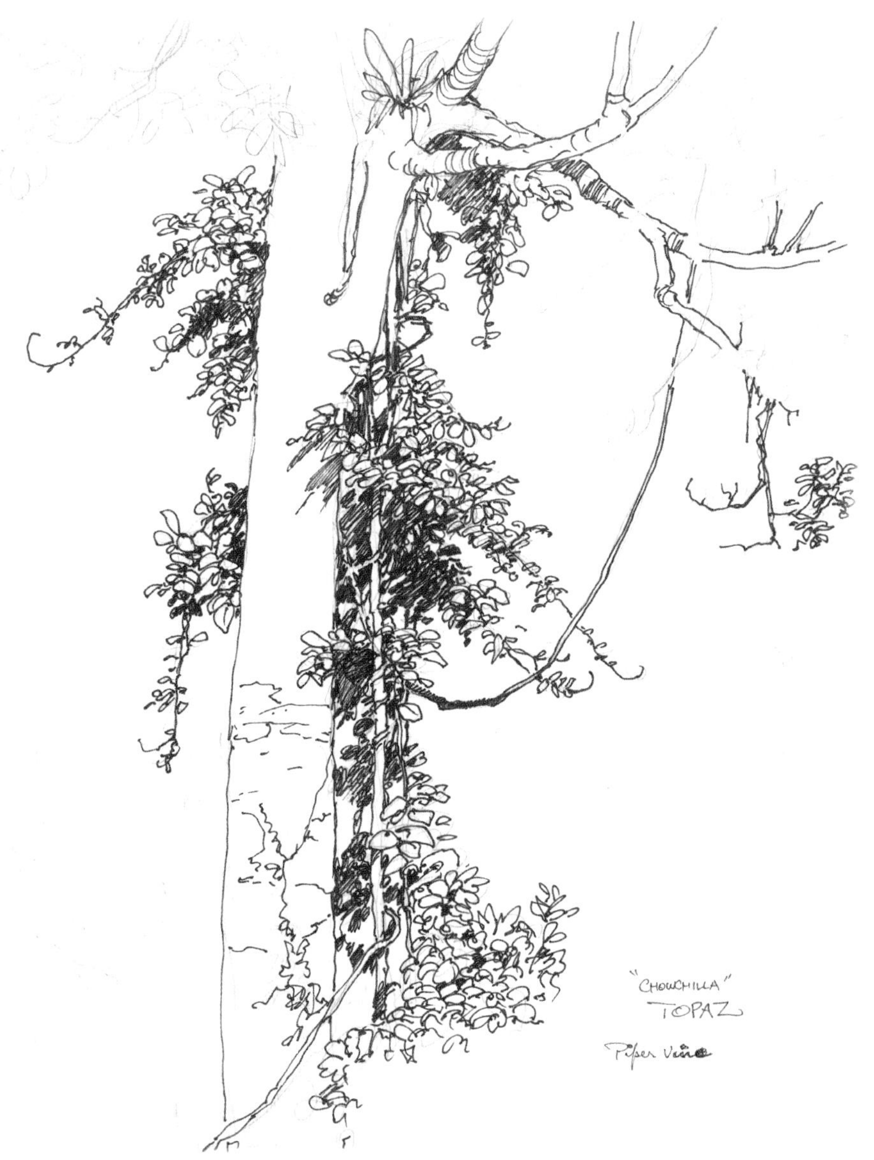

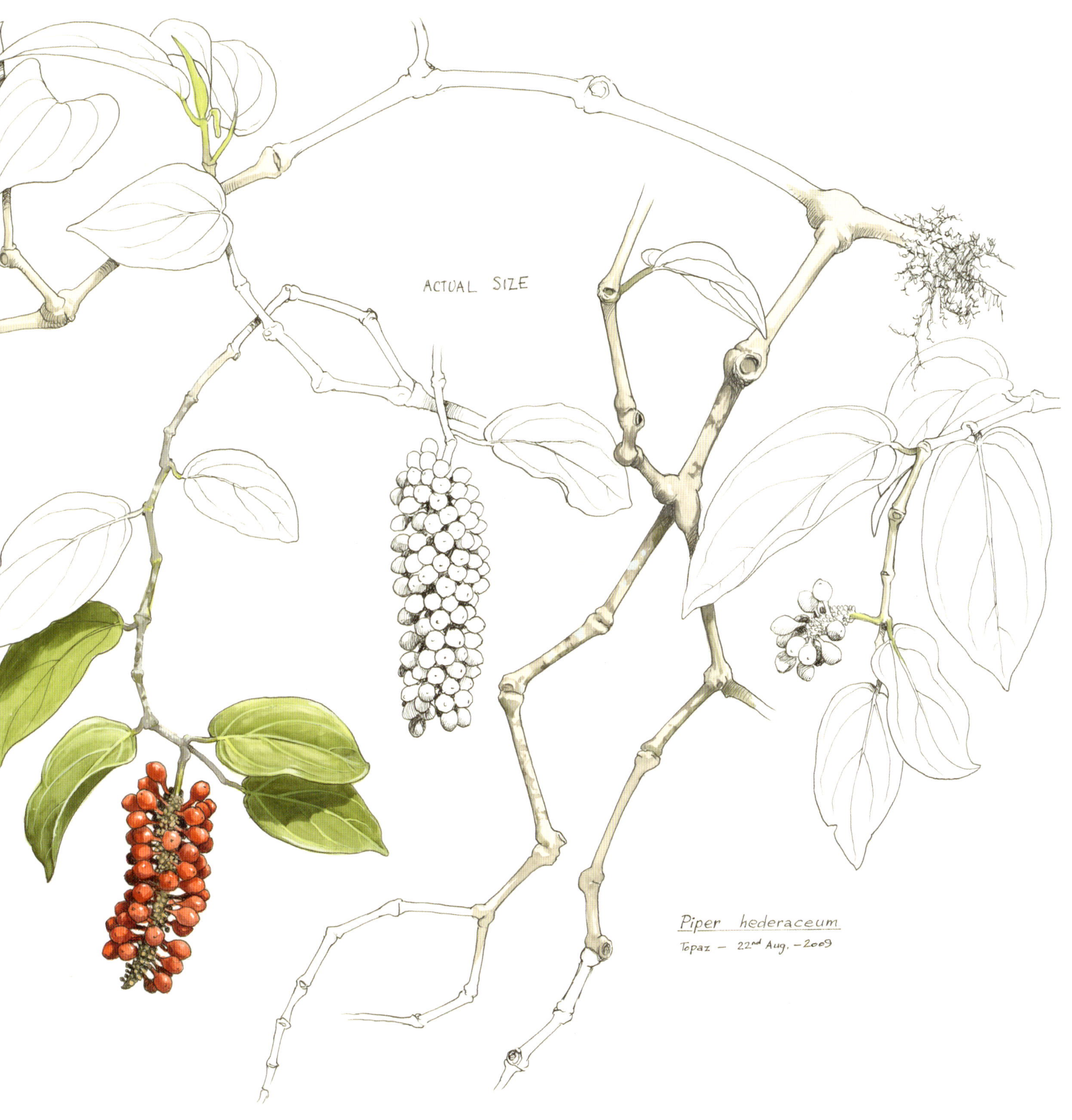

↑ 2009; pen, pencil and watercolour; 420 x 595mm
National Library of Australia, 8070419

← 1988; pen and pencil; 270 x 205mm
State Library of New South Wales

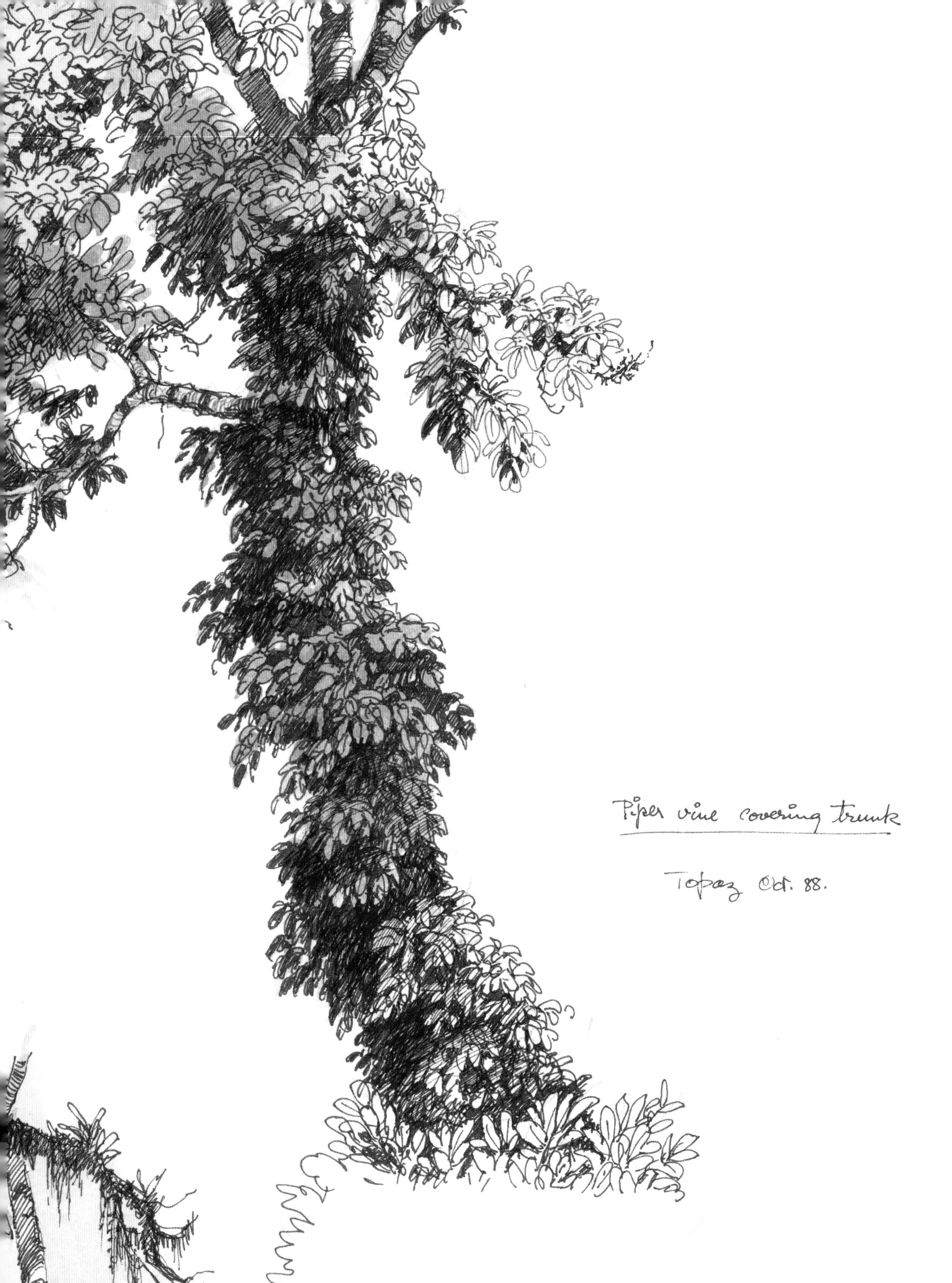
Piper vine covering trunk
Topaz Oct. 88.

← undated; pen and pencil
State Library of New South Wales

↘ Giant Pepper Vine with Riflebirds
undated (c.1985); pen, pencil and wash
430 x 380mm
Private collection

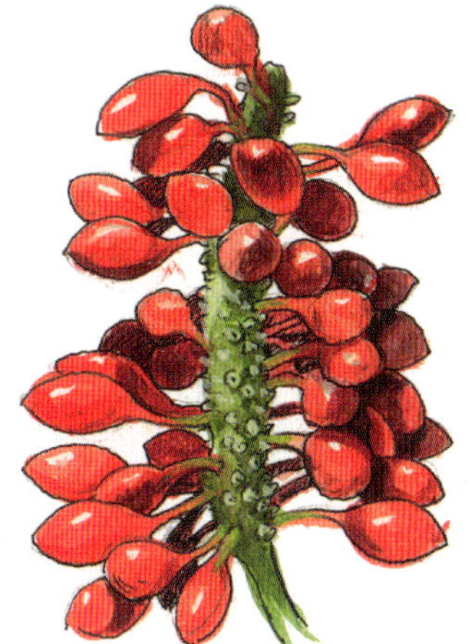

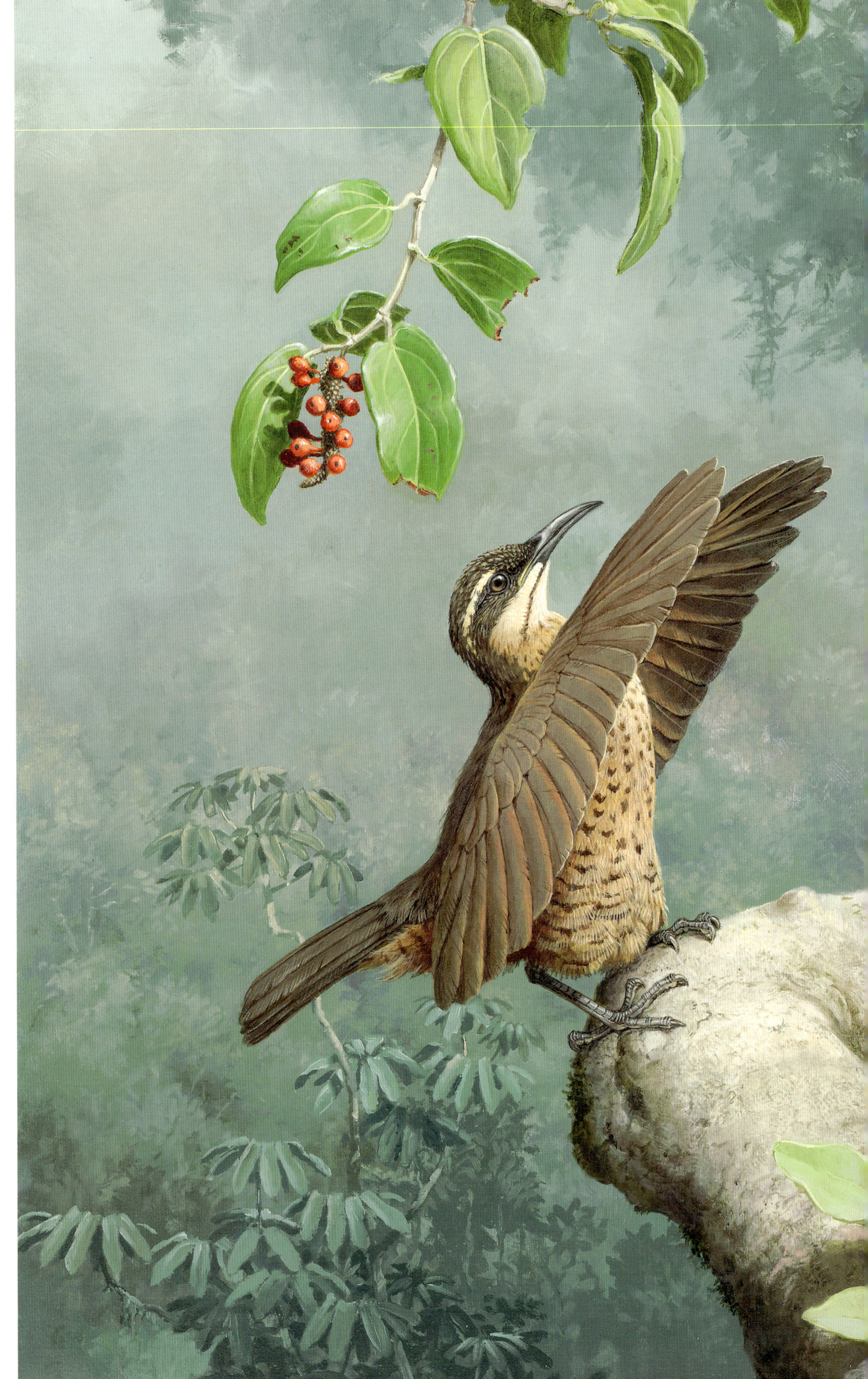

→ Victoria's Riflebirds
2013; acrylic on board
500 x 655mm
Private collection

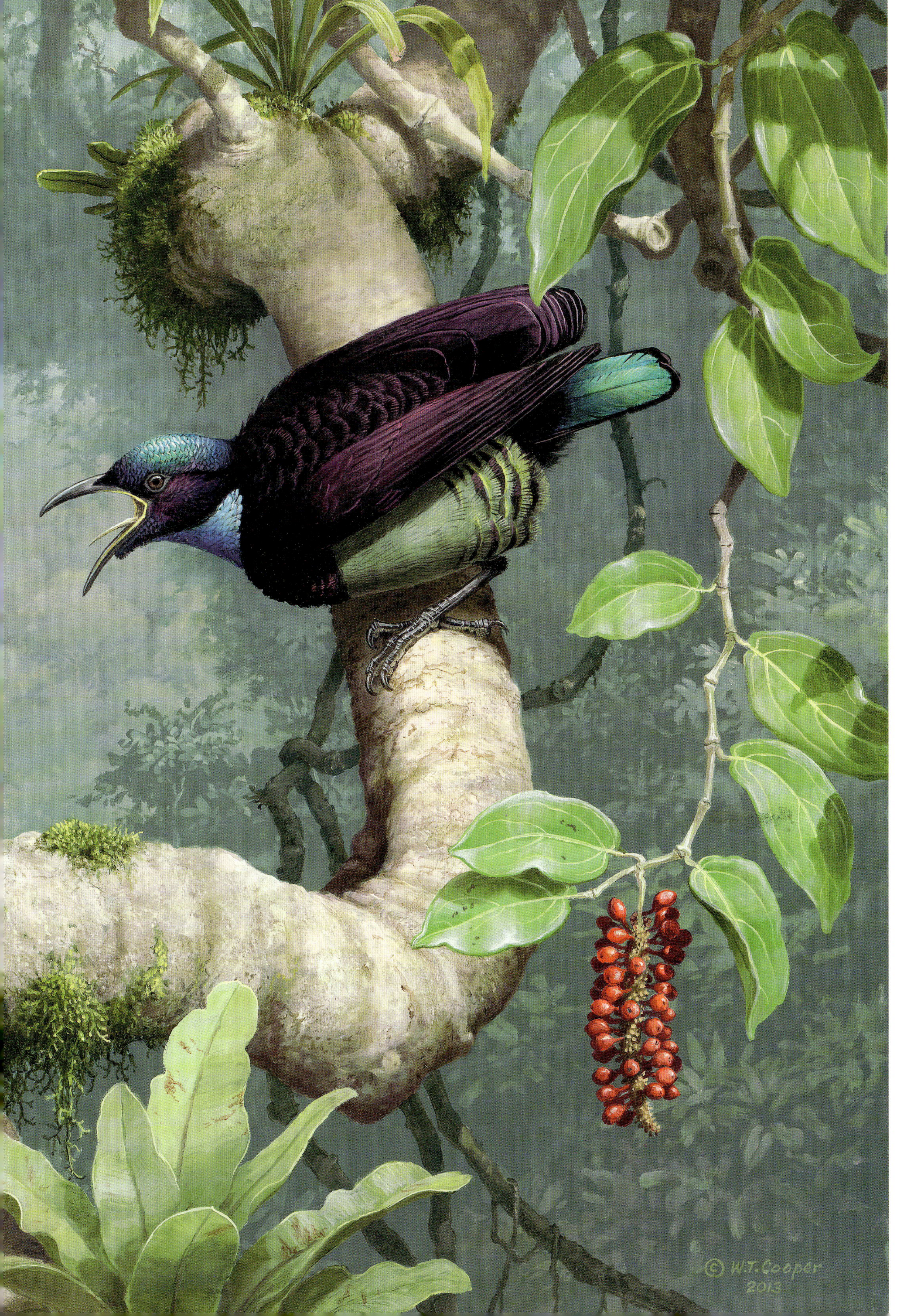
© W.T.Cooper
2013

Pothos longipes

Pothos

ARACEAE

Pothos is from *Potha*, a Sinhalese name for the genus.

longipes is from *longus* (long) and *pes* (foot), referring to the elongated and flattened leaf stalk that looks like the lower half of the leaf.

Pothos climbs with adventitious roots that grip tree trunks or rocks. It grows in wet rainforests from Kendall in New South Wales to Cape York Peninsula in north Queensland. The single-seeded fruit is eaten by many bird species as well as Spectacled Flying-foxes.

The plant's unusual leaves were worthy of inclusion in a painting. This sketch was done when Bill lived in central New South Wales, where the plant does not occur. He wasn't to realise that one day *Pothos longipes* would be common a few steps from our back door in North Queensland.

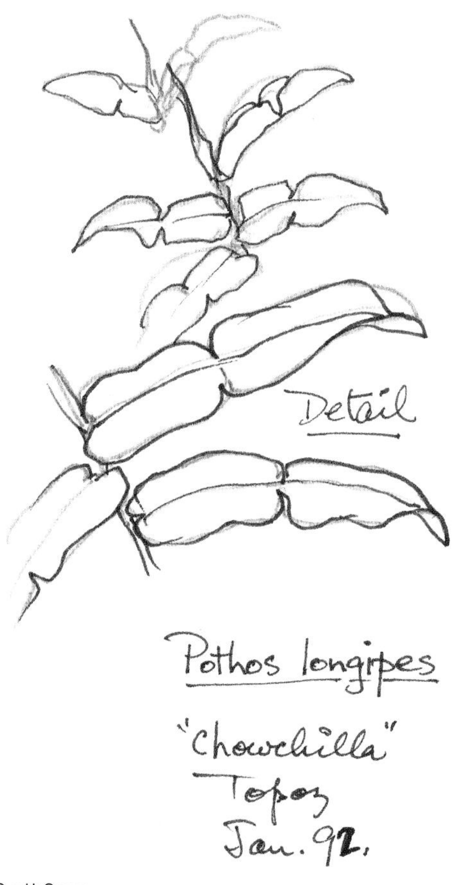

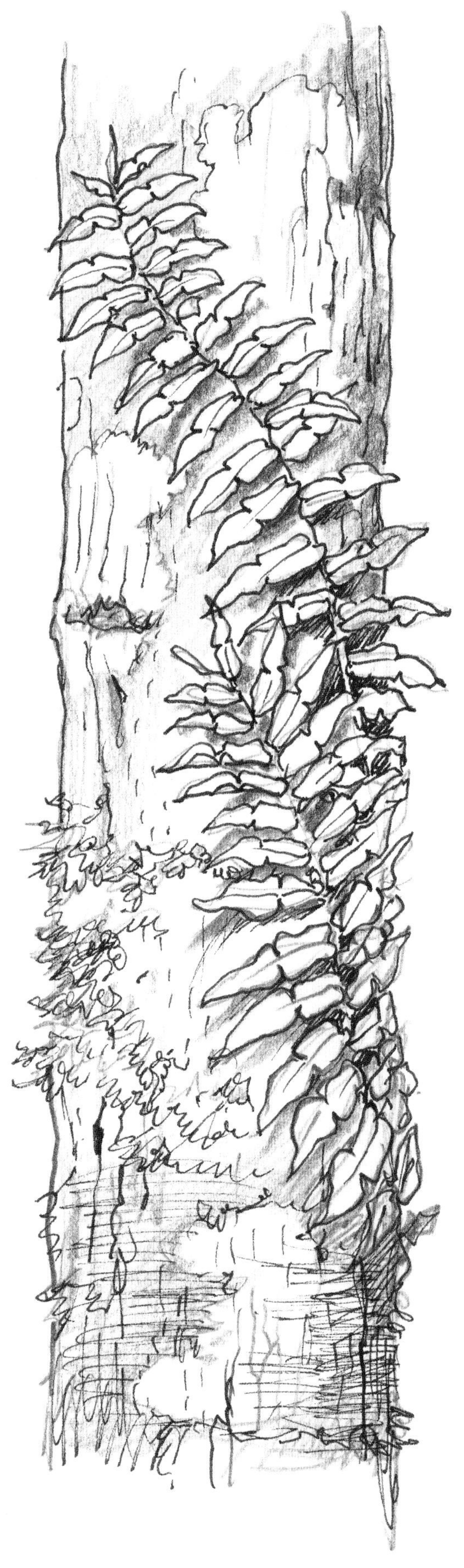

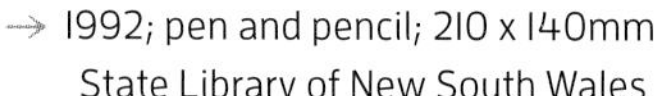

→ 1992; pen and pencil; 210 x 140mm
State Library of New South Wales

↠ 1971; pen, pencil and gouache; 260 x 275mm
State Library of New South Wales

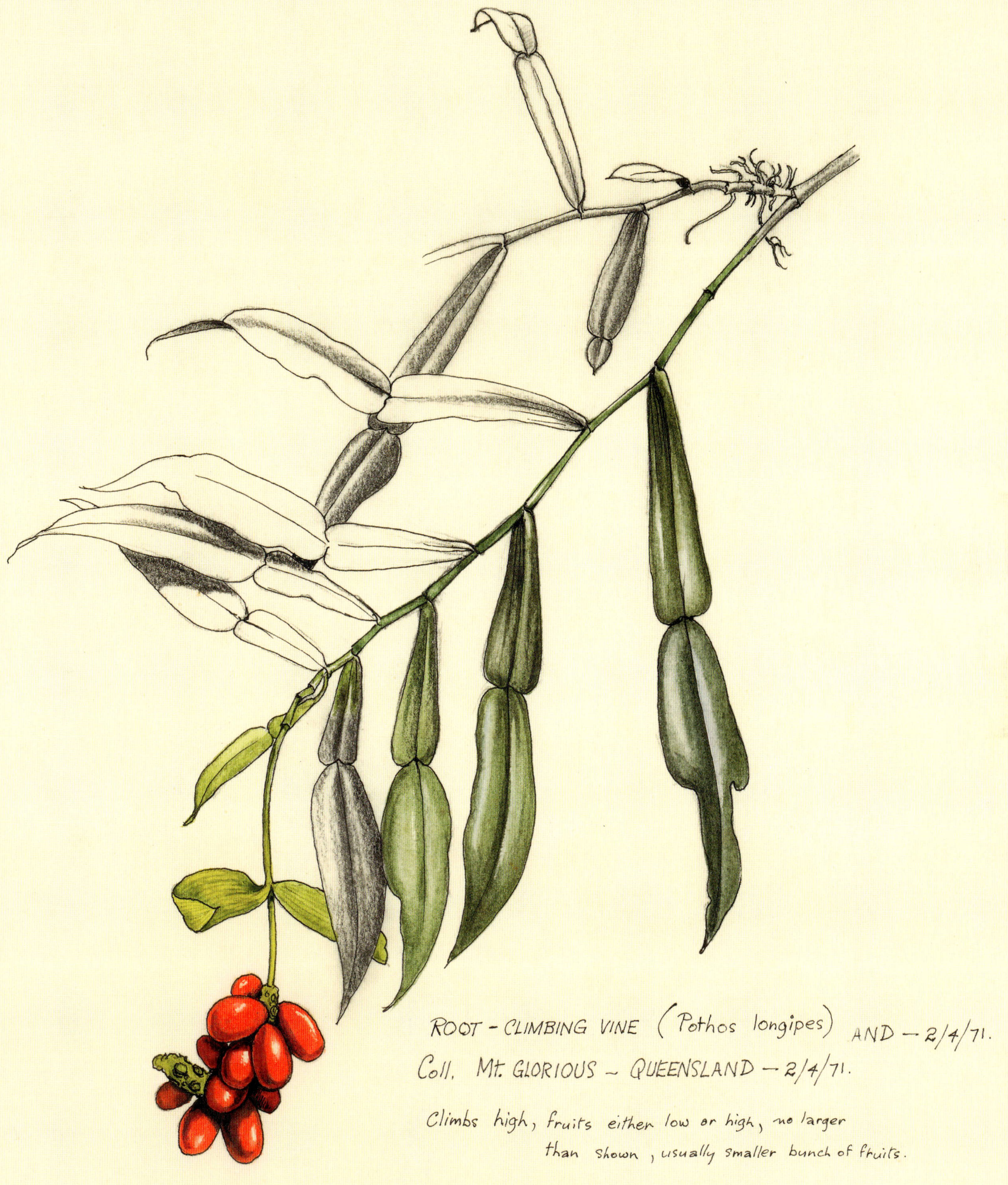
ROOT - CLIMBING VINE (Pothos longipes) AND - 2/4/71.
Coll. Mt. GLORIOUS - QUEENSLAND - 2/4/71.
Climbs high, fruits either low or high, no larger
than shown, usually smaller bunch of fruits.

Pyrrosia confluens

Silvery Felt Fern

POLYPODIACEAE

Pyrrosia is from *pyros* (red), referring to the reddish star-shaped hairs on some species.

confluens (confluent or merging), referring to the sori spots on the leaf undersides, which often merge into a mass.

The variety *Pyrrosia confluens* var. *confluens* commonly adorns branches in disturbed rainforest on the Atherton Tablelands in north Queensland. The leaf tracings below show the spore pattern on the underside of Silvery Felt Fern.

The illustration opposite shows the southern variety, *dielsii* (named in honour of German botanist Friedrich Ludwig Emil Diels (1874–1945)), clustered on a fallen rotting branch at Seal Rocks in New South Wales.

→ 1969; pen, pencil and gouache; 280 x 230mm
State Library of New South Wales

↓ Branch Engulfed by Silvery Felt Fern
1991; pen, pencil and watercolour; 520 x 630mm
State Library of New South Wales

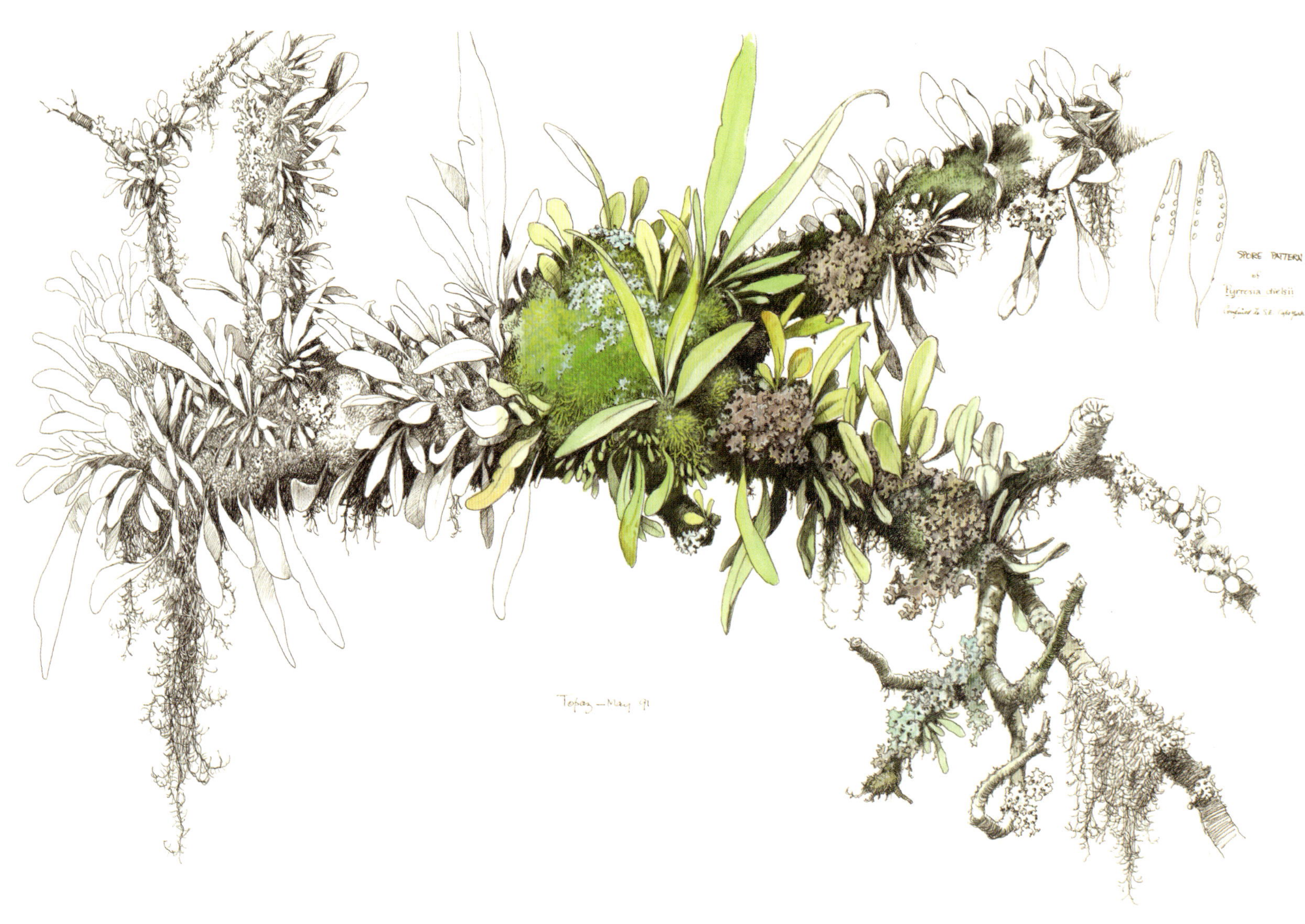

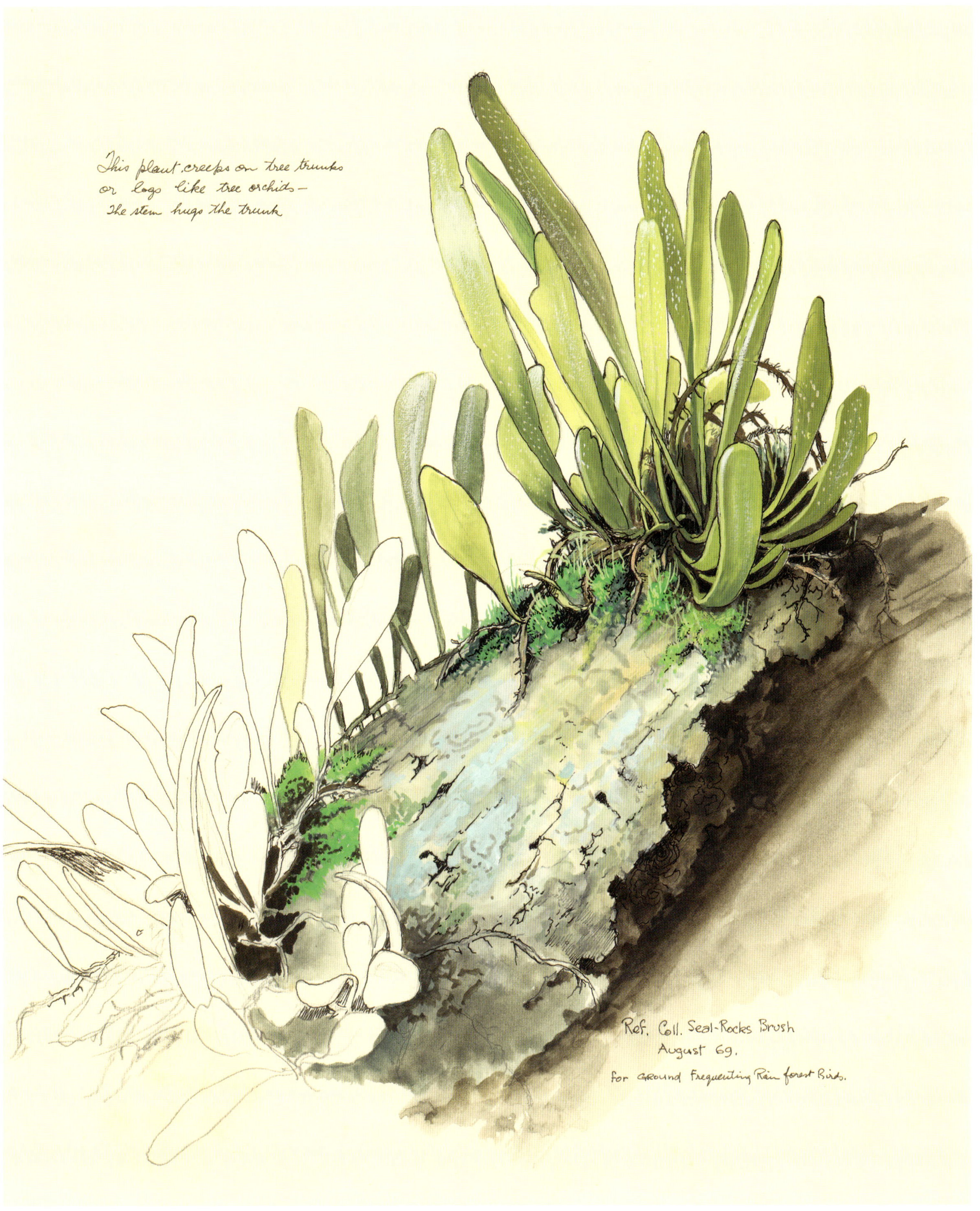
This plant creeps on tree trunks
or logs like tree orchids -
The stem hugs the trunk
Ref. Coll. Seal-Rocks Brush
August 69.
for Ground Frequenting Rain forest Birds.

Ripogonum

RIPOGONACEAE

Ripogonum is from *rhips* (wickerwork) and *gonia* (angle or corner), referring to the jointed older stems, which look like basket-making cane.

Ripogonum are restricted to six species occurring in Australia, New Zealand and New Guinea. Five species occur in Australia, and four of these are endemic.

Ripogonum album

White Supplejack

album is from *albus* (white), referring to the flowers.

White Supplejack is a stiff robust climber occurring in various rainforest types from Victoria to north Queensland and in New Guinea. The prickly stems can wind great distances along the ground or into the sub-canopy. Ripe fruits are red and are eaten by many birds.

This painting is labelled as *Ripogonum papuanum*, which is an old name for *Ripogonum album*.

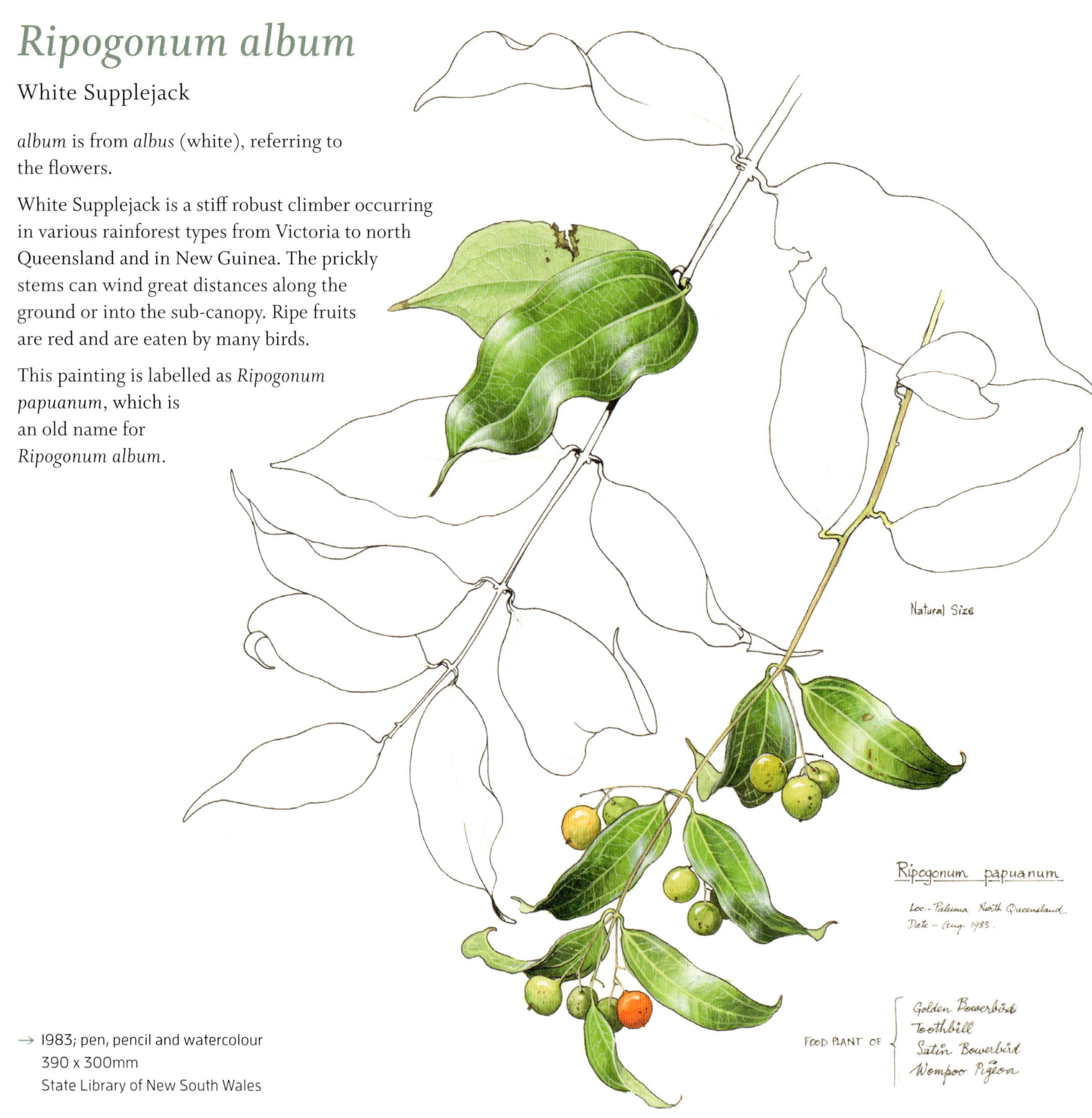

→ 1983; pen, pencil and watercolour
390 x 300mm
State Library of New South Wales

Ripogonum fawcettianum

Small Supplejack

fawcettianum is after Hugh Charles Fawcett (1812–1890), plant collector and police magistrate at Richmond River, New South Wales.

Small Supplejack is an understorey climber and scrambler, with somewhat wiry stems, in the wetter rainforests of New South Wales. Unlike most Supplejacks, the prickles are small and sparse. The plant illustrated grew on our property at Bungwahl. Bill painted it during that exhilarating time when we were first becoming interested in learning about rainforest plants.

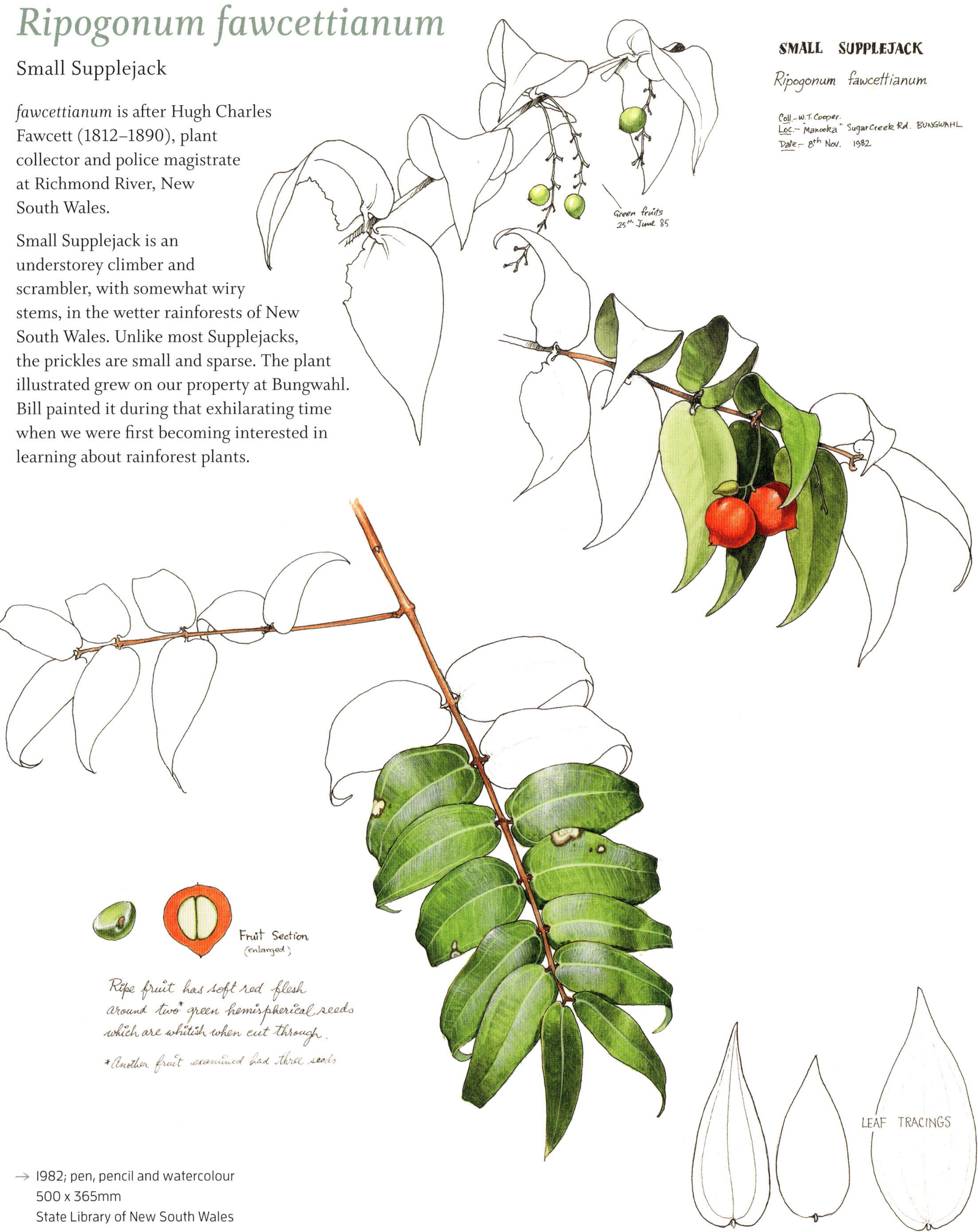

→ 1982; pen, pencil and watercolour
500 x 365mm
State Library of New South Wales

Salacia

CELASTRACEAE

Salacia is named after the Greek goddess of the ocean, referring to the type species, which most commonly occurs on the beachfront.

Salacia is a pantropical genus with about 150 species worldwide. Three species occur in Australia, within tropical Queensland.

Salacia chinensis

Lolly Berry

chinensis is from China, where the type specimen was collected.

Lolly Berry is a widely distributed vine or scrambling shrub, occurring in coastal areas from Australia through South-East Asia to India. Inside the fruit is an edible and insipidly sweet white flesh that remains strongly attached to the seed, giving rise to the common name. This sketch was done for a plate in *Cockatoos: A Portfolio of All Species* featuring the Salmon-crested Cockatoo, which is endemic to Maluku, Indonesia.

Salacia disepala

disepala is from *di-* (two) and *sepalus* (having sepals), referring to the calyx, which splits into two lobes as the flower bursts open.

Salacia disepala is a tendril-climbing vine that occurs in rainforests north from Tully in Queensland. The fruit of this vine was unknown to botanists until Bill and I found it. That fruit contained caterpillars, which we took home and bred through to adults, using the *Salacia* fruit. The caterpillars in fact eat the seeds in the fruit, and the butterflies turned out to be the beautiful Australian Plane, or Sword-tailed Flash.

Bill wrote about the experience in his diary:

> *21st Feb., 1992. Reared an Australian Plane butterfly. The caterpillar was found in the seed of Salacia disepala. It pupated 9/2/92 and emerged (female) on 21/2/92. Kept it as a specimen. The pupa case had no girdle. Pupated in curled dead leaf. The caterpillar had a wide pale band across middle.*

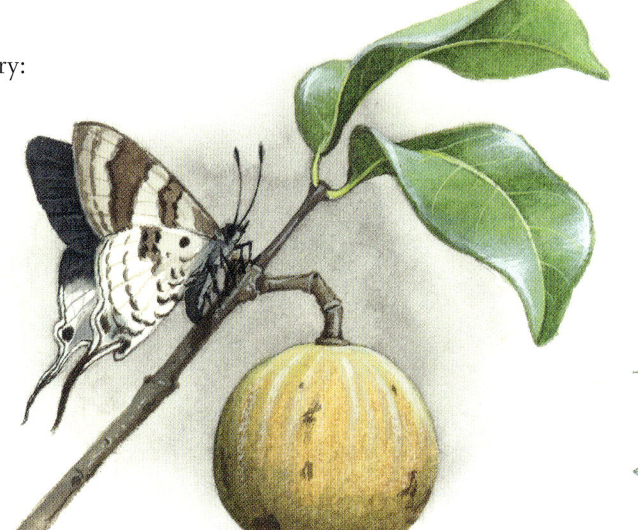

→ 1989; pen and watercolour
520 x 665mm
Collection of Bruce Gray

← undated (c.1991); watercolour
150 x 190mm
Private collection

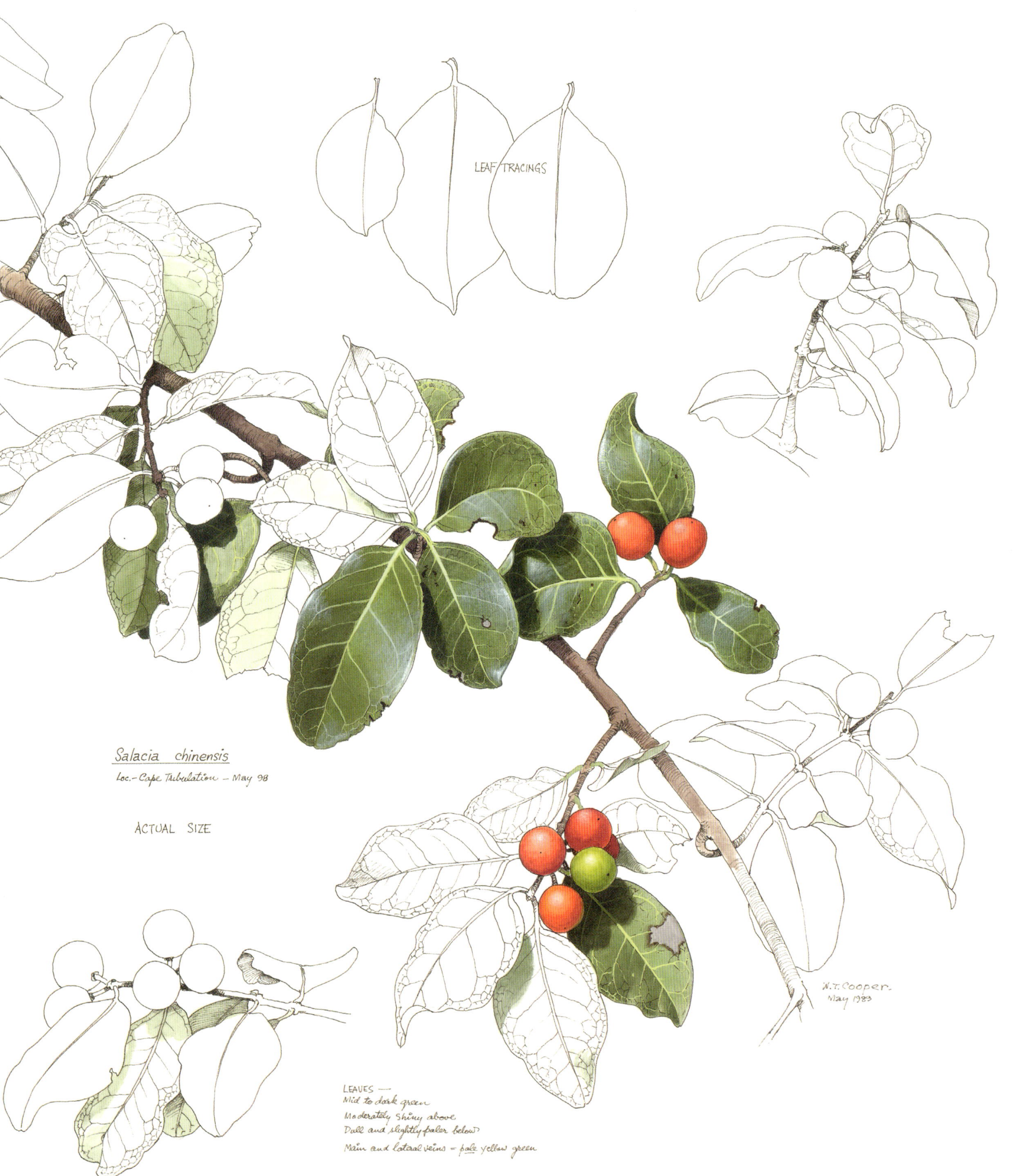
LEAF TRACINGS
Salacia chinensis
Loc. - Cape Tribulation - May 98
ACTUAL SIZE
W.T. Cooper.
May 1983
LEAVES —
Mid to dark green
Moderately Shiny above
Dull and slightly paler below
Main and lateral veins = pale yellow green

Tecomanthe burungu

Tecomanthe

BIGNONIACEAE

Tecomanthe is after the related genus *Tecoma* and from *-anthus* (-flowered).

burungu is the Eastern Kuku Yalanji name for the area where the plant occurs.

This plant is endemic to rainforest in a small area north of the Daintree River. Even though it was first collected on Thornton Peak in 1937 by Leonard Brass, it has only recently been described and named by botanists Frank Zich and Andrew Ford. The showy pink flowers are up to 86 millimetres long, arising in large clusters on leafless woody stems.

The painting opposite shows some unidentified moss-adorned vine stems from our property at Topaz.

→ Unidentified Vine Stems at Topaz
1988; pen, pencil and watercolour
490 x 290mm
State Library of New South Wales

↓ 2000; watercolour; 33.5 x 29mm
Australian National Herbarium, Centre for Australian National Biodiversity Research, National Research Collections Australia, CSIRO

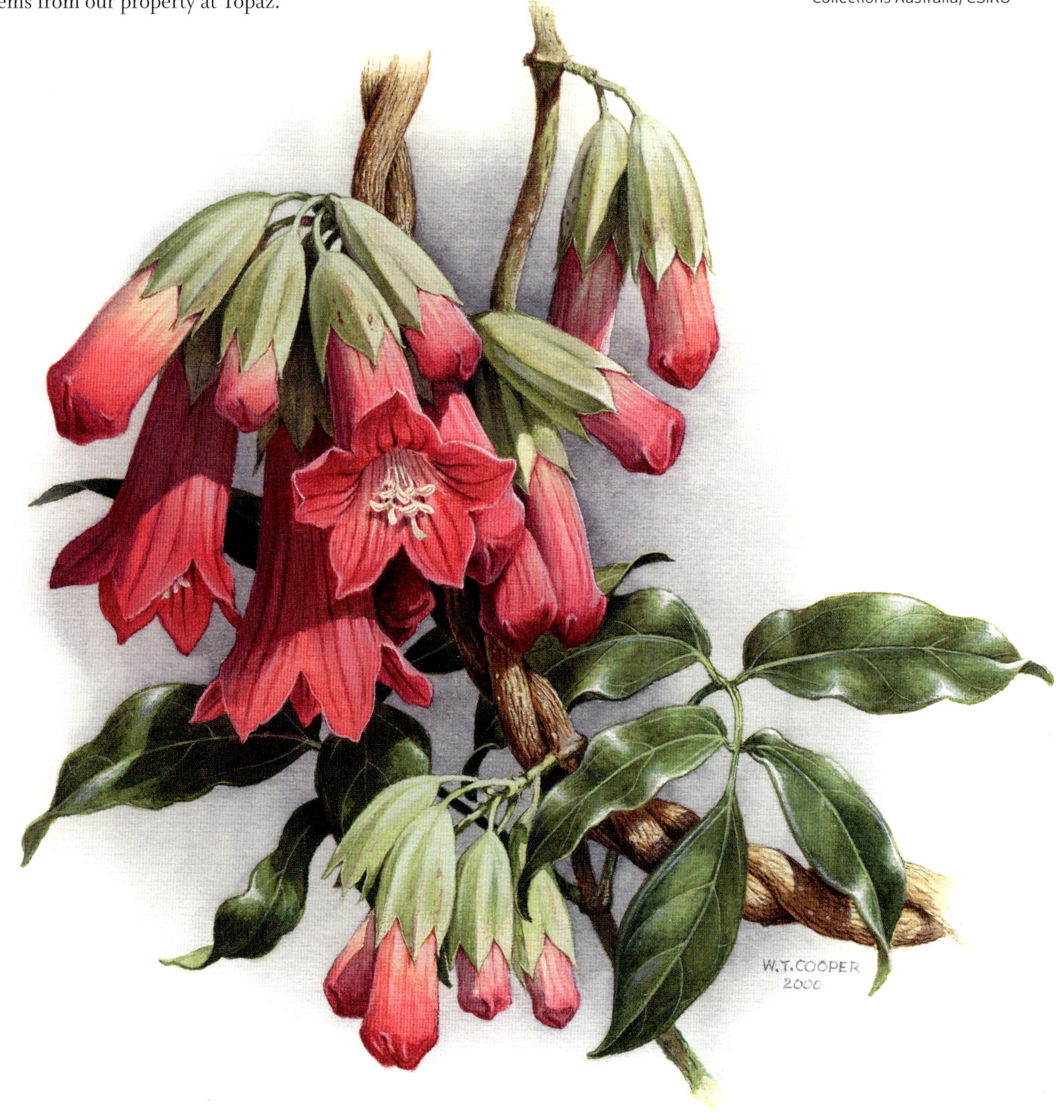

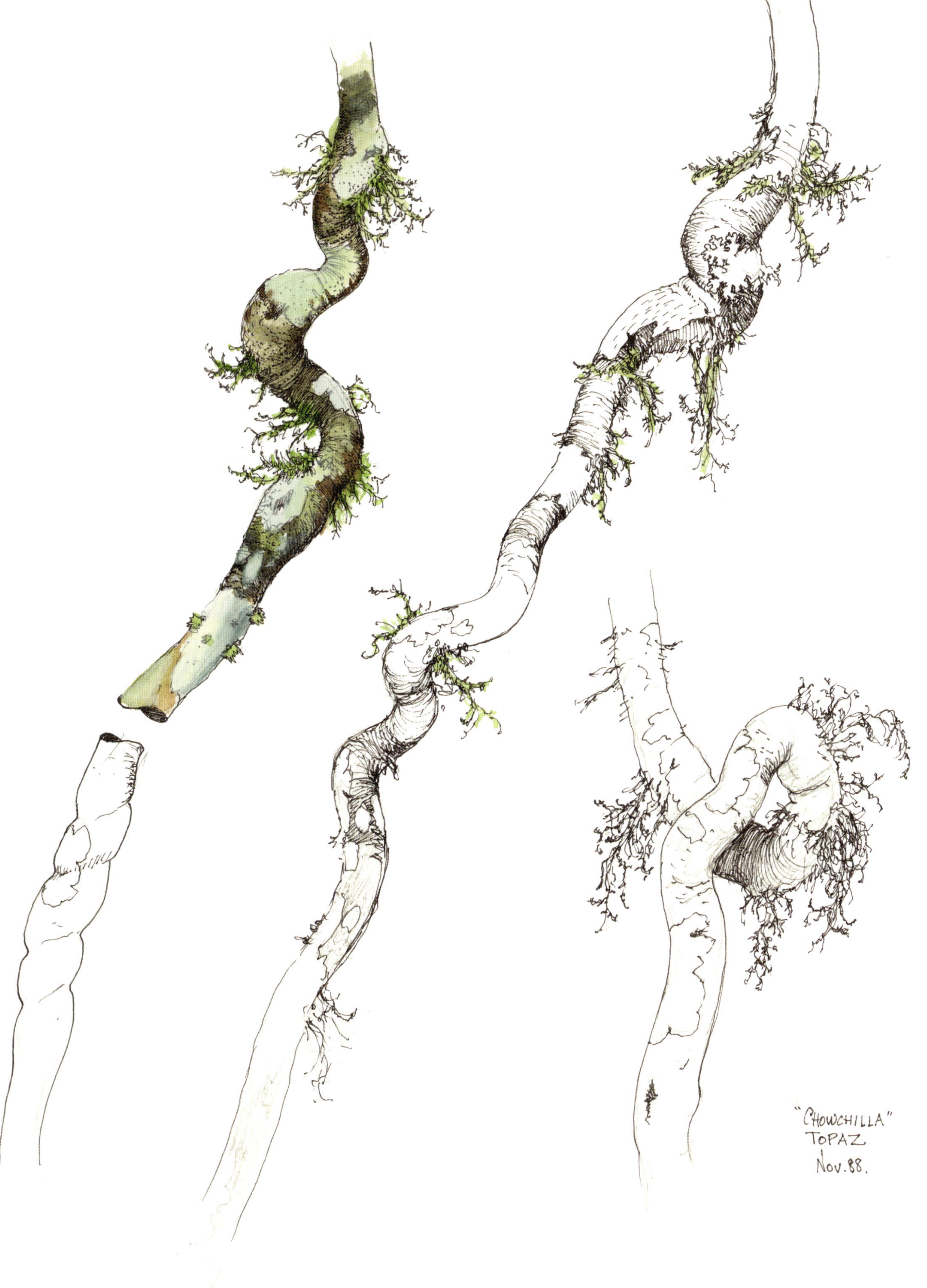

"CHOWCHILLA"
TOPAZ
Nov. 88.

Trichosanthes

CUCURBITACEAE

Trichosanthes is from *trichos* (hair) and *-anthus* (-flowered), referring to the fringed petals.

Trichosanthes occurs from India and China through to Australia and the eastern Pacific. There are approximately 100 species within that area, with six in Australia.

→ undated (c.2010); pen; 450 x 360mm
Private collection

↓ Rainforest at Topaz
1993; pen, pencil and gouache
State Library of New South Wales

Trichosanthes odontosperma

Rainforest Gourd

odontosperma is from *odonto* (tooth) and *-sperma* (seed), referring to the toothed seeds.

Rainforest Gourd is a tendril climber, mostly seen on rainforest margins, including around our garden. The delicately frilled flowers open at night but it was a convenient subject just a few steps from the house, and the unfurling flowers could be observed and described as part of naming this species. As mentioned elsewhere, Bill was not keen on overly detailed work but kindly illustrated this, my first taxonomic paper with co-author Andrew Ford, to describe and name a new species.

Bill observed in his diary:

> *5th March 1990 … Within the last week have seen riflebird feeding on Trichosanthes sp (orange fruit). It ate out the inside leaving the red inner flesh—must have just eaten the seeds.*

We later observed that the birds either flick the seeds from the flesh or regurgitate the seeds, only needing to consume the flesh that surrounds the seeds.

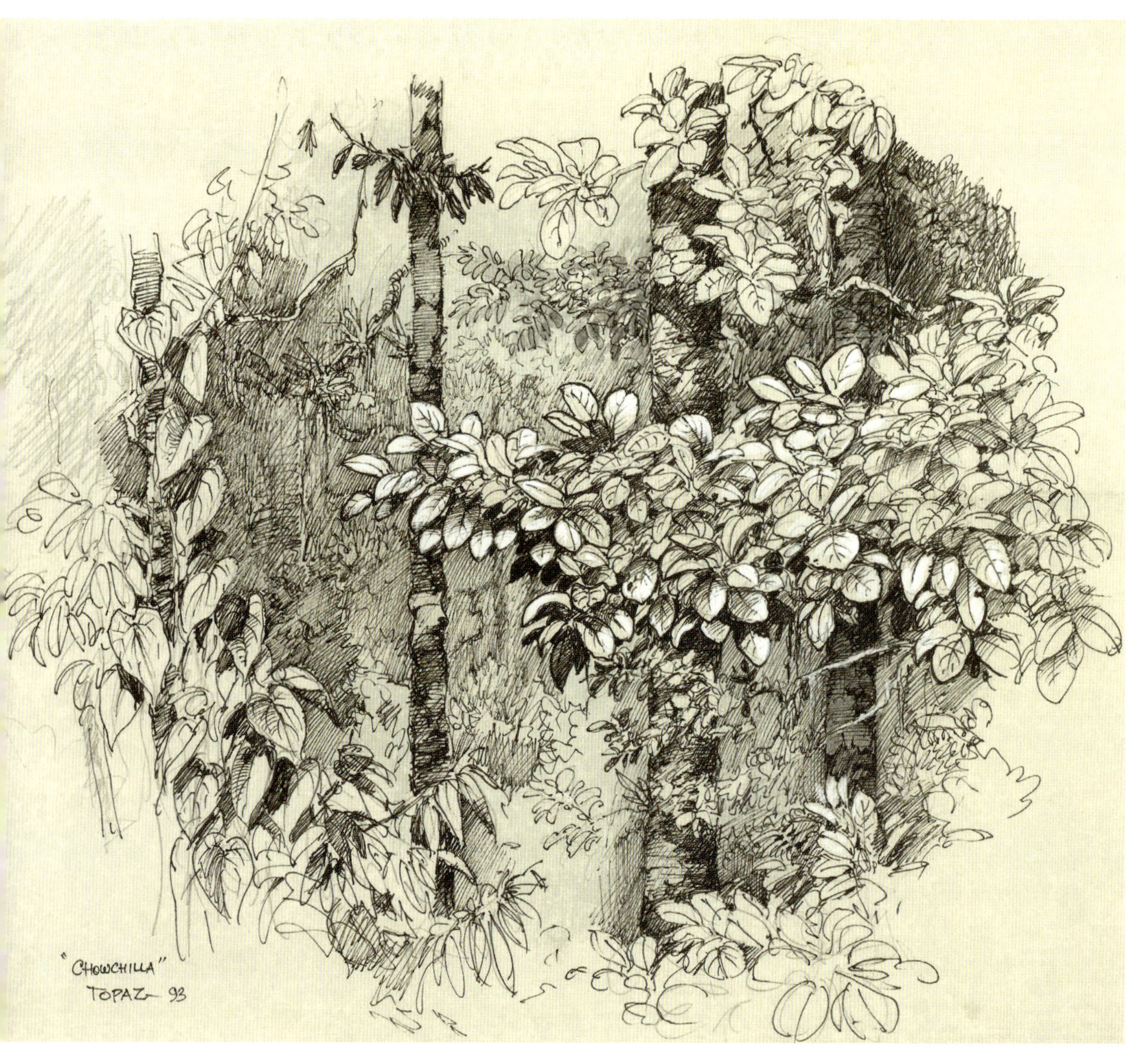

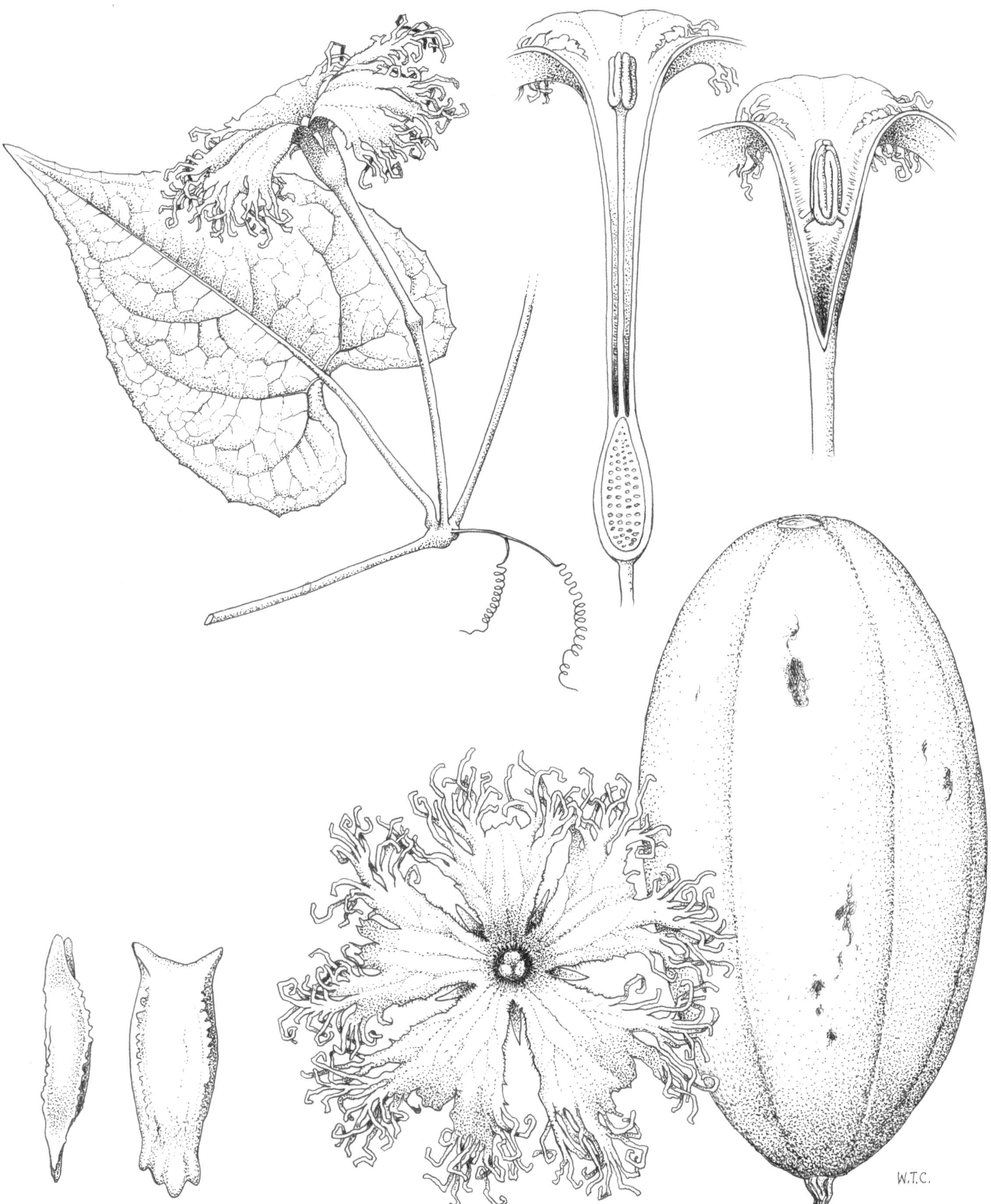
W.T.C.

Trichosanthes pentaphylla

Red Gourd

pentaphylla is from *penta* (five) and *-phyllus* (-leaved), referring to the compound leaf with five leaflets.

Red Gourd is a widespread tendril climber, mostly of lowland rainforest in tropical Queensland. This sketch was done in 1981, when we had no idea what species the plant was. At that time, the species was not known to occur at Iron Range. We knew that both the unidentified plant and Palm Cockatoos occurred in the same forest, so Bill created this painting soon after our observations.

Thirty years later, I did taxonomic work on this group, so we returned to the area to collect specimens in order to officially confirm that the plant does indeed occur in those forests.

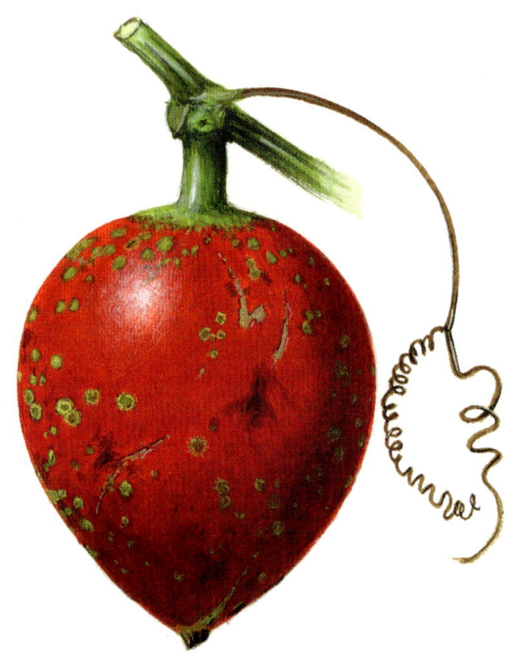

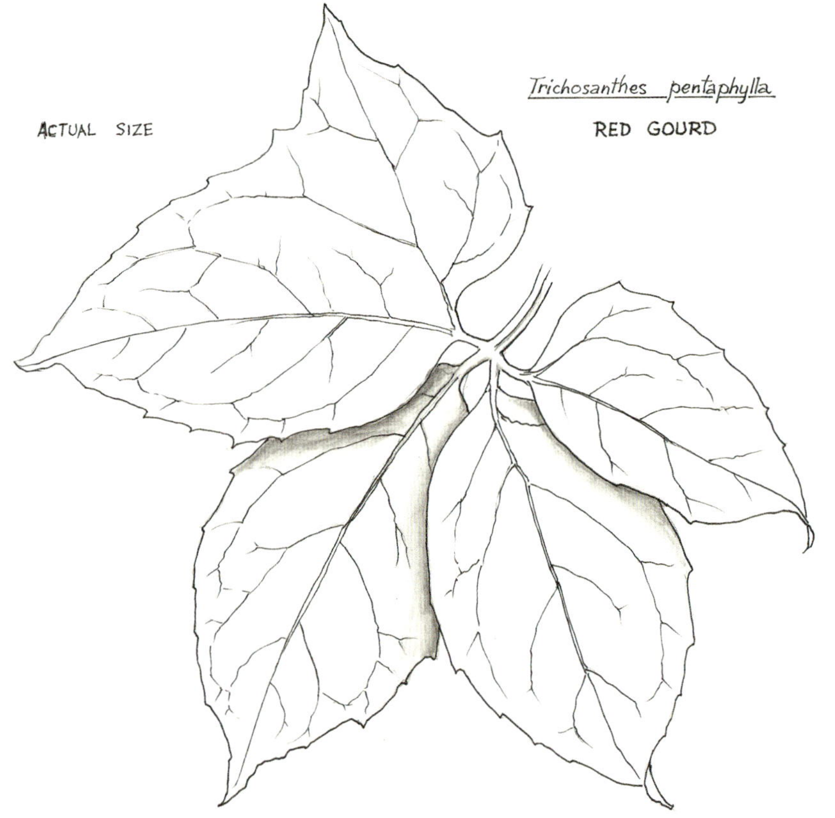

↑ Rainforest Gourd
1994; watercolour
State Library of New South Wales

← 2009; pen and pencil; 400 x 295mm
State Library of New South Wales

→ 1981; pen, pencil and watercolour; 510 x 370mm
State Library of New South Wales

↓ 2004; watercolour
State Library of New South Wales

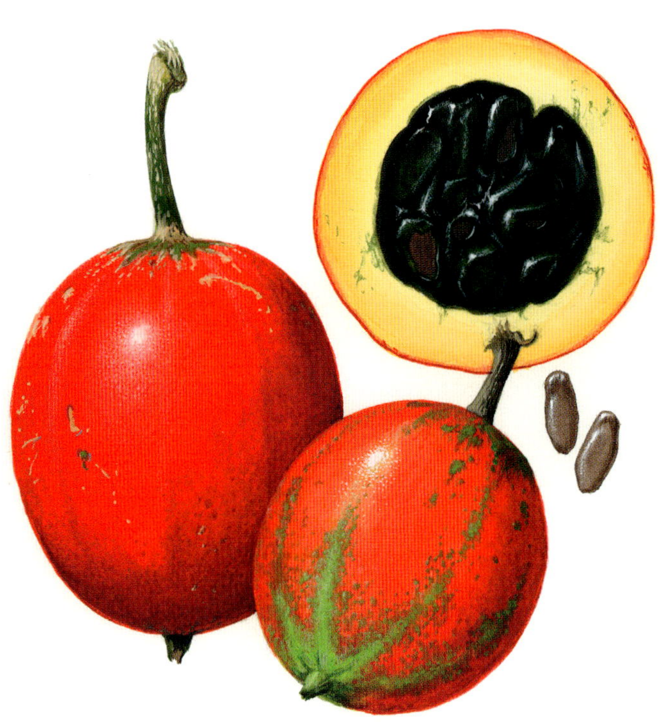

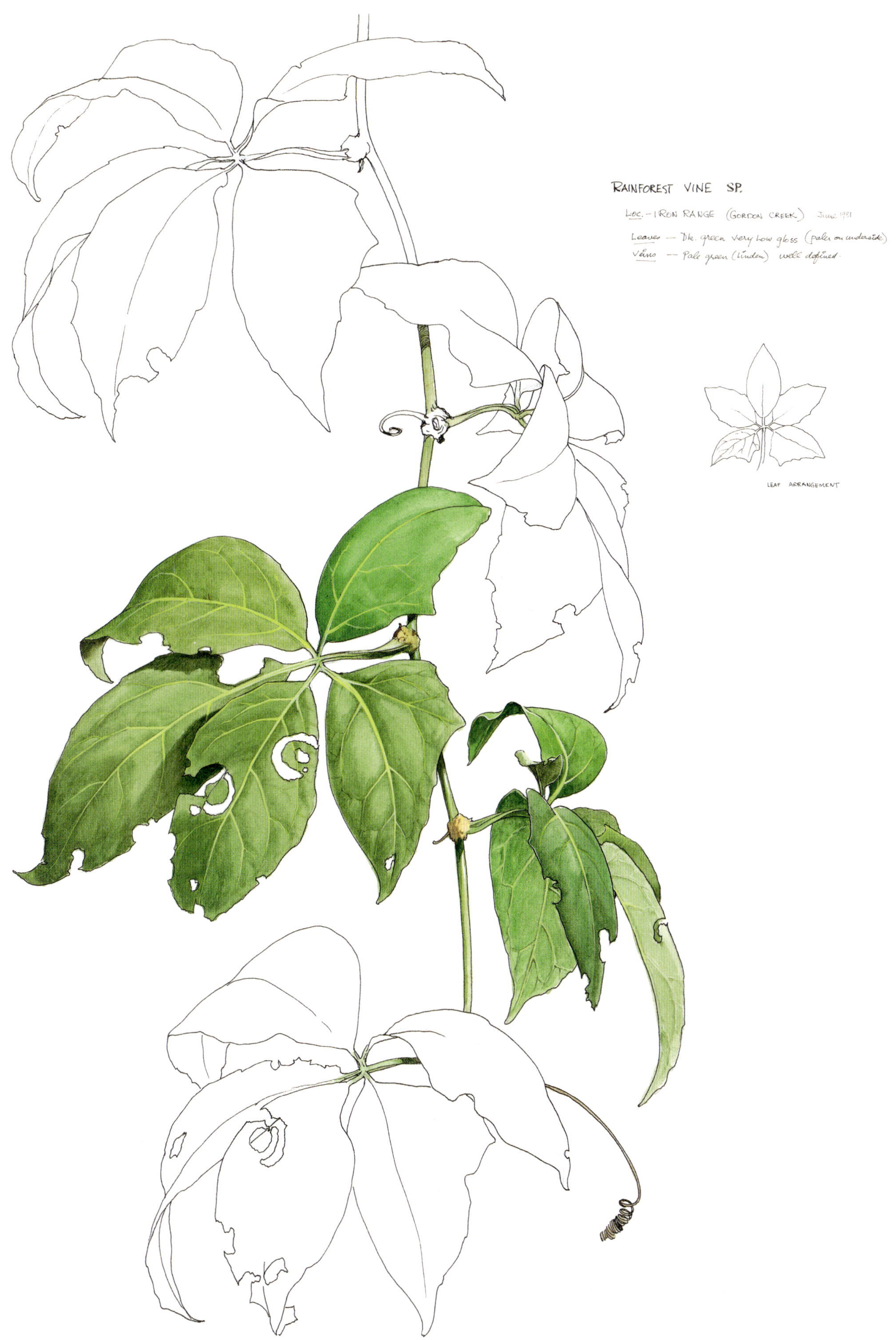
RAINFOREST VINE SP.
LOC. – IRON RANGE (GORDON CREEK) June 1981
Leaves — Dk. green very low gloss (paler on underside)
Veins — Pale green (linden) well defined.
LEAF ARRANGEMENT

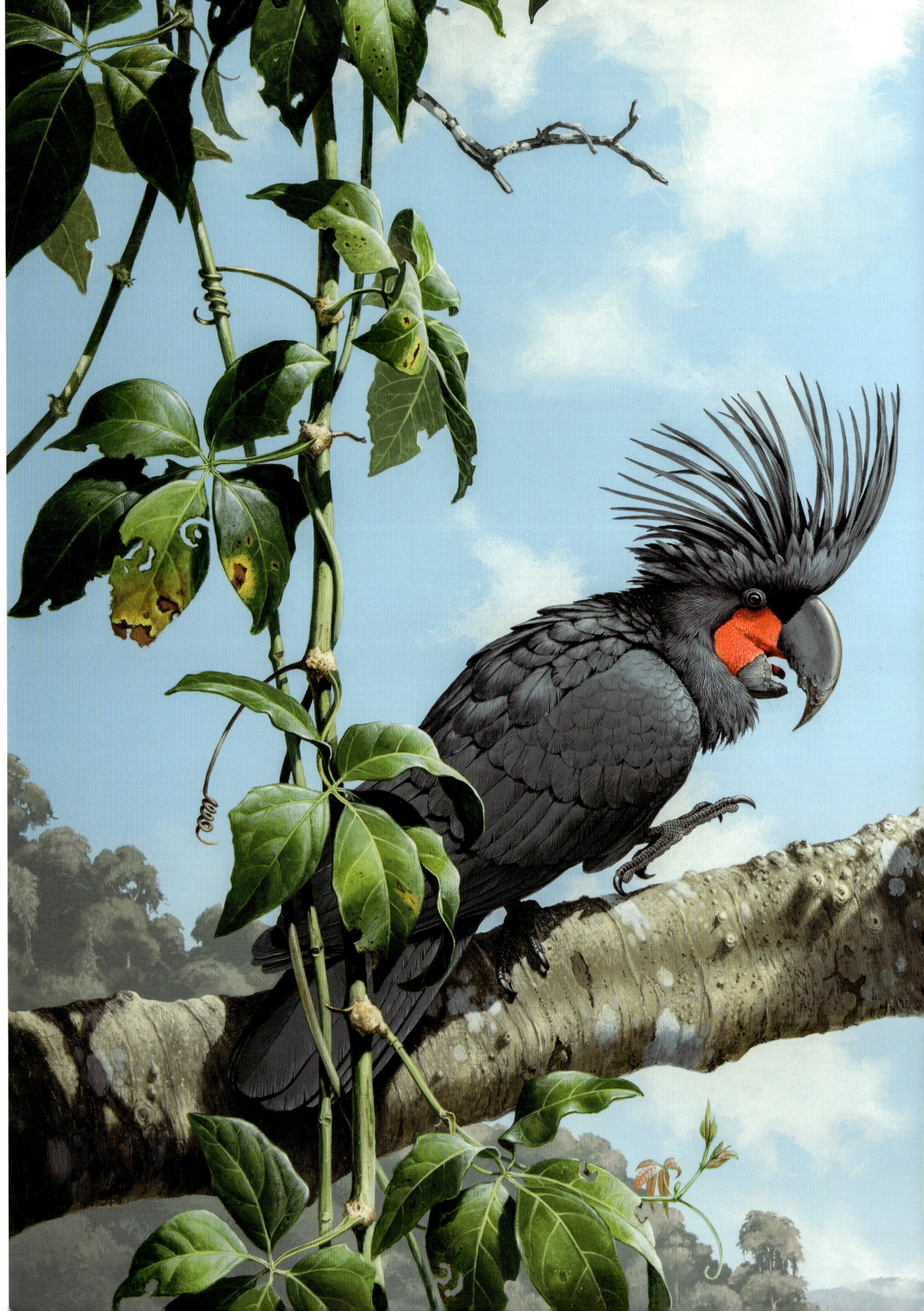

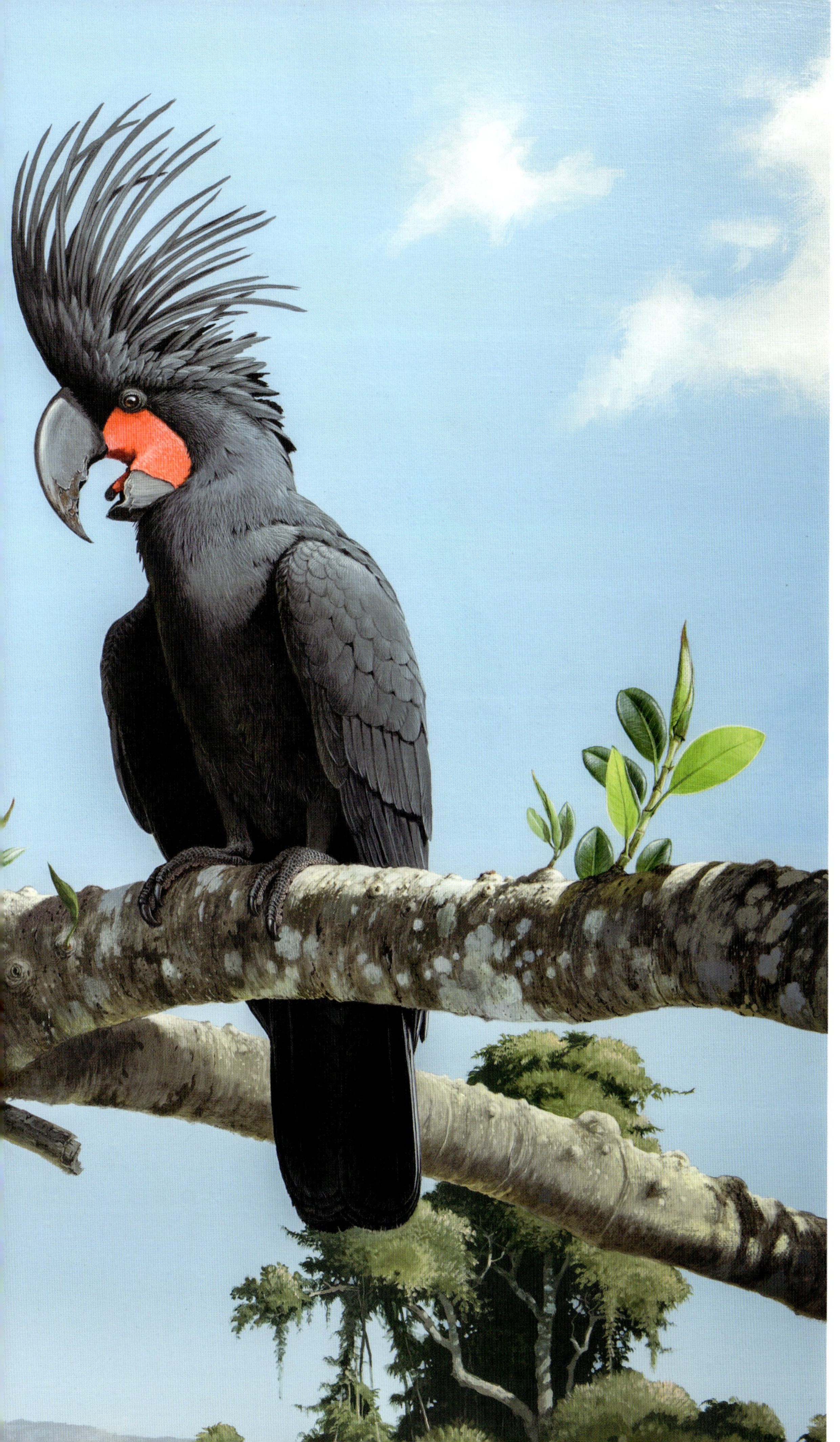

← Palm Cockatoos
undated (c.1986);
acrylic on canvas on board
101 x 126cm
Collection of Peter Boag

Chapter 3

Dry Country

→ *Senna artemisioides* subsp. *helmsii*
undated; pen, pencil and watercolour
290 x 420mm
State Library of New South Wales

↙ Bill Sketching and Making Notes at Noonbah near Longreach, c.2012

Bill's childhood was shaped by a father attuned to the outdoors and a mother who was an accomplished pencil artist. From a very young age, he found inspiration in the family's supply of art materials and natural history books, and in landscapes populated by shrieking birds speeding among eucalypts, wattles and banksias.

Initially, Bill painted birds from published illustrations, so his earliest paintings are of species not native to his surrounds. With the sights, scents and sounds of his childhood environment cherished and stored, the young man became a notable landscape painter. Later, when birds began once more to occupy his attention and to take form on paper, he painted the birds he could see with his monocular, a low-powered telescope that he kept as a treasured possession throughout his life.

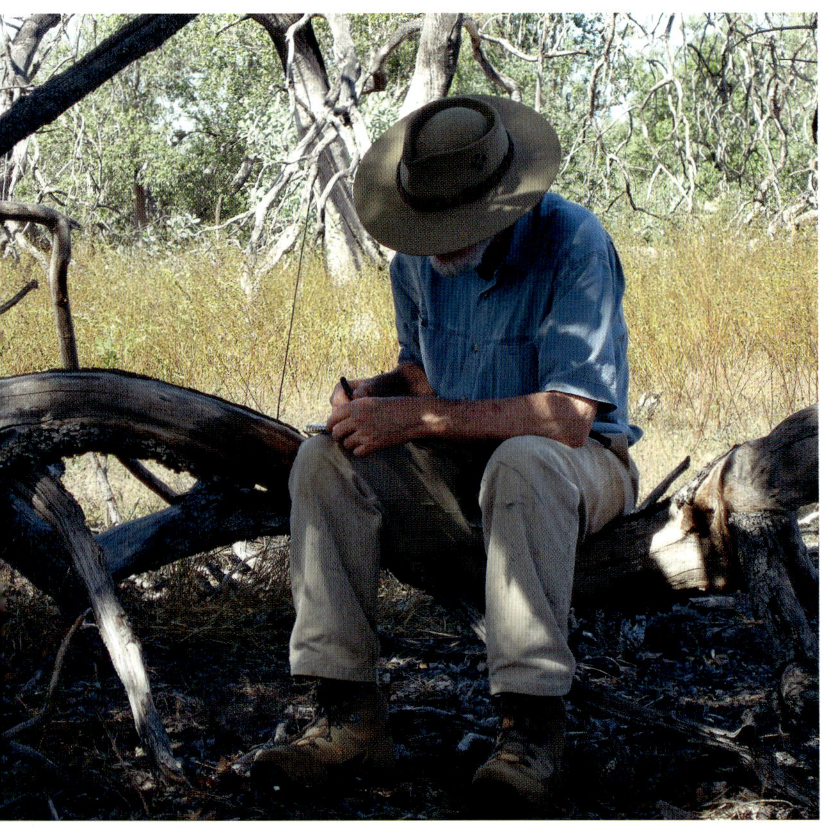

Plants were not much included in early compositions, but as Bill's confidence in his ability grew, vegetation details relevant to each work became more prominent. With close observation and determined accuracy, he began to draw and paint those plants on which birds, particularly parrots and cockatoos, were seen feeding. Bill saw the detail without necessarily knowing the scientific names for a swelling at the base of a leaf or calyx lobes persisting at the tip of a fruit. It was the complexity of the subject that attracted Bill's attention, and this complexity, perfectly depicted, is what enthrals the viewer gazing at any of his paintings.

The hardy, gnarly Old Man Banksia (*B. serrata*), its seeds a magnet for Black Cockatoos, appears in several exquisite studies done as reference, the way a writer might make notes to refer to. But where, for a writer, a quick scribble might be enough, the work that Bill put into multiple studies is evident in the wrinkled branches, saw-tooth leaves, both fresh and ageing flower spikes, and fruiting spikes with wide open seed cases, recalling May Gibbs' Big Bad Banksia Men.

The Coast Banksia (*B. integrifolia*) of eastern Australia also occupied Bill's attention. On a page showing several details for this tree, Bill has included notes on the leaves:

> *Under side—edge (margins) strongly recurved so as to be visible from below. Main vein as shown lat[eral] veins netted very pale greenish—underside [of leaf] appears almost white.*

In a painting showing Coast Banksia twigs, there is the comment: 'Branches get smooth and shiny where rubbed. Note! This purplish colour also extends to some of the smaller twigs'.

Bill enjoyed the scent of eucalypt forests, and he admired the twists and turns of trunk and branch, and the colours as bark peeled, all of which he set down on paper. His River Red Gum (*Eucalyptus camaldulensis*) drawings illustrate some of this, with a page added to show the delicate lines of twigs carrying long leaves and rotund flower buds.

In one of several paintings of the Red-winged Parrot, Bill has them perched on a Silver-leaved Ironbark (*Eucalyptus*

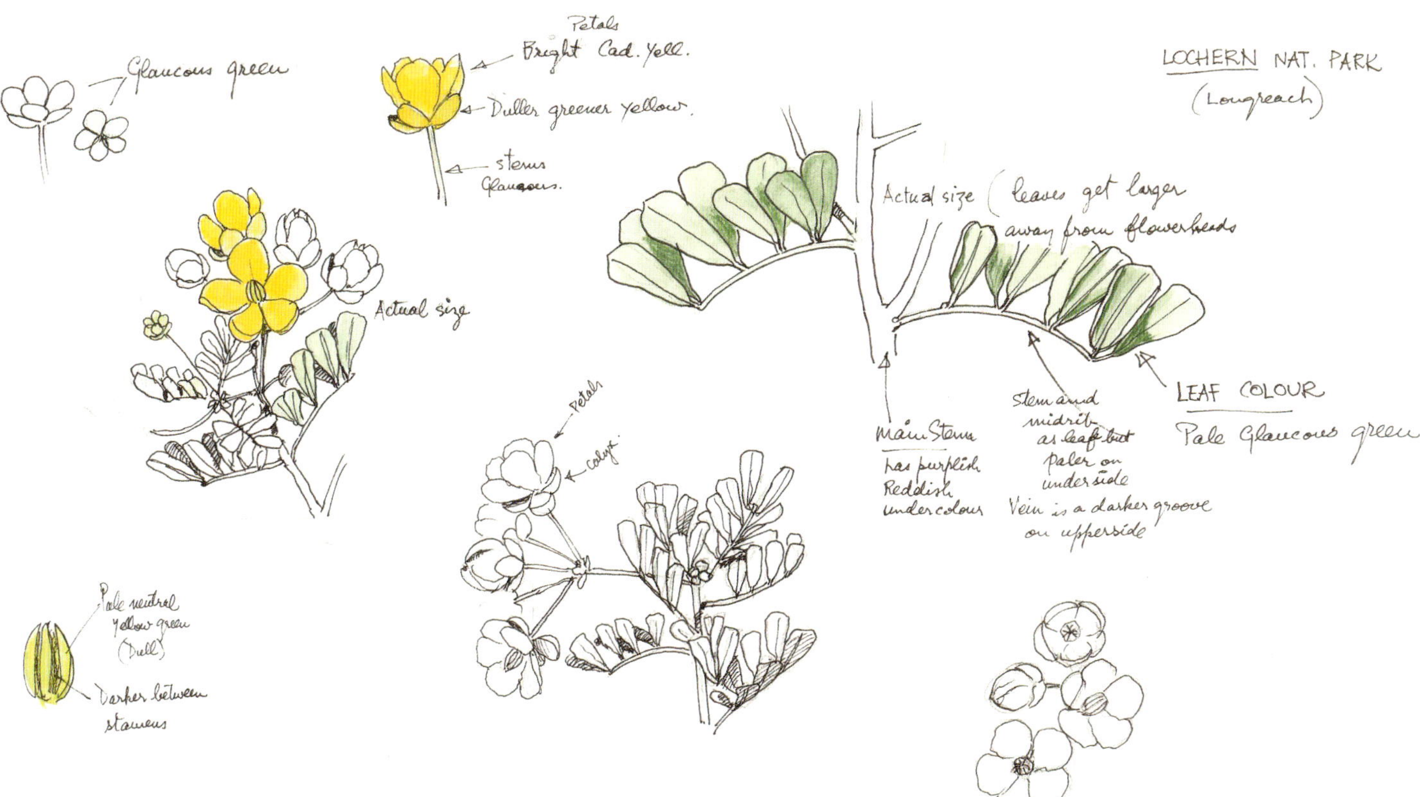

shirleyi), a central part of the habitat frequented by these beautiful, strong-flying birds. With attractive silvery-blue leaves around him and his mate, the male bird reaches for a seed capsule. Before Bill began the final work, he had covered several pages with drawings showing the exact shapes and colours of a range of leaves and their juxtaposition with other leaves on the twig. On one twig showing the lovely pink of newly sprouted leaves, there is a jotting to show that a young stem is square in cross-section. The colour given to the leaves in one study did not satisfy Bill, so there is a second study beside it. Alongside this is written: 'Note!! This is closer to the true colour of the leaves'. Next to the leaves is another note: 'Veins are almost pure white, laterals less so. Leaves have no shine'.

One of the Silver-leaved Ironbark drawings carries Bill's initials—W.T.C.—in tiny letters. It was rare for Bill to initial a study, but perhaps he was particularly satisfied with how he had captured the complex arrangement of the eucalypt's foliage along a twig.

In open forest, Bill valued the light of the early morning or late evening. Cloudy days were also welcome, because to Bill's eye redness was accentuated in overcast conditions. With a composition in mind and enough light left in the day, Bill and Wendy might take off, driving from their home in the very wet Topaz rainforest to travel west of Herberton or out to Chillagoe, equipped with billy and binoculars. Once a branchlet or leafy twig was deemed just right for what Bill had in mind, there would be a hurried trip home to convey the plant-loaded esky to the studio.

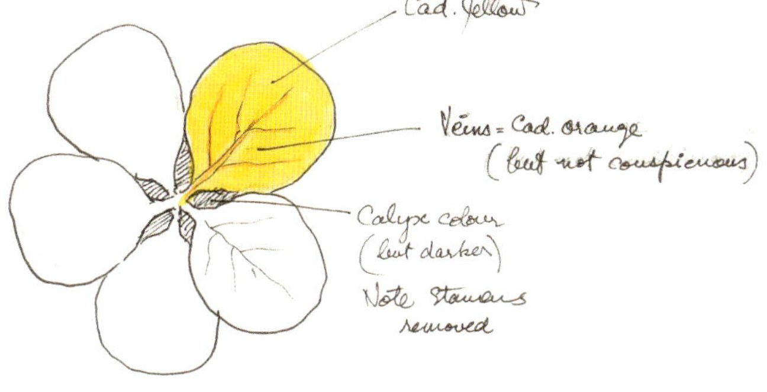

On one Chillagoe trip, Bill chose a distinctive northern eucalypt, the Cabbage Gum or Broad-leaved Carbeen (*Corymbia confertiflora*), as a subject for several studies. Carbeen leaves, sandpapery to touch, are unusual because the mature tree retains the juvenile eucalypt habit of leaves set opposite each other. Bill's note reads: 'Leaf colour closest to 371 U ... more yellow when light comes through the leaf'. This distinctive eucalypt is used in a grand painting, *Above the Gorge*, of Red-tailed Black-Cockatoos, which Bill saw in Quincan country around Laura. Bill noted:

> *They came into view flying quite close, but on seeing me veered off and sailed on down the gorge … I imagined how, if they had not seen me, they might have landed on the Eucalyptus branch in front of where I was sitting, and how wonderful they would have looked against the escarpment behind.*

Rupert Russell

Acacia aneura

Mulga

FABACEAE

Acacia is from *akis* (a sharp point), referring to the spines on the type species, *Acacia nilotica*.

aneura is from *a* (without) and *neuron* (nerve), referring to the almost-veinless phyllodes (false leaves).

Mulga is usually a tree that grows to about 18 metres. It is widespread in arid places, commonly forming dense stands, which may dominate large areas. The long taproot gives the tree extraordinary drought tolerance; its leaves are regarded by pastoralists as useful stock fodder in dry times.

The reference drawing opposite was done for use in the foreground of this Princess Parrot plate for *Australian Parrots*. The trees in the background are also Mulgas. While the birds have not been recorded feeding on Mulga, they are known to occur in these habitats.

→ 1979; pen and watercolour; 410 x 300mm
State Library of New South Wales

↓ Princess Parrots
1989; watercolour; 310 x 400mm
Private collection

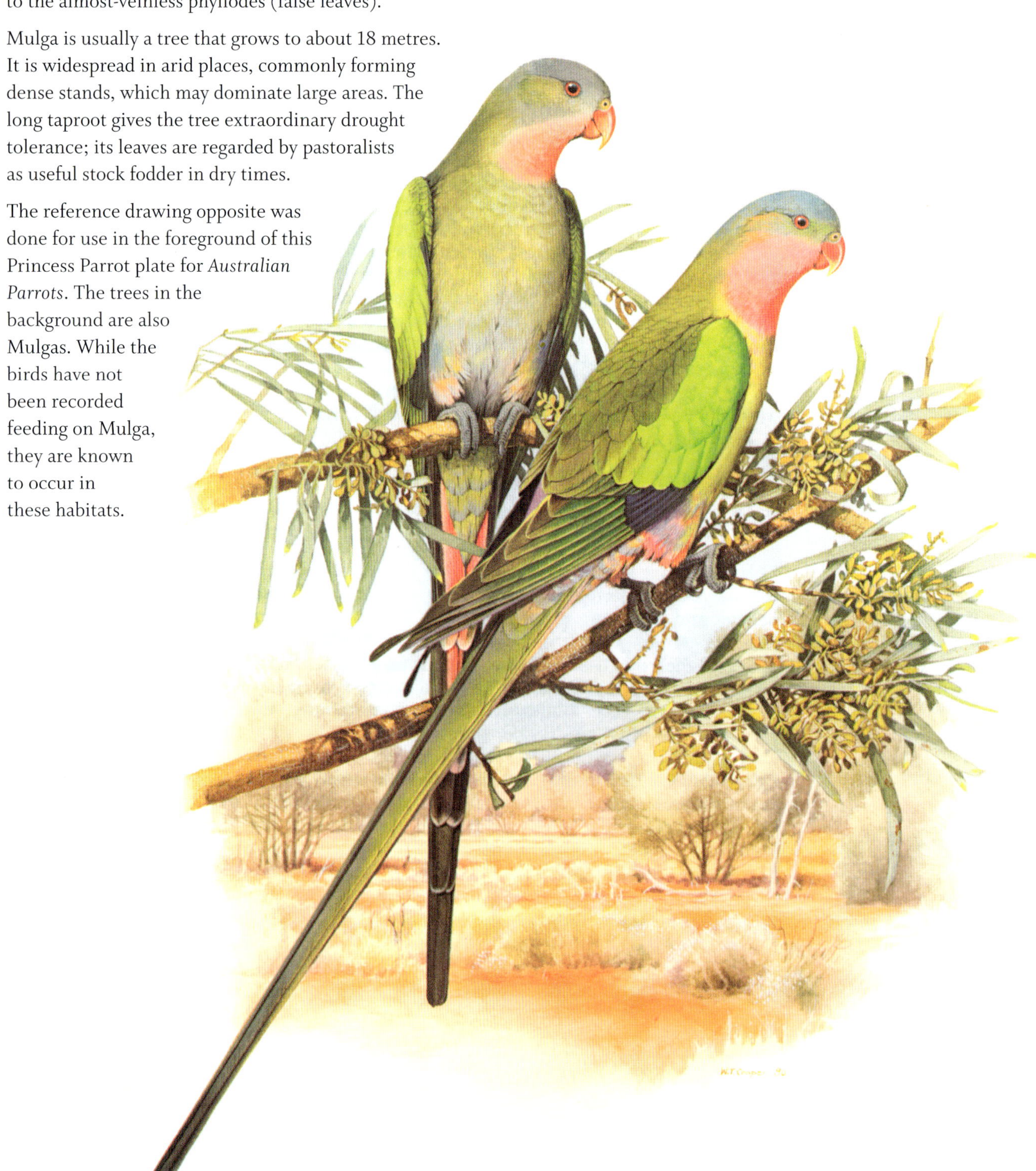

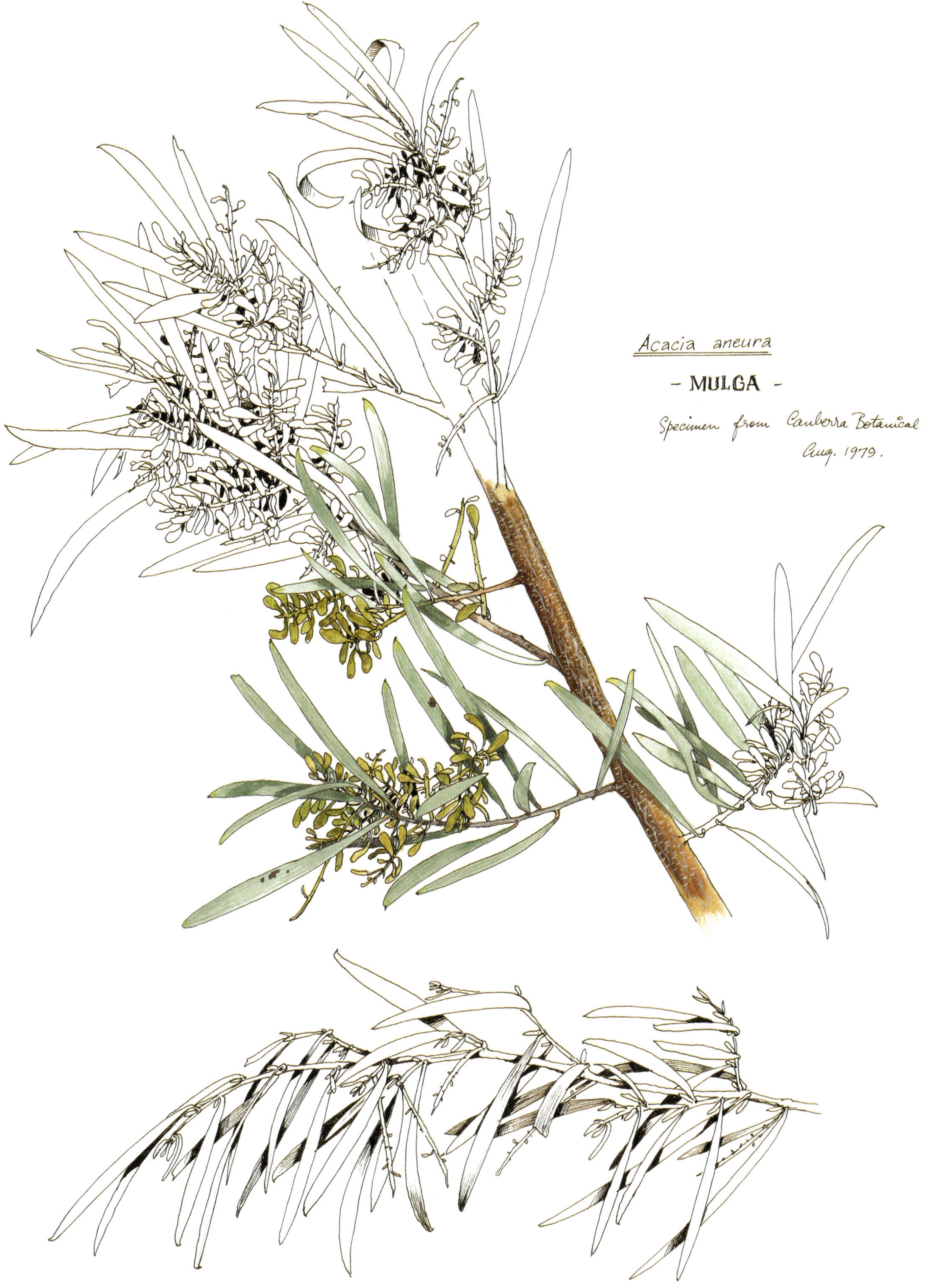
Acacia aneura
- MULGA -
Specimen from Canberra Botanical
Aug. 1979.

Actinotus helianthi

Flannel Flower

APIACEAE

→ undated (c.1978); oil on board; 490 x 380mm
Private collection

↓ undated; pencil; 250 x 235mm
State Library of New South Wales

Actinotus is from *actino-* (star-like or radiating from a centre) and *-otus* (possession), referring to the ray florets, which are petal-like and radiate around the central pin-cushion of tiny flowers.

helianthi is from *helios* (sun) and *anthos* (flower), referring to the similarity of the floral structure of Sunflowers.

Flannel Flowers are common on the sandy heath country around Myall Lakes in New South Wales. During the years following soil disturbance or fires, these plants flower in abundance for a long time. A few flowers can be found at any time of the year, but just before spring they usually flower en masse. The show was always a delight, especially as a backdrop to the numerous more colourful heath plants. Bill painted Flannel Flowers as the main feature in quite a few oil landscapes. This one is a study, painted mostly in a broad style with a palette knife. It was gifted to me for a birthday when we lived at Bungwahl.

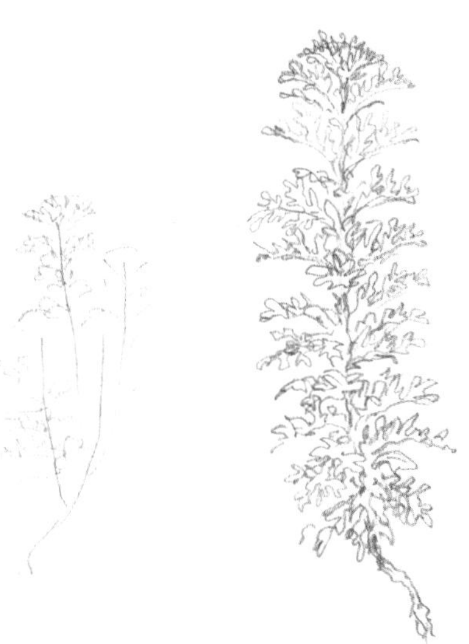

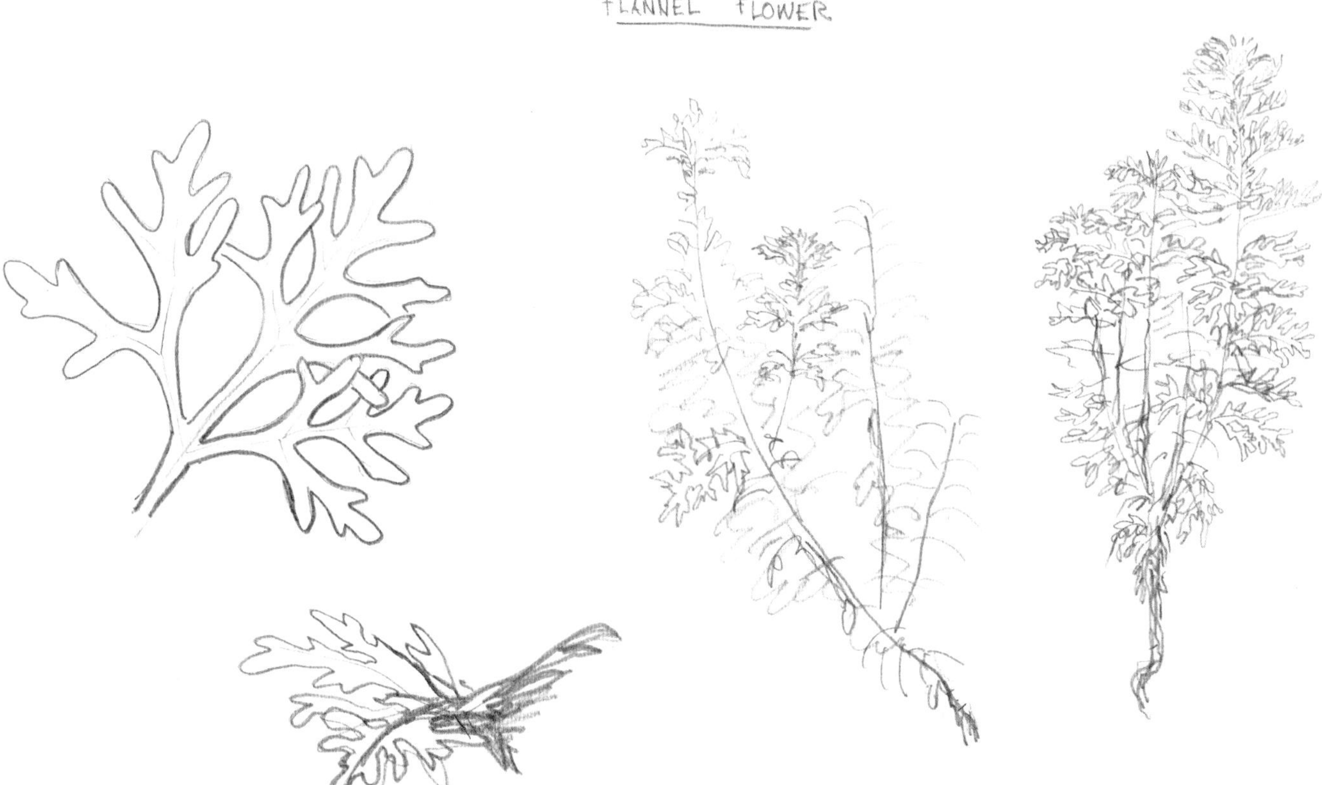

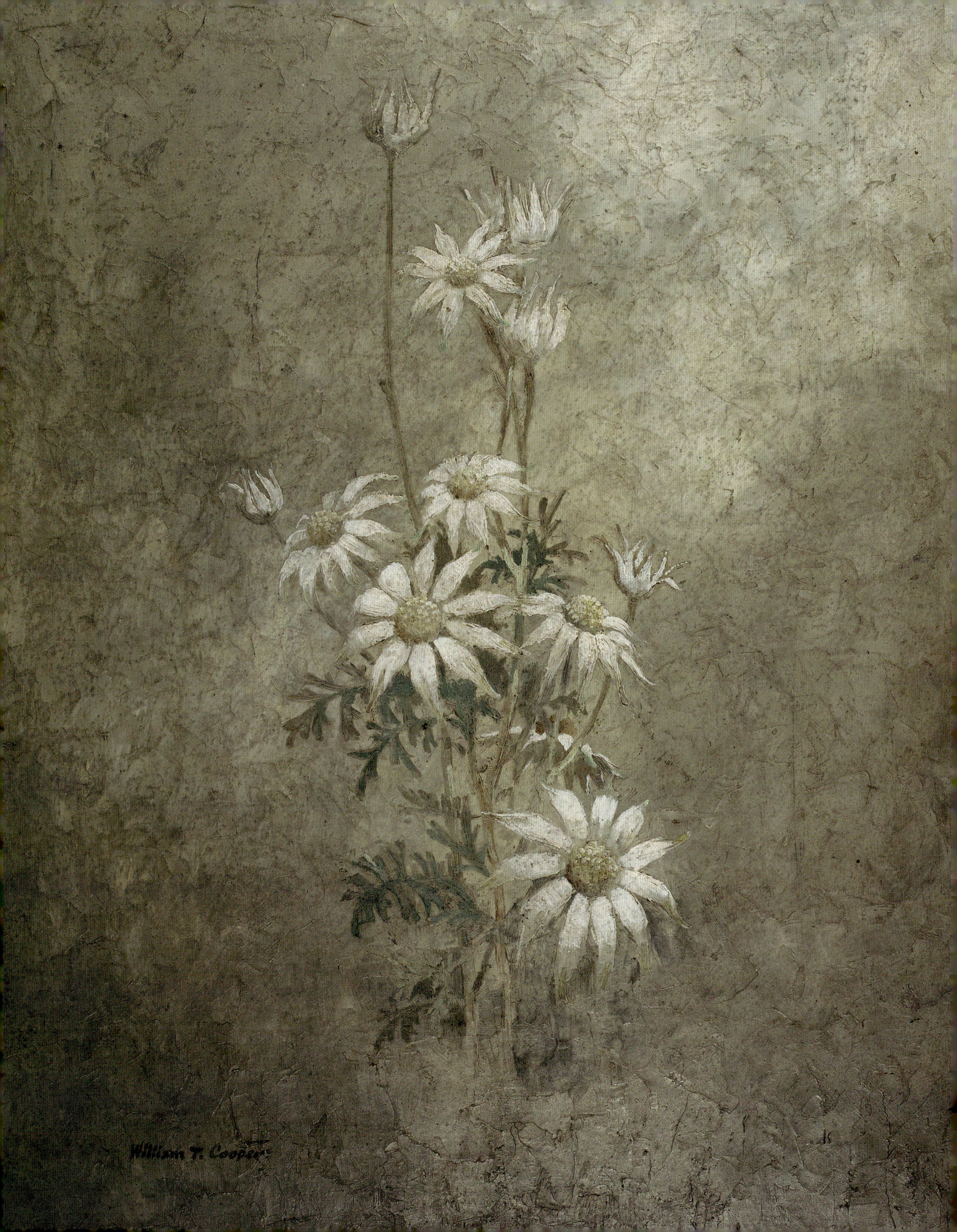
William T. Cooper

Banksia

PROTEACEAE

Banksia is named in honour of botanist Joseph Banks (1743–1820), who sailed with James Cook on the *Endeavour*.

There are 76 species of *Banksia* and they are almost solely Australian. All species occur here, with one (*Banksia dentata*) also occurring in New Guinea.

→ 2000; pen, pencil and watercolour; 297 x 410mm
National Library of Australia, 8068755

↓ 2000; pen, pencil and watercolour; 297 x 410mm
National Library of Australia, 8068757

Banksia integrifolia

Coast Banksia

integrifolia is from *integer* (entire) and *folium* (leaf), referring to the untoothed leaves on this species, an unusual feature for a *Banksia*.

Coast Banksia is a tree that grows up to 25 metres, occurring on sand dunes and beachfronts from southern Victoria to south-east Queensland. *Banksia integrifolia* was the first *Banksia* specimen to be collected by Joseph Banks at Botany Bay in 1770.

The seeds are taken by Yellow-tailed Black-Cockatoos, a bird that Bill painted many times. It was important for him to have drawings of their various food trees to avoid repetition in his paintings.

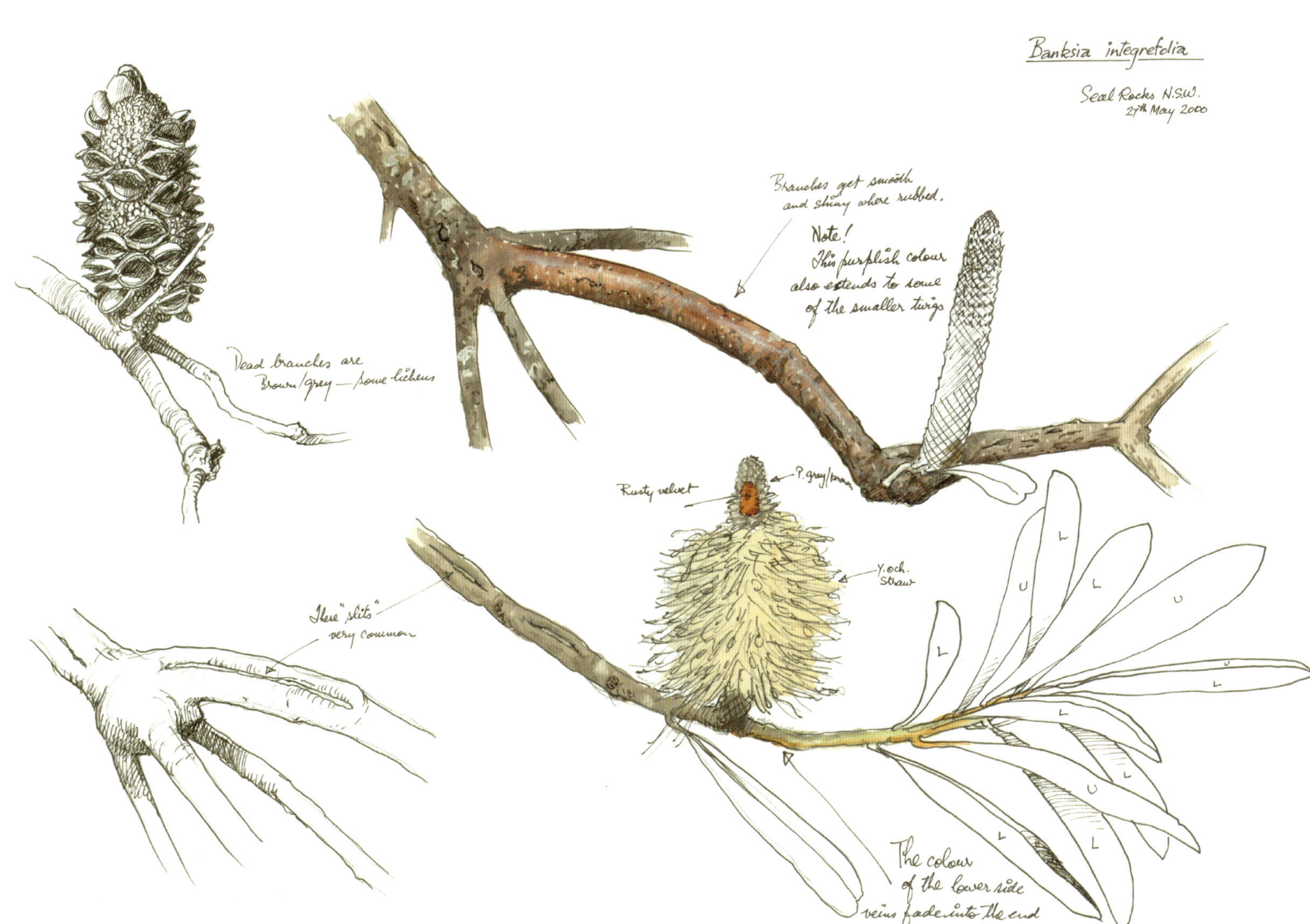

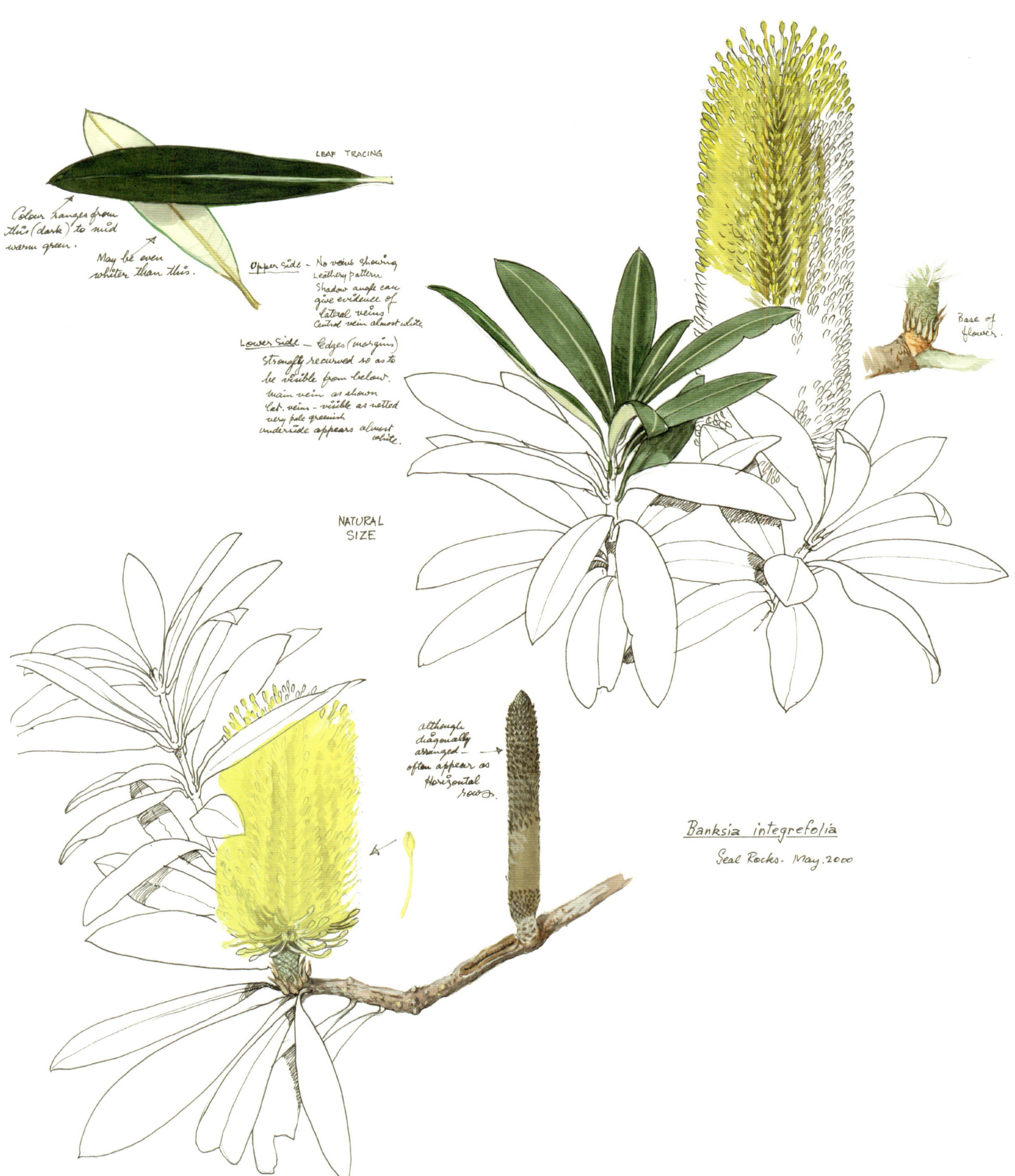
LEAF TRACING
Colour ranges from this (dark) to mid warm green.
May be even whiter than this.
Upper side - No veins showing
Leathery pattern
Shadow angle can give evidence of lateral veins
Central vein almost white.
Lower side — Edges (margins) strongly recurved so as to be visible from below.
Main vein as shown
Lat. veins - visible as netted very pale greenish
underside appears almost white.
Base of flower.
NATURAL SIZE
although diagonally arranged — often appear as Horizontal rows.
Banksia integrefolia
Seal Rocks. May. 2000

Banksia menziesii

Menzies' Banksia or Firewood Banksia

menziesii is named in honour of Archibald Menzies (1754–1842), botanist, surgeon and naturalist on HMS *Discovery*.

Menzies' Banksia is a shrubby tree that grows to 10 metres, occurring mostly in coastal areas of open heath, or sand plains on deep sandy soil north and south of Perth, from the Murchison River to the Carbunup River.

We travelled to south-west Western Australia to see cockatoos and collect plant and habitat material for *Cockatoos: A Portfolio of All Species*. On our first afternoon at Mogumber, we were excited to see Carnaby's Black-Cockatoos actively feeding on this *Banksia*. When the birds departed, Bill pruned a few pieces to take away for drawing and painting.

→ undated (c.1999); pen, pencil and watercolour; 297 x 410mm
National Library of Australia, 8068763

↓ 1999; pen and pencil; 220 x 220mm
State Library of New South Wales

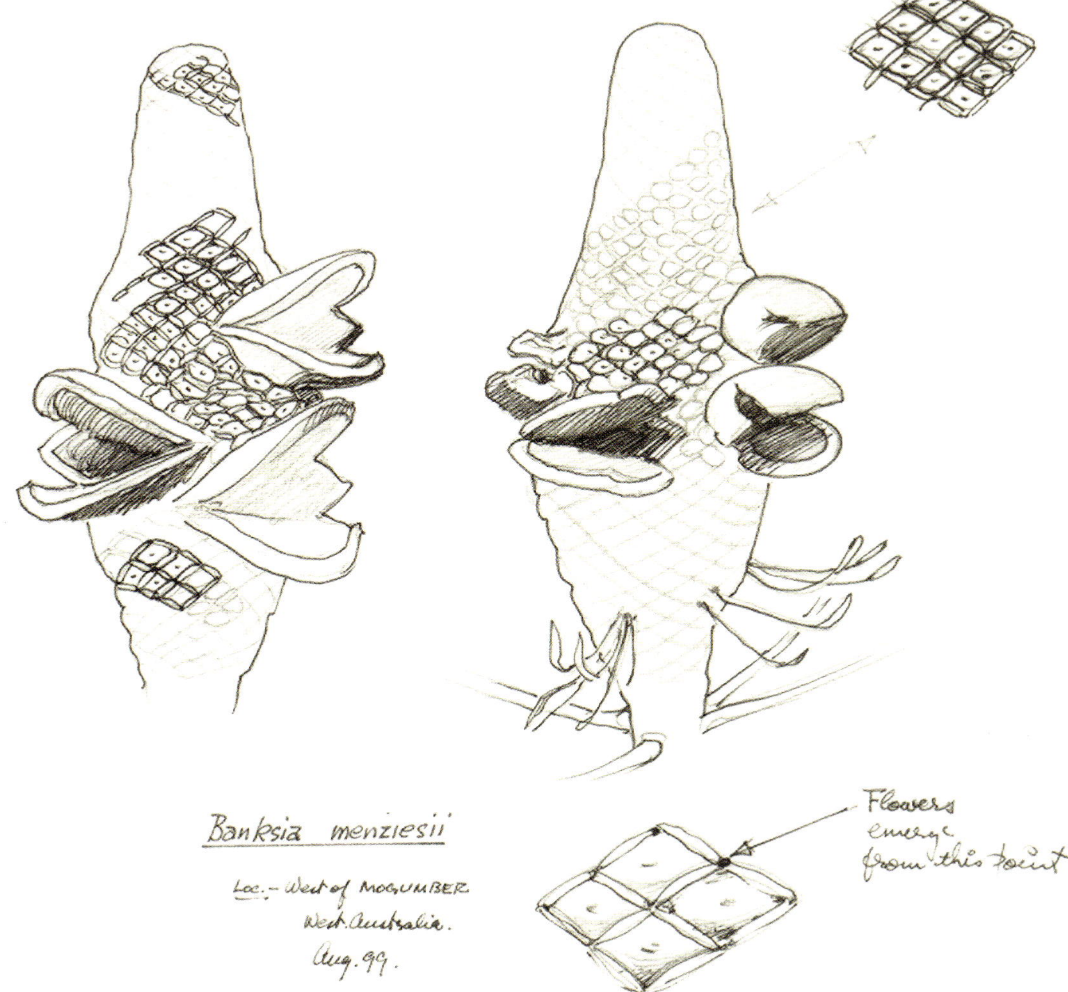

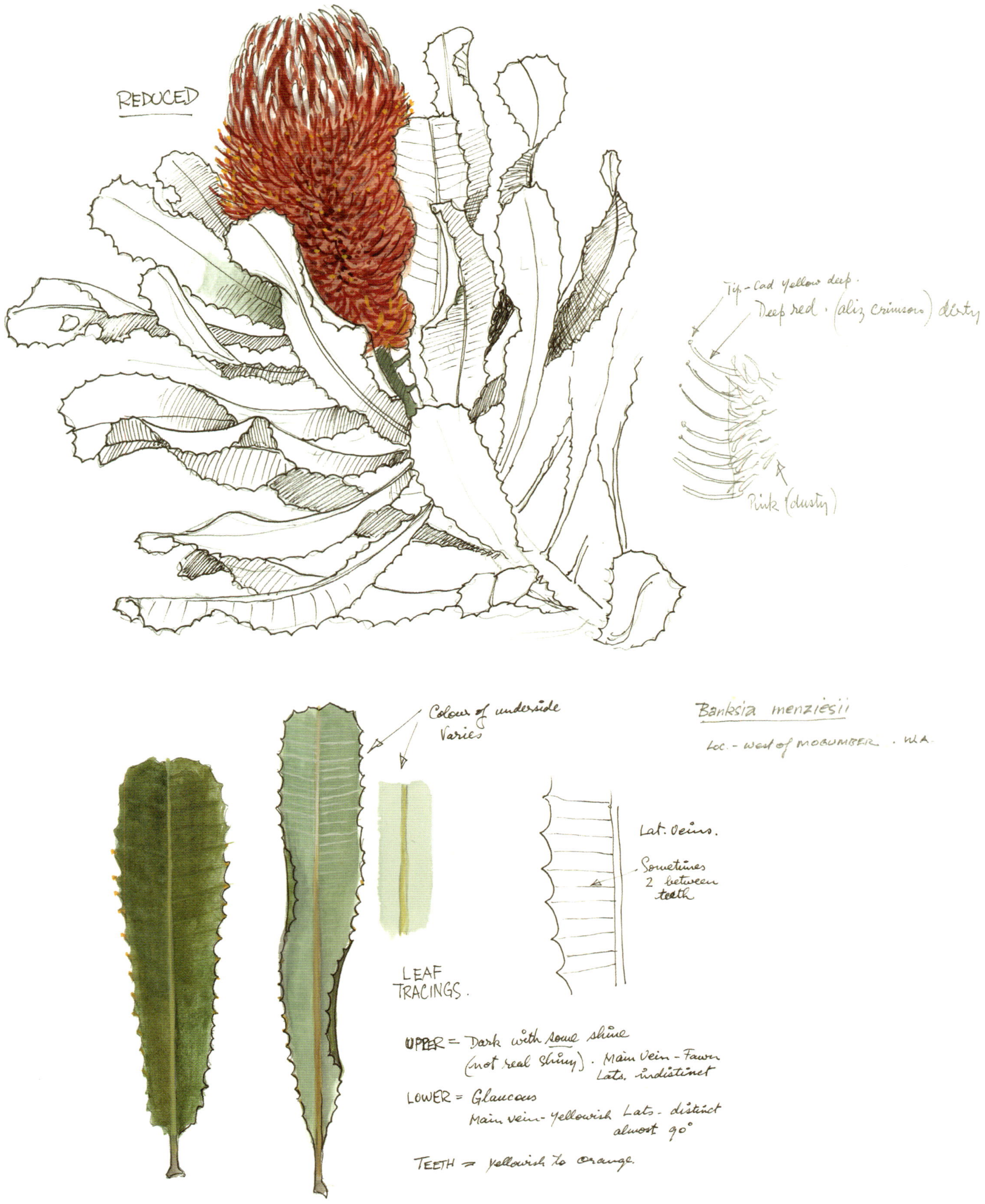

REDUCED

Tip - Cad yellow deep.

Deep red. (aliz crimson) dirty

Pink (dusty)

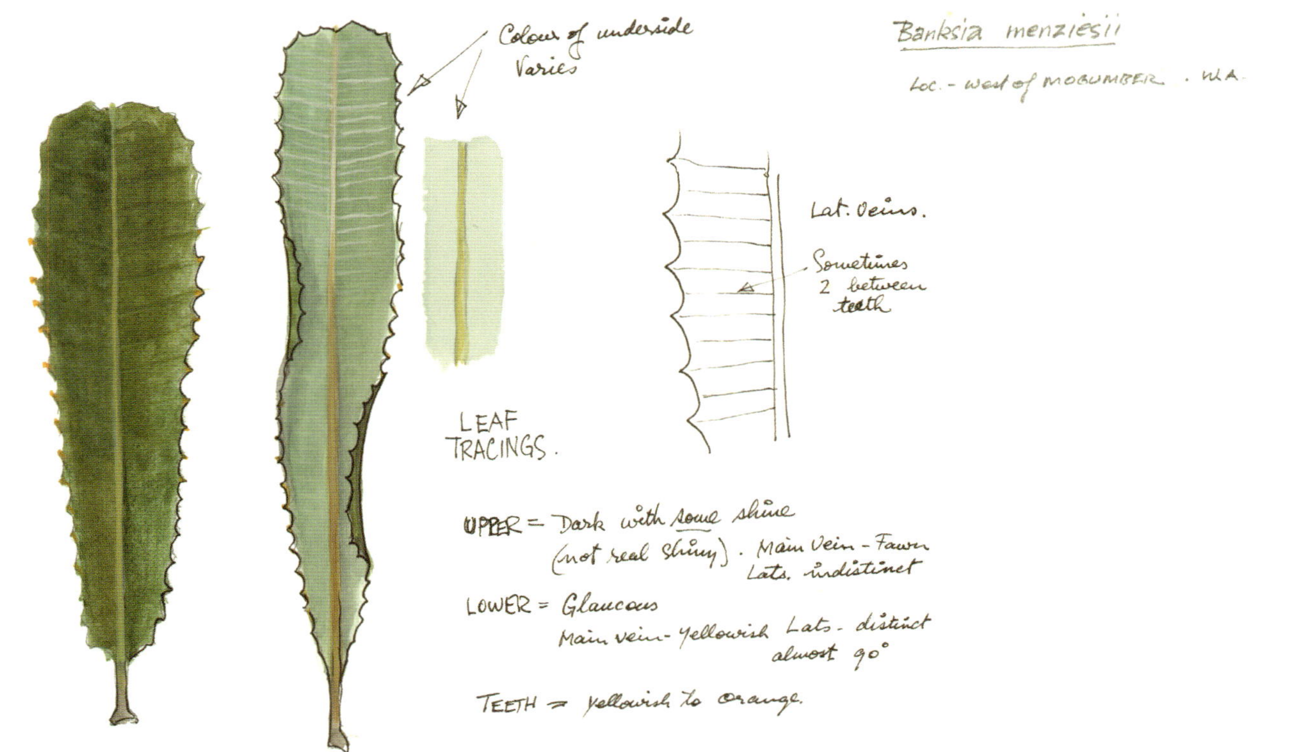

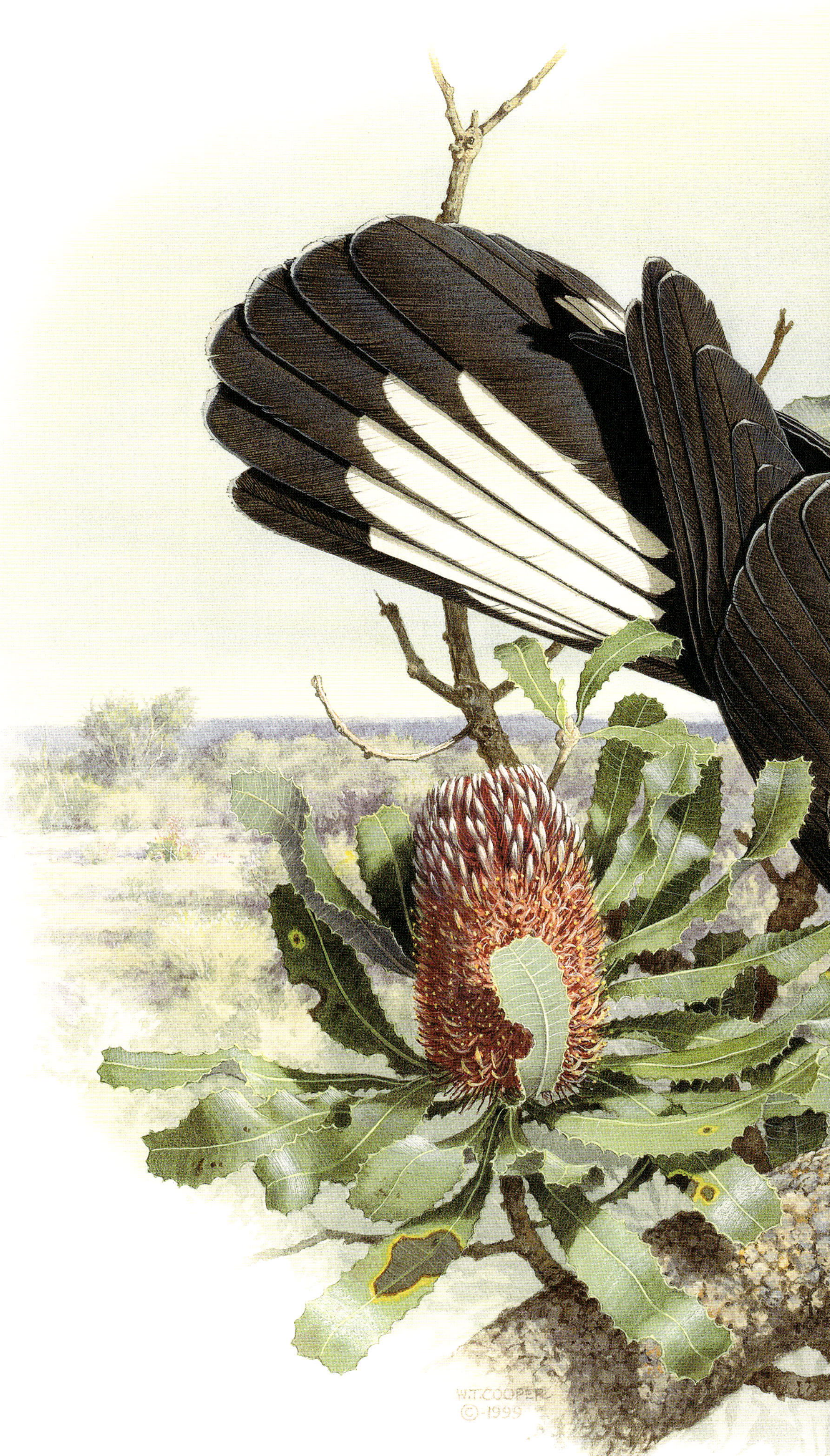

→ Carnaby's Black-Cockatoos
1999; watercolour
c.550 x 700mm
Private collection

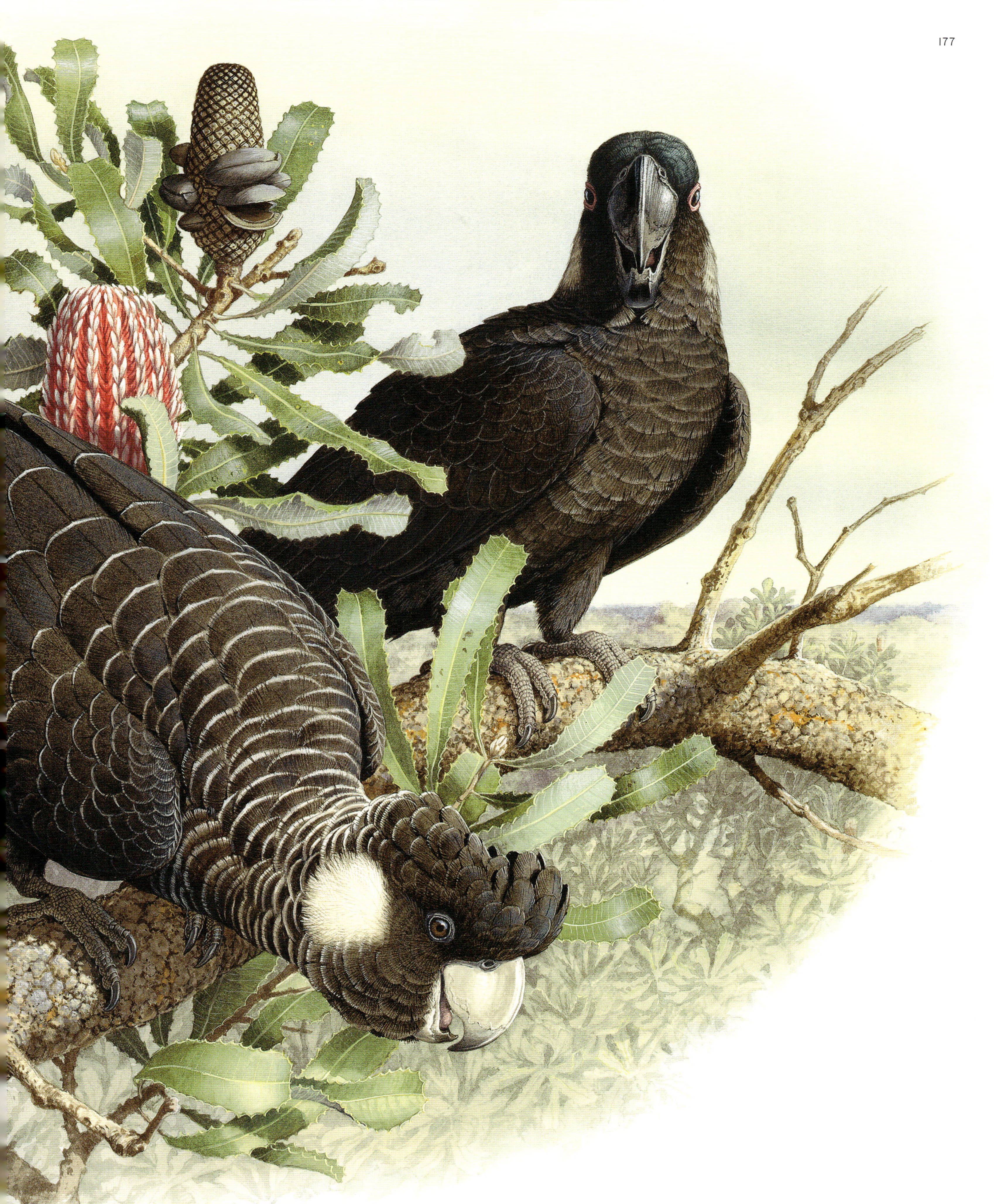

Banksia serrata

Old Man Banksia

serrata is from *serratus* (serrated), referring to the toothed leaves.

Old Man Banksias, with their warty bark and gnarled branches, are characteristic trees of coastal forest, from Tasmania and southern Victoria to southern Queensland. In the sand dune forest that surrounded our first home at Bungwahl, these banksias were interspersed with *Angophora* trees, *Xanthorrhoea* and numerous wildflowers.

The flowers are full of nectar, attracting many honeyeaters and noisy, cackling wattlebirds. We delighted in watching Yellow-tailed Black-Cockatoos delicately extracting the slim seeds from the cones. Bill noted one experience in his diary:

> *12th Dec. 1977. Manooka, Scarlet Honeyeaters everywhere. As usual there seems to be territory butting territory. Birds (males only) calling from top branches of their particular trees. Watched one this morning (male) feeding on a Banksia serrata cone, hanging upside down and probing into flowers. A beautiful sight.*

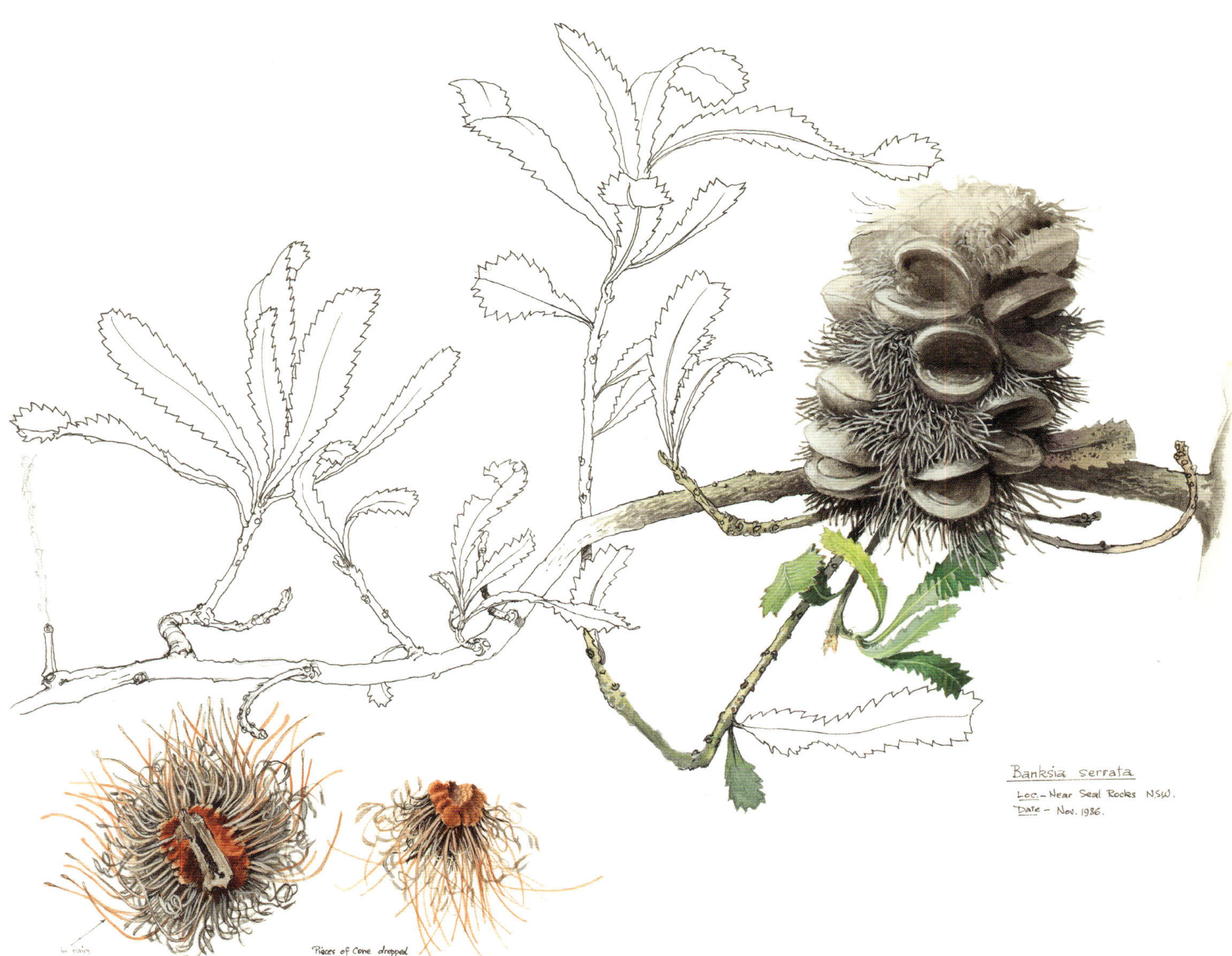

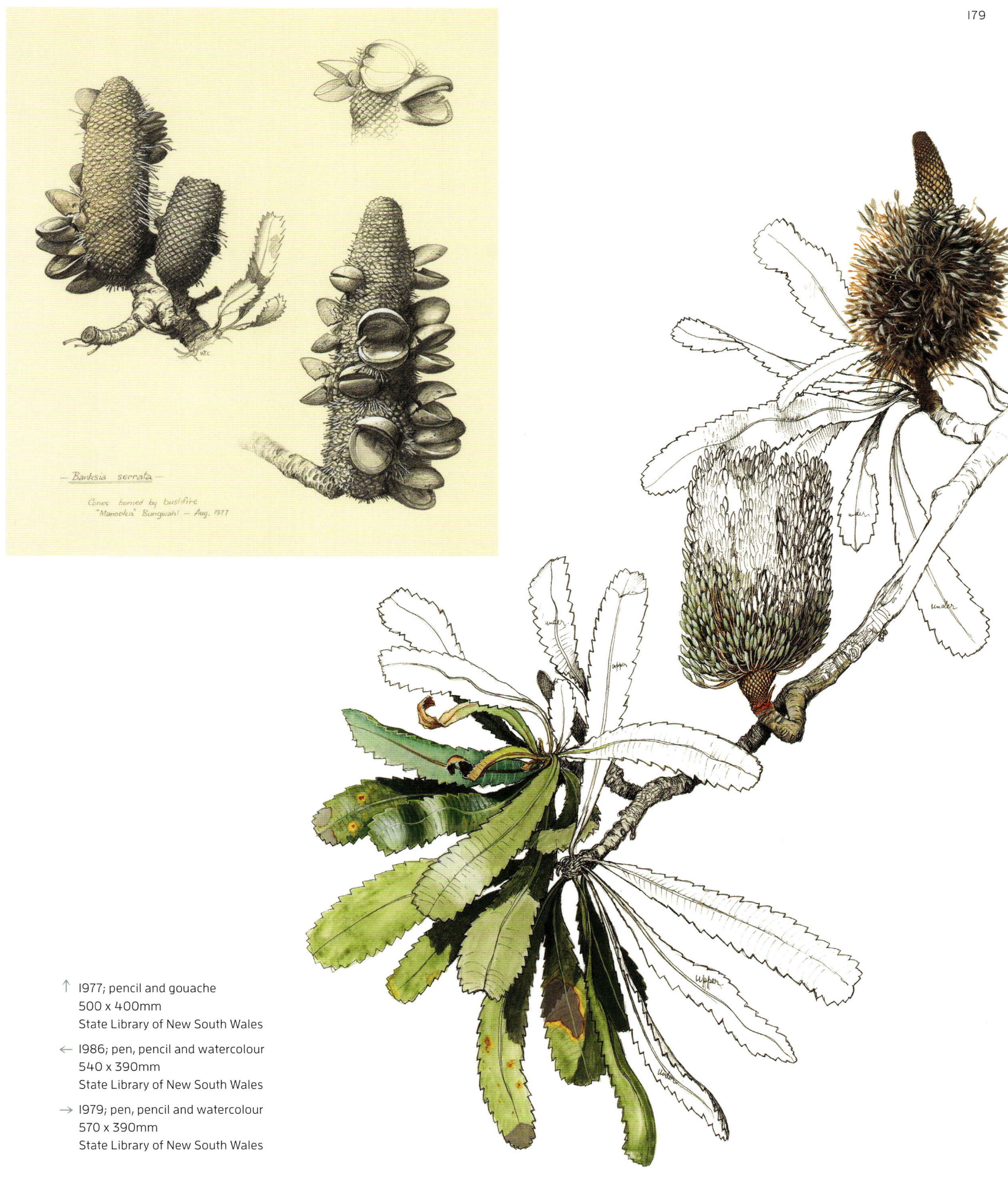

↑ 1977; pencil and gouache
500 x 400mm
State Library of New South Wales

← 1986; pen, pencil and watercolour
540 x 390mm
State Library of New South Wales

→ 1979; pen, pencil and watercolour
570 x 390mm
State Library of New South Wales

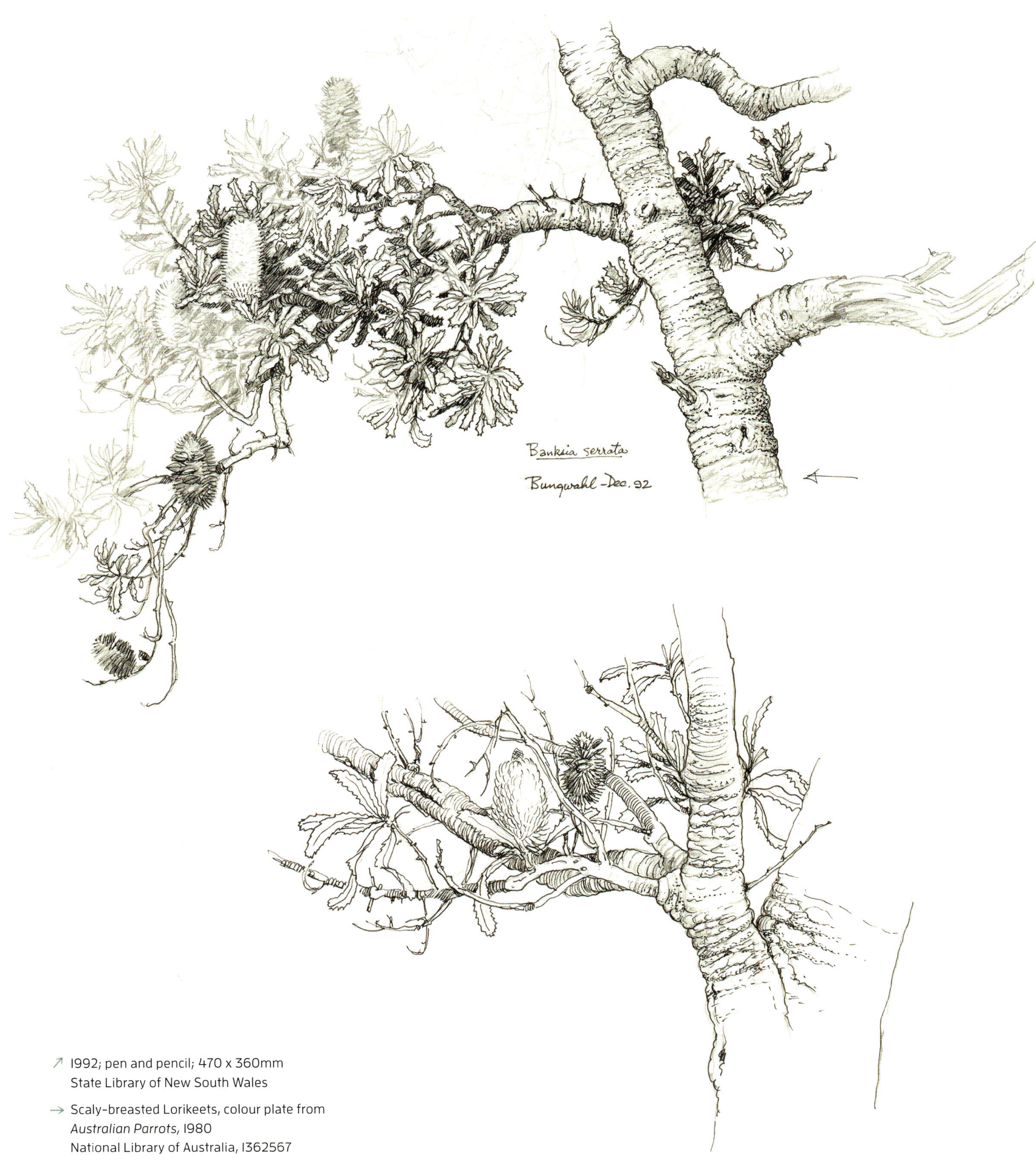

↗ 1992; pen and pencil; 470 x 360mm
State Library of New South Wales

→ Scaly-breasted Lorikeets, colour plate from *Australian Parrots*, 1980
National Library of Australia, 1362567

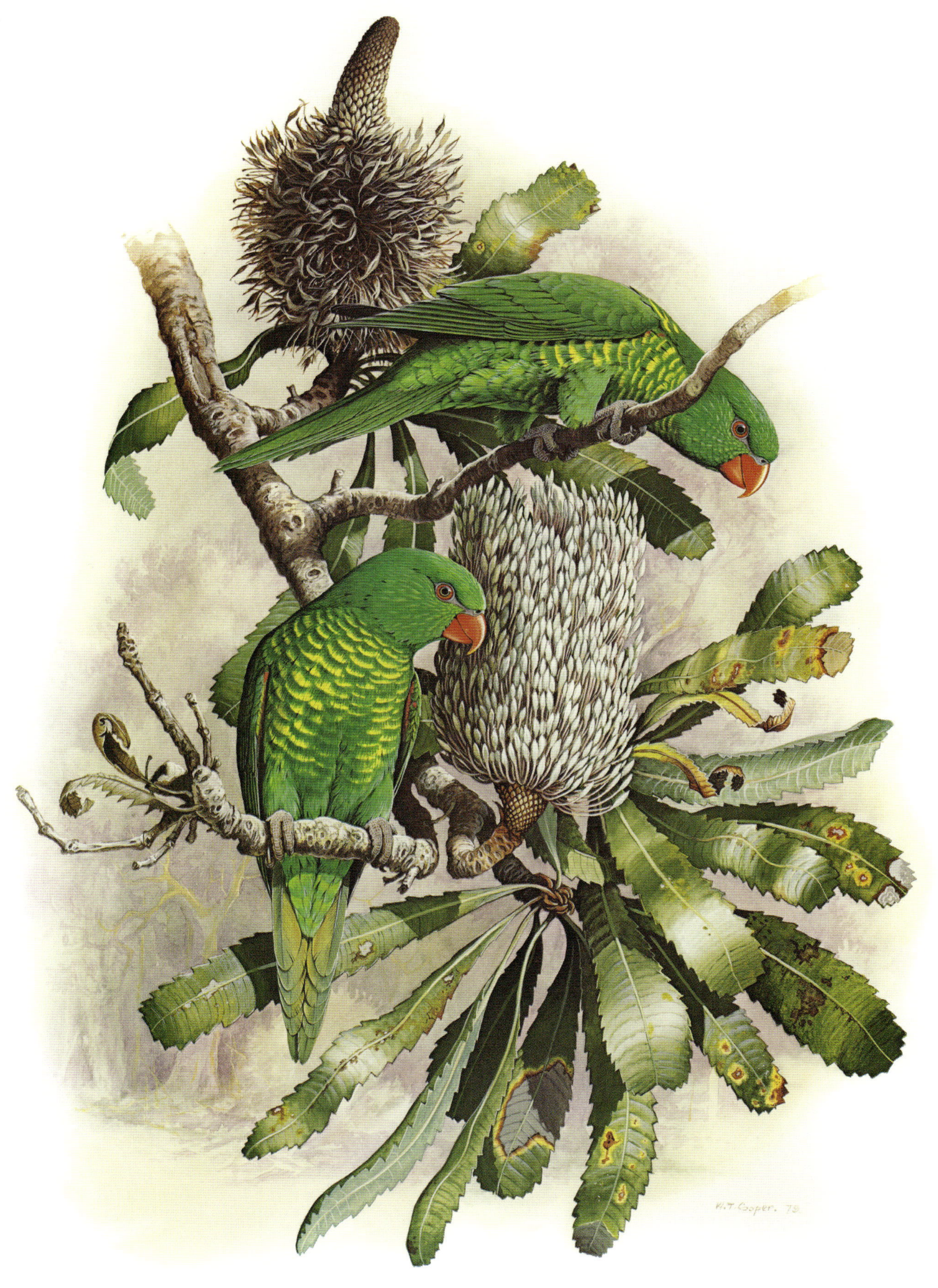
W.T. Cooper. 78

Callitris glaucophylla

White Cypress

CUPRESSACEAE

→ 1979; pen, pencil and watercolour
480 x 460mm
State Library of New South Wales

↓ 1979; pen and pencil; 270 x 320mm
State Library of New South Wales

Callitris is from *callos* (beauty) and *tris* (three), referring to the leaves, which are usually in whorls of three.

glaucophylla is from *glaucus* (blue-grey) and *-phyllus* (-leaved).

White Cypress is a tree that grows to 20 metres, with fine glaucous foliage like pine needles. It is very widely distributed in all states except Tasmania, occurring in a range of habitats, including open forest, woodland and dry rainforest.

We were on a camping trip in search of Major Mitchell Cockatoos in the mallee at Round Hill in western New South Wales. Bill was particularly struck by the beauty and soft colours of the mallee country, as he wrote in his diary:

> *I was surprised by the amount of lichens in this dry country. The Cypress trees have beautiful orange lichens and the ordinary greens too. Rocks are covered with them and dead branches have lots of patches of it.*

These colours were captured in the soft pink and white of the Major Mitchells against the equally soft bluish foliage of the White Cypress trees, used in the plate for *Australian Parrots*.

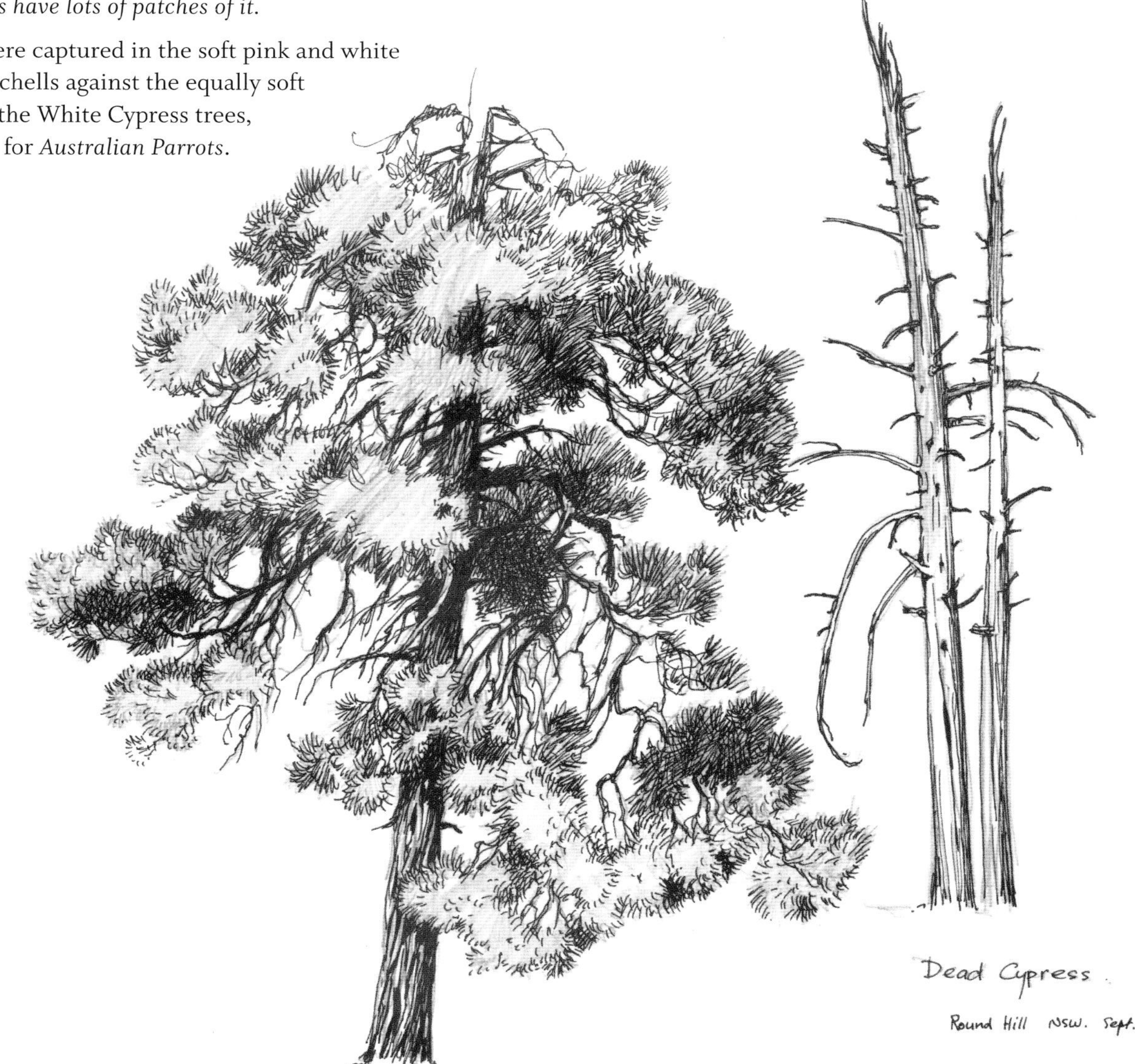

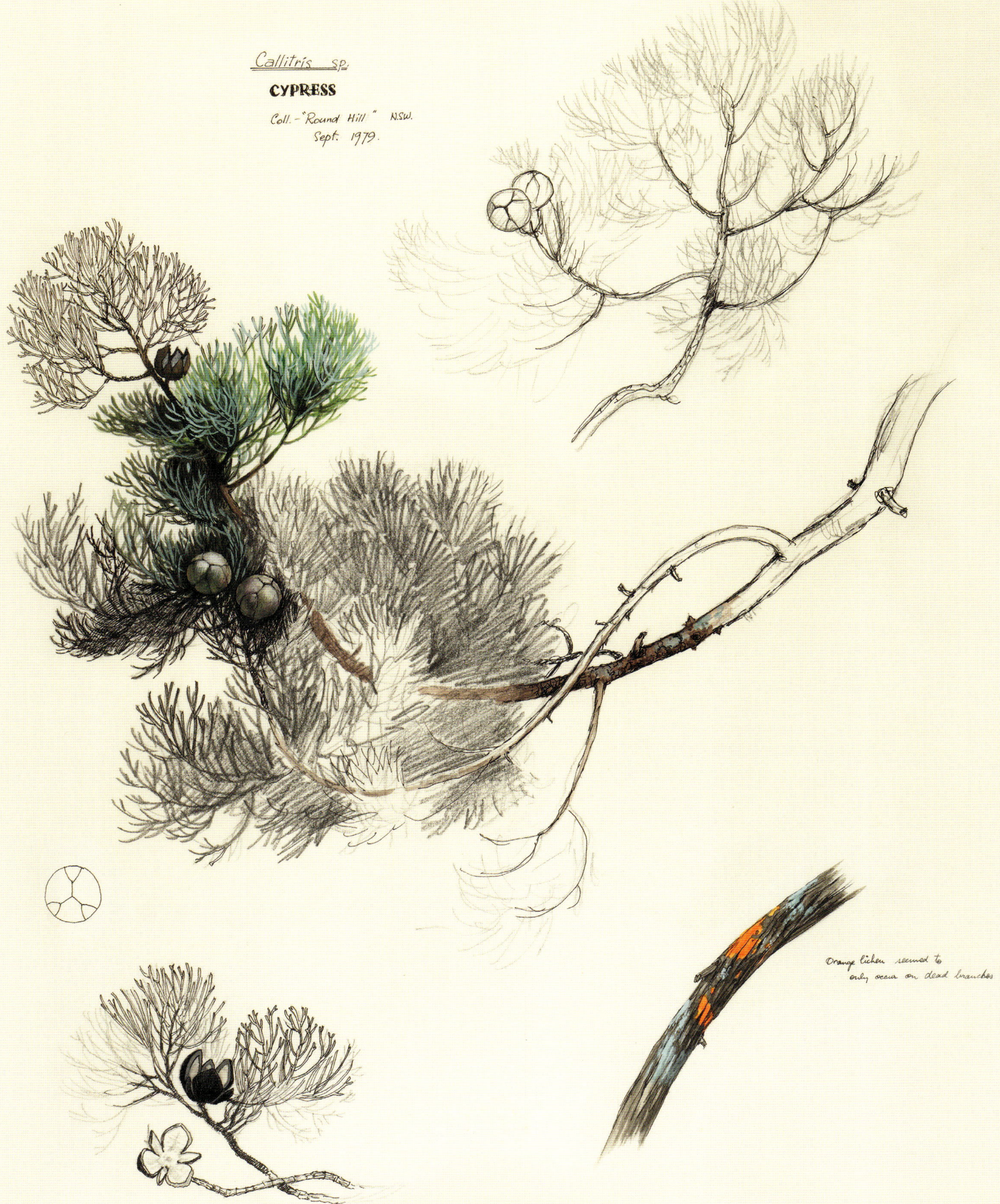

Callitris sp.
CYPRESS
Coll. - "Round Hill" N.S.W.
Sept. 1979.
Orange lichen seemed to
only occur on dead branches

Corymbia

MYRTACEAE

Corymbia is from corymb, a flat or round-topped inflorescence with lower pedicels longer than those in the upper section.

Corymbia are the bloodwoods, ghost gums and spotted gums that were once part of the genus *Eucalyptus*.

Corymbia calophylla

Marri

calophylla is from *calo-* (beautiful) and *-phyllus* (-leaved).

Marri is a bloodwood that grows to 40 metres and very occasionally as high as 60 metres, in the forest of south-west Western Australia. The large gumnuts of Marri trees produce seeds that are relished by Red-capped Parrots.

This reference sketch was used for the Red-capped Parrot in *Australian Parrots* and appeared on the cover of one of the editions.

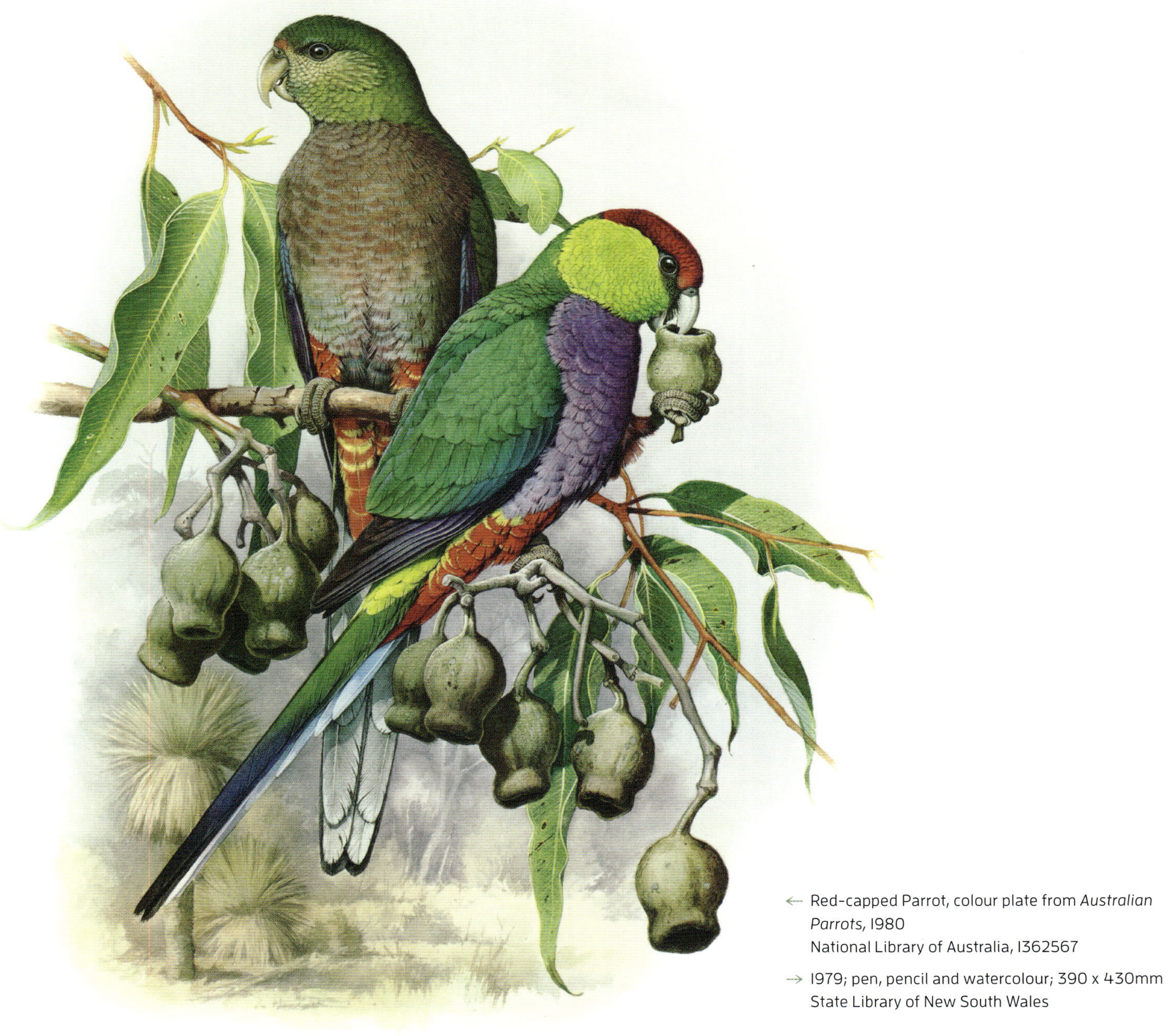

← Red-capped Parrot, colour plate from *Australian Parrots*, 1980
National Library of Australia, 1362567

→ 1979; pen, pencil and watercolour; 390 x 430mm
State Library of New South Wales

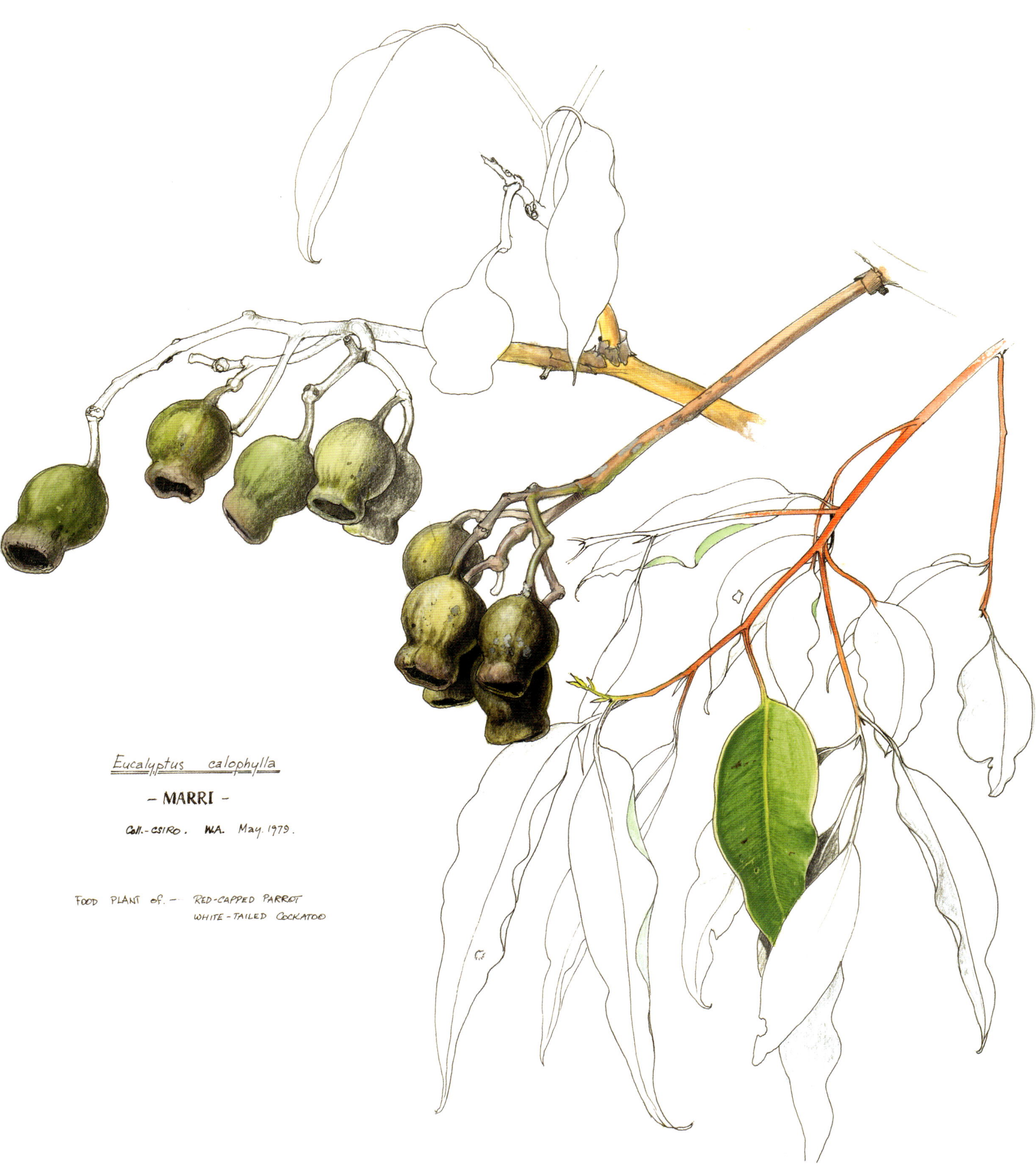
Eucalyptus calophylla
- MARRI -
Coll.- CSIRO. W.A. May. 1979.
FOOD PLANT of.- RED-CAPPED PARROT
WHITE-TAILED COCKATOO

Corymbia confertiflora

Broad-leaved Carbeen or Cabbage Gum

confertiflora is from *confertus* (crowded or densely) and *-flora* (flowered).

Broad-leaved Carbeen trees are unusual in being deciduous gumtrees. The sandpapery leaves fall late in the tropical dry season. The trees are usually leafless for a few weeks before new growth emerges.

This sketch was done for a painting of Red-tailed Black-Cockatoos, which inhabit the area where Broad-leaved Carbeen grows.

Most *Eucalyptus* or *Corymbia* trees have narrow sickle-shaped leaves. The wide leaves with short stalks of this species held Bill's interest as they provided an unusual appearance for a painting with a savannah background where gums dominate.

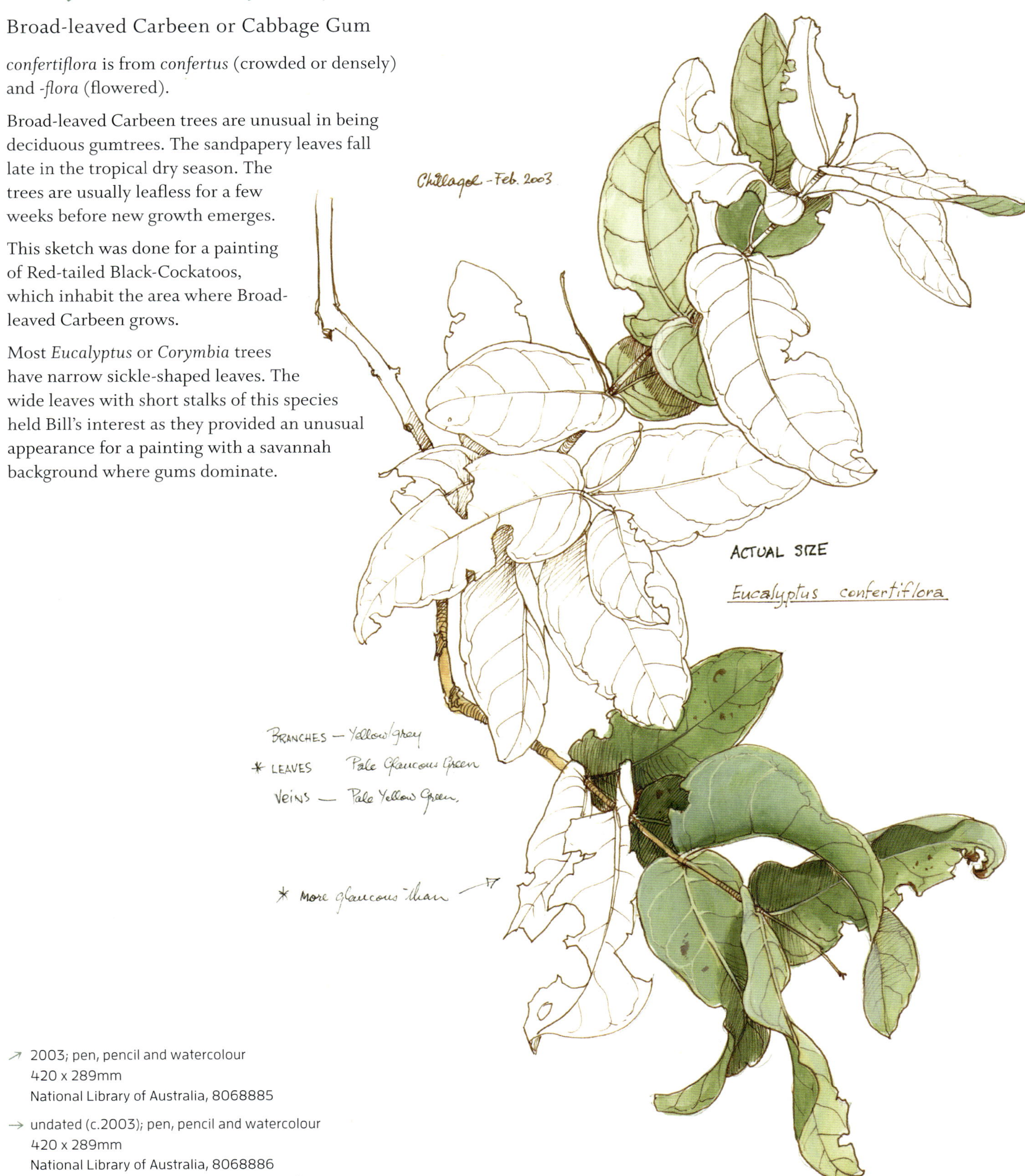

↗ 2003; pen, pencil and watercolour
420 x 289mm
National Library of Australia, 8068885

→ undated (c.2003); pen, pencil and watercolour
420 x 289mm
National Library of Australia, 8068886

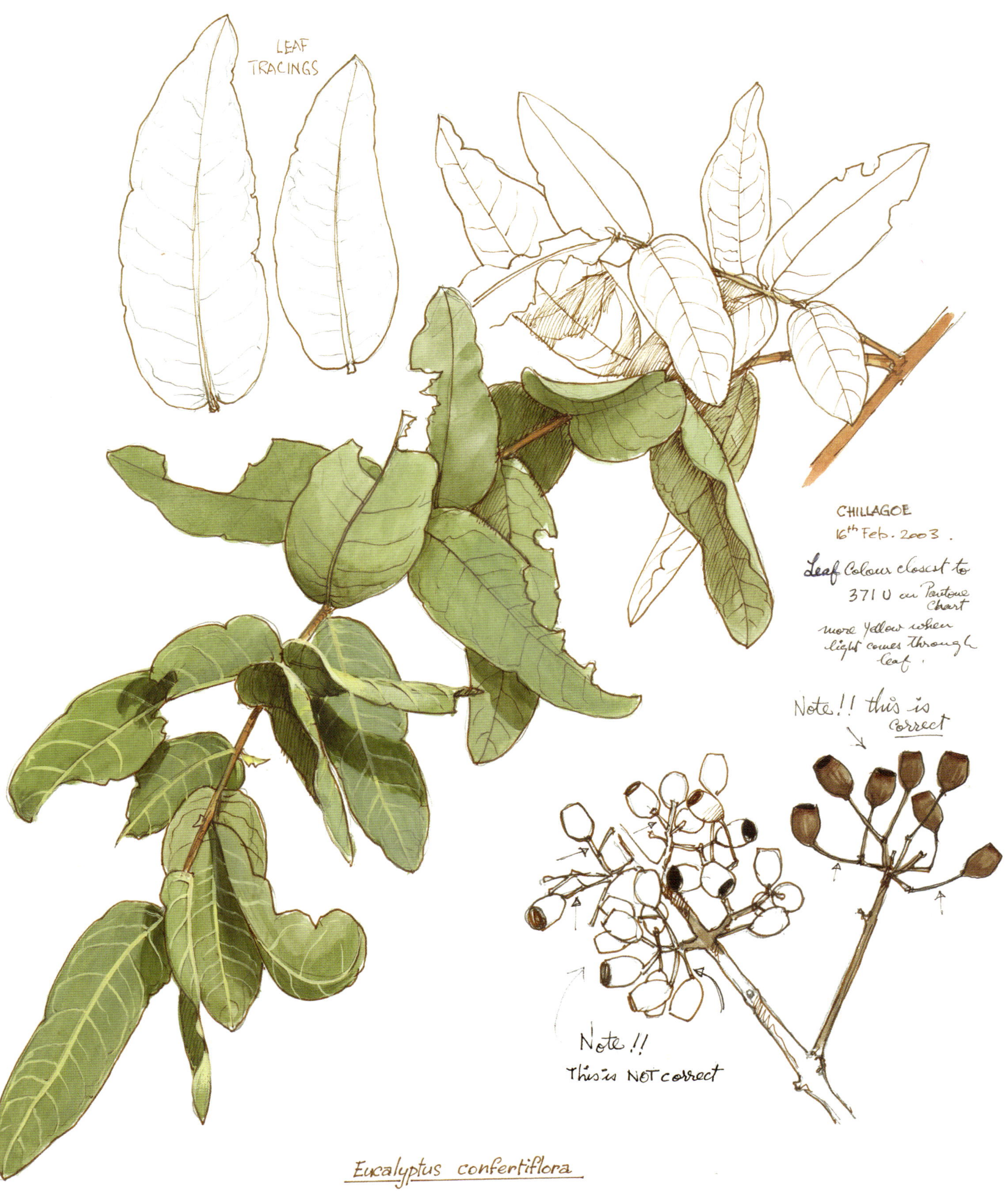
LEAF
TRACINGS
CHILLAGOE
16th Feb. 2003.
Leaf Colour closest to
371 U on Pantone
Chart
more Yellow when
light comes through
leaf.
Note!! this is
correct
Note!!
This is NOT correct
Eucalyptus confertiflora

Corymbia intermedia

Pink Bloodwood

intermedia is from *intermedius*, referring to the leaf oils being similar to and intermediate between *C. gummifera* and *C. eximia*.

Pink Bloodwood is a tall gum tree that grows to 35 metres. The bark is rough and tessellated throughout, including the smaller branches. It occurs north from Gloucester, New South Wales, to Cooktown in Queensland.

This specimen was collected at Tumoulin and came with an assurance by local foresters that it was the closely related Red Bloodwood, *Corymbia gummifera*, from New South Wales and southern Queensland, as annotated on the sketch. However, the leaves and gumnuts of both species are indistinguishable, which is a relief, as Bill used this reference sketch in a painting of Yellow-tailed Black-Cockatoos, a southern species.

→ 1991; pen, pencil and watercolour
590 x 430mm
State Library of New South Wales

↓ Eucalyptus Forest Sketch
undated; watercolour

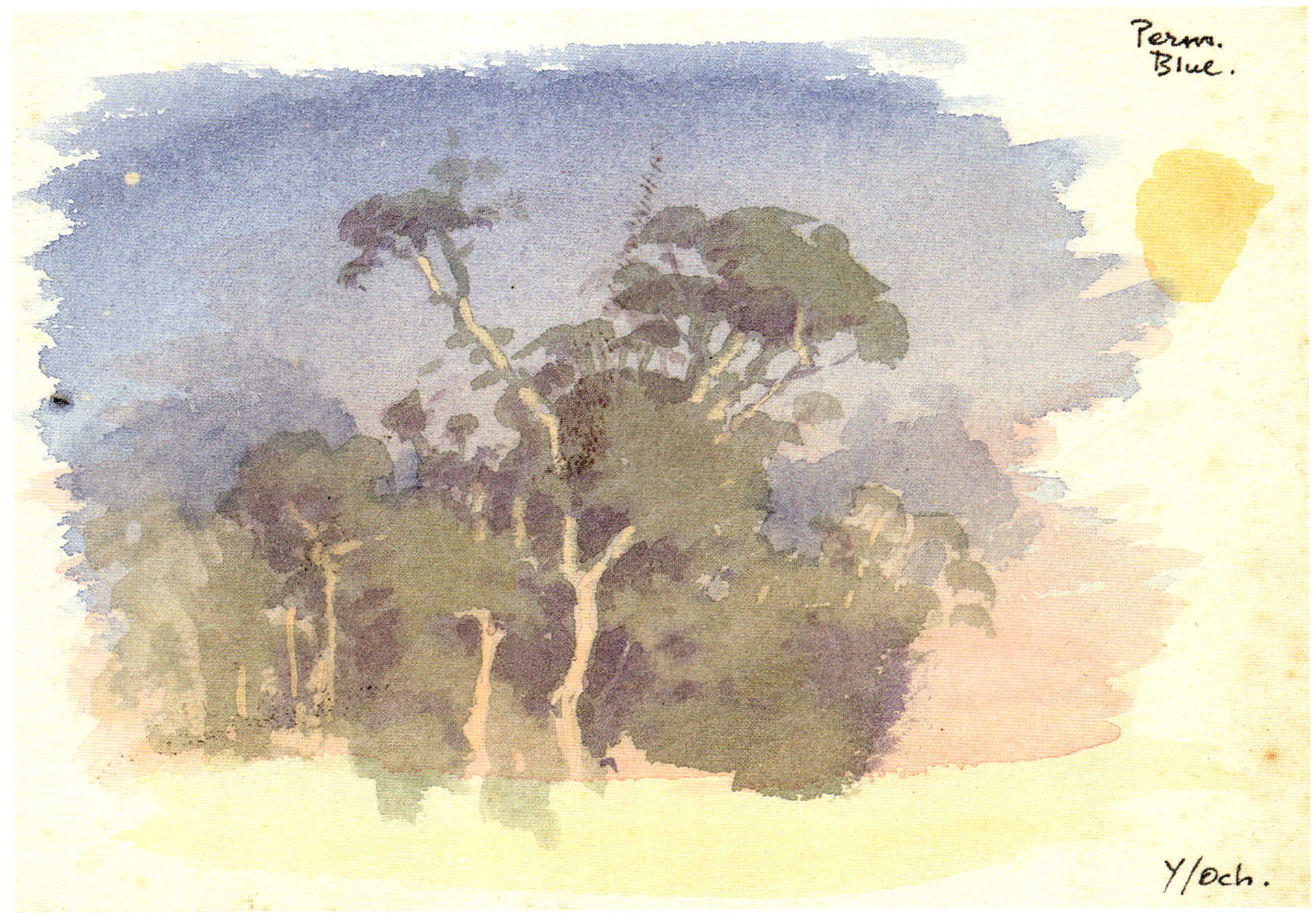

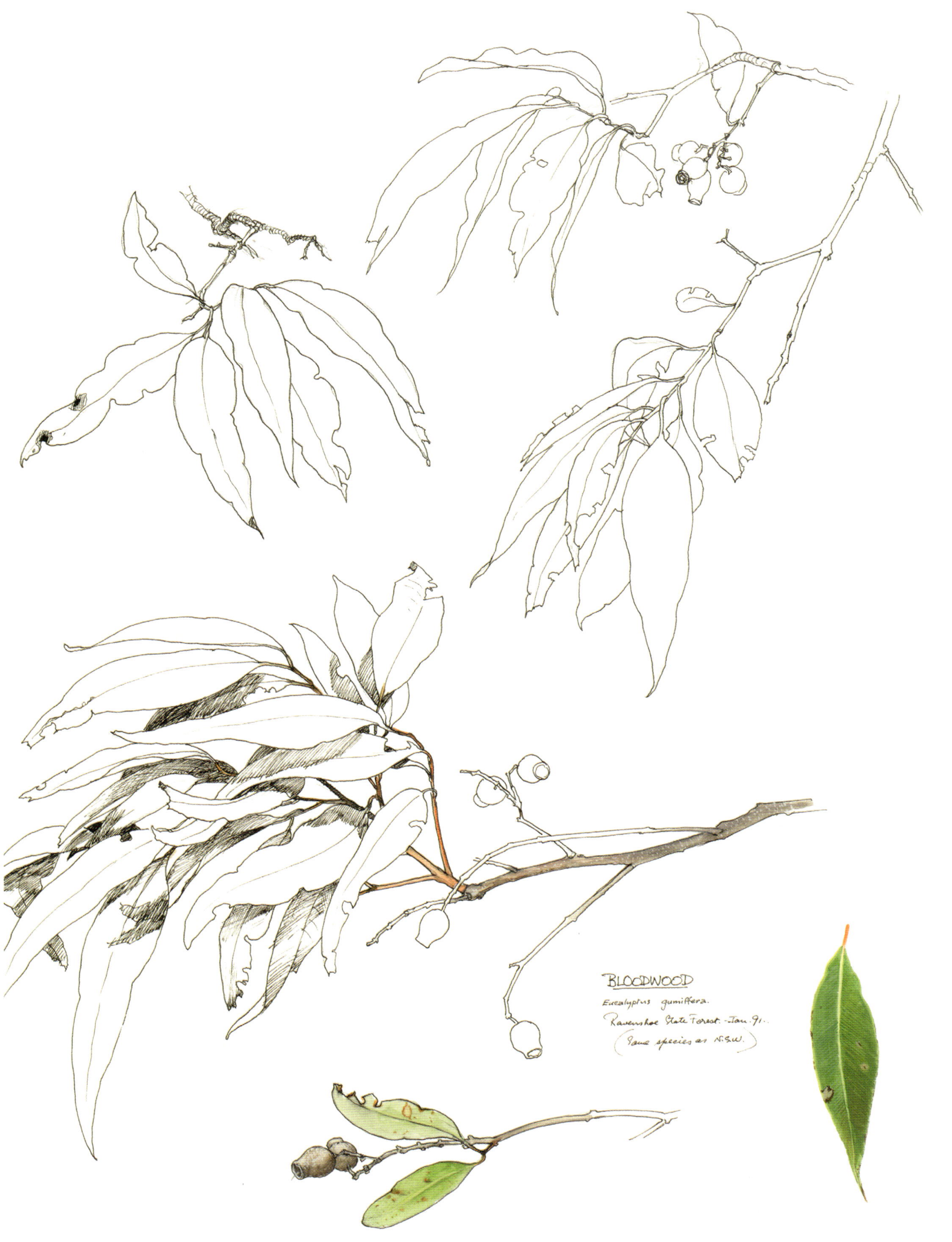
BLOODWOOD
Eucalyptus gumiffera.
Ravenshoe State Forest. - Jan. 91.
(Same species as N.S.W.)

Corymbia ptychocarpa

Swamp Bloodwood

ptychocarpa is from *ptycho* (grooves or folds) and *carpus* (fruited), referring to the strongly ribbed fruit capsules or gumnuts.

Swamp Bloodwood is a native gumtree of northern Australia, from the Kimberley in Western Australia to north-western Queensland. The beautiful flowers and attractively ribbed gumnuts have made it a popular plant in cultivation.

Drawings were made of the leaves and gumnuts from street trees near Chillagoe, west of the Atherton Tablelands in north Queensland. The Chillagoe area was always an interesting place to visit, being close to home but with a dry and starkly different habitat. The vegetation there is varied, from savannah to deciduous vine thickets and monsoon forest clustered around limestone karsts.

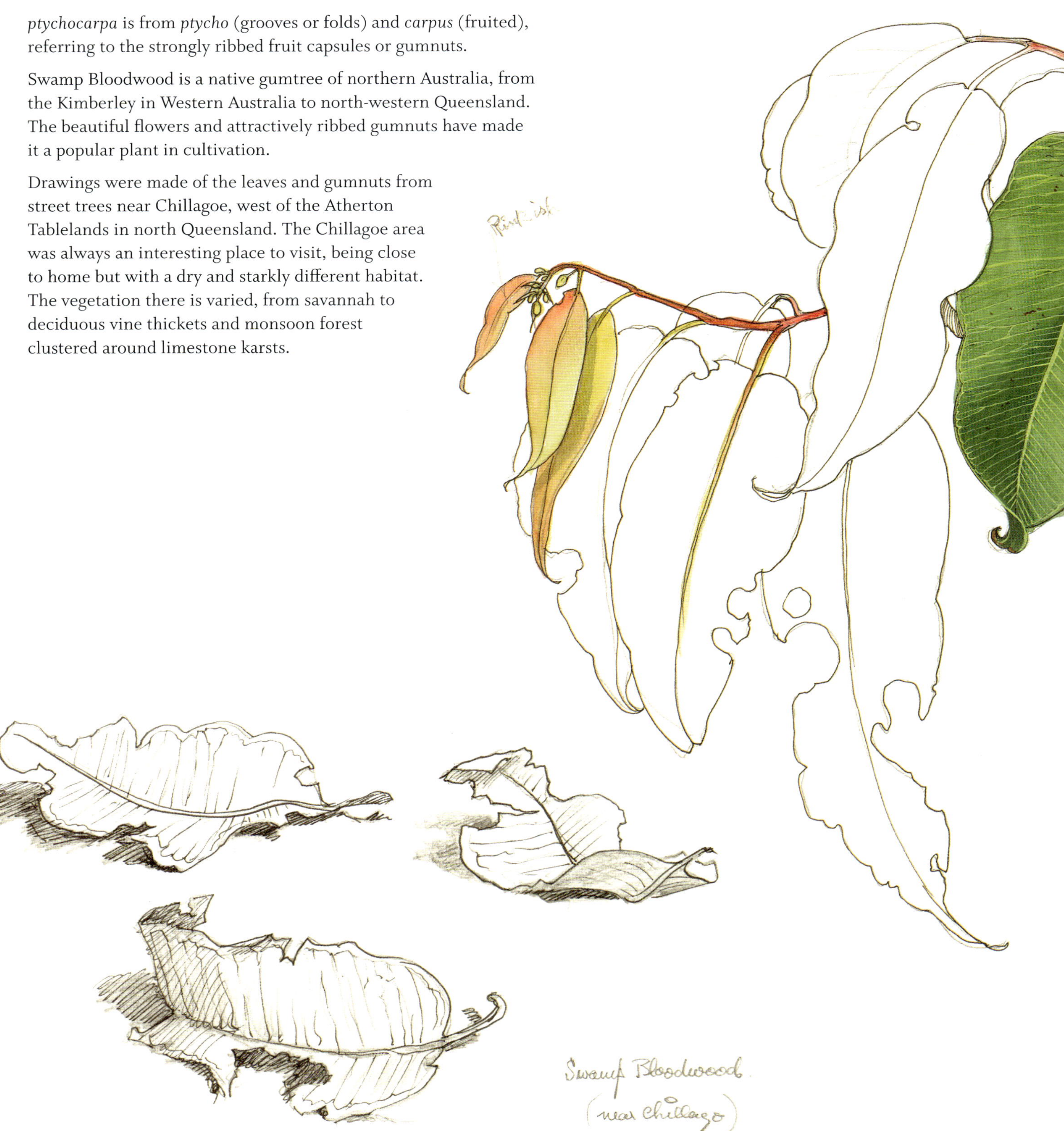

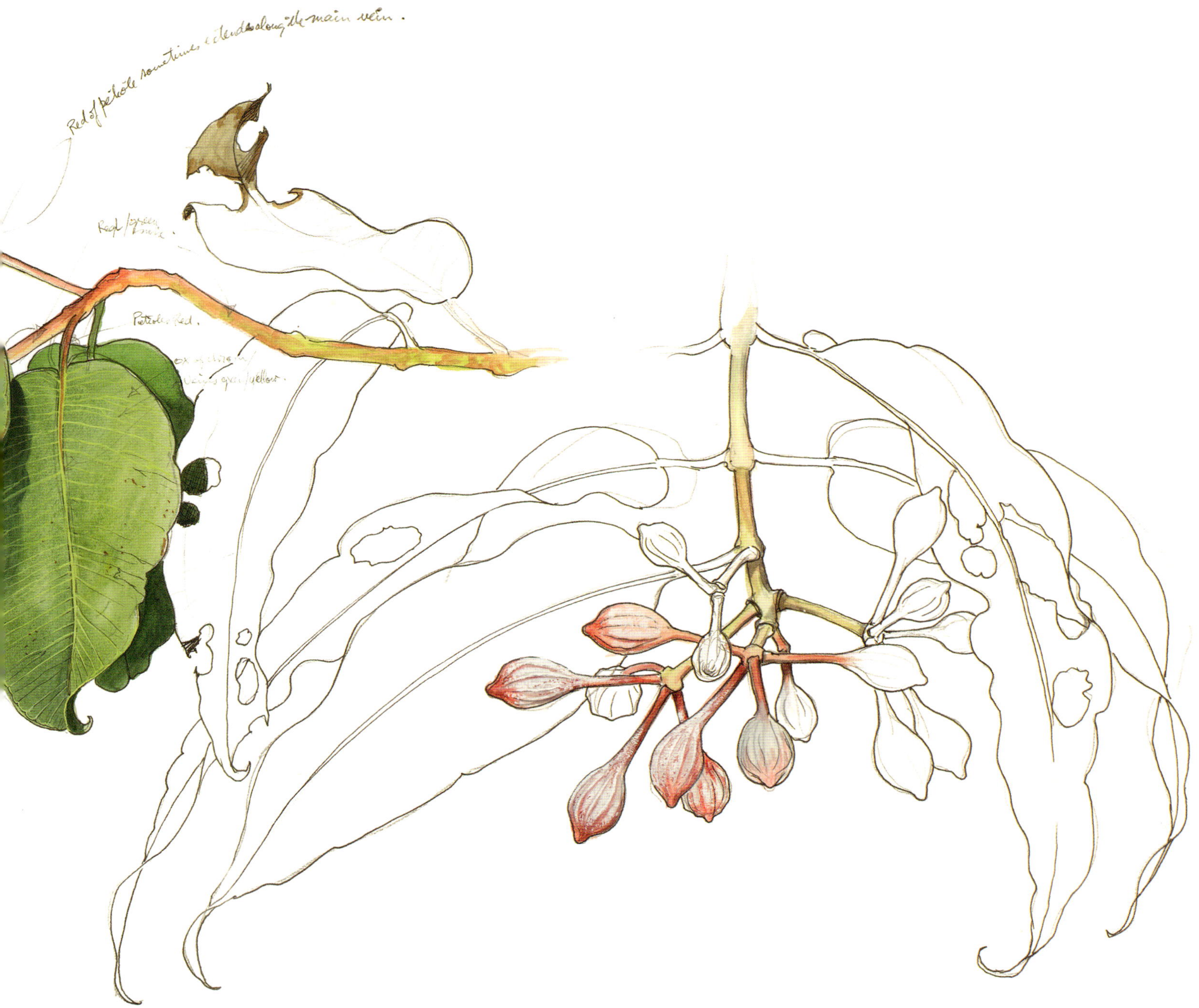

↑ undated; pen, pencil and watercolour; 460 x 410mm
State Library of New South Wales

← undated; pen and pencil; 130 x 210mm
State Library of New South Wales

Eucalyptus

MYRTACEAE

→ 1979; pen and watercolour; 490 x 350mm
State Library of New South Wales

↓ 1997; pencil; 285 x 420mm
State Library of New South Wales

Eucalyptus is from *eu-* (well) and *calyptra* (covered), referring to the calyptra or hood that covers the flower until it pops off as the flower unfurls.

Eucalyptus trees are quintessentially Australian. About 620 of the 622 species worldwide occur on this continent.

Eucalyptus camaldulensis

River Red Gum

camaldulensis is after an abandoned Italian monastery in Camalduli in Italy, where River Red Gum seeds collected from South Australia were cultivated by botanist and chief gardener Frederick Dehnhardt. It was specimens from these plants that he used for describing the species.

River Red Gums are magnificent gumtrees that commonly line rivers across Australia, mostly throughout the inland. The bark is beautifully smooth and usually two-toned in greys and browns. The large old trees contain numerous hollows, which are important as nest sites for parrots and possums.

The pencil drawings were done on one of our many trips to Noonbah Station in south-west Queensland. With its arid landscape, it was a favourite place to visit because of its extreme contrast with the wet tropical rainforest of home in Topaz.

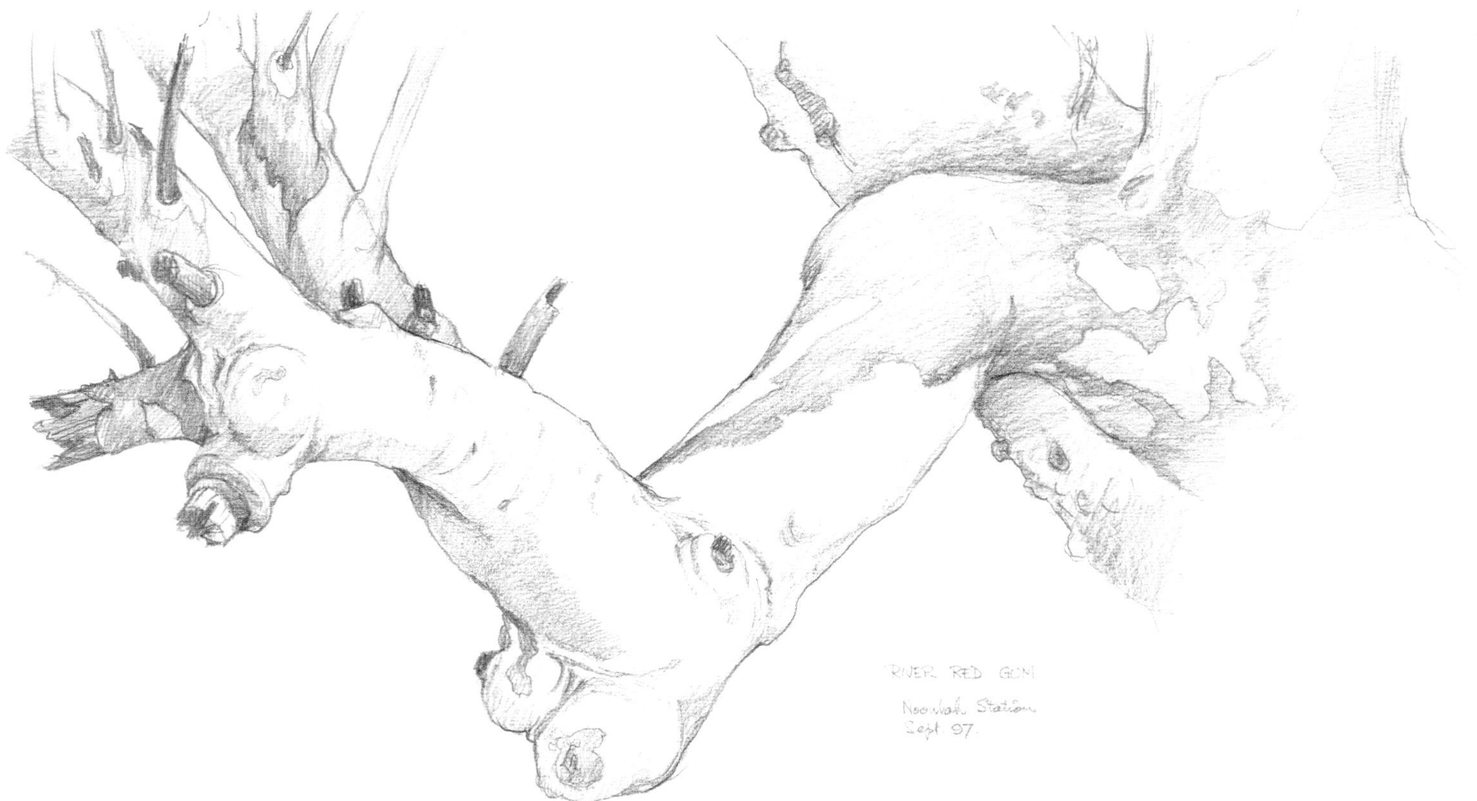

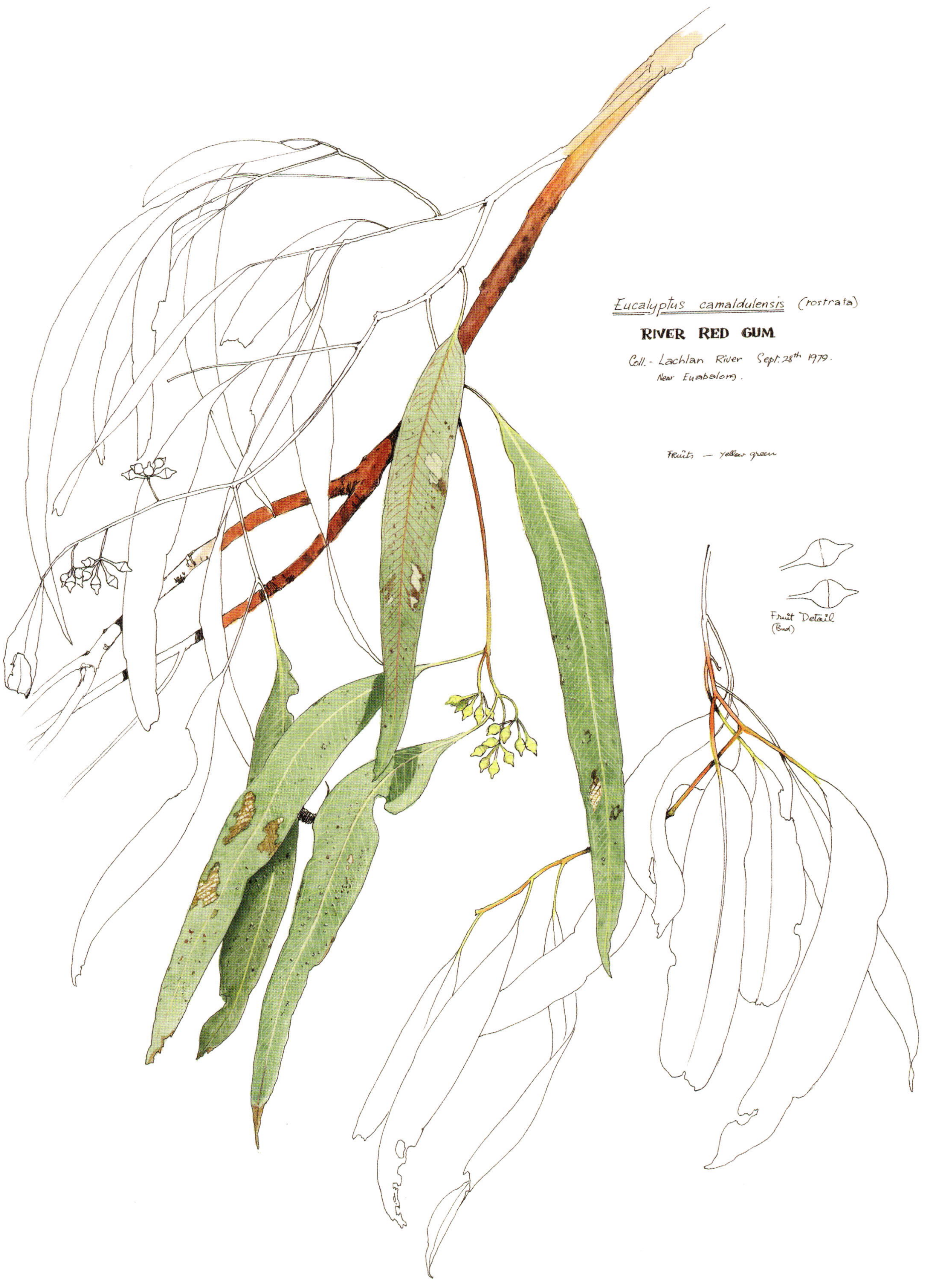
Eucalyptus camaldulensis (rostrata)
RIVER RED GUM
Coll. - Lachlan River Sept. 28th 1979.
Near Euabalong.
Fruits — yellow green
Fruit Detail
(Bud)

Eucalyptus chartaboma

Woollybutt

chartaboma is from *charte* (paper) and *boma* (base), referring to the papery bark.

Woollybutt is a tree that grows to about 18 metres. The lower part of the trunk is clothed in rough papery bark, while the upper branches are smooth and whitish. It occurs in savannah and woodland west of Cairns, north Queensland.

The vivid orange blossoms are nectar-rich, attracting numerous honeyeaters and lorikeets. Their noisy bickering as they compete for nectar makes a flowering Woollybutt obvious from a distance. When the Woollybutts were in bloom west of Atherton, we knew that we would see the gaudy spectacle of two Australian icons together: parrots and *Eucalyptus*.

→ Rainbow Lorikeets
2004; watercolour; c.290 x 420mm
Private collection

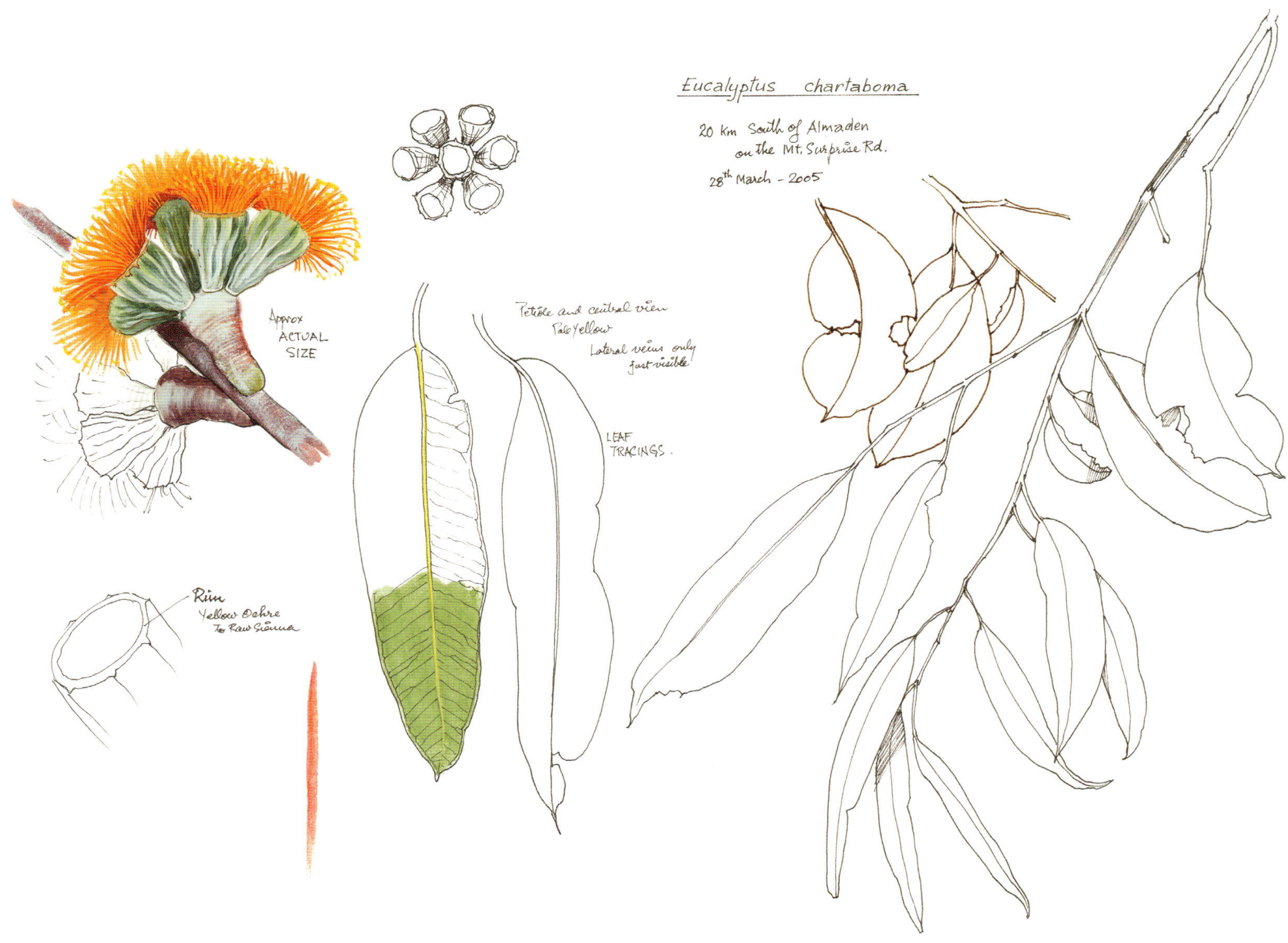

↓ 2005; pen, pencil and watercolour; 285 x 410mm
State Library of New South Wales

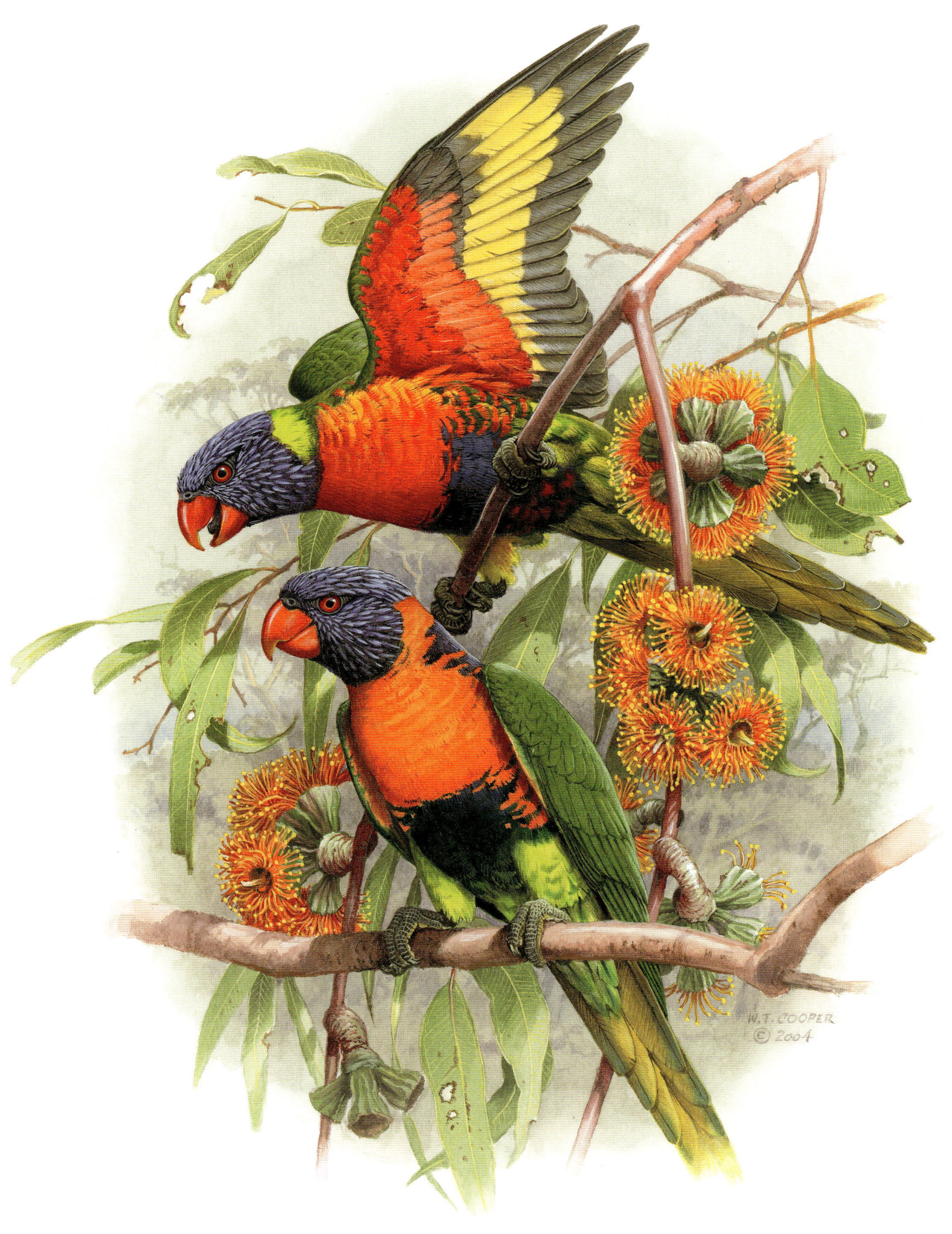
W.T. COOPER
© 2004

Eucalyptus cinerea

Argyle Apple

cinerea is from *cinereus* (ash-grey), referring to the leaf colour.

Argyle Apple is also known as Silver Dollar for its round, silvery juvenile leaves. Although most mature leaves are narrow and more typical of *Eucalyptus*, round leaves are often present on older trees with flowers or gumnuts. The tree occurs in temperate forest from north-eastern Victoria to the Southern Tablelands in New South Wales.

In the winter of 2000, we drove down eastern Australia gathering material for various cockatoos for the plates in *Cockatoos: A Portfolio of All Species*. In strikingly chilly Canberra, with snow on the surrounding hills, Bill was able to draw Argyle Apple for use in the Gang-gang Cockatoo plate. Gang-gangs deftly remove the small seeds before dropping the empty capsule.

→ 2000; pen, pencil and watercolour
295 x 410mm

↘ 2000; pen, pencil and watercolour
297 x 405mm
National Library of Australia, 8069057

↓ 2000; pen, pencil and watercolour
297 x 410mm
National Library of Australia, 8068917

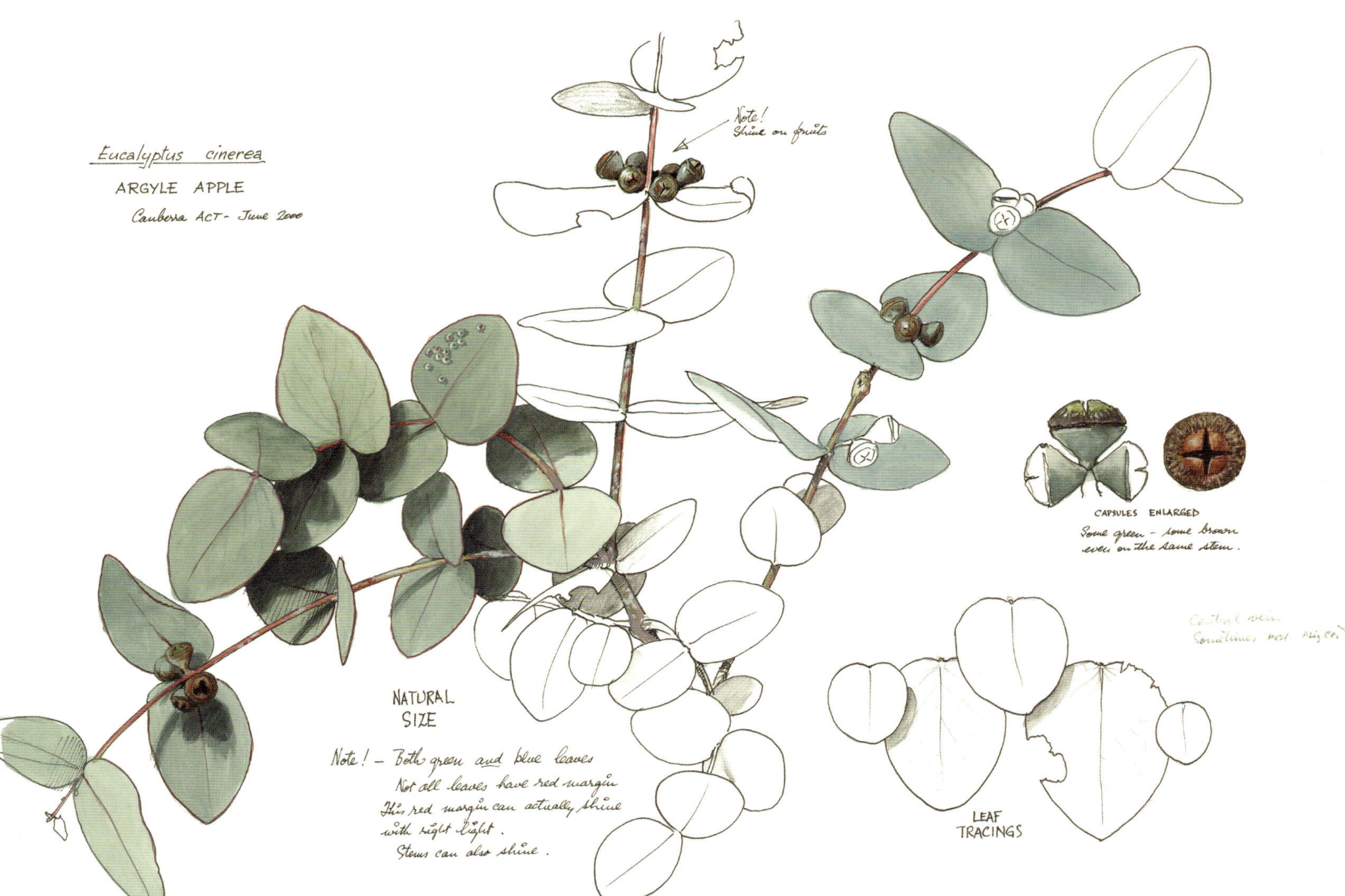

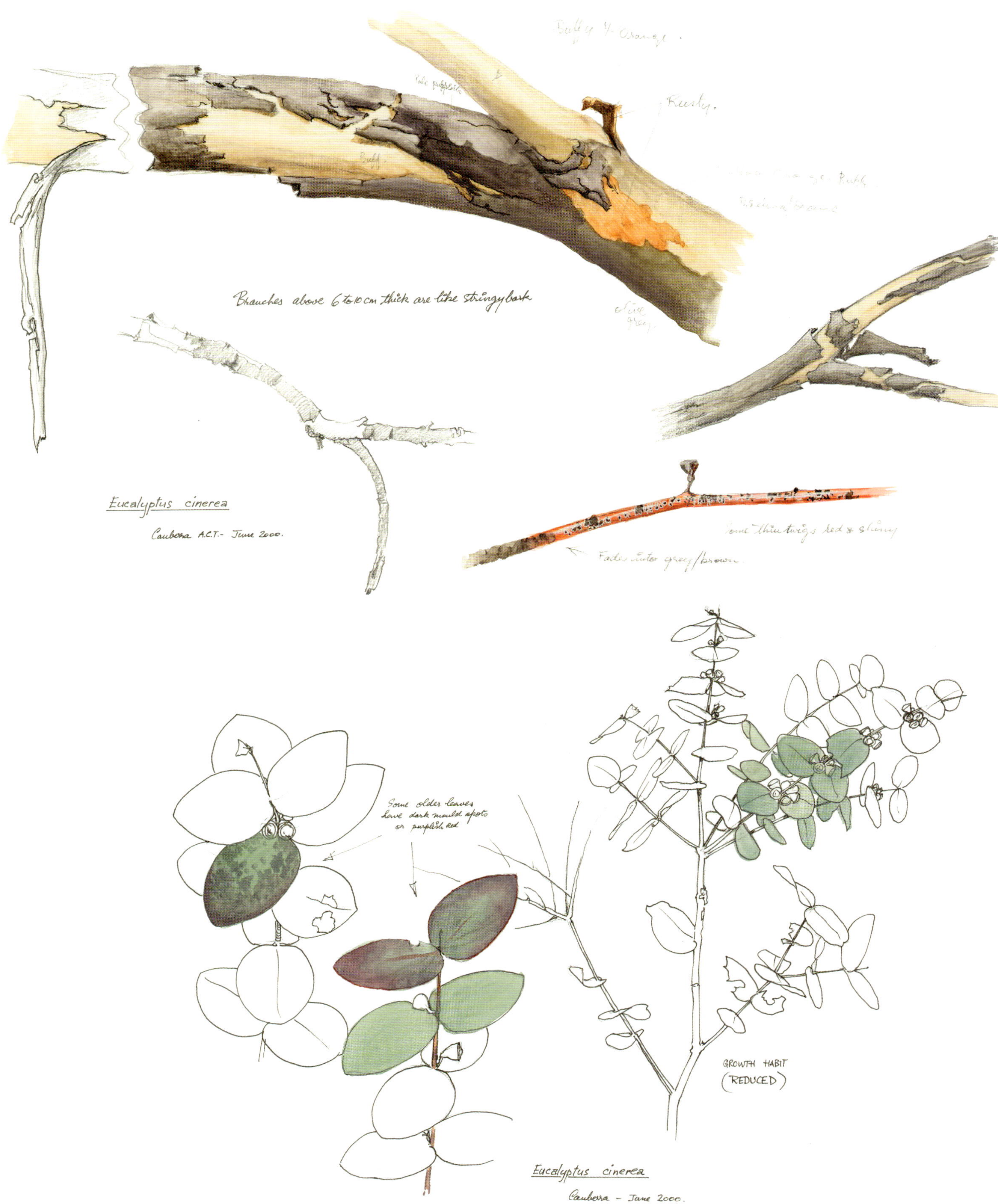
Rusty.
Branches above 6 to 10 cm thick are like stringybark
Eucalyptus cinerea
Canberra A.C.T.- June 2000.
Some thin twigs red & shiny
Fades into grey/brown.
Some older leaves have dark mould spots or purplish red
GROWTH HABIT (REDUCED)
Eucalyptus cinerea
Canberra - June 2000.

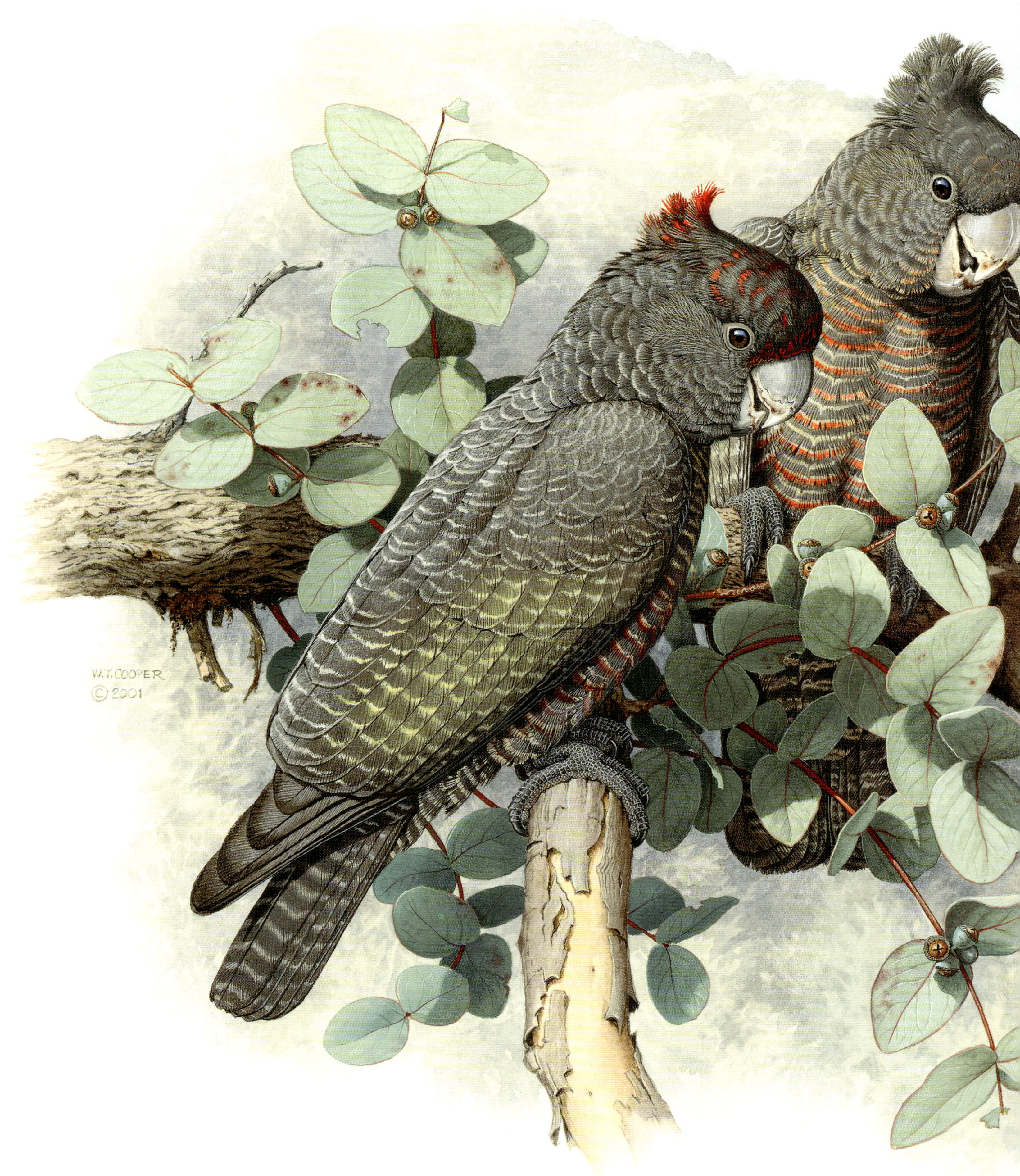
W.T.COOPER
© 2001

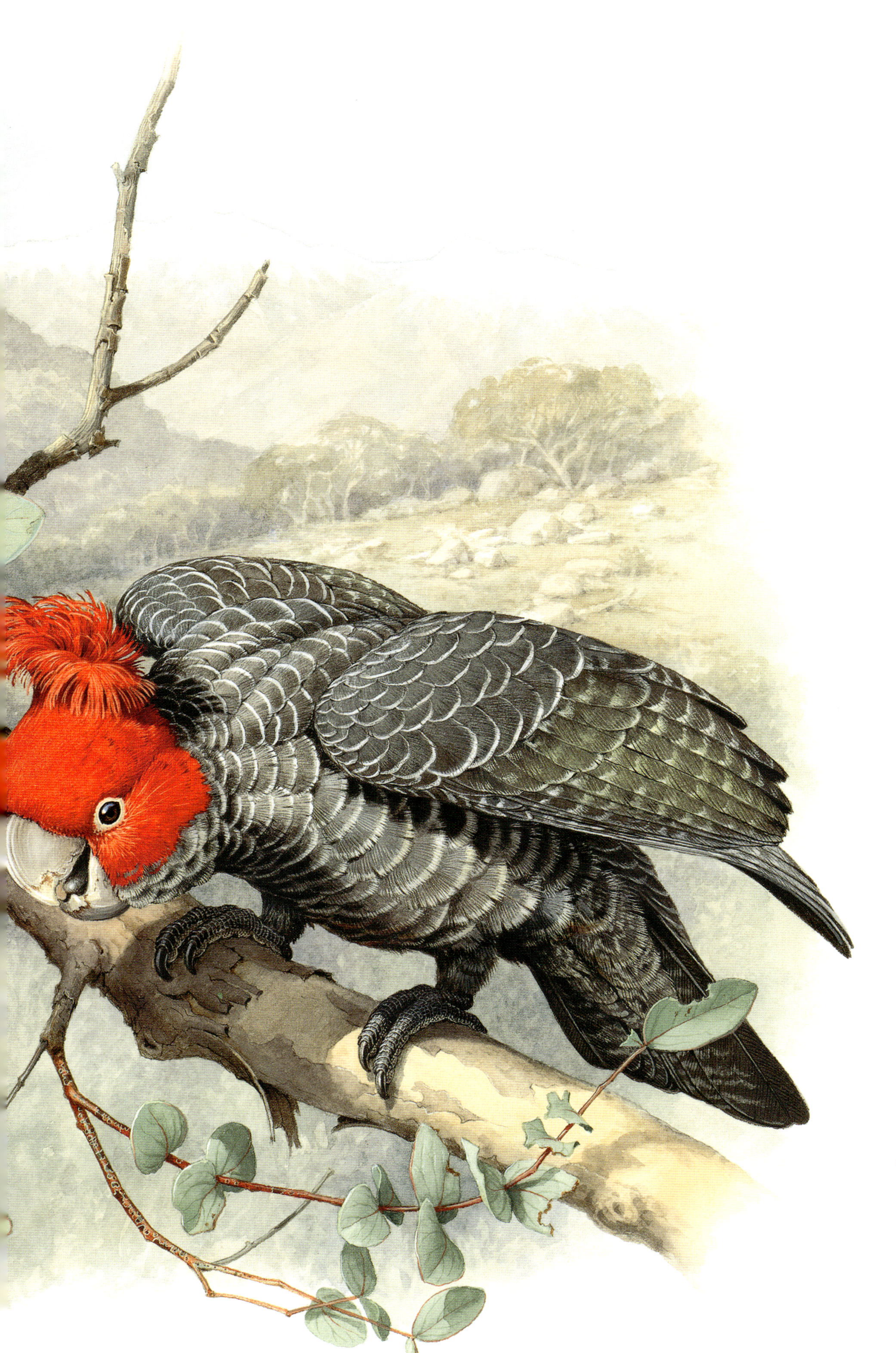

← Gang-gang Cockatoos
2001; watercolour
c.550 x 700mm
Collection of Peter
and Marilyn Chapman

Eucalyptus shirleyi

Silver-leaved Ironbark

shirleyi is named in honour of John Shirley (1849–1922), an educator from London with an interest in botany, especially lichens.

Silver-leaved Ironbark is mostly a mallee, with multiple stems arising from the base. It grows to about 7 metres tall in savanna and woodland habitats on poor soils in north-eastern Queensland.

We drove west of Herberton in search of a suitable *Eucalyptus* for inclusion in a Red-winged Parrot oil painting. This was the plant chosen, as typical of a stunted *Eucalyptus* in an area of low soil nutrients. It was an especially cold misty morning, which is reflected in the final painting (opposite). The pink new leaf growth and the red feathers added some warmth to an otherwise bleak day.

The painting has been annotated as *Eucalyptus melanophloia*, a similar species, but the two differ slightly.

→ Red-winged Parrots
2015; oil on board; 530 x 460mm
Private collection

↙ undated (c.2015); pen, pencil and watercolour
297 x 420mm
National Library of Australia, 8069068

↓ undated (c.2015); pen, pencil and watercolour
297 x 420mm
National Library of Australia, 8069065

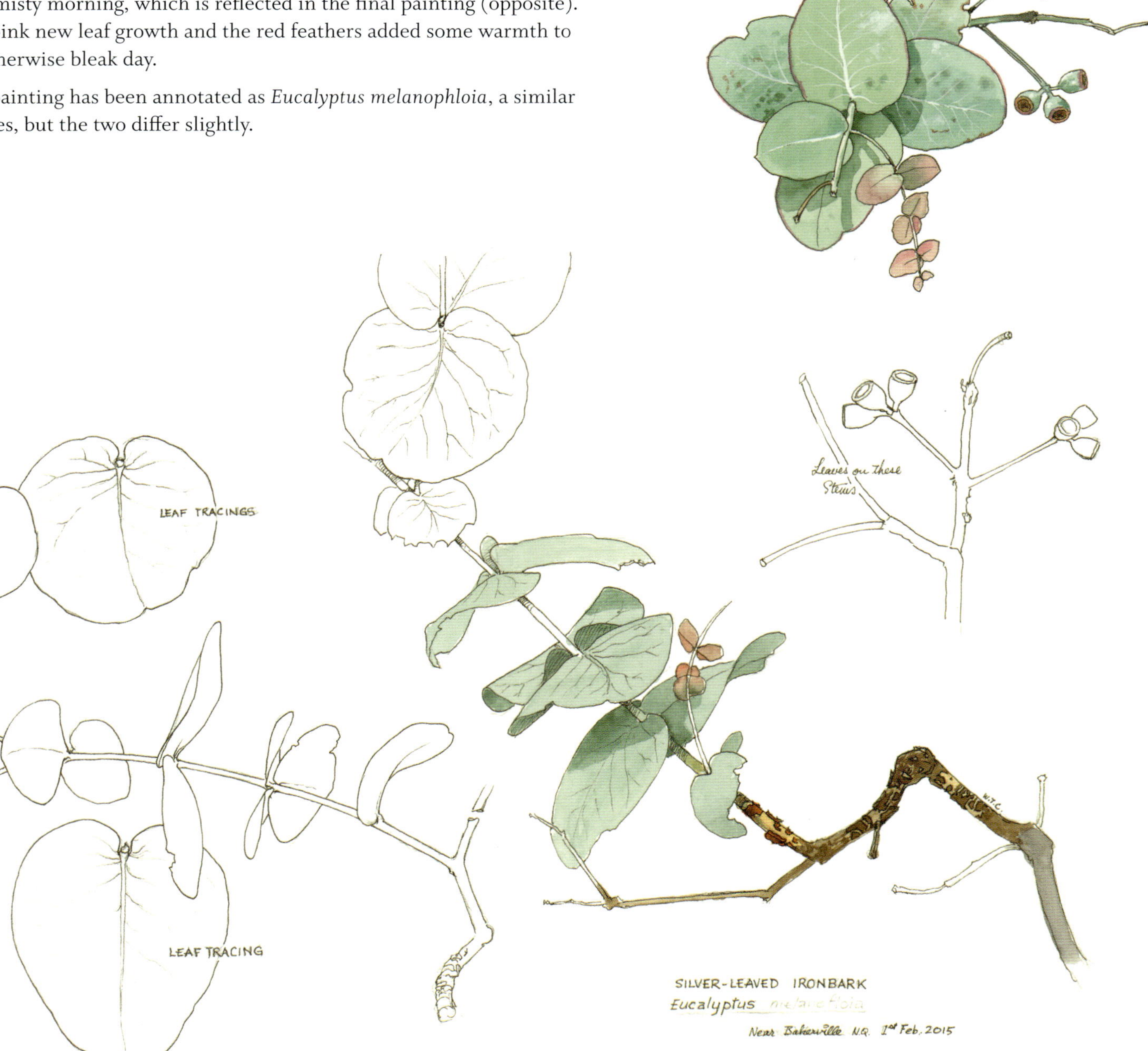

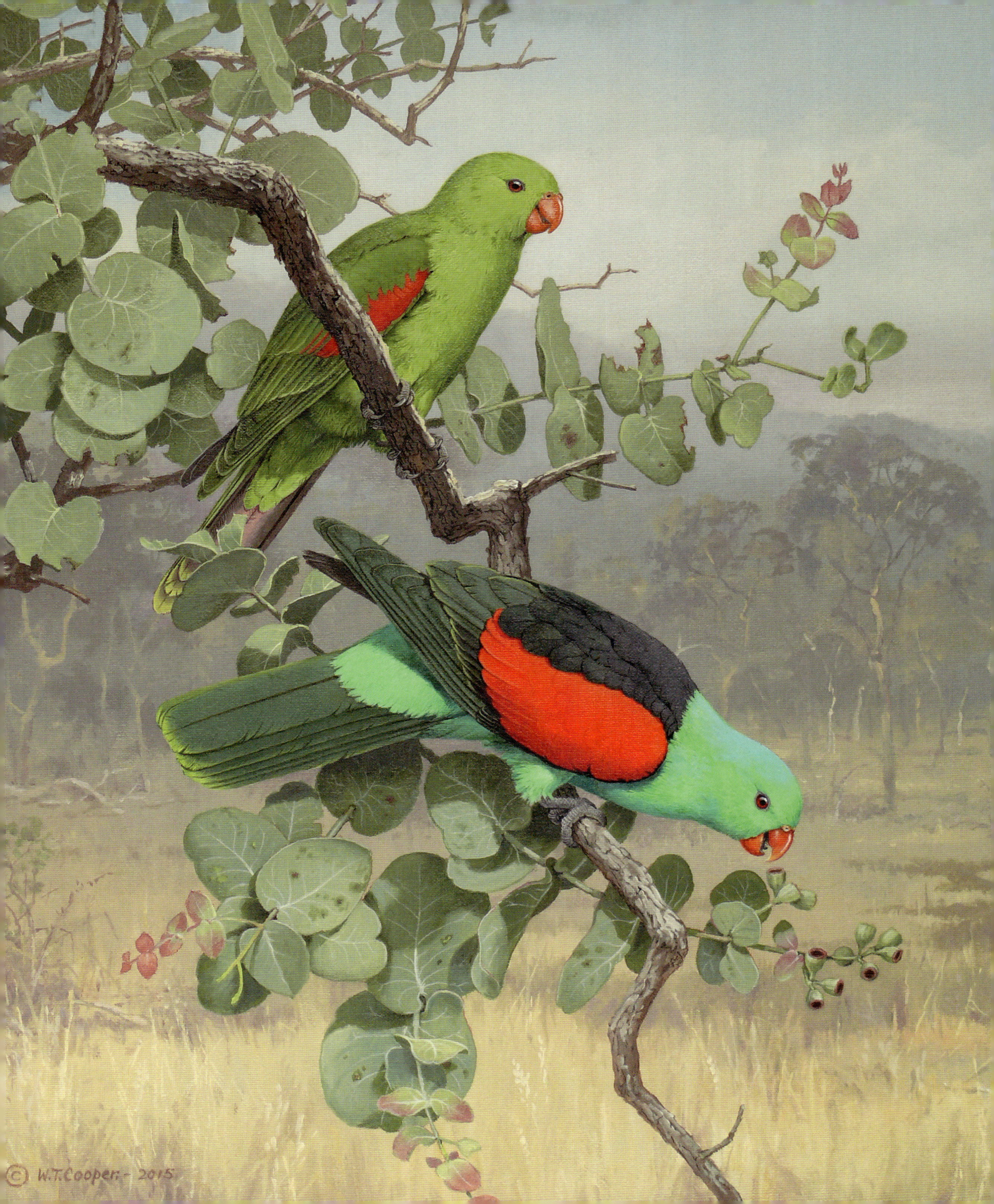
© W.T.Cooper - 2015

Grevillea pteridifolia

Fern-leaved Grevillea

PROTEACEAE

→ Rainbow Lorikeets, colour plate from *Australian Parrots*, 1980
National Library of Australia, 1362567

↓ 1979; pen, pencil and watercolour; 360 x 510mm
State Library of New South Wales

Grevillea is named in honour of Charles Francis Greville (1749–1809), British politician, Fellow of the Royal Society and the Linnaean Society, and a keen gardener, who cultivated *Vanilla planifolia* in Britain.

pteridifolia is from *pterido* (fern) and *-folius* (leaved).

Fern-leaved Grevillea is a spindly small tree that tends to grow in clusters of single-species colonies, mostly in areas of poor soils and seasonal waterlogging. It occurs from Cape York to Central Queensland, the Northern Territory and Western Australia. During flowering season, the trees are seen as swathes of orange flowers across low-lying areas in the landscape.

The specimen for this sketch was mailed in a tube from the Northern Territory in the days before express couriers. The leaves and flowers, carefully wrapped in moistened paper and plastic bags, arrived looking very fresh. The material was included in the plate of Rainbow Lorikeets in *Australian Parrots*.

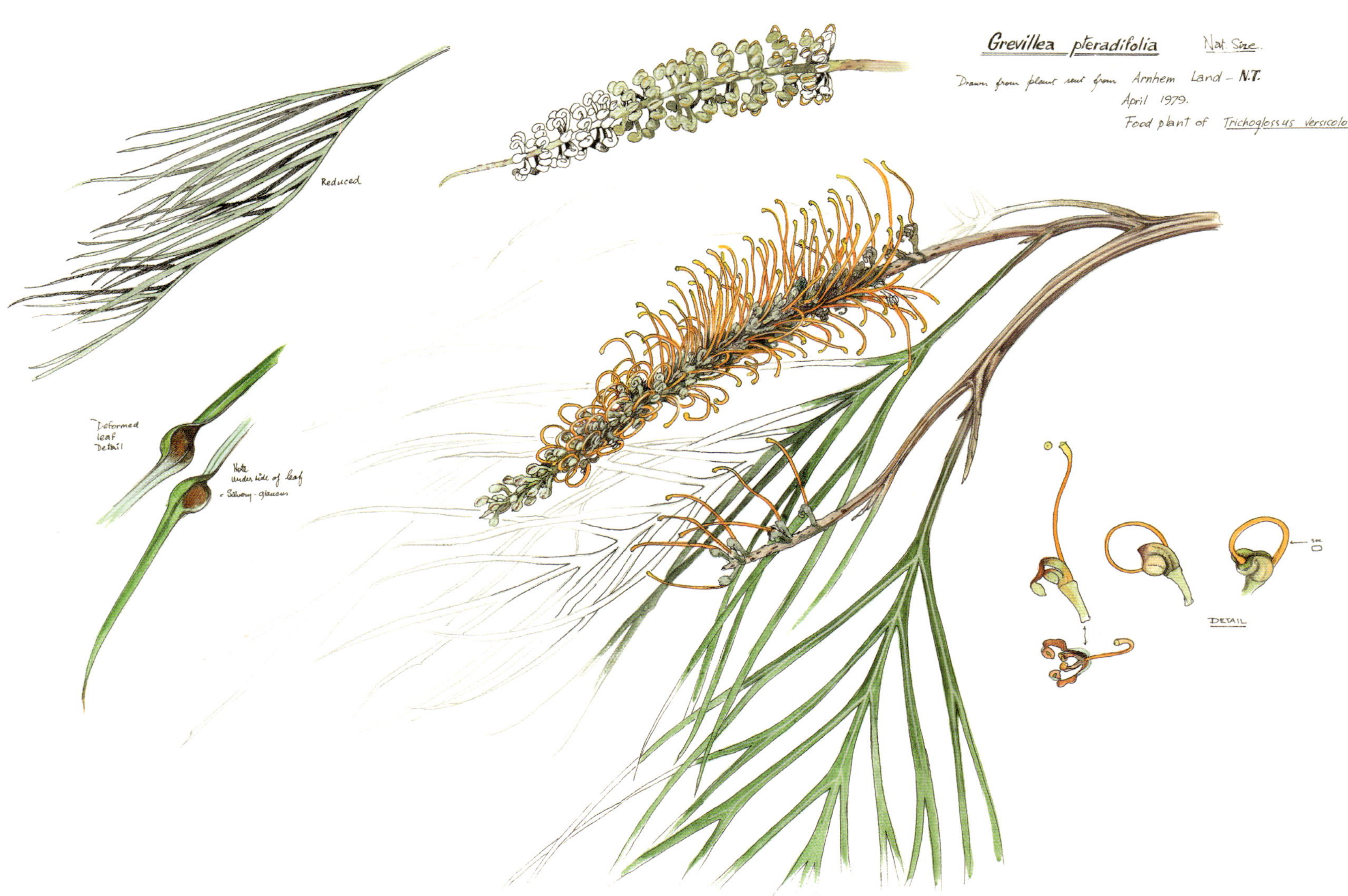

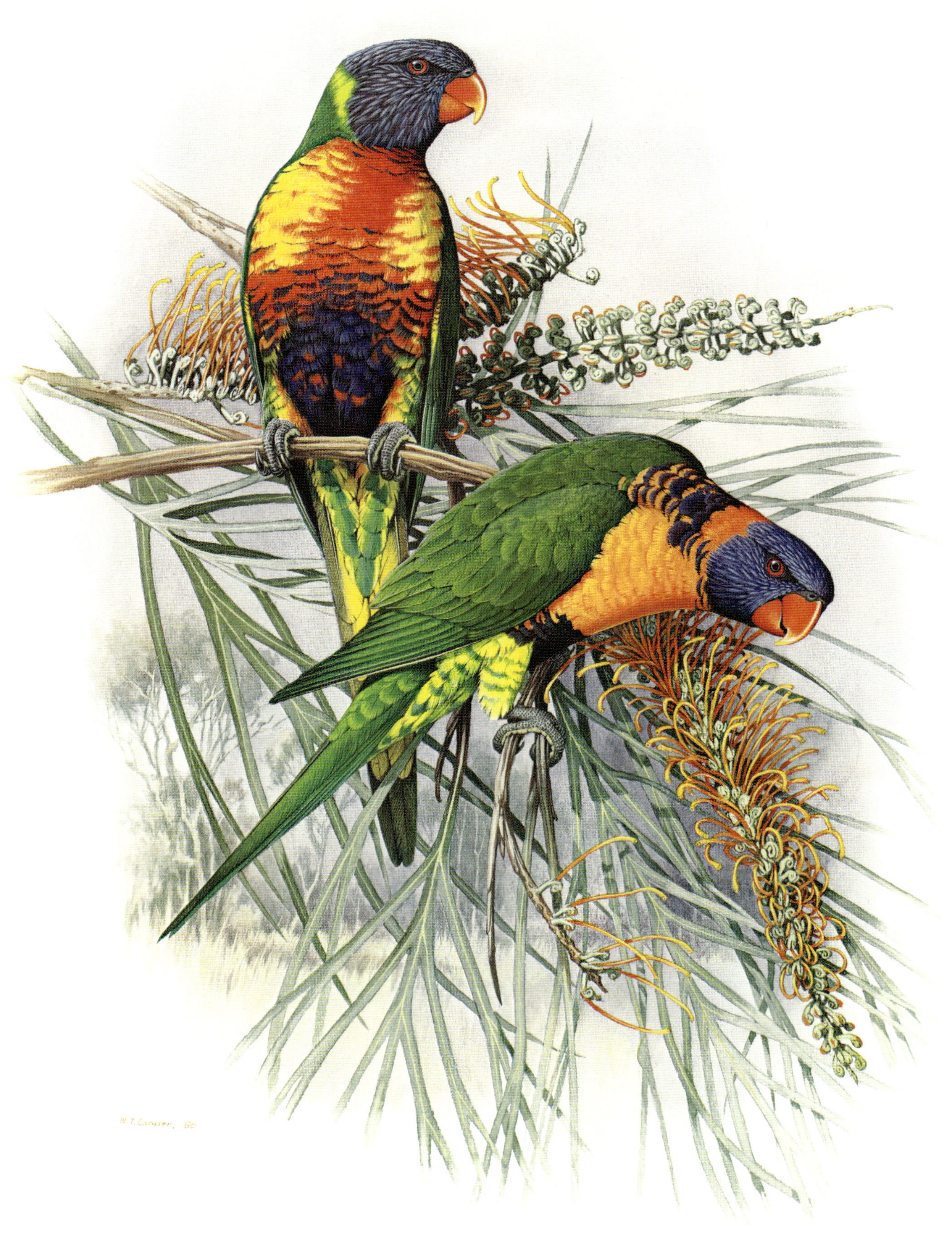

Lambertia formosa

Mountain Devil

PROTEACEAE

Lambertia is named in honour of Aylmer Bourke Lambert (1761–1842), British botanist and member of the Linnaean Society, who published the multi-volume *A Description of the Genus Pinus* between 1803 and 1824.

formosa is from *formosus* (beautiful), referring to the flowers.

A shrub of heathland and sclerophyll forest along the coast and ranges of New South Wales, this would have been a familiar plant for Bill. He had frequently walked this country since childhood, to go rock fishing at Redhead. It is an area that often experienced bushfires, for which this plant is well adapted, resprouting from a lignotuber (woody swelling). The leaves are stiff and the flowers are favoured by honeyeaters. Mountain Devil derives its name from the woody two-horned fruit.

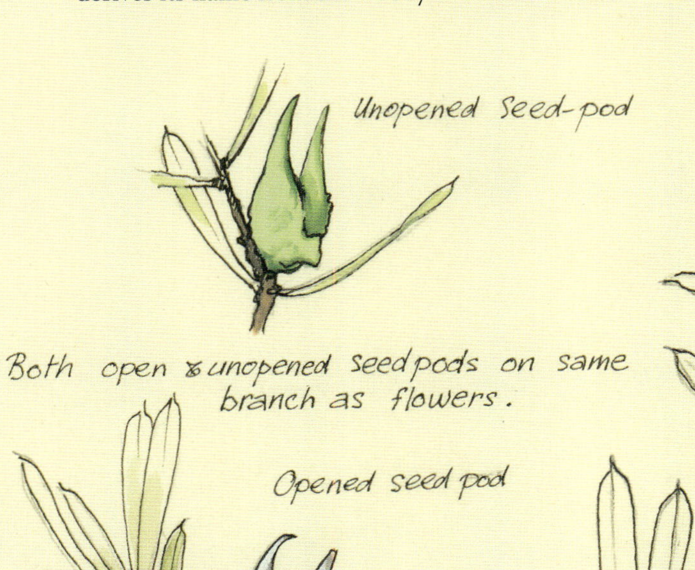

HONEY FLOWER
(Lambertia formosa)
Collected – Redhead, May 68.
also blooms – Spring & summer.

→ 1968; pen, pencil and gouache
280 x 220mm
State Library of New South Wales

Lomatia silaifolia

Crinkle Bush or Wild Parsley

PROTEACEAE

Lomatia is from *loma* (fringe or border), referring to the funicle, which is a border around the seed wing.

silaifolia is after the genus *Silaus* and *folia* (leaves), referring to the similar leaves.

Wild Parsley is a shrub that grows to 2 metres tall, widespread in heathland and woodland from Jervis Bay in southern New South Wales to the Blackdown Tableland in Queensland.

This illustration was done from the plants growing in the heath country around Bungwahl, among numerous other wildflowers, Blackbutt and *Angophora* trees. Wild Parsley is well adapted to fire-prone areas (as this area was) in its ability to quickly resprout from a lignotuber.

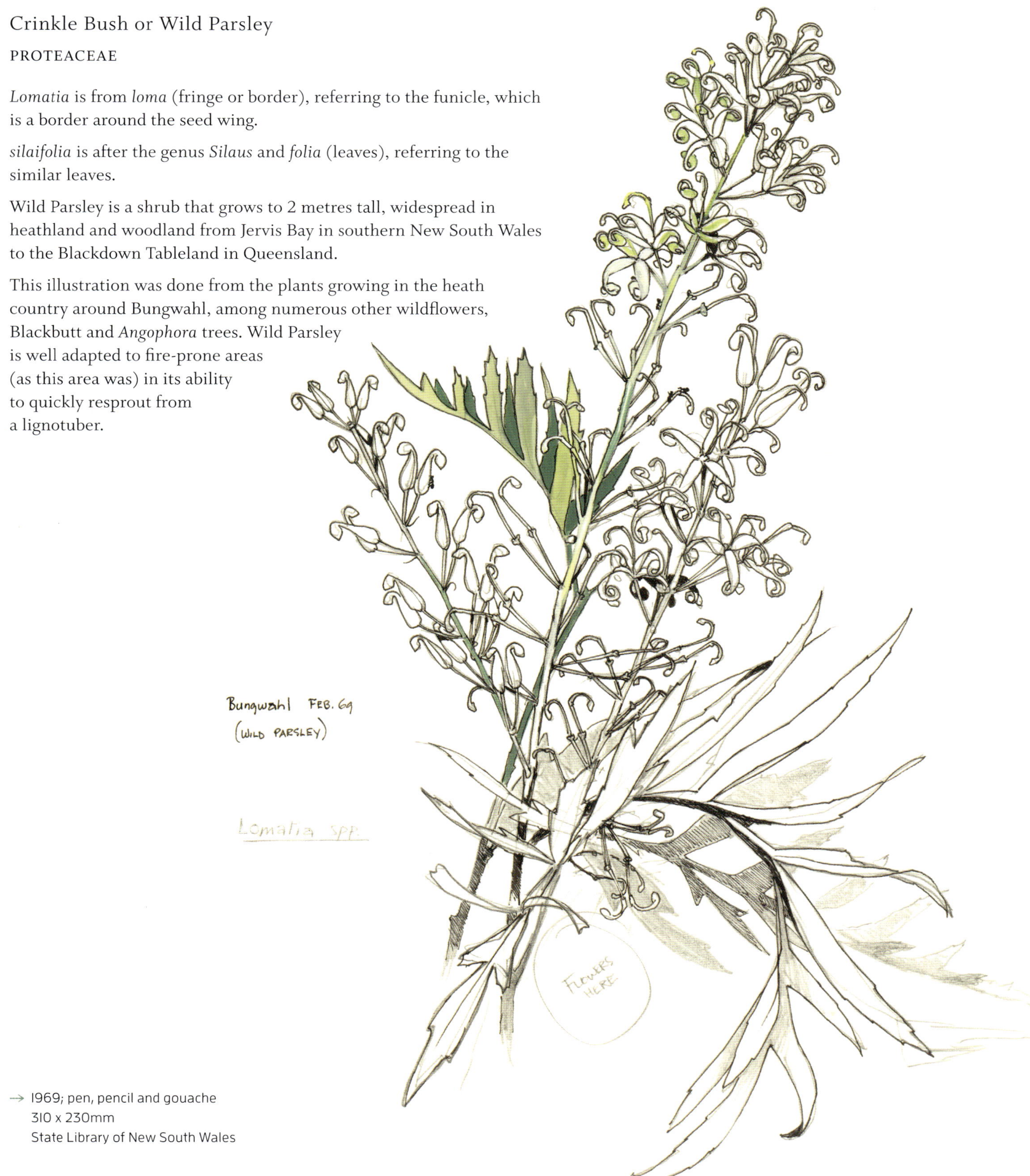

→ 1969; pen, pencil and gouache
310 x 230mm
State Library of New South Wales

Lophostemon confertus

Brush Box

MYRTACEAE

→ undated; pen and watercolour; 280 x 490mm
State Library of New South Wales

↓ 1992; pencil and watercolour; 400 x 455mm
State Library of New South Wales

Brush Box is an especially attractive tree with smooth pinkish flaky bark. It is widespread, occurring in rainforest and wet sclerophyll forest in eastern Australia, from the Hunter Valley to the Cooktown area. The glossy dark-green leaves mostly grow in whorls. The woody capsules house tiny seeds up to 3 millimetres long, which are delicately removed and eaten by Red-tailed Black-Cockatoos and Crimson Rosellas.

Angophora costata trees (below) also have salmon-coloured bark. Both species often occur together, especially in coastal New South Wales.

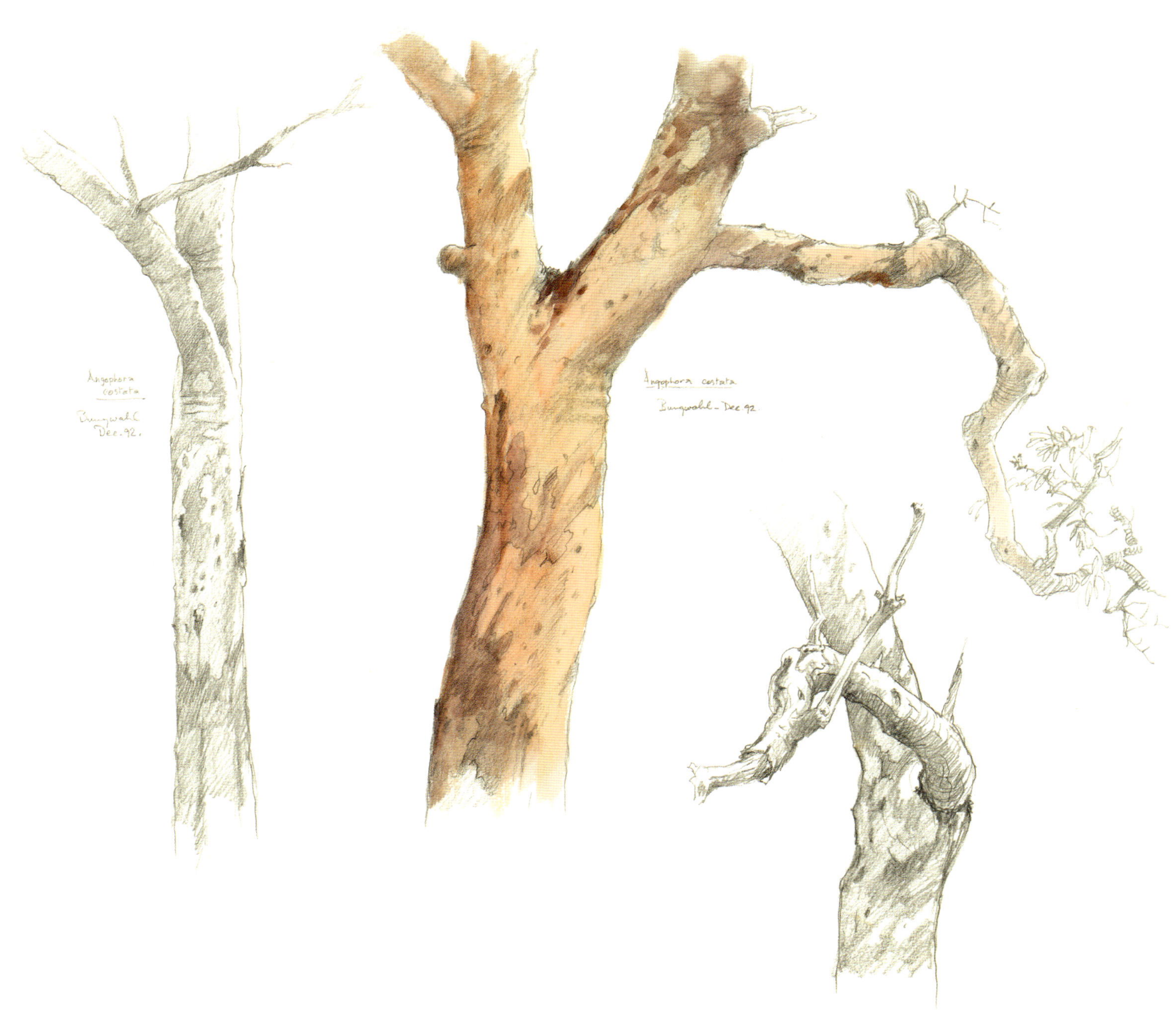

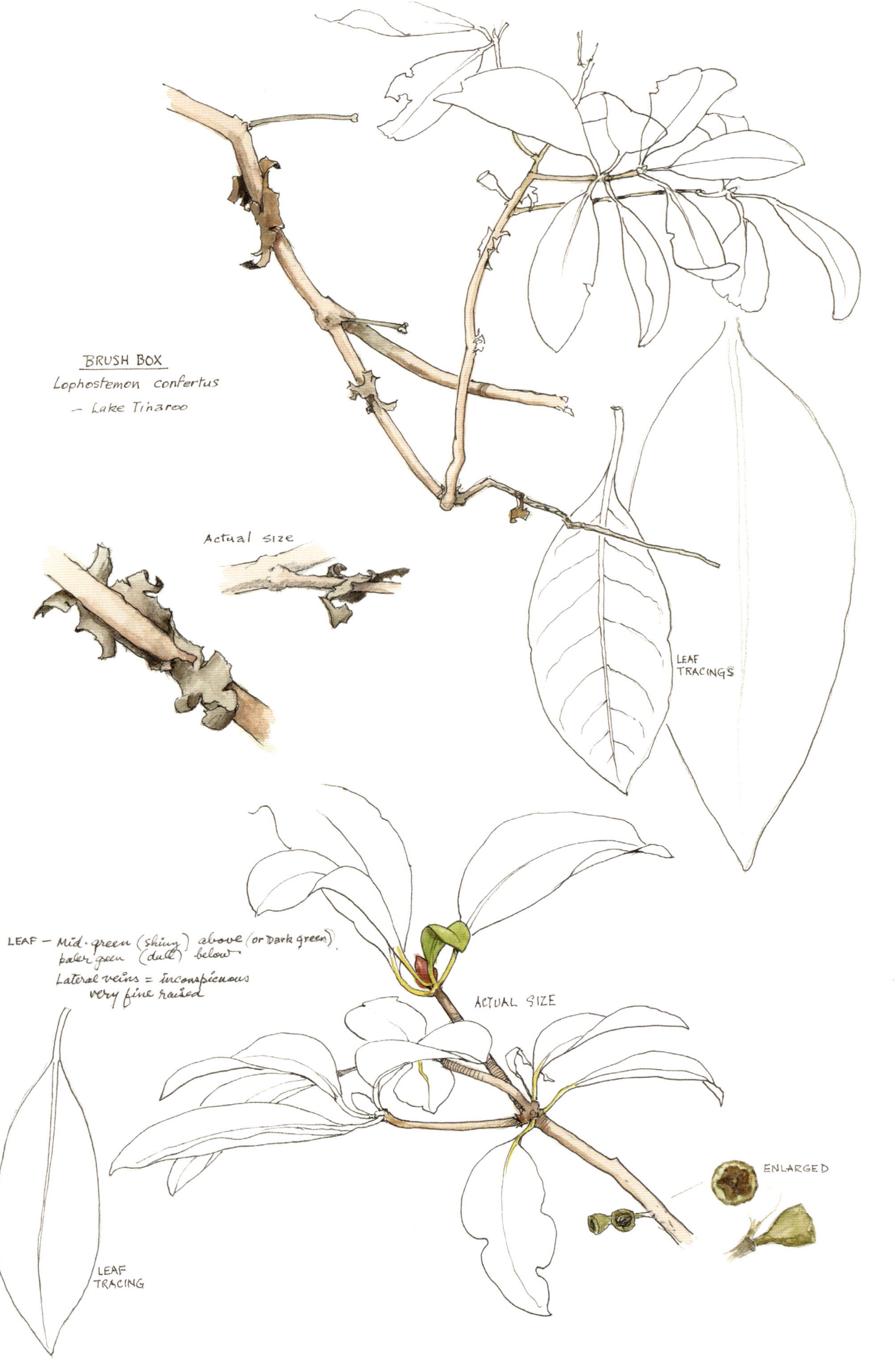

BRUSH BOX
Lophostemon confertus
– Lake Tinaroo
Actual size
LEAF TRACINGS
LEAF – Mid-green (shiny) above (or Dark green)
paler green (dull) below
Lateral veins = inconspicuous
very fine raised
ACTUAL SIZE
ENLARGED
LEAF TRACING

Lysiphyllum cunninghamii

Bauhinia or Bean Tree

FABACEAE

→ 2004; pen and watercolour; 420 x 290mm
State Library of New South Wales

↓ 1997; pen and pencil; 400 x 290mm
State Library of New South Wales

Lysiphyllum is from *lysis* (loose or free) and *phyllon* (leaf), referring to the compound leaves with two leaflets, differing from the related genus *Bauhinia*, which has deeply lobed simple leaves.

cunninghamii is named in honour of English botanical collector Allan Cunningham (1791–1839).

The Bean Tree is a deciduous species that grows to 10 metres, occurring in woodland and savannah as well as various dry forms of rainforest in tropical Queensland, the Northern Territory and Western Australia. The butterfly-like leaves are compound, with two separate leaflets. Rusty-coloured calyces holding red flowers are followed by large flat bean-like fruit of up to 210 millimetres in length.

In the winter of 2004, we were on a trip through western Queensland, the Northern Territory and the Kimberley to gather material for *Pigeons and Doves in Australia*. The study opposite was never used, even though it is annotated with the idea to pair it with Spinifex Pigeons. These birds spend most of their time on the ground and the branches from this tree also hang near to the ground.

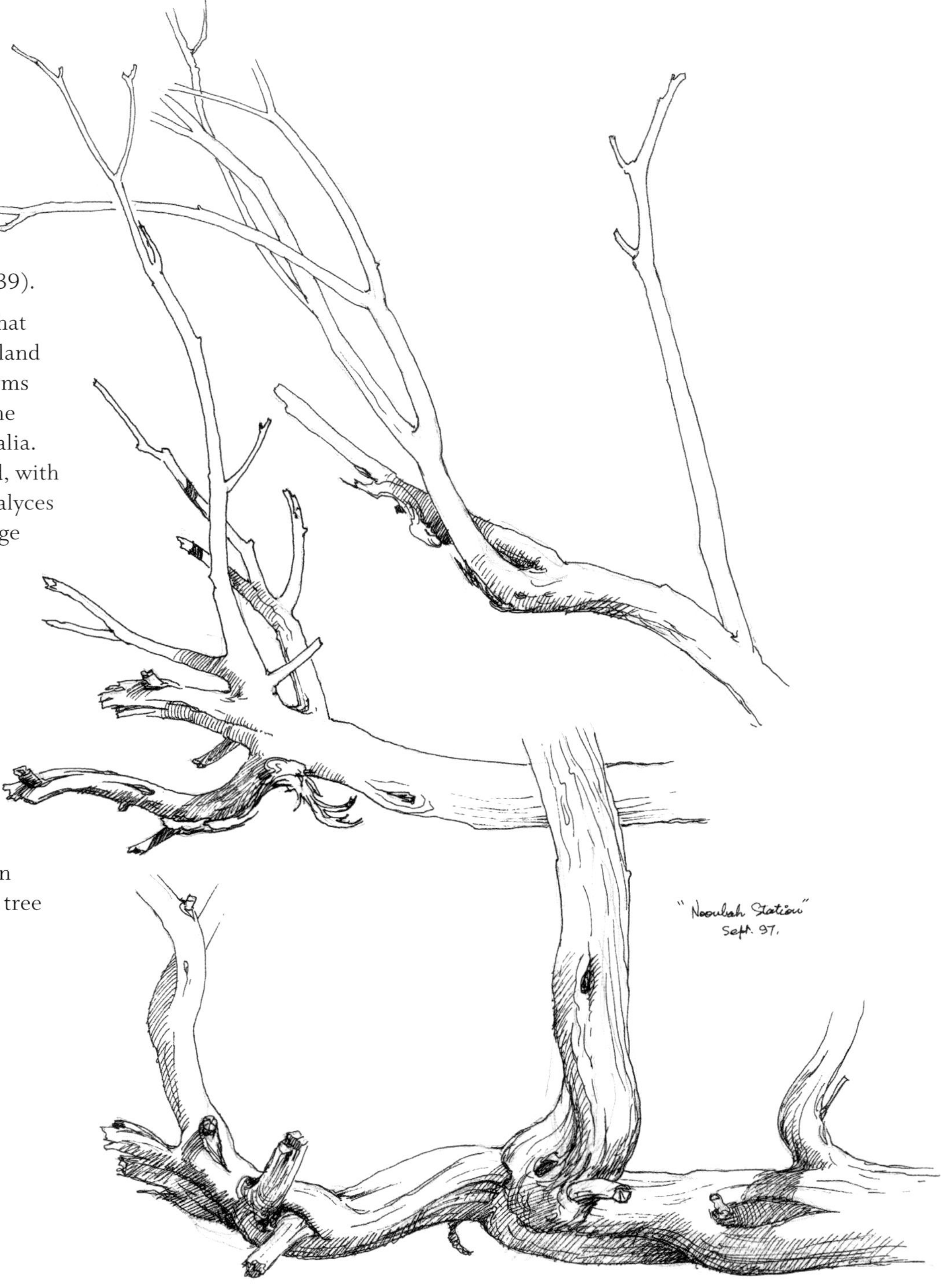

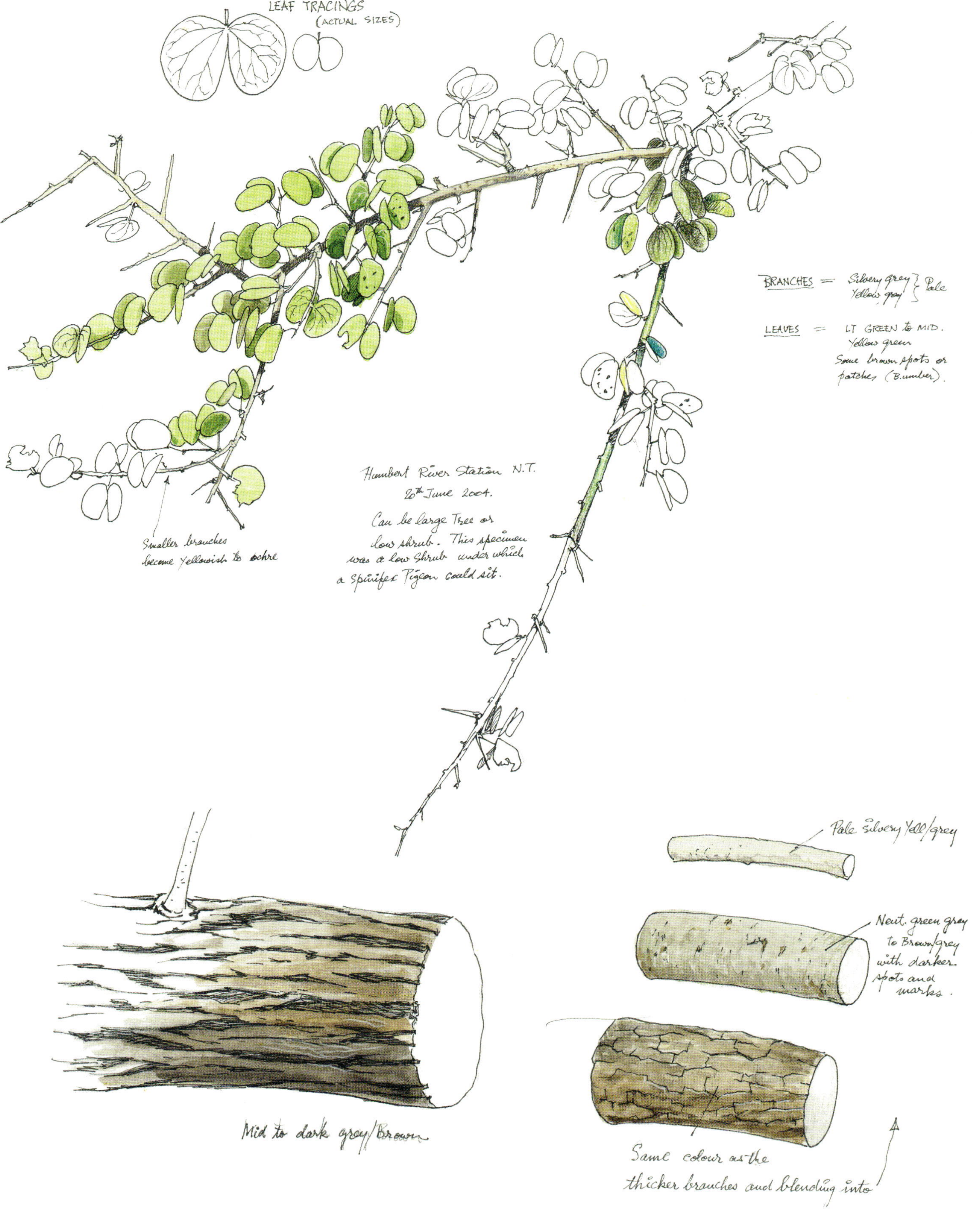
LEAF TRACINGS
(ACTUAL SIZES)
BRANCHES = Silvery grey
Yellow grey } Pale
LEAVES = LT GREEN to MID.
Yellow green
Some brown spots or
patches (B.umber).
Humbert River Station N.T.
20th June 2004.
Can be large Tree or
low shrub. This specimen
was a low Shrub under which
a Spinifex Pigeon could sit.
Smaller branches
become Yellowish to ochre
Pale Silvery Yell/grey
Neut. green grey
to Brown/grey
with darker
spots and
marks.
Mid to dark grey/Brown
Same colour as the
thicker branches and blending into

Melaleuca

MYRTACEAE

Melaleuca is from *mel-* (black) and *leucos* (white), referring to the often burnt black trunks and white unburnt upper branches.

Melaleuca trees are most conspicuous as Australian paperbarks, even though not all species have such bark. There are about 220 species in Australia and another 10 species in New Caledonia, New Guinea and Asia.

Melaleuca quinquenervia

Broad-leaved Paperbark

quinquenervia is from *quinque* (five) and *nervus* (veined), referring to the leaves usually having five veins.

Broad-leaved Paperbark grows to 25 metres, often occurring in dense stands in coastal forests and swamps from the Sydney area north to New Guinea. It is an important tree ecologically, as its regular and prolific flowering supports many invertebrates, birds and mammals.

This specimen was collected from Sugar Creek Road, near Myall Lakes. When in flower, these trees were alive with parrots by day and flying-foxes at night.

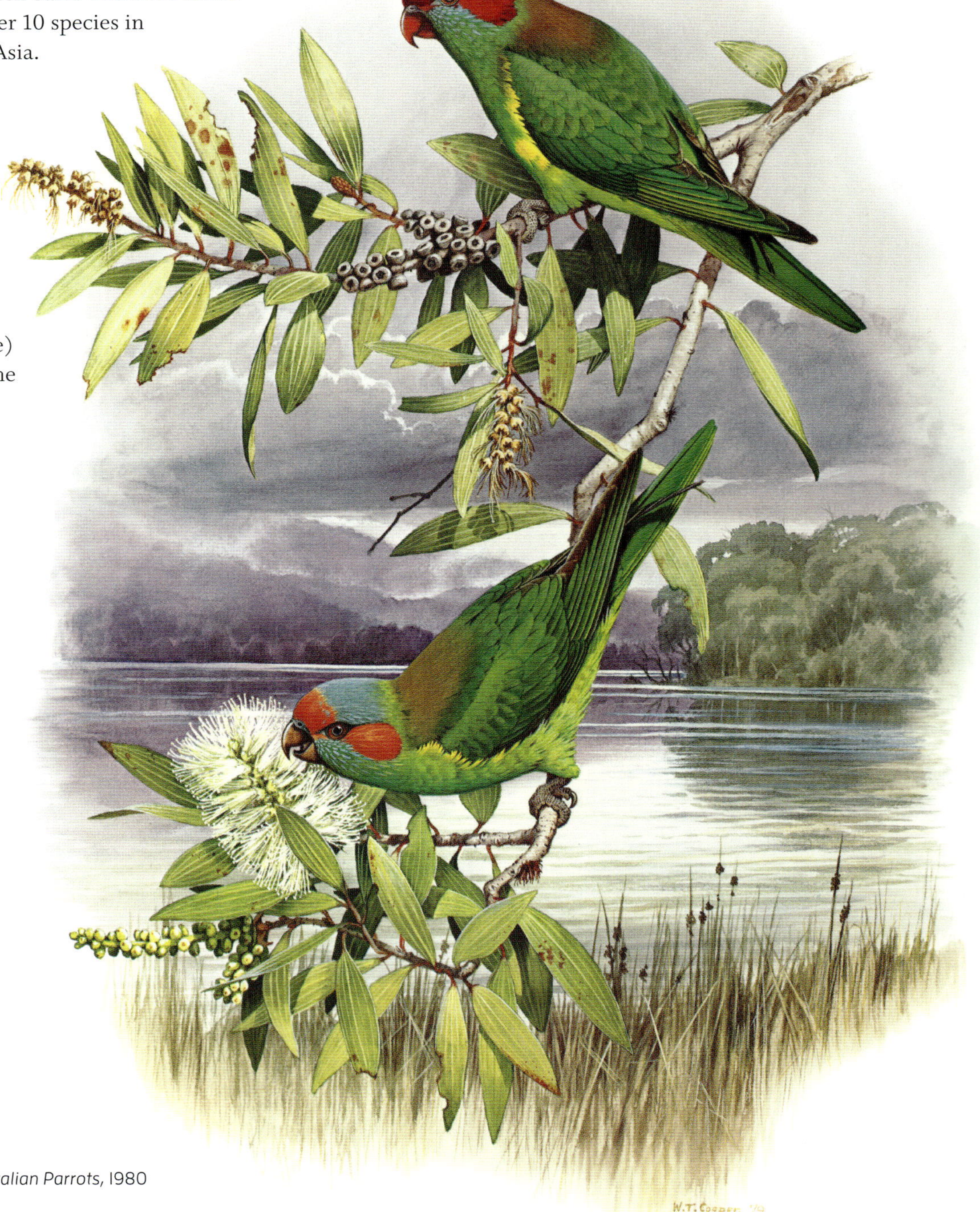

↗ Musk Lorikeets, colour plate from *Australian Parrots*, 1980
National Library of Australia, 1362567

→ 1979; pen, pencil and watercolour; 450 x 350mm
State Library of New South Wales

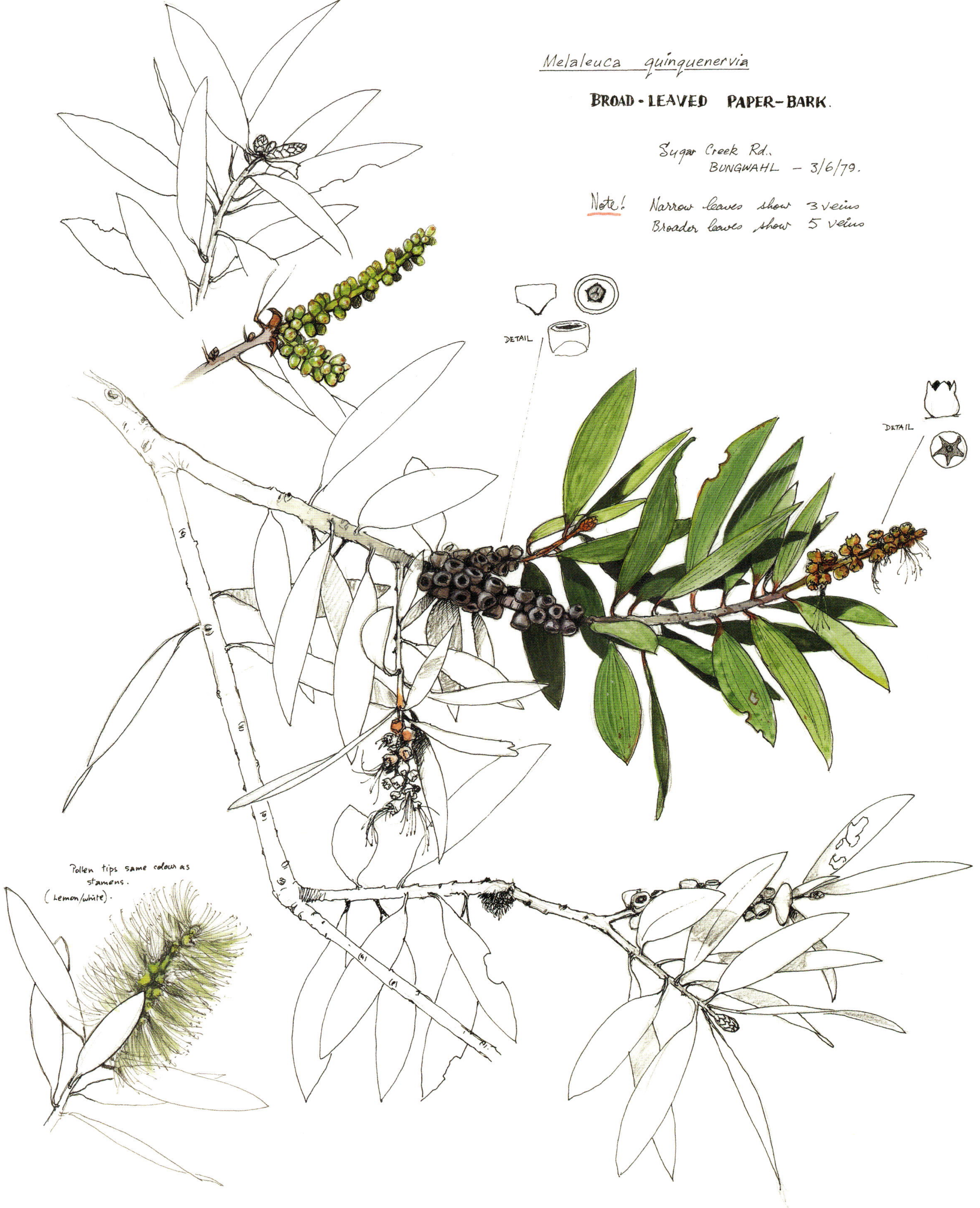

Melaleuca quinquenervia
BROAD-LEAVED PAPER-BARK.
Sugar Creek Rd.,
BUNGWAHL – 3/6/79.
Note! Narrow leaves show 3 veins
Broader leaves show 5 veins
DETAIL
DETAIL
Pollen tips same colour as
stamens.
(Lemon/white).

Melaleuca viridiflora

Broad-leaved Tea-tree

viridiflora is from *viridis* (green) and *-florus* (flowered), referring to the flowers, which are usually green but can also be pink, as in this painting.

Broad-leaved Tea-tree, with its papery bark, grows in swampy areas, mostly in open forest and rarely in rainforest, from southern Queensland to New Guinea, Western Australia, the Northern Territory and Indonesia. It can be found in flower at almost any time of year, with the nectar-rich flowers visited by numerous honeyeaters.

This specimen was collected between Iron Range and the Wenlock River on Cape York, in July 1981, just before our return home from the first of many trips to that area. Bill noted in his diary:

> *1100 hrs—much activity in blossoming Melaleucas growing in grassland savannah. There were hundreds of birds, mostly Little Friarbirds, Blue-faced Honeyeaters, Rainbow Lorikeets also Pied Butcherbirds and Dusky Honeyeaters.*

Bill later completed a watercolour of Blue-faced Honeyeaters with this sketch.

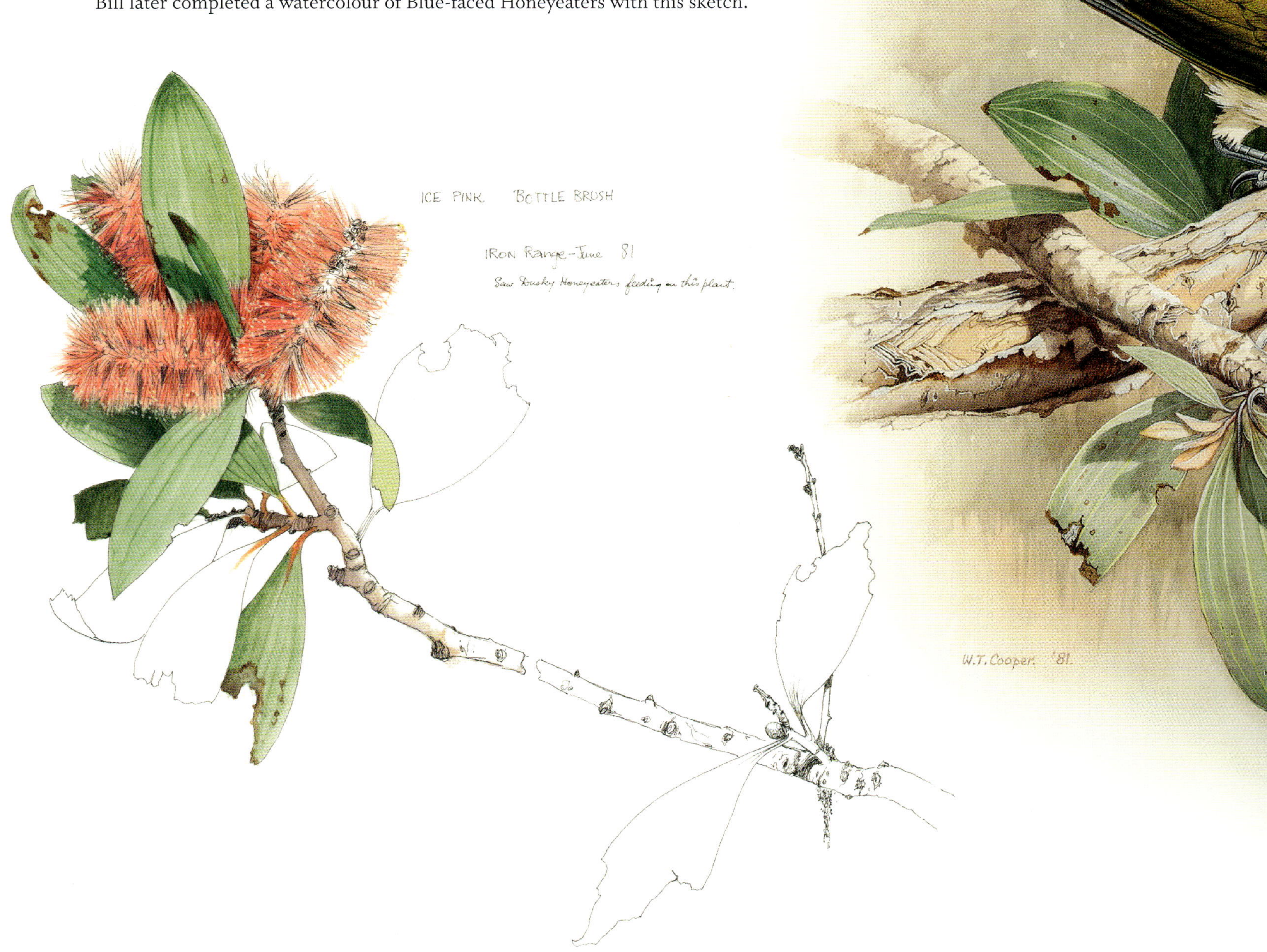

← 1981; pen, pencil and watercolour; 465 x 450mm
State Library of New South Wales

↓ Blue-faced Honeyeaters
1981; watercolour; 460 x 620mm
Private collection

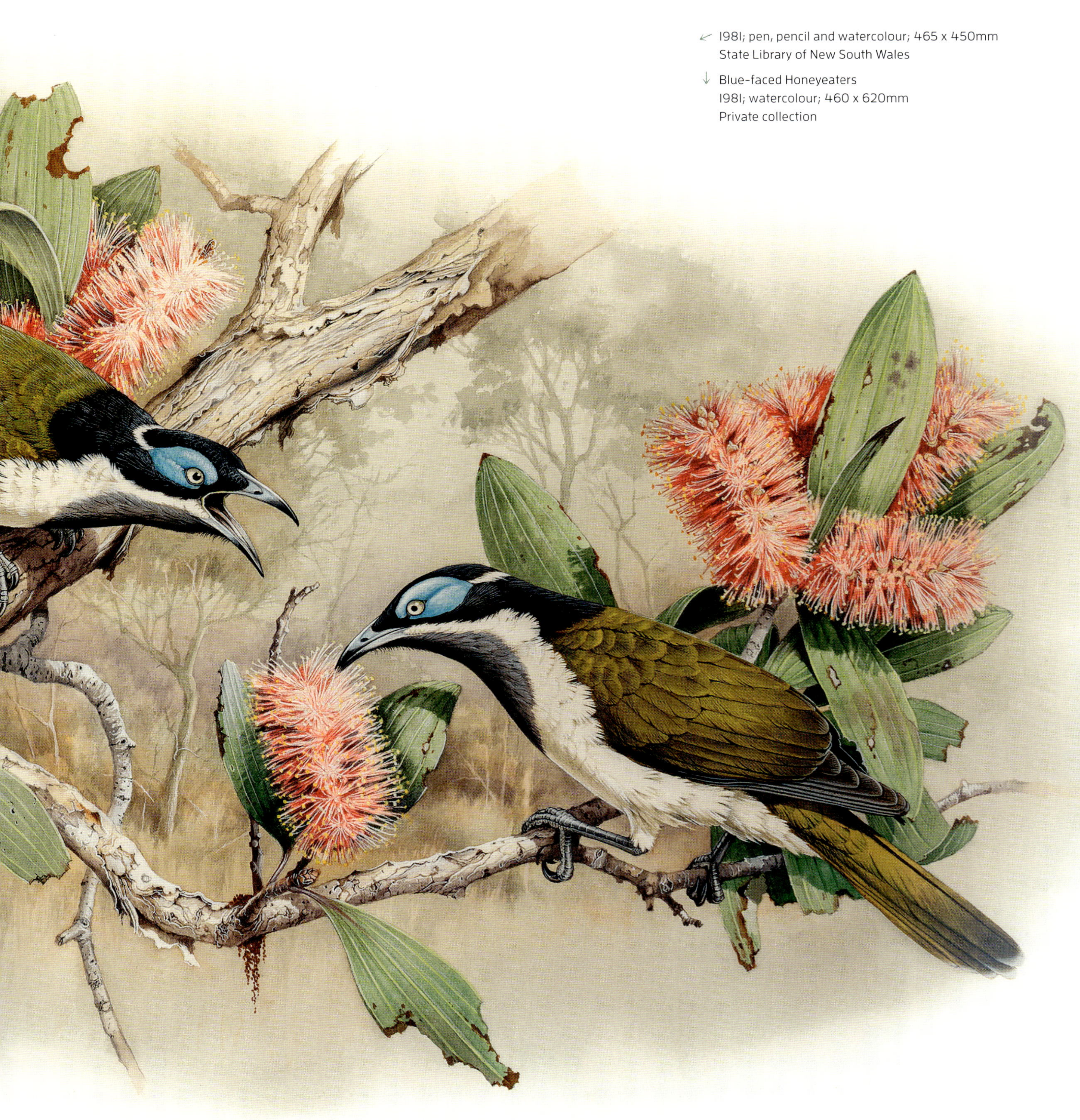

Pandanus cookii

→ 2005; pencil and watercolour; 710 x 590mm
Collection of Bruce and Joy Gray

↓ 1988; pen and pencil; 400 x 320mm
State Library of New South Wales

Cook's Pandan

PANDANACEAE

Pandanus is from a Malaysian word meaning conspicuous.

cookii is named in honour of English navigator and explorer Captain James Cook (1728–1779).

Pandanus is an interesting group of plants in somewhat of a taxonomic mess, partly because the species are variable and they hybridise. In 2005, we decided to begin work on this group, which involved Bill painting a full-page illustration for each species. *Pandanus* species have many overlapping features and are not a simple group to work on. Alas, I was new to taxonomy and DNA methods were in their infancy, so I was soon defeated and no more *Pandanus* plates were completed. Today, other botanists are getting close to untangling *Pandanus* species concepts.

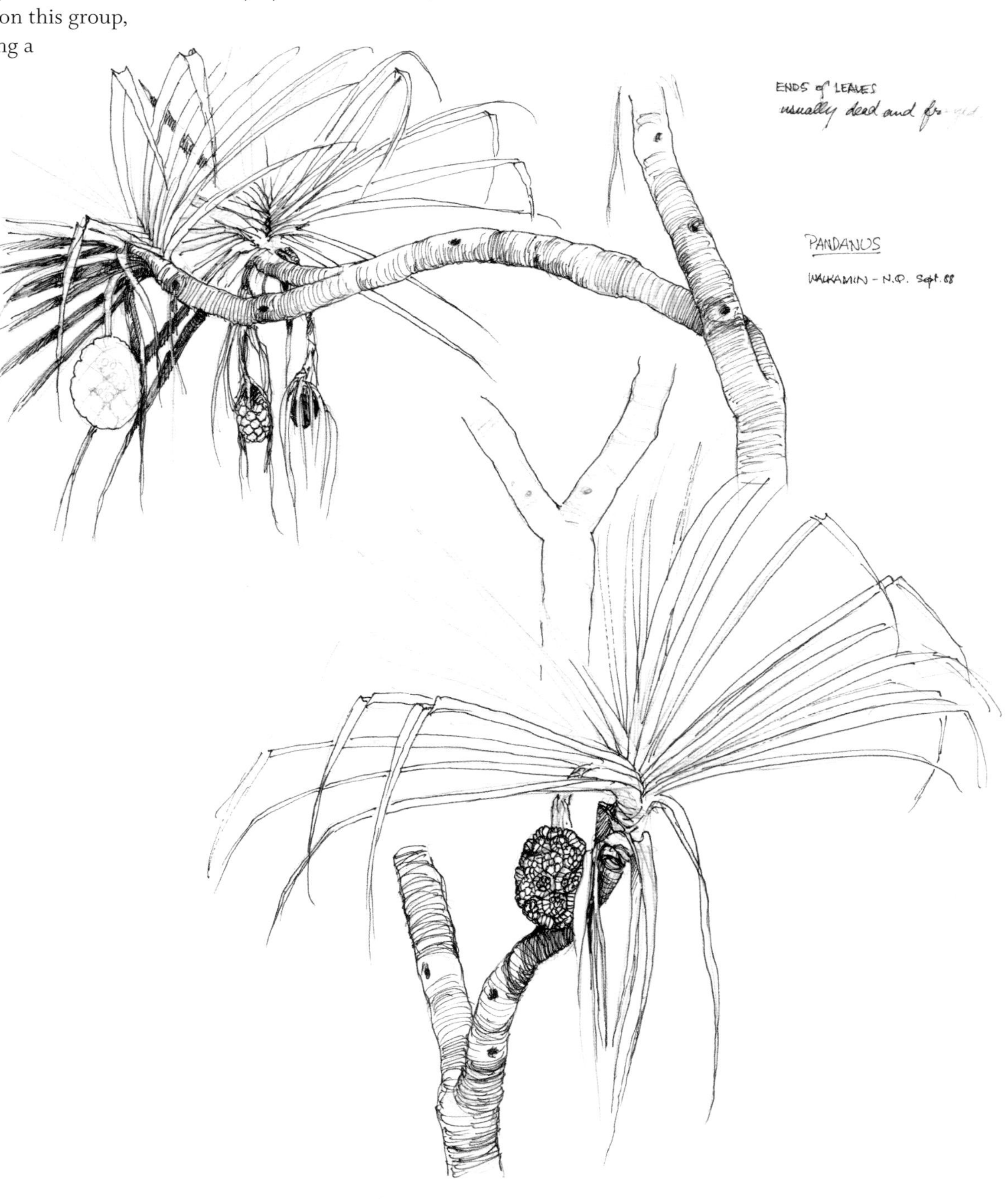

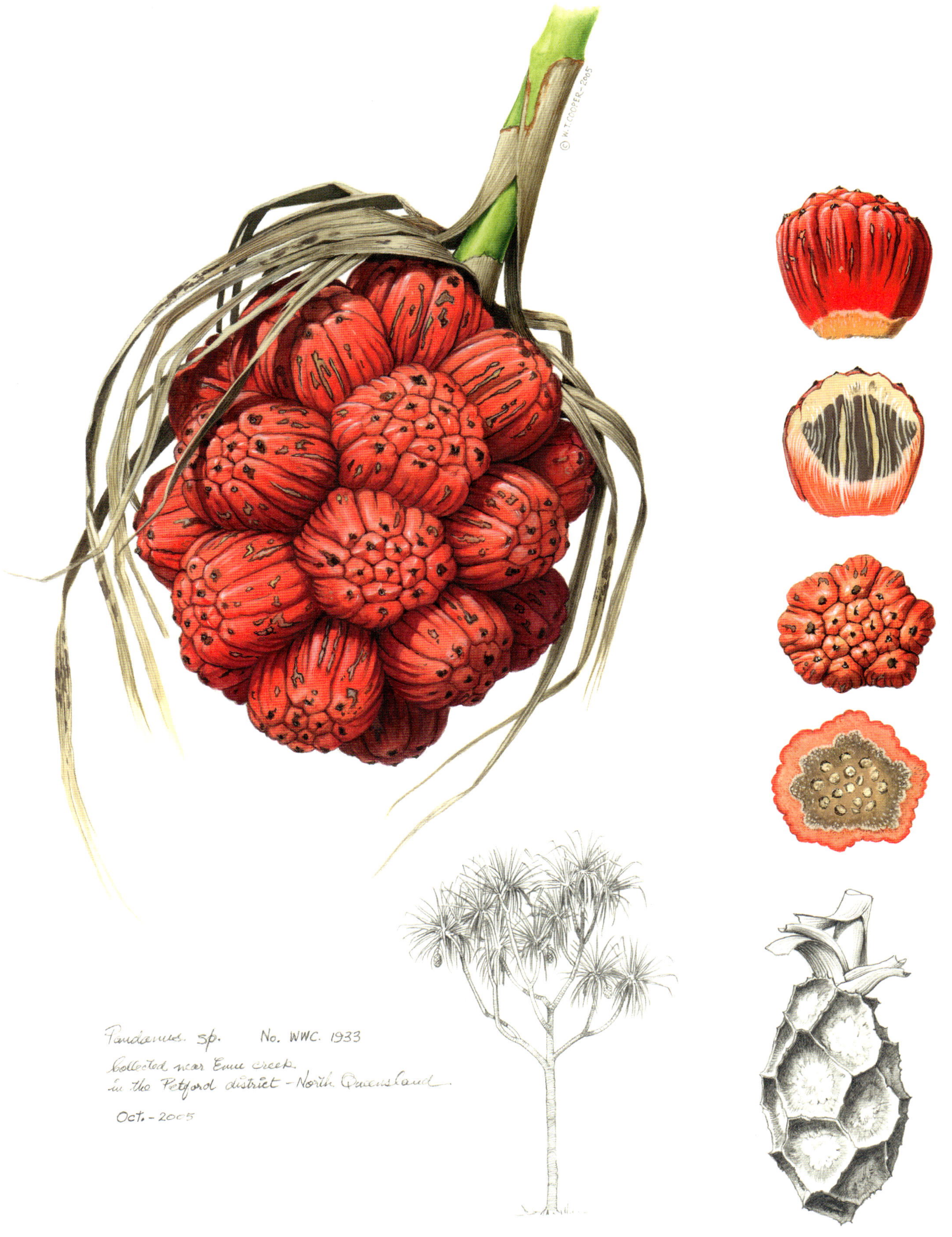
© W.T. COOPER – 2005
Pandanus sp.
No. WWC. 1933
Collected near Emu creek.
in the Petford district – North Queensland
Oct. – 2005

Persoonia linearis

Narrow-leaved Geebung

PROTEACEAE

Persoonia is named in honour of Christiaan Hendrik Persoon (1761–1836), a mycologist.

linearis is from *linearis* (linear), referring to the slender leaves.

Narrow-leaved Geebung is a common plant among *Angophora*, *Banksia*, Blackbutt and Grass trees in sclerophyll forests on old sand dunes at Bungwahl, coastal New South Wales. The tree has a different appearance from other species it grows beside, mostly because of the dense narrow leaves and twisted trunk with black papery bark that occasionally peels away exposing a red layer beneath. The succulent green-and-purple blotched fruit are eaten by birds and mammals.

The name Geebung is an Aboriginal name from the Dharug language of the Sydney area.

↓ 1978; pen, pencil and watercolour
285 x 400mm
State Library of New South Wales

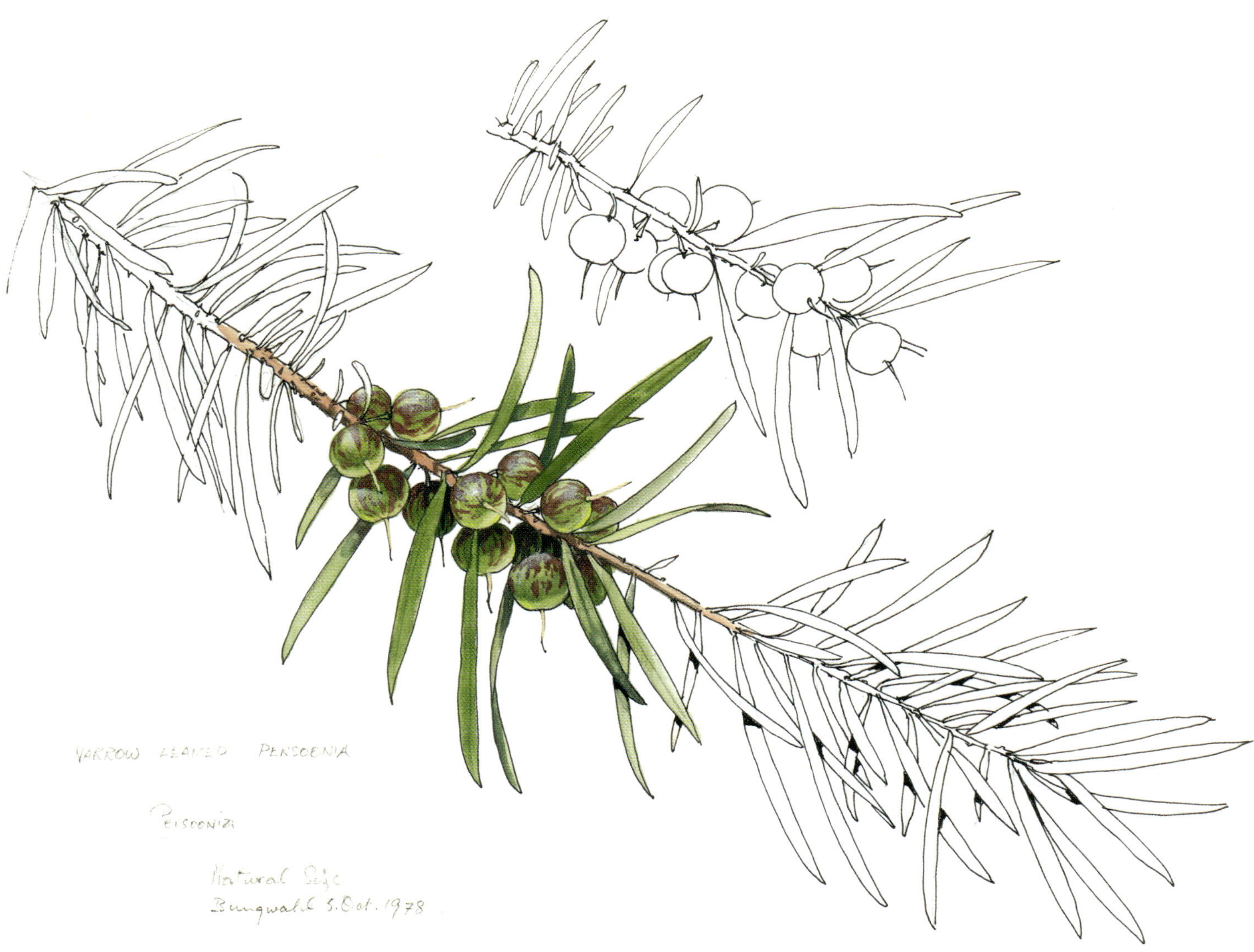

Planchonia careya

Cocky Apple

LECYTHIDACEAE

Planchonia is named in honour of French botanist Jules Émile Planchon (1823–1888).

careya is after the genus *Careya*, which is the name taxonomists initially gave this plant.

Cocky Apple is a deciduous shrub or small tree, mostly occurring in woodland. Red-tailed Black-Cockatoos and Palm Cockatoos are known to extract the seeds from these hard fruits. Bill thought its popularity with cockatoos and its attractive leaves, especially those turning red, were good reasons to have a sketch on hand, ready for a painting with either of the above-mentioned parrots. However, the plant is also common in the area where Red-winged Parrots occur and was used for that plate on the cover of *Australian Parrots*.

↓ 1988; pen, pencil and watercolour
430 x 580mm
State Library of New South Wales

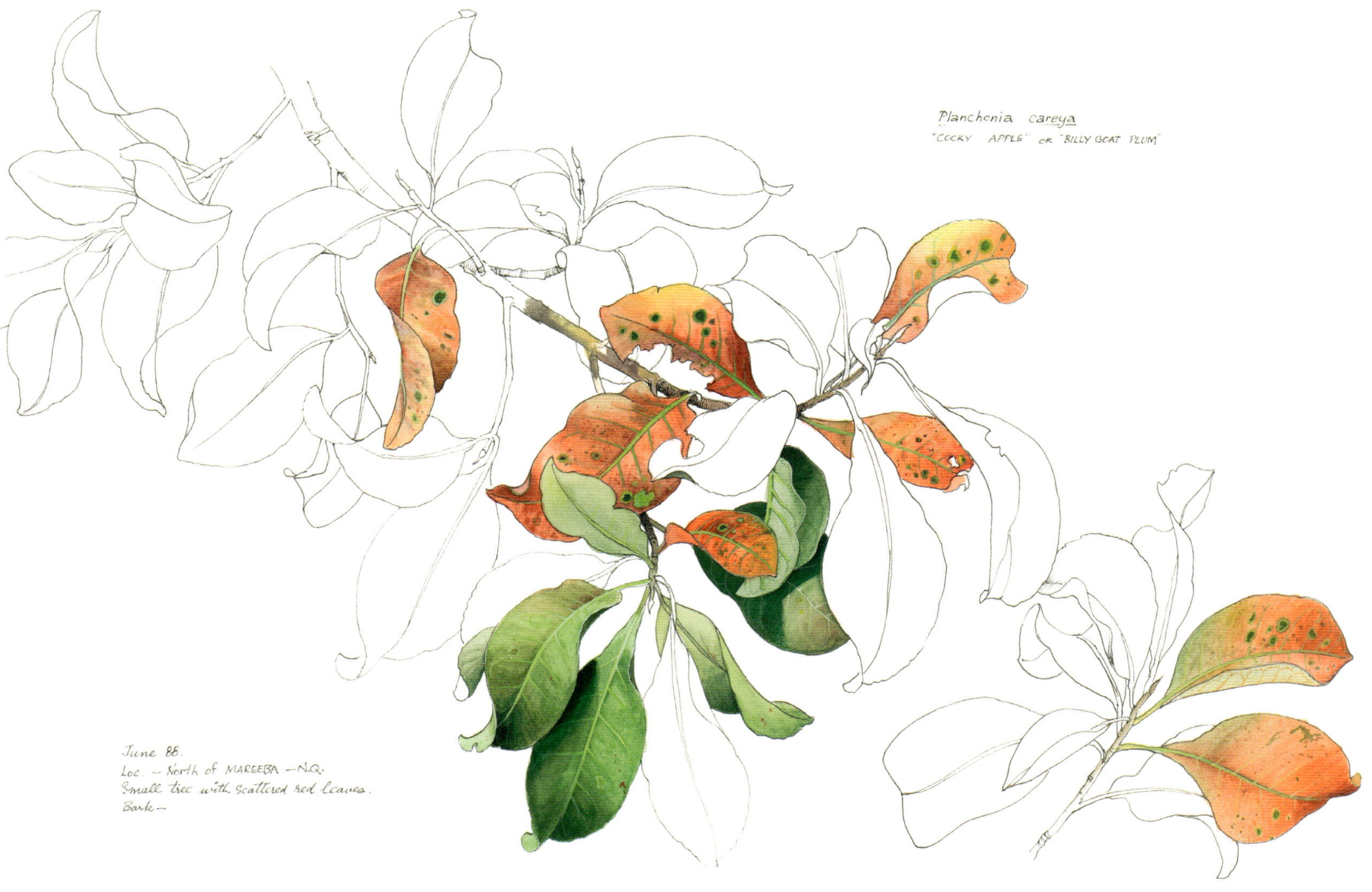

Syncarpia glomulifera

Turpentine

MYRTACEAE

→ 1979; pen, pencil and watercolour
370 x 285mm
State Library of New South Wales

↓ undated; pen and watercolour
400 x 275mm
State Library of New South Wales

Syncarpia is from *syn-* (together) and *carpos* (fruit), referring to the joined fruit segments.

glomulifera is from *glomulus* (little ball) and *-fer* (bearing), referring to the flowers.

Turpentines are grand trees with a particularly attractive branching habit, especially in New South Wales; they are less flamboyant in Queensland. The trees are related to *Eucalyptus* and grow in the wet sclerophyll forests of eastern Australia.

One of these sketches is from our former home at Bungwahl and the other from Mount Baldy near Atherton in north Queensland. Turpentine forest is home to Yellow-tailed Black-Cockatoos, one of Bill's most favourite birds. The distant sound of their mournful calls would usually have Bill running outside to watch these remarkable and acrobatic cockatoos. The older sketch from Bungwahl was used for a painting of Yellow-tails in *Australian Parrots*.

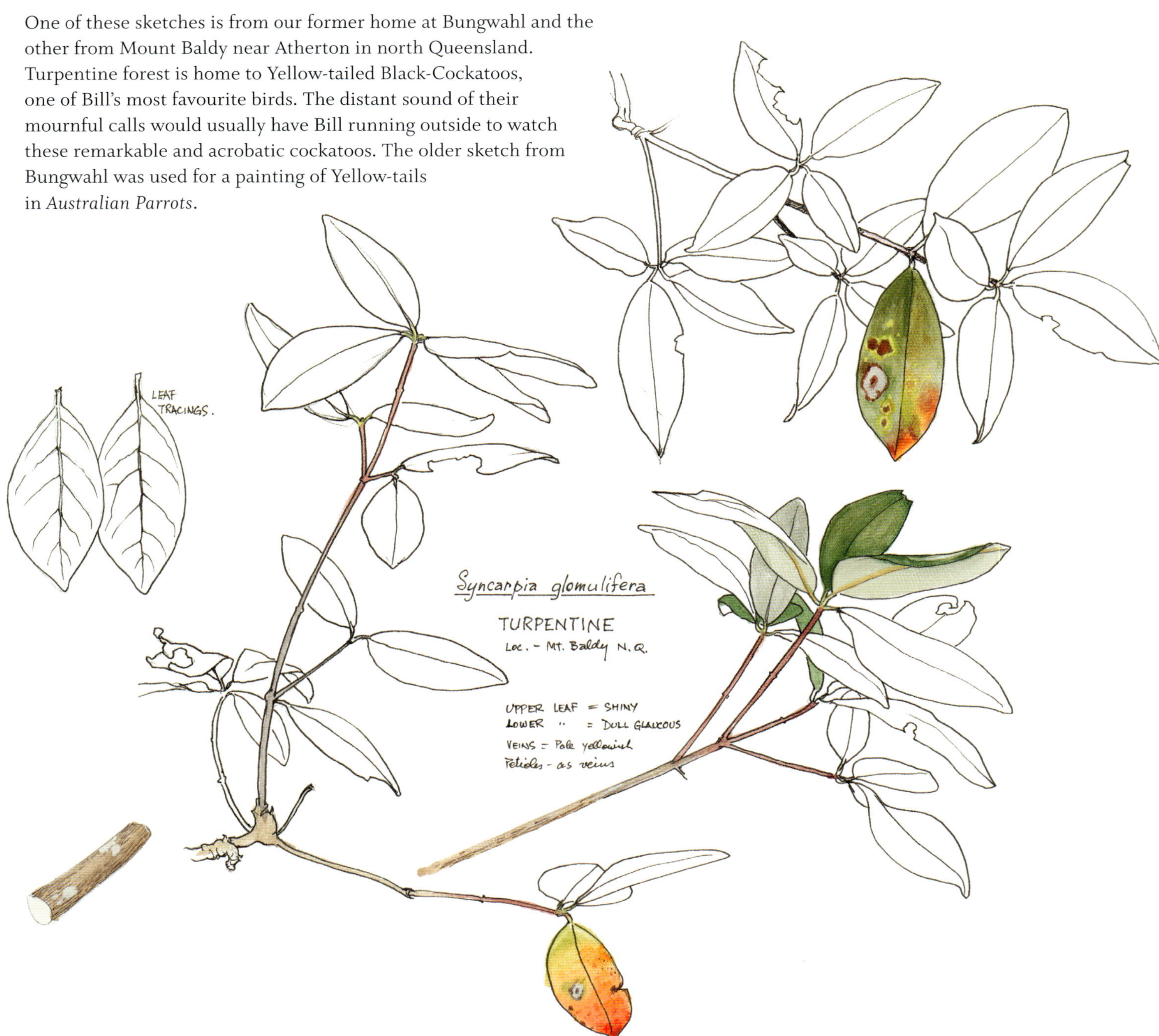

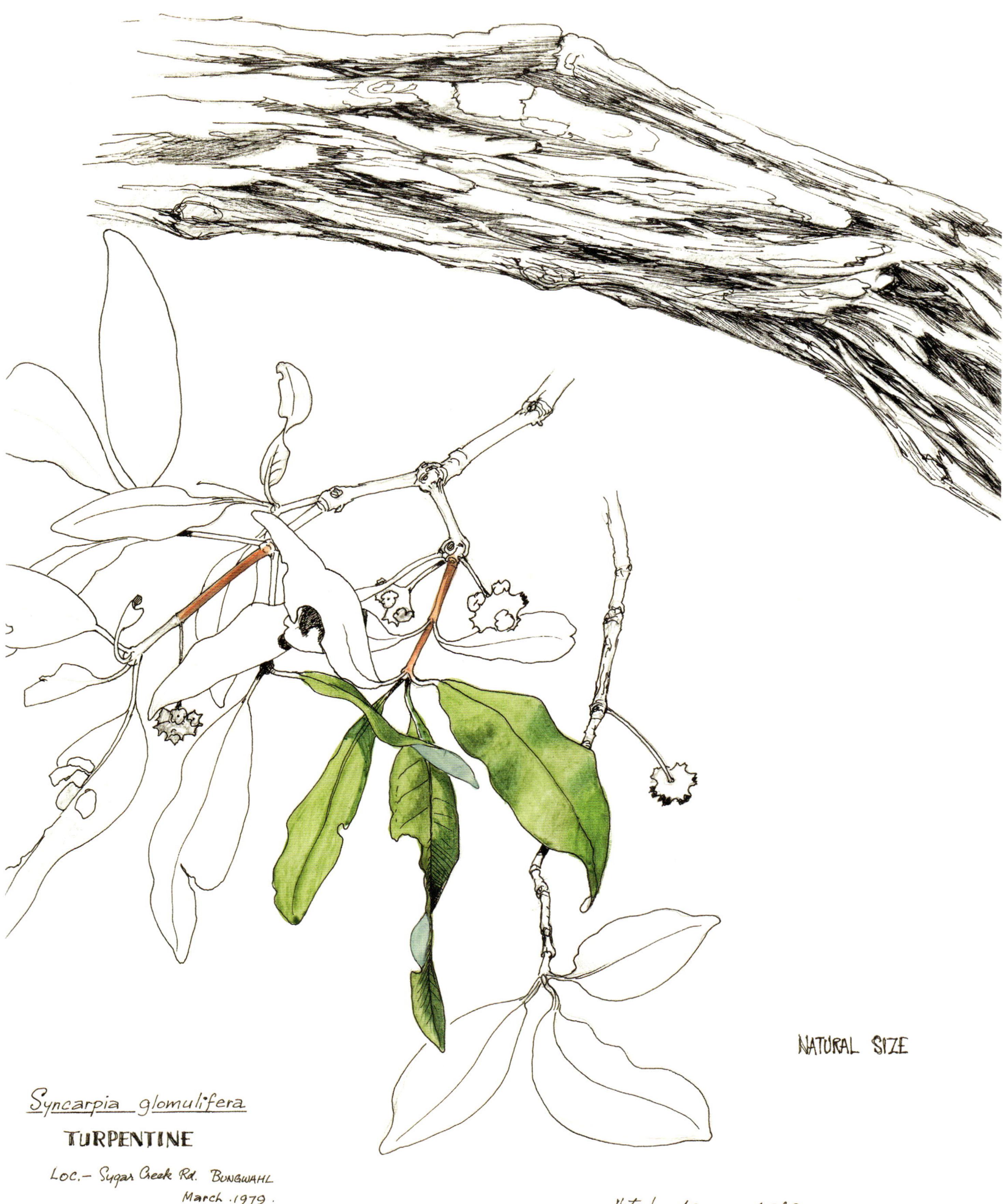
NATURAL SIZE
Syncarpia glomulifera
TURPENTINE
Loc.– Sugar Creek Rd. BUNGWAHL
March .1979 .
Note! Leaves not shiny

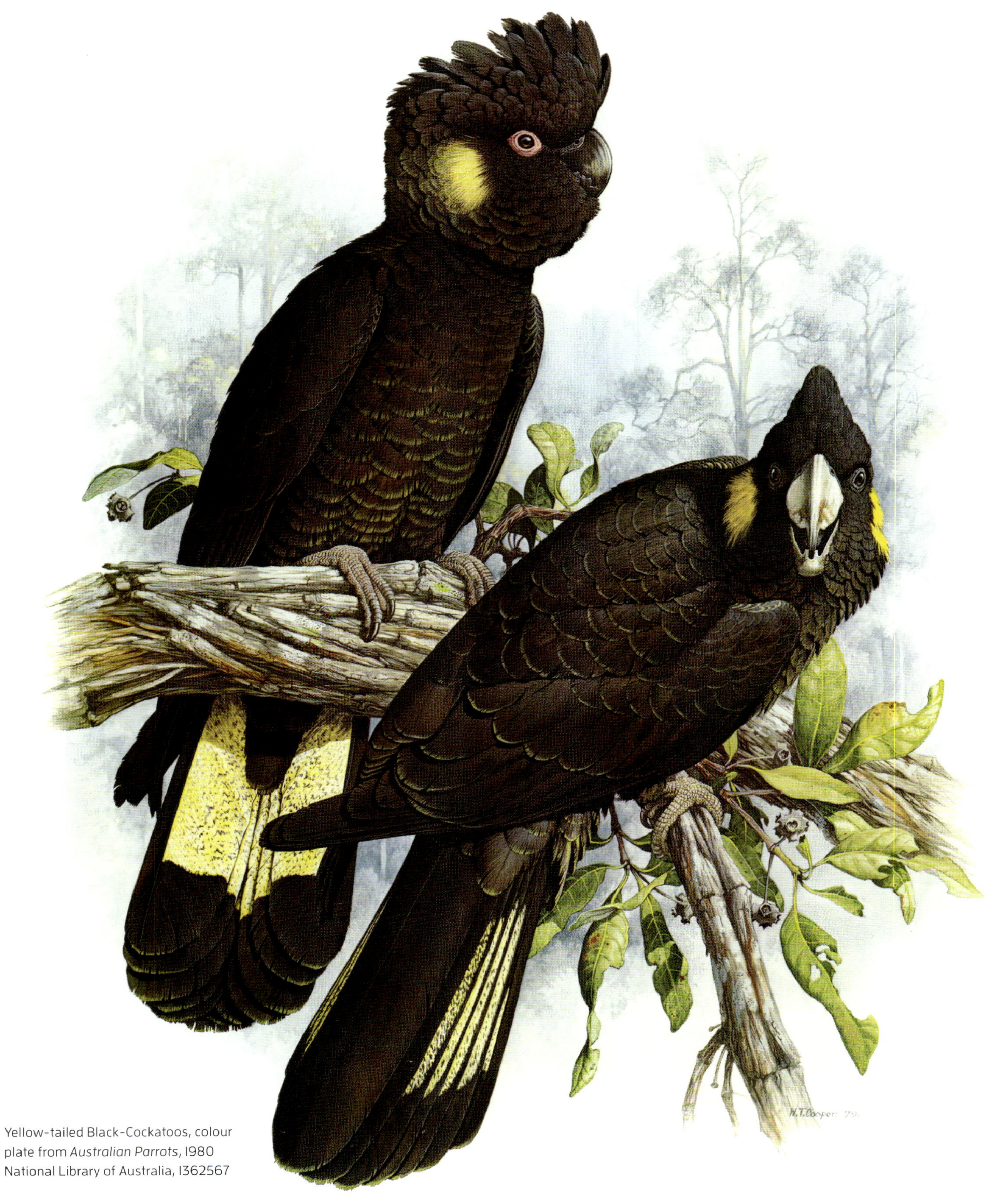

→ Yellow-tailed Black-Cockatoos, colour plate from *Australian Parrots*, 1980
National Library of Australia, 1362567

Terminalia platyphylla

Wild Plum

COMBRETACEAE

Terminalia is from *terminalis* (terminal), referring to the many species having cluster leaves at the tips of the branches.

platyphylla is from *platy-* (broad-) and *-phyllus* (-leaved).

Wild Plum is a deciduous tree that grows to 20 metres, mostly along creeks and rivers in the savannah of northern Australia. This sketch was made near Halls Creek on the O'Donnell River in Western Australia. With friends, we were on a field trip across the Top End to gather material for the book *Pigeons and Doves in Australia*. While this plant is not especially associated with pigeons, Bill saw Red-tailed Black-Cockatoos feeding in this tree so took advantage of an attractive plant with potential for use in a painting. He noted in his diary on Monday 12 July 2004: 'All started walk upstream. I turned back to do some work rather than walk for walking's sake. Drew some Terminalia—Black Cocky foodplant'.

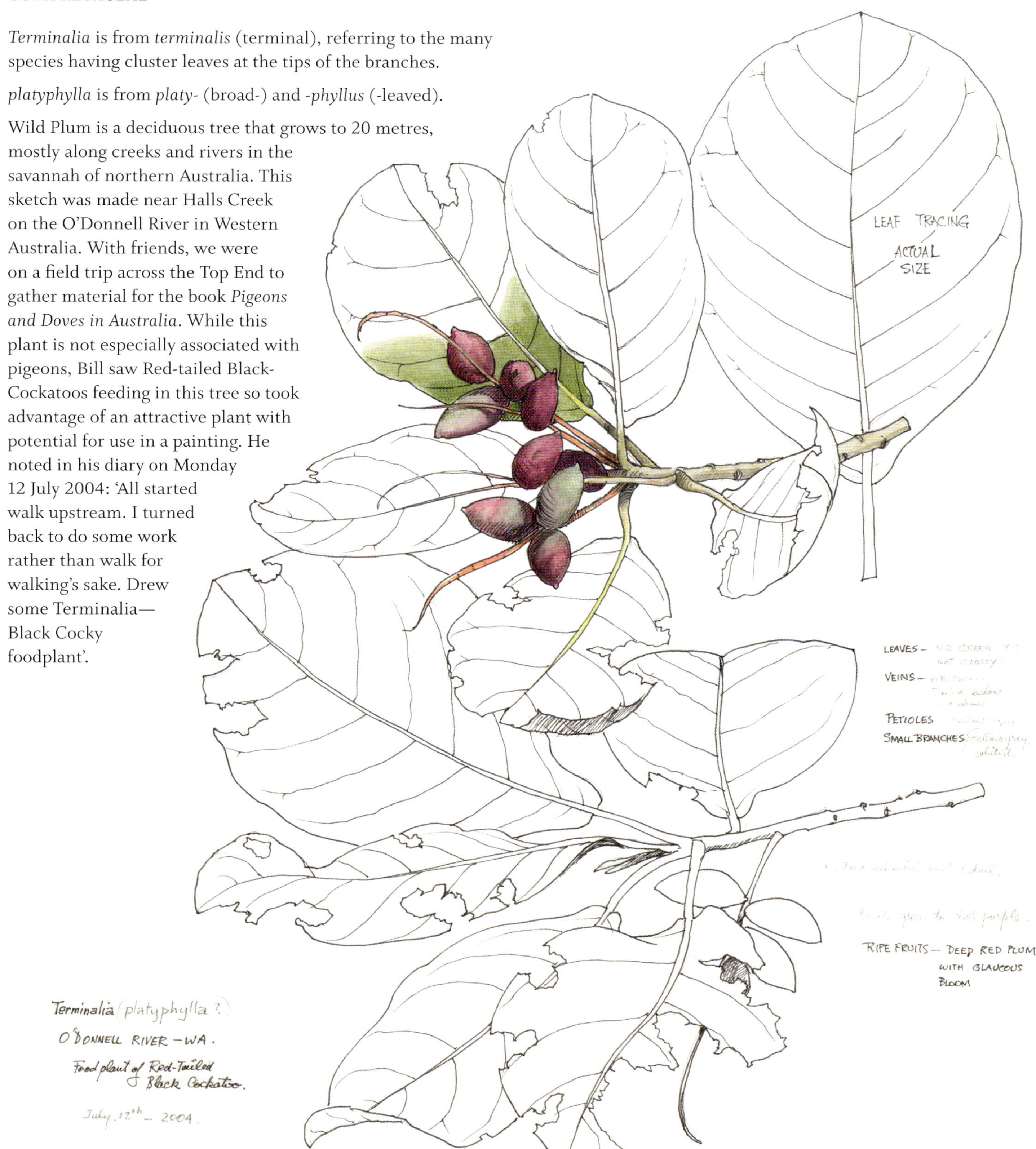

↓ 2004; pen, pencil and watercolour; 415 x 290mm
State Library of New South Wales

Chapter 4

Swamps, Mangroves & Beach Forest

Using his elbows to wriggle patiently forward between tussocks, as he slithered over the mud of a swamp, the boy Bill moved past a quizzical heron that watched his progress but did not lift off. A parting squawk from the heron would have alarmed ducks resting at the edge of the open water, the birds he wanted to take a close look at.

In keeping with Wordsworth's line—'the Child is father of the Man'—Bill retained his fondness for swamps and lagoons, and for their plants, birds and beasts. As an adult, he would sit by a lagoon recording mauve-flowered waterlilies floating on a placid surface, delighting when a Comb-crested Jacana trailing its chicks came lily-trotting through the scene. The complexity of the venation on the underside of a waterlily leaf absorbed Bill's attention; he included notes on the veins alongside his painted study. There is also a note on the colour of the comb in a Jacana sketch. Bill could tolerate mosquitoes in huge numbers as they feasted on an exposed arm, simply pausing occasionally to wipe them off. If a snake came by, it would be admired and allowed to pass, though in boyhood some of the snakes encountered were added to his live collection.

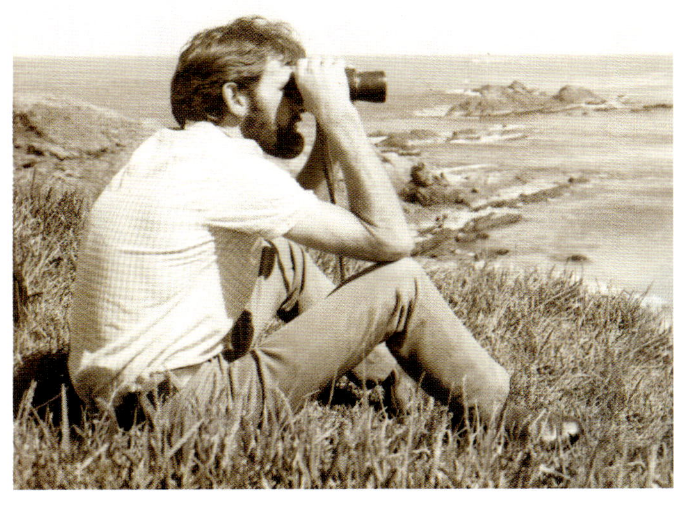

Sure-footed and nimble on rocks, Bill loved fishing from headlands, revelling in a stormy ocean's colours and turbulence. Paradoxically, the man who loved the ocean did not swim, nor did he have any patience for idling on a beach. Yet Bill took an interest in many of the trees that adorn the foreshore. A beach exposed to buffeting winds might be dominated by spindly *Casuarina*, but a bay protected by a rocky headland can nurture a variety of trees with broad leaves, several of which appeared in Bill's paintings.

To a study of the pink-centred flowers of Dog Bane (*Cerbera manghas*) a note was added: 'flowers have lovely perfume'. Bill showed the distinctive branching pattern of twigs on the tree known as a Sea Almond (*Terminalia catappa*) along with the bunched arrangement of leaves at the end of a twig. This handsome tree has leaves that turn an attractive brick-red with the advance of cold weather. It carries quite large and very tough fruit, encasing a small but tasty kernel, hence almond as part of its name. Bill admired the gatherings of Red-tailed Black-Cockatoos padding solemnly about under Sea Almonds, scissoring the fruit in order to neatly extract the rewarding kernel.

There is a tree with the imaginative but puzzling name of Sea Hearse (*Hernandia nymphaeifolia*), which bears nuts concealed within strange, almost translucent, shrouds. Bill depicted it with exquisite attention to the light and shadow on a cluster of long-stemmed leaves. The small hard nuts have given this plant traction on the shores of Africa, Sri Lanka, China and Hawaii, so its country of origin is uncertain. In Australia, the tree is known along the north-east coastline, and on a page in Bill's work.

→ *Barringtonia asiatica*
c.2003; pencil
State Library of New South Wales

↙ Bill beside Sugarloaf Point Lighthouse near Seal Rocks, above His Favourite Fishing Spot, c.1975

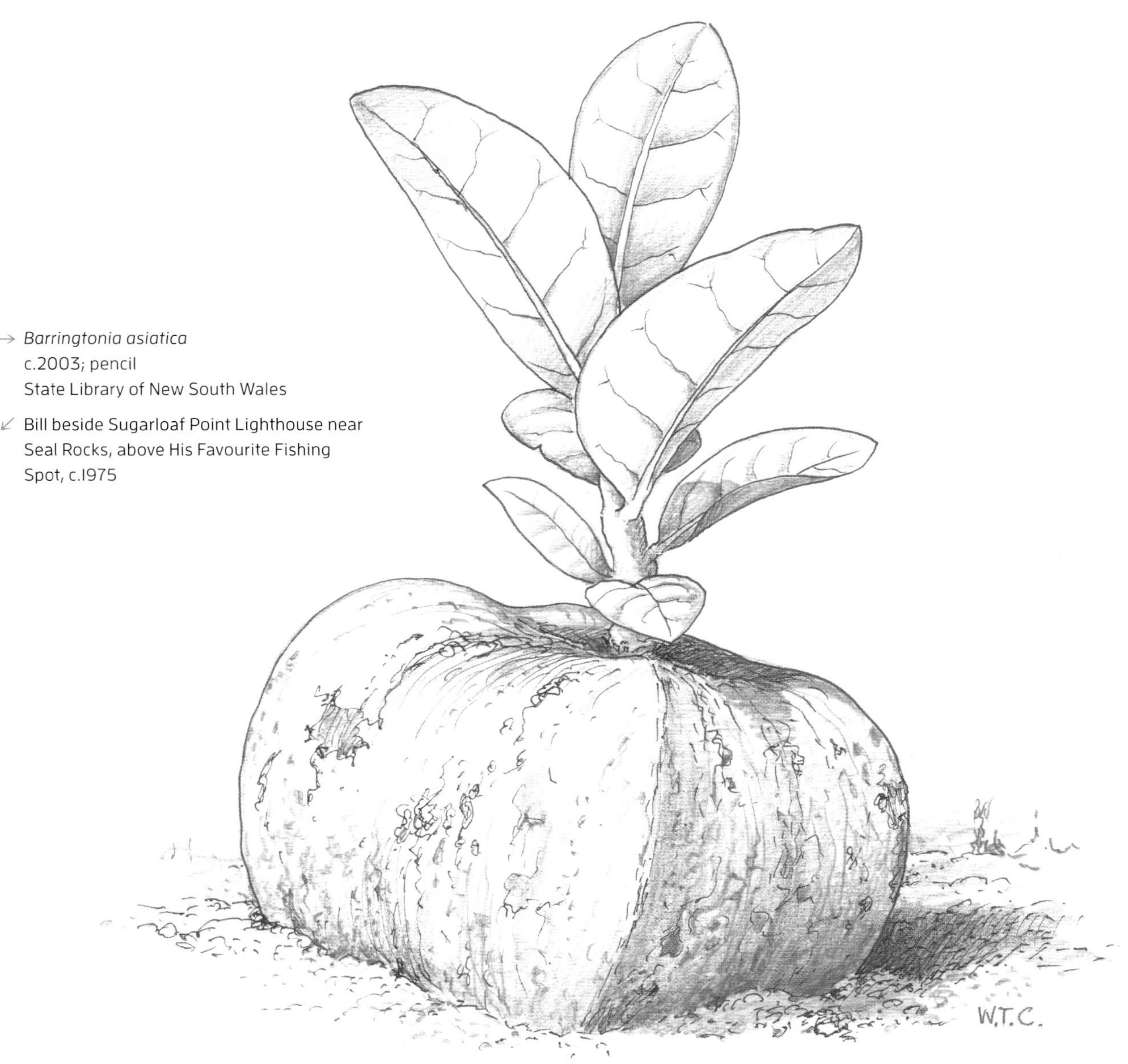

A small coastal tree produces large, fragile, pink and white flowers that open in the evening and drop off the following morning. The beauty of its flower is not reflected in the tree's name: Box Fruit (*Barringtonia asiatica*). The fruit, though, is an intriguing package, a life concealed within a dense mass of fibre. A Box Fruit might float across the oceans of the world, surviving months at sea, before being thrown ashore by a wave, at which point the embryo within can germinate, shooting out through the fibrous barrier to begin another lovely tree. The beachcomber in Bill recorded a Box Fruit well founded in sand, with a sturdy shoot aiming for the sky.

For many people, mangroves are a frightening, bewildering habitat—a fence of close stems, entangled and entangling roots, and the haunt of mosquitoes and biting midges. Yet playing within the mangroves is the comedy of fiddler crabs and goggle-eyed mudskippers, along with notable birds, such as the Shining Flycatcher. Flower, fruit and leaf details of the classic Red Mangrove (*Bruguiera gymnorhiza*) occupy a page, with its knobbly breathing-roots sketched on another.

Many of Bill's paintings may have been his response to beauty, a way of caressing the object and sharing it with the viewer. His flowering branch of a Golden Guinea Tree (*Dillenia alata*), for instance, is an exquisite vignette that both embraces the plant and presents it to the viewer. Bill particularly delighted in irregularities of form and colour. A Sea Almond could offer leaf after leaf chewed by insects, fretted by wind or coloured by autumn, whereas depicting the disciplined patterns on shells found along a beach was drudgery. Birds, the plants on which they depend and the unceasing sea were perpetual gifts to an artist who found beauty in complexity and in imperfect nature.

Rupert Russell

Avicennia marina

White Mangrove or Grey Mangrove

ACANTHACEAE

Avicennia is in honour of abu-Ali al-Husayn ibn-Sina (980–1037), Persian philosopher, physician and polymath.

marina is from *marinus* (occurring in the sea), referring to the mangrove habitat.

White Mangrove is the most widespread mangrove species in Australia. Like most other mangroves, it occurs predominantly in the tropical north. However, it also extends to Victoria and South Australia, much further south than any of the other mangrove species in the world.

Most mangrove trees are strict in their occurrence within precise tidal delimitations, either high, low or intermediate. White Mangrove occurs in all areas of the system. It has pencil-like above-ground respiratory roots (knee-roots), usually clustered near the tree base.

This reference was sketched in 1970 for use in the Blue-rumped Parrot plate in *Parrots of the World*. The bird was known to occur in the mangroves of Malaysia, so White Mangrove was a safe plant to use with this bird. The plant was also used later for the Paradise Crow plate in *Birds of Paradise and Bower Birds* because, again, the bird was known from the mangroves, this time on Halmahera, the largest island within North Maluku province, Indonesia.

→ Four Species of Parrot (*Psittinus* and *Bolbopsittacus*)
1970; gouache; 455 x 355mm
National Library of Australia, 1812925

↘ Paradise Crow, colour plate from *Birds of Paradise and Bower Birds*, 1977
National Library of Australia, 69465

↓ 1970; pen, pencil and gouache; 255 x 350mm
State Library of New South Wales

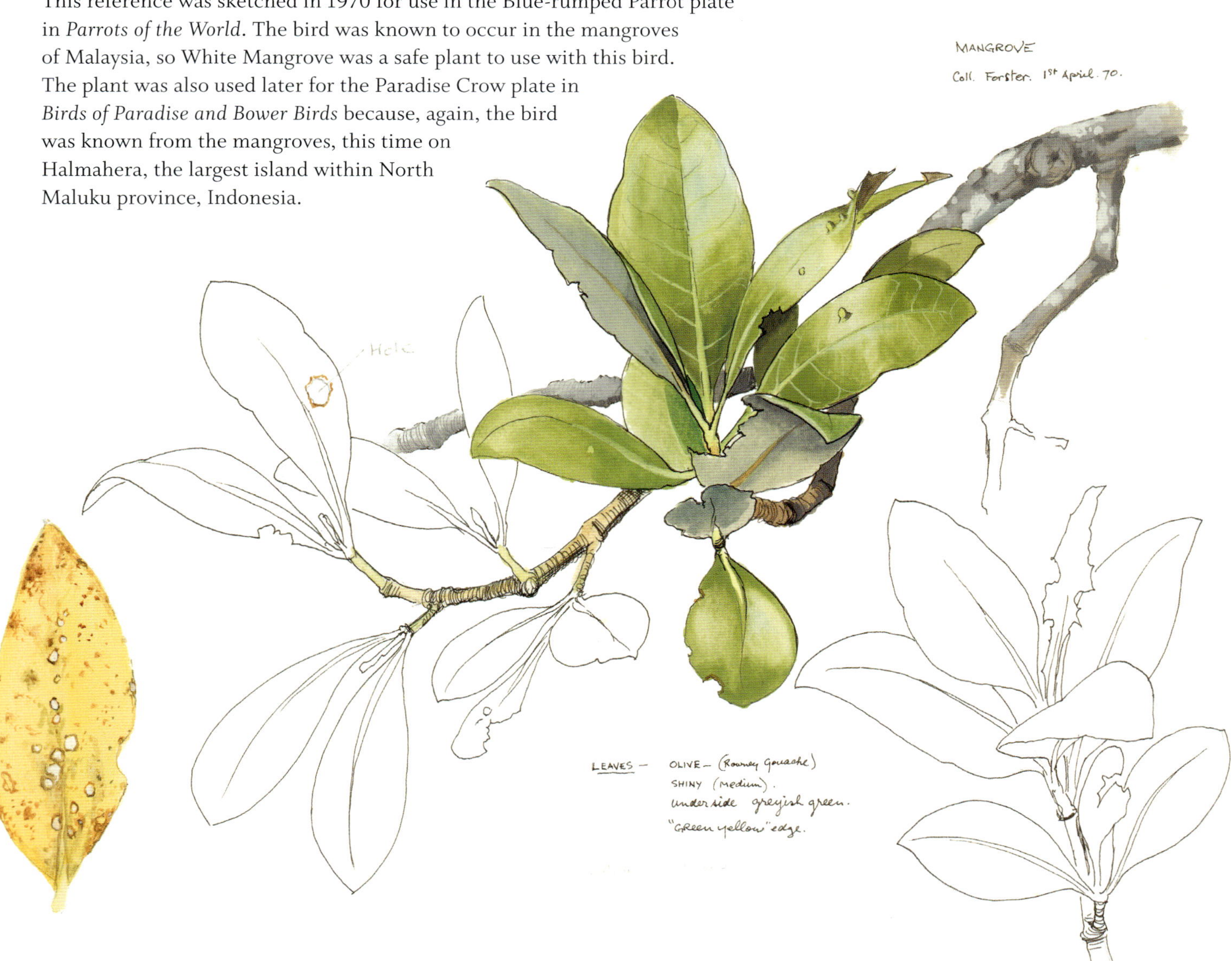

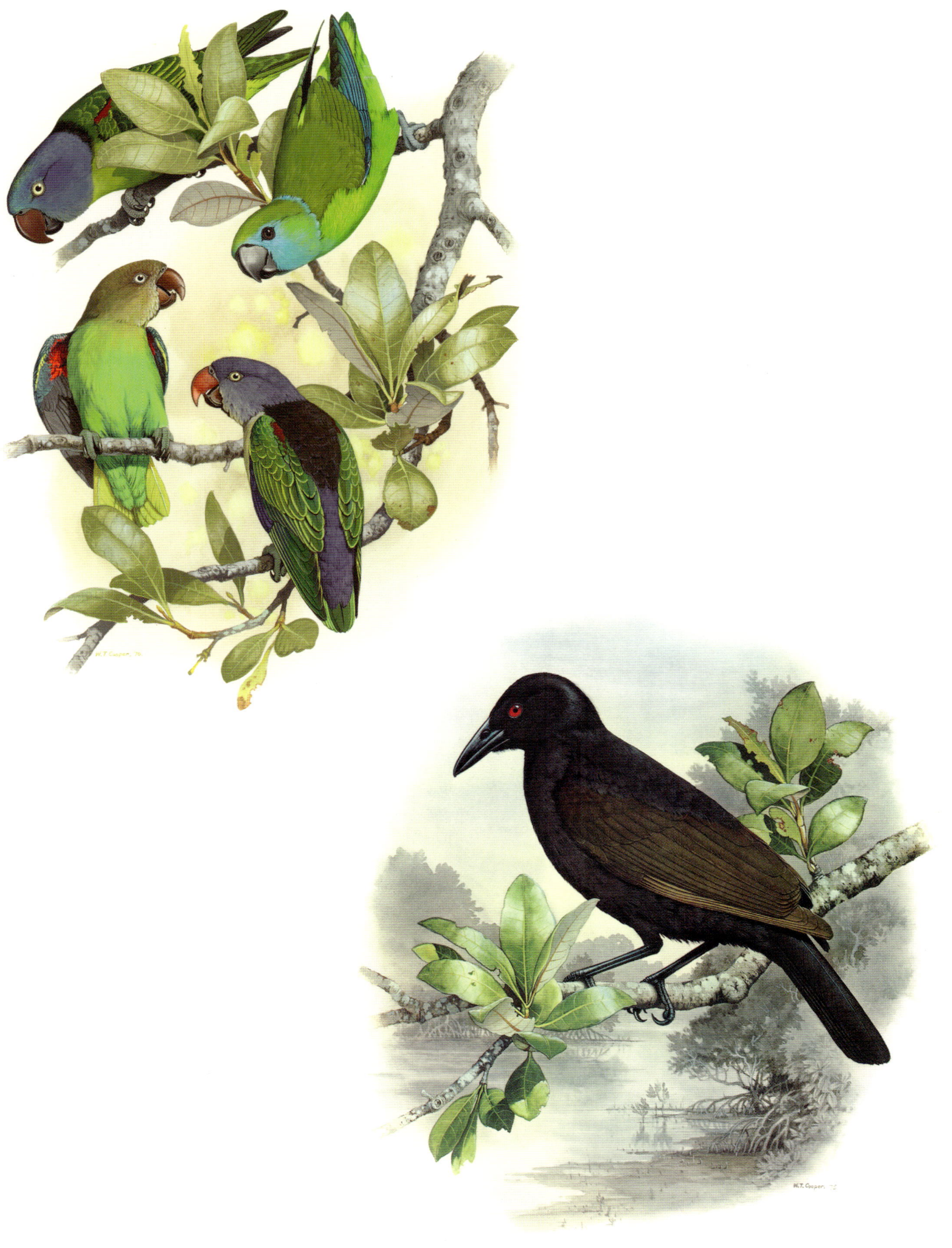

Bruguiera gymnorhiza

Red Mangrove

RHIZOPHORACEAE

Bruguiera is named after Jean Guillaume Bruguière (1750–1798), French explorer and biologist.

gymnorhiza is from *gymnos* (naked) and *rhiza* (root), referring to the numerous above-ground knee roots.

Red Mangroves can become large trees of up to 35 metres, occurring in most tropical areas of the world, with the exception of the Americas and West Africa. Curious raised knobs, or knee-roots, are formed where the roots branch. The flower-like red fruit is often found along tropical beaches at the high tide mark. Seeds germinate while still attached to the parent plant, forming fleshy, bean-like green propagules. Many of these drift with the ocean tides but those that drop and spear into the mangrove mud become seedlings.

Mangroves are rich habitats for many species, providing nectar, fruit, seeds, insects, nesting sites for birds and protection for mud-dwellers. Seedling plants are often used by kingfishers as perches from which to plunge for prey. The reference drawing opposite was used for the painting of Trumpeter Hornbills (on page 229), known to inhabit mangroves in Kenya where the plant also occurs.

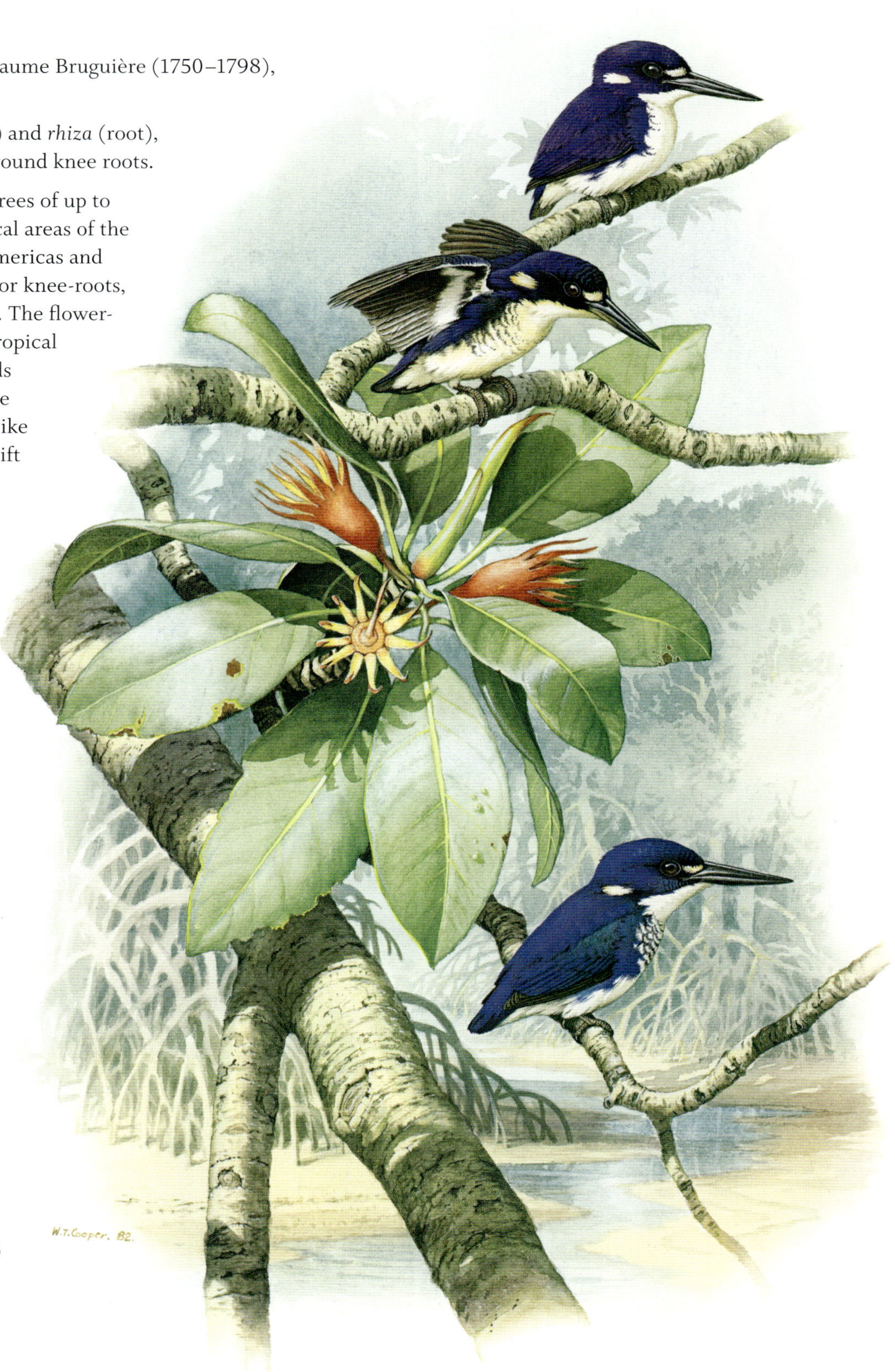

→ Little Kingfisher, colour plate from *Kingfishers and Related Birds*, 1982
National Library of Australia, 2494045

→→ undated; pen, pencil and watercolour
520 x 367mm
National Library of Australia, 8068773

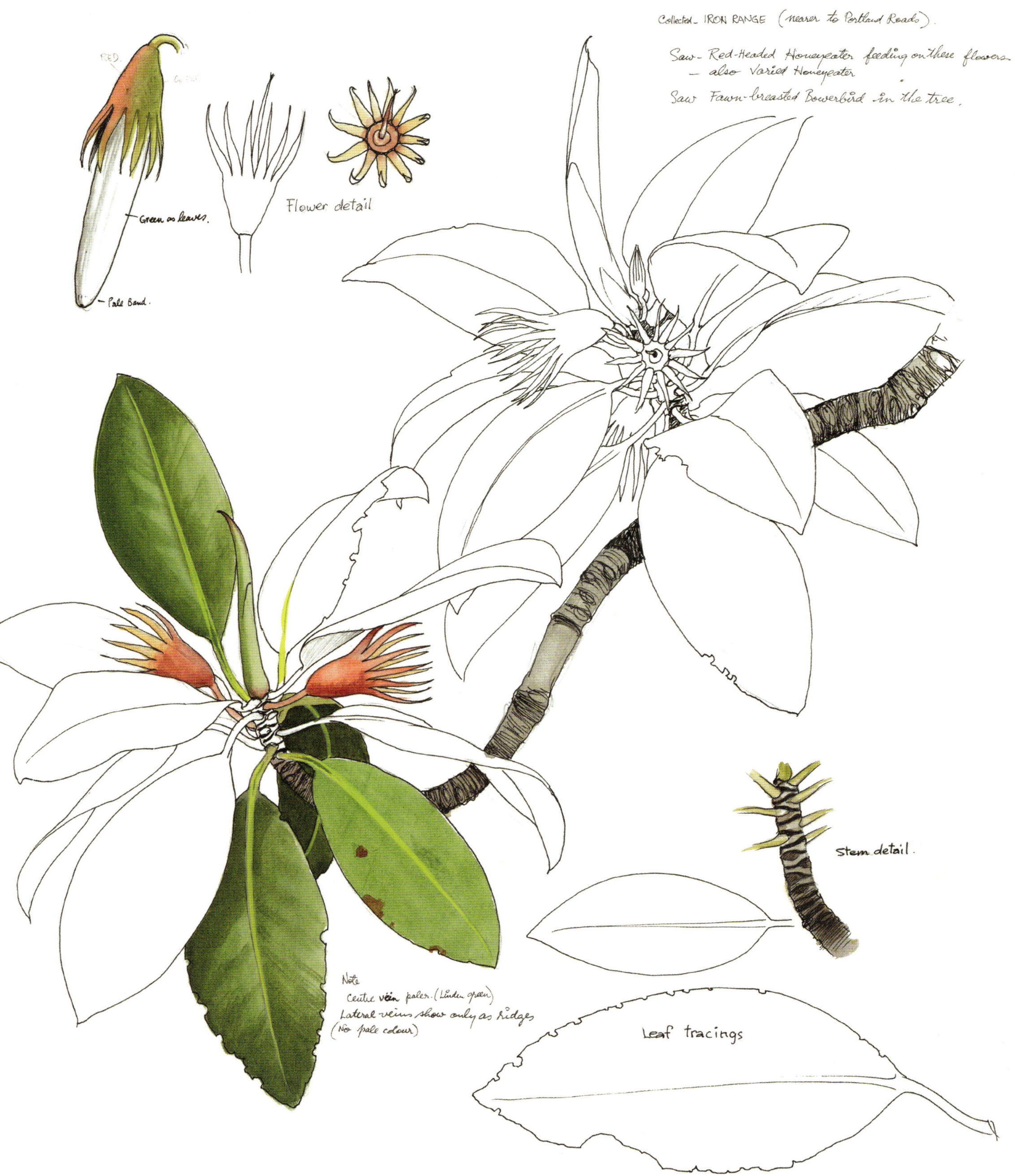
LARGE-LEAFED ORANGE MANGROVE
Bruguiera gymnorrhiza
Collected - IRON RANGE (nearer to Portland Roads).
Saw - Red-Headed Honeyeater feeding on these flowers.
- also Varied Honeyeater
Saw Fawn-breasted Bowerbird in the tree.
RED
Green as leaves.
Pale Band.
Flower detail
Stem detail.
Note
Centre vein paler. (Linden green)
Lateral veins show only as ridges
(No pale colour)
Leaf tracings

→ Trumpeter Hornbill, colour plate from *Kingfishers and Related Birds*, 1993
National Library of Australia, 2494045

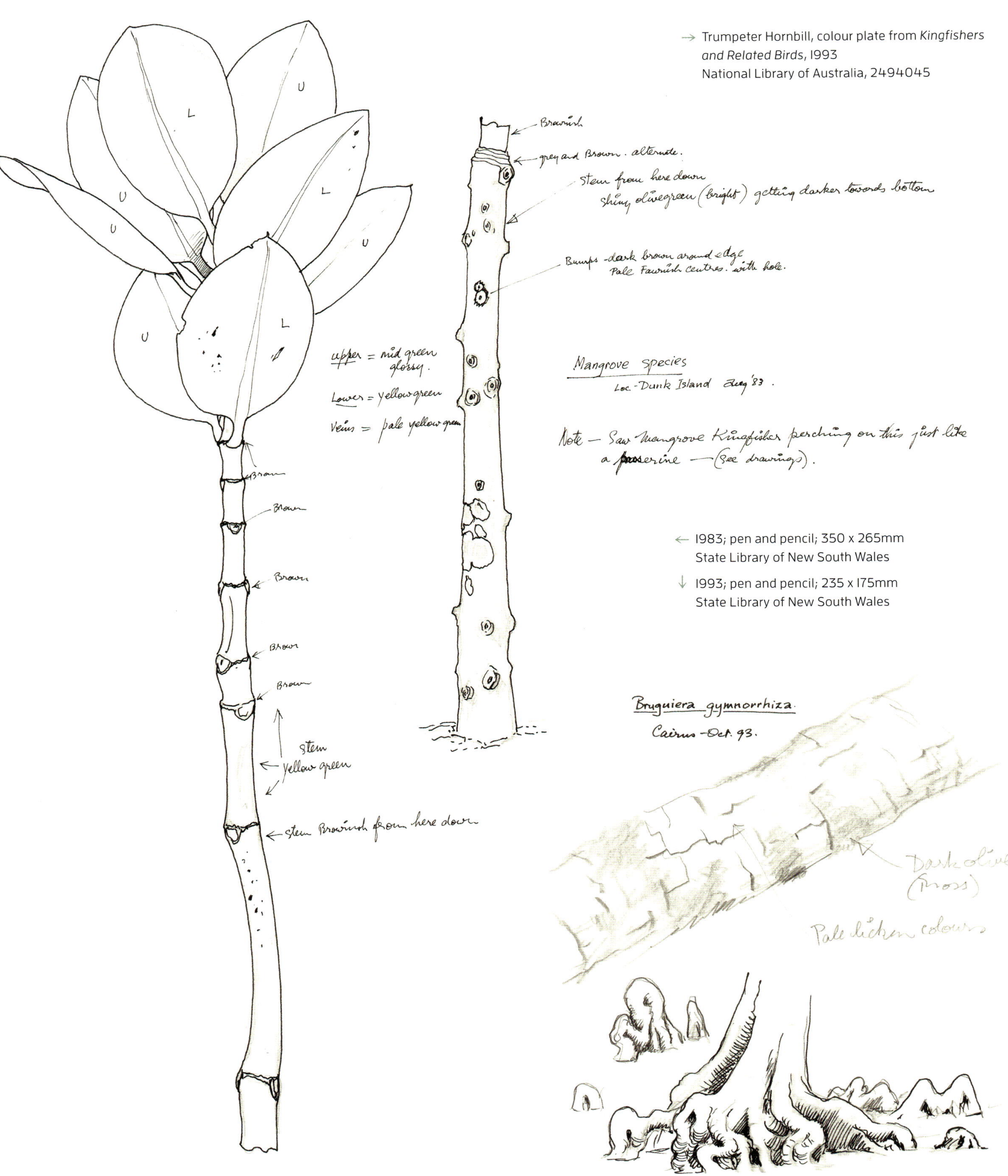

← 1983; pen and pencil; 350 x 265mm
State Library of New South Wales

↓ 1993; pen and pencil; 235 x 175mm
State Library of New South Wales

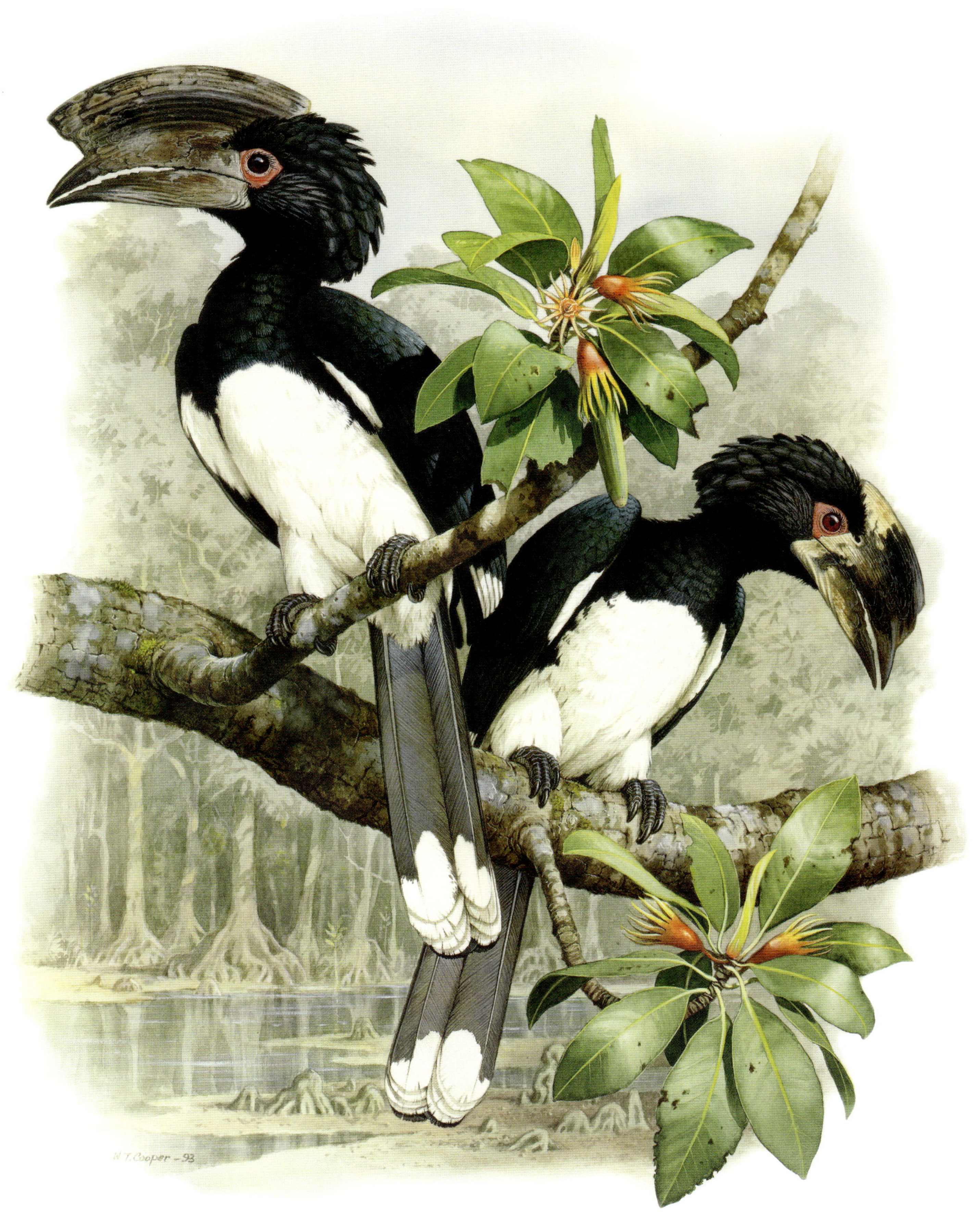
W.T.Cooper -93

Carpobrotus glaucescens

Pigface

AIZOACEAE

Carpobrotus is from *carpos* (fruit) and *brota* (edible), referring to the edible fruit.

glaucescens is from *glaucus* (blue-grey), referring to the leaves.

Pigface is a common plant that grows on beaches and sand dunes around the Australian coast. It is a wonderful sand stabiliser as it forms roots at each node. The bright pink-purple flowers are always a delight in the sunny exposed positions that they favour. The succulent leaves are salty but edible.

This painting was done at Treachery Head near our old home at Bungwahl, New South Wales. We were on our way home to north Queensland after Bill had had a highly successful exhibition of paintings at Morpeth near Newcastle, and we were enjoying the nostalgia and familiarity of our old home turf.

↓ 2003; pen and watercolour; 190 x 280mm
State Library of New South Wales

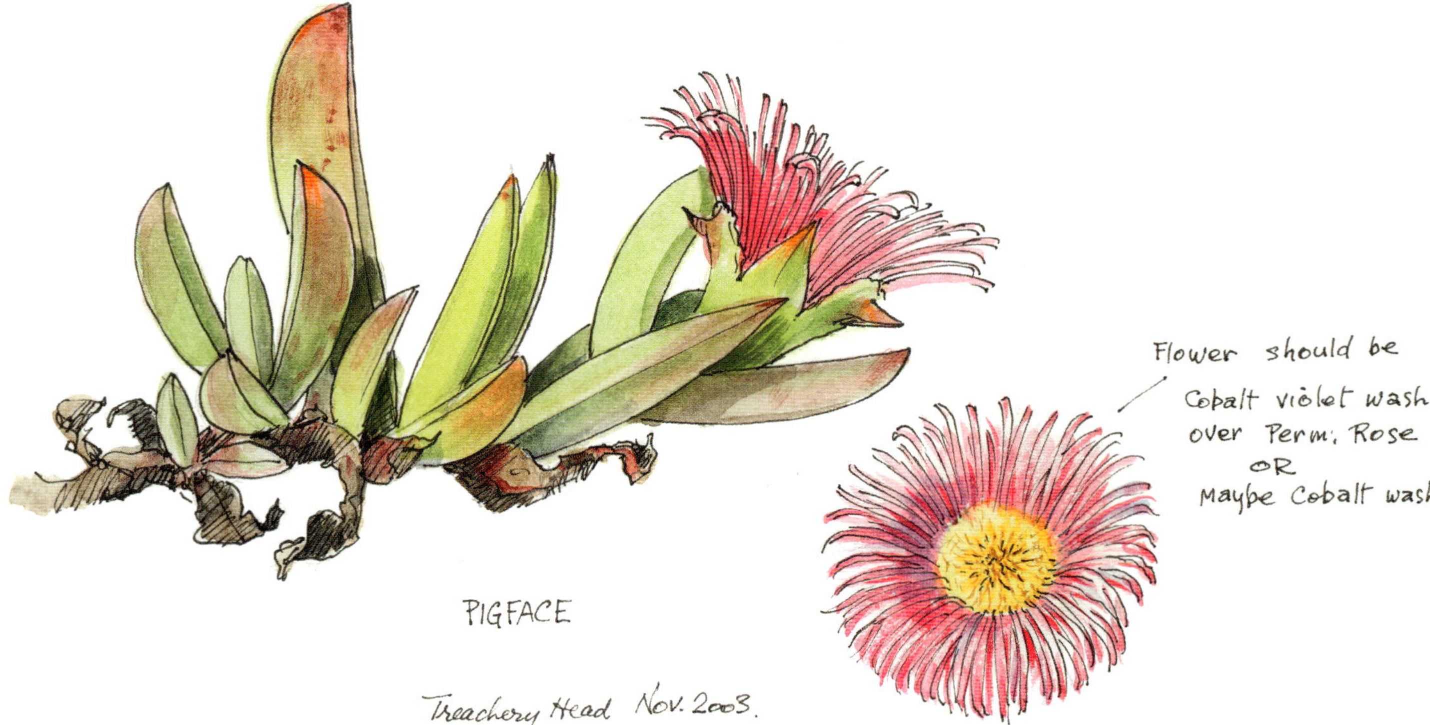

Cerbera manghas

Rubber Tree, Dog Bane or Pink-eyed Cerbera

APOCYNACEAE

Cerbera is named in honour of the mythological Greek dog Cerberus, referring to the fruit, which is often called Dog Bane, presumably because it is poisonous.

manghas is after the genus *Manghas* (mango), referring to the mango-like fruit.

Pink-eyed Cerbera is named for its pink-centred and delicately fragrant white flowers with a diameter to 35 millimetres. The tree is spectacular in every way, with beautiful flowers, large glossy dark-green leaves and large red mango-shaped fruit. It occurs in coastal rainforest north from Noosa, Queensland, to the Northern Territory, New Guinea and Asia.

↓ 1993; pen and pencil; 200 x 290mm
State Library of New South Wales

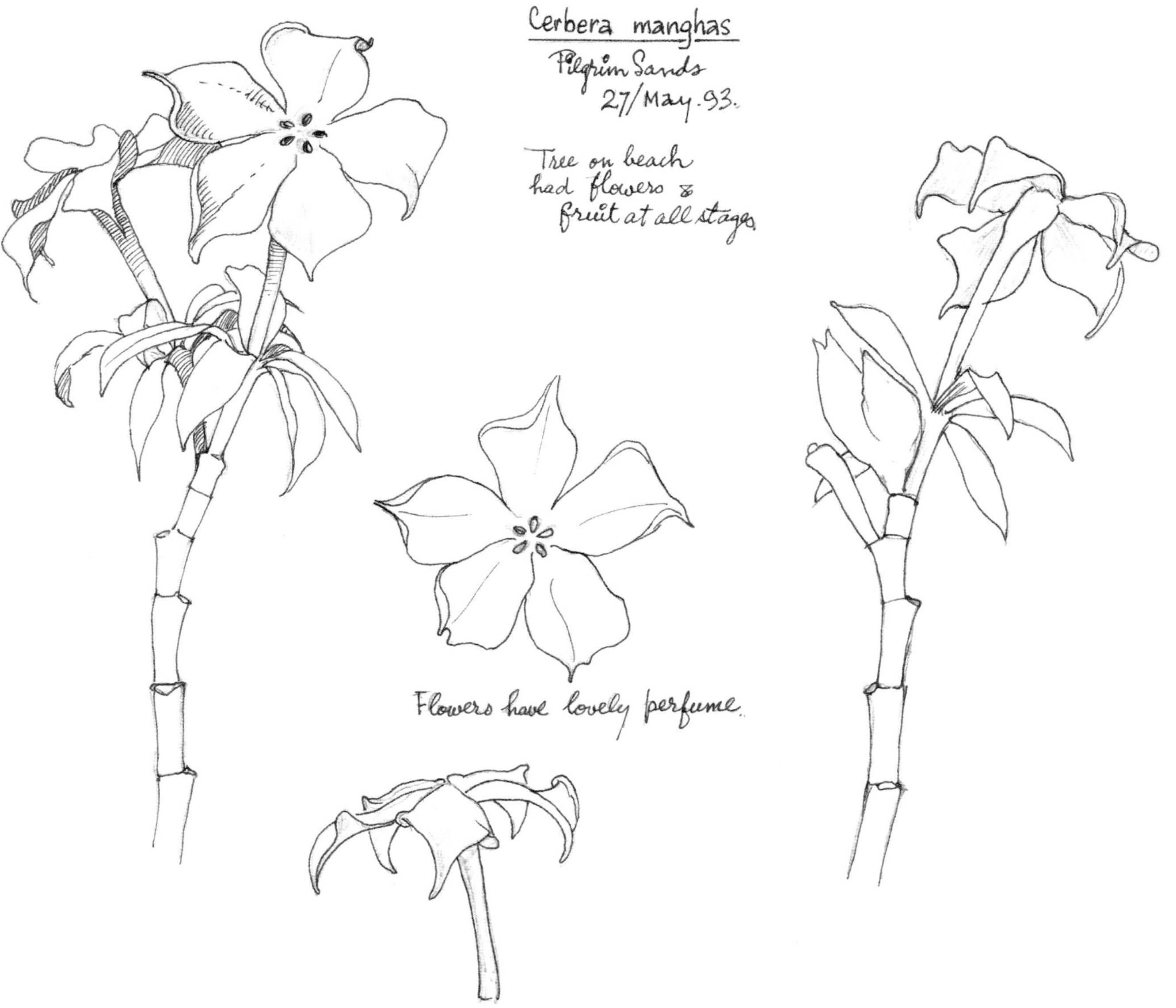

Cocos nucifera

Coconut Palm

ARECACEAE

Cocos is from the Spanish and Portuguese *coco* (grimace or mask), referring to the monkey face pattern on the seed.

nucifera is from *nucifer* (nut-bearing).

Coconut is a familiar fruit to most people, as is the Coconut Palm to those who have visited tropical beaches. Coconut Palms are widely dispersed along tropical coastlines, usually growing at the top of the beach. The huge seeds are enclosed in a buoyant husk, which can last for some time in the salt water.

After *Fruits of the Rainforest* was completed, Bill and I began work on *A Beachcomber's Guide*. The Coconut was one of the first pieces for the book but not long into the project we abandoned it. Although Bill's paintings are considered detailed, he disliked immensely the monotony of painting the precise geometric patterns and subtleties in many shells.

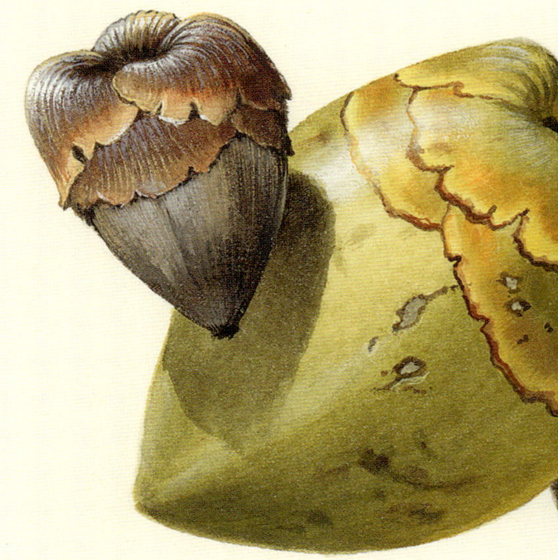

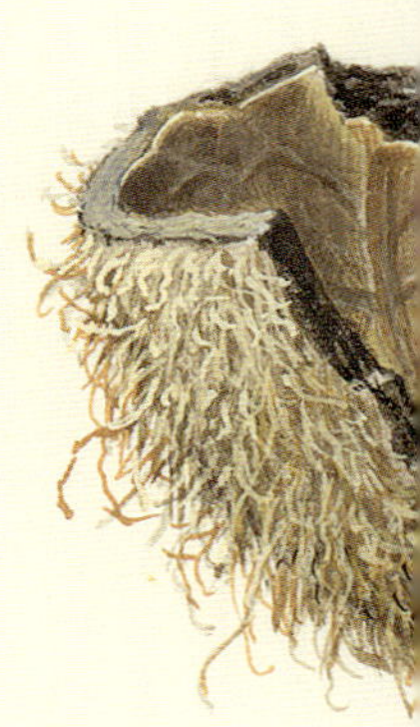

→ 2005; watercolour; 375 x 550mm
State Library of New South Wales

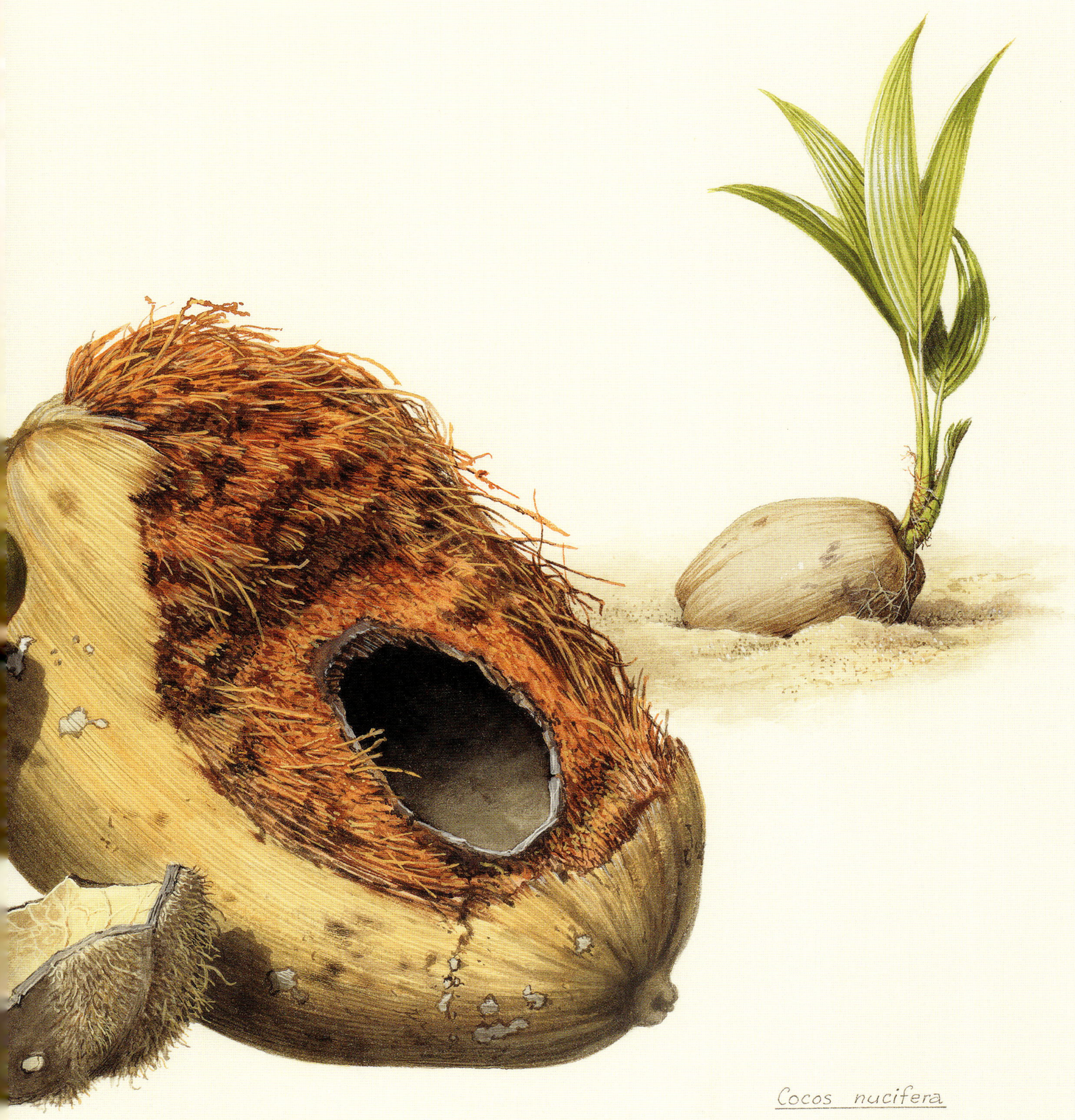
Cocos nucifera
Thornton beach Q – 16th June 2005

Dillenia alata

Red Beech or Golden Guinea Tree

DILLENIACEAE

Dillenia is named after Johann Jacob Dillenius (1684–1747), a German botanist who became a Professor of Botany at Oxford.

alata is from *alatus* (winged), referring to the leaf stalks or petioles.

Red Beech or Golden Guinea Tree is a distinctive tree with red papery bark, mostly found growing in coastal swamp forests from Cardwell to the Torres Strait in Queensland, the Northern Territory and New Guinea.

A few bright yellow flowers are present at most times of the year. The fruit are red flower-shaped capsules with black seeds enclosed in fleshy white arils. Most rainforest trees have small, insignificant white flowers but this species has such large showy flowers that Bill made them the subject of complete botanical paintings on a couple of occasions.

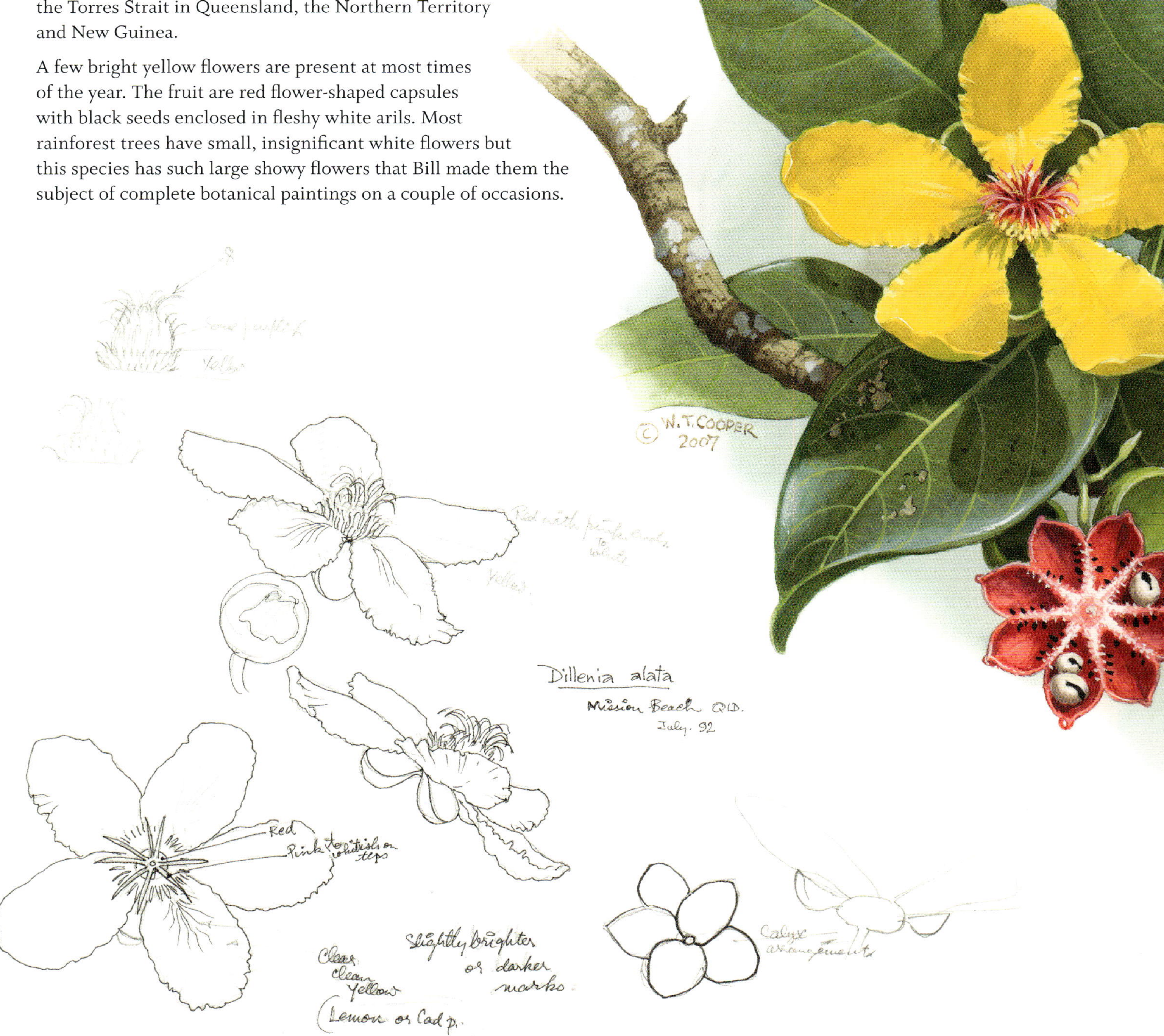

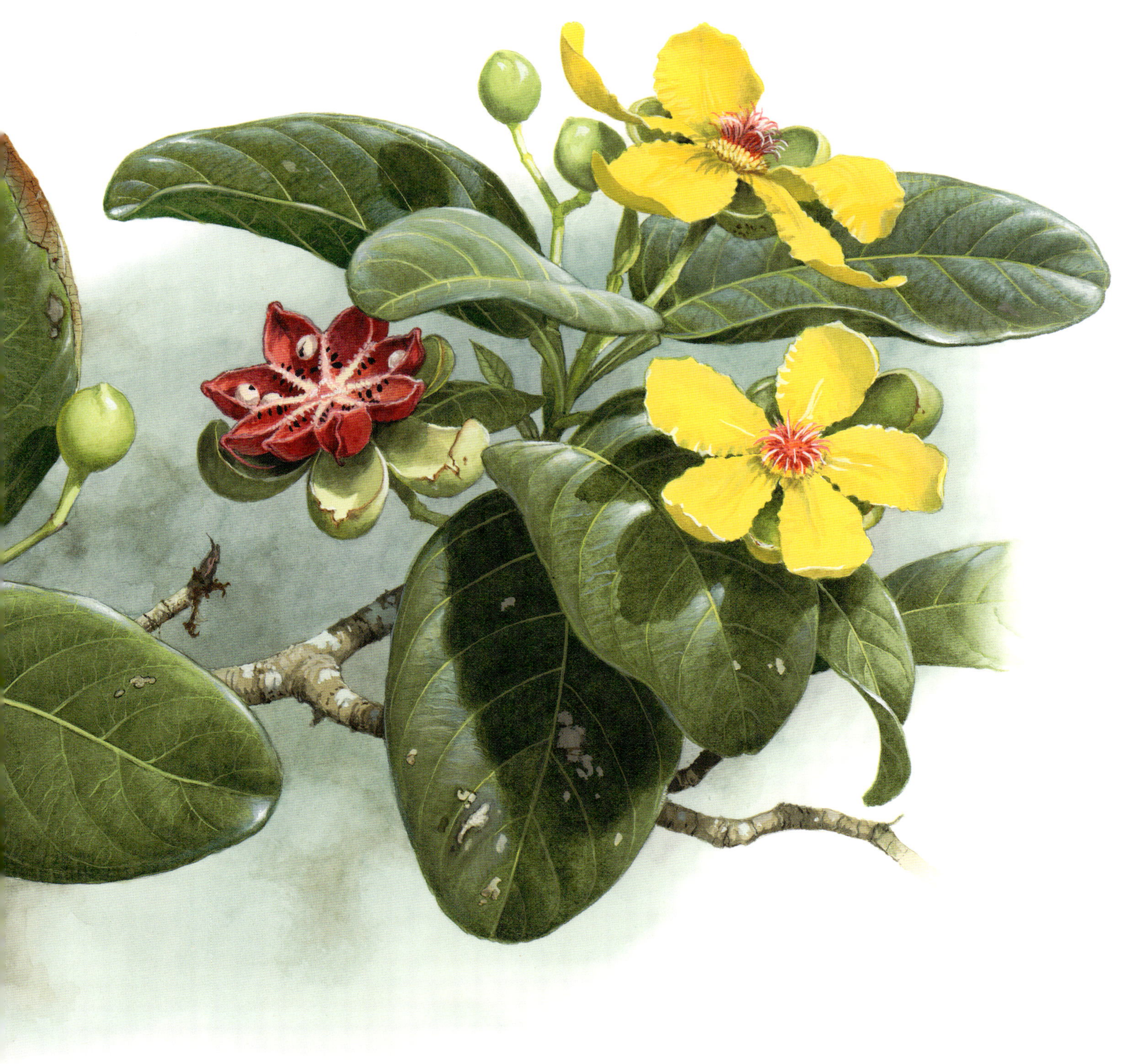

GOLDEN GUINEA TREE

—*Dillenia alata*—

↑ 2007; watercolour; c.560 x 390mm
Collection of Nigel and Tracey Tucker

← 1992; pen and pencil; 280 x 210mm
State Library of New South Wales

Erythrina variegata

Indian Coral Tree

FABACEAE

→ undated; pen and watercolour; 550 x 450mm
State Library of New South Wales

↓ undated; pen; 295 x 405mm
State Library of New South Wales

Erythrina is from *erythros* (red), referring to the red flowers on the type species.

variegata is from *variegatus* (variegated), referring to the variously shaped leaves.

Indian Coral Tree is a striking sight when in flower beside quietly lapping waves on beach sand. As a widespread species occurring in coastal areas in the world's tropics, including north Queensland, this plant was ideal for use in the painting of Ducorp's Corella, an endemic species to the Solomon Islands. The red flowers and thorny branches also added colour and contrast to the painting of the White Cockatoo for *Cockatoos: A Portfolio of All Species*.

Near Mission Beach there is a lovely Indian Coral Tree overhanging a secluded tropical beach, a beautiful place that we frequently visited. Flowers were collected from this tree for the illustrations shown here.

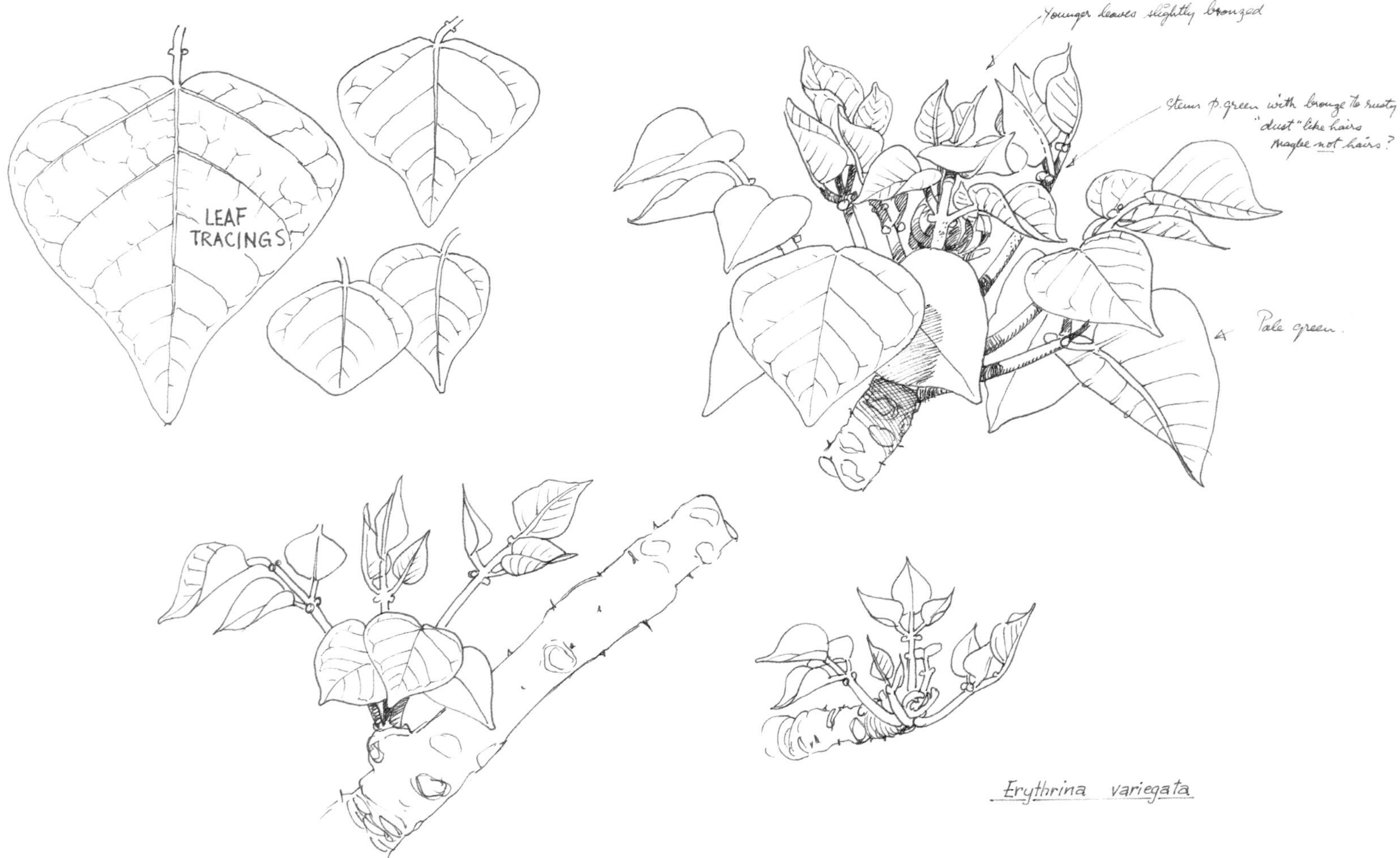

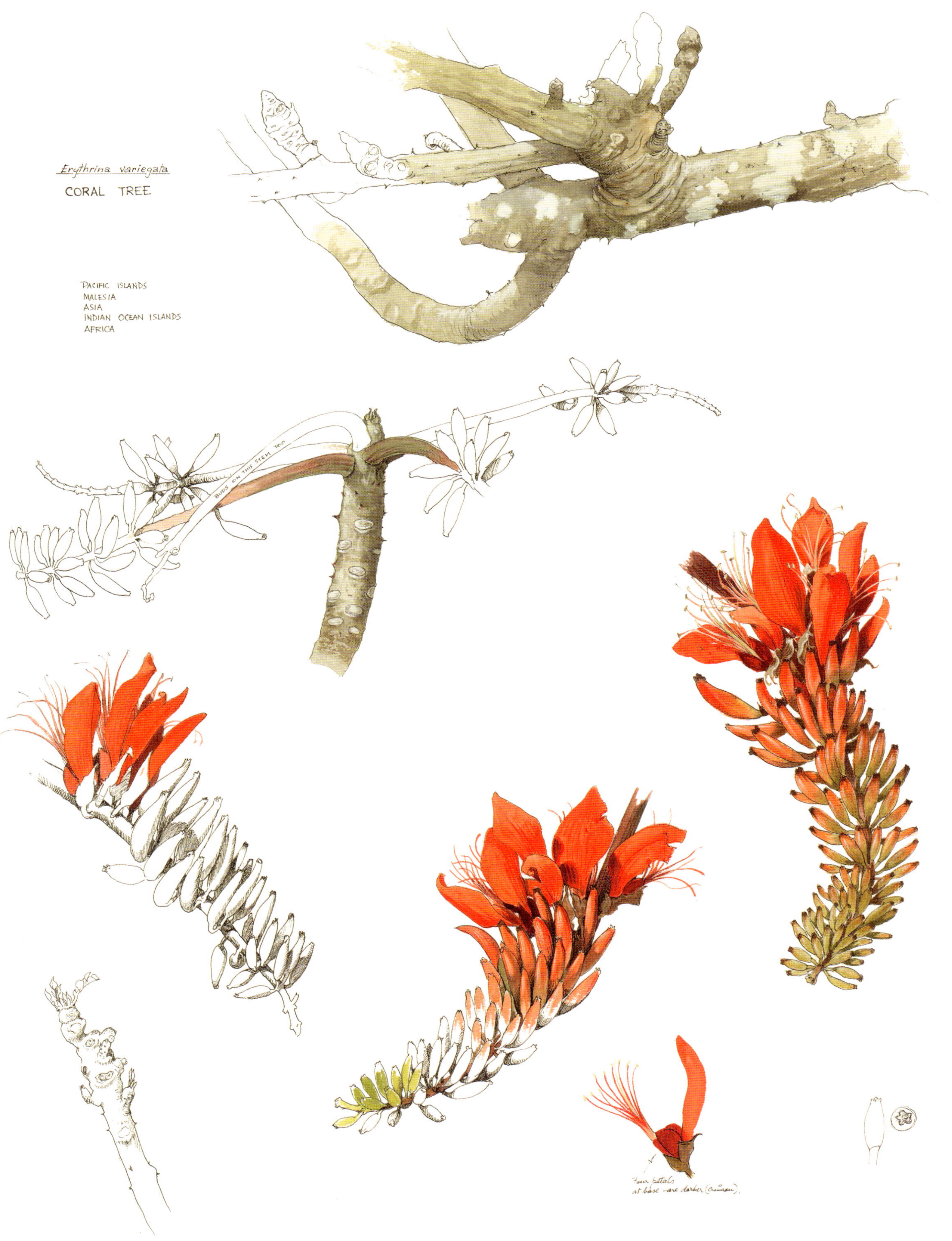

Erythrina variegata
CORAL TREE
PACIFIC ISLANDS
MALESIA
ASIA
INDIAN OCEAN ISLANDS
AFRICA
BUDS ON THIS STEM TOO
Four petals
at base – are darker (crimson).

Hernandia nymphaeifolia

Sea Hearse

HERNANDIACEAE

Hernandia is named in honour of Francisco Hernández de Toledo (1514–1587), physician and naturalist to King Philip II of Spain and author of *Historia Naturalis Mexicana*.

nymphaeifolia is after the genus *Nymphaea* (waterlily) and *-folius* (-leaved), referring to the similarly shaped leaves.

Sea Hearse is a lush tree occurring on tropical beaches and in mangrove areas from north Queensland through South-East Asia to East Africa. The fruit is the black nut surrounded by lampshade-like green or reddish bracts.

Bill needed a plant for a painting of Goffin's Corella, a species from the Tanimbar Islands of Indonesia. We knew that Sea Hearse occurred with the bird in the wild, so we were able to use a familiar tree growing at Emmagen Beach near Cape Tribulation. This painting was published in *Cockatoos: A Portfolio of All Species.*

→ 1998; pen, pencil and watercolour; 570 x 710mm
National Library of Australia, 8069150

↓ undated (c.1998); pencil; 420 x 540mm
State Library of New South Wales

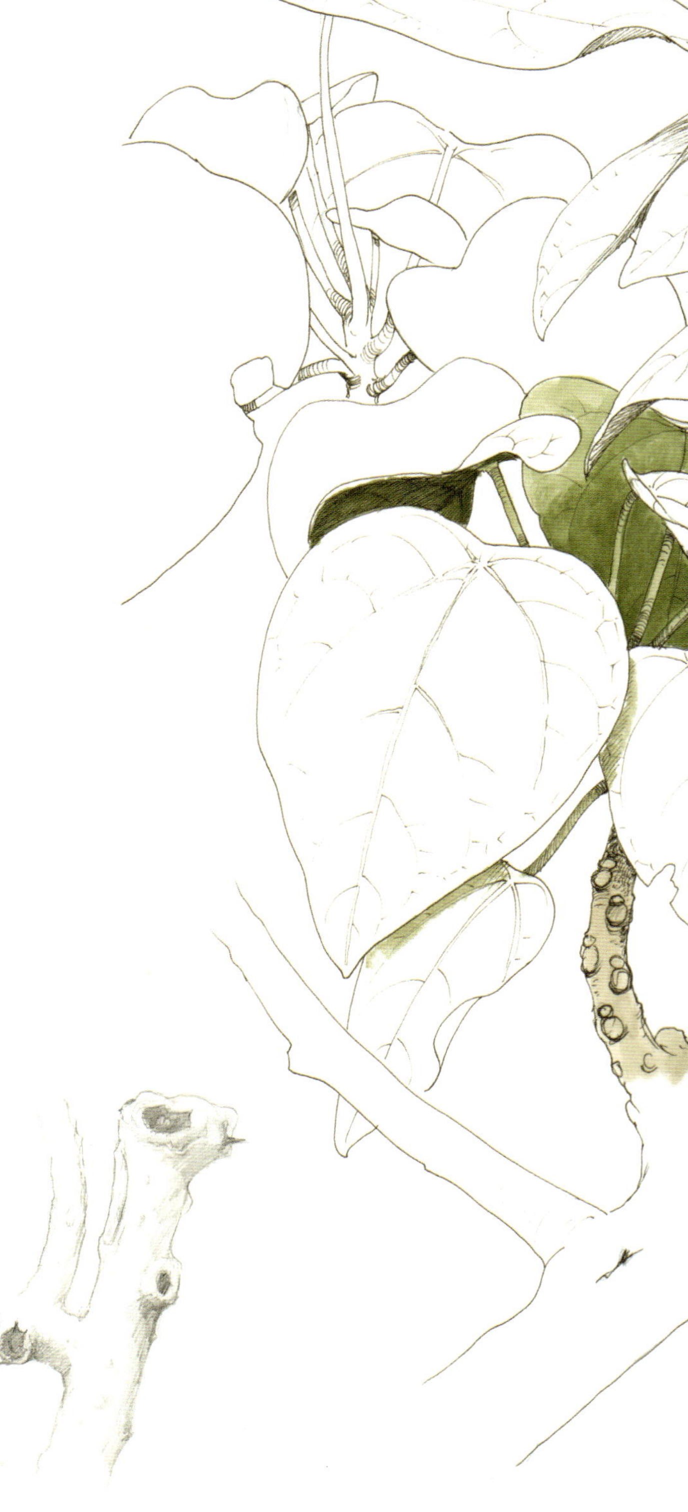

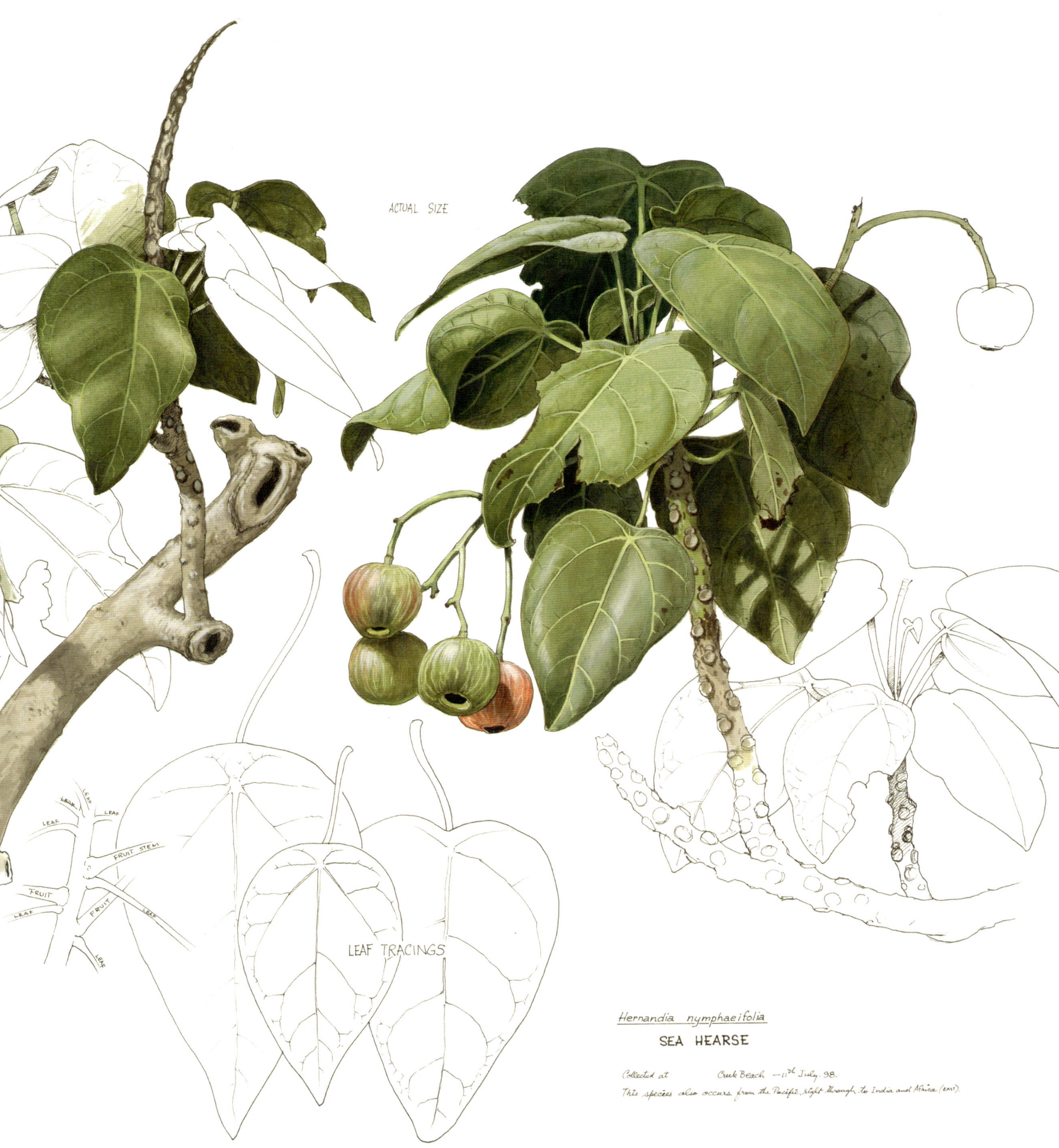
ACTUAL SIZE
LEAF
LEAF
LEAF
LEAF
FRUIT STEM
FRUIT
LEAF
FRUIT
LEAF
LEAF
LEAF TRACINGS
Hernandia nymphaeifolia
SEA HEARSE
Collected at Creek Beach – 11th July. 98.
This species also occurs from the Pacific, right through to India and Africa (EAST).

Intsia bijuga

Kwila

FABACEAE

Intsia is probably from *Intsy*, a Malagasy word that means 'there it is'.

bijuga is from *bi-* (two) and *jugus* (paired), referring to the paired leaflets.

Kwila is an elegantly shaped rainforest tree famous for its timber quality. It occurs from Australia and the Pacific through Asia to islands in the Indian Ocean.

Again, this plant was done as a reference sketch because of its wide distribution, but this one was never used in that way. Somehow Bill's artistic eye was often prone to choose aberrant samples for drawing. Here, he was attracted to a most unusual flower with two labellum petals rather than the usual one. He was always faithful to his subject, so the botanist in me was frequently checking on the pieces he chose to draw, but I wasn't quick enough to see the aberrant flowers being depicted here.

→ 1998; pen and watercolour
282 x 420mm
National Library of Australia, 8069173

↓ 1998; pen, pencil and watercolour
282 x 420mm
National Library of Australia, 8069167

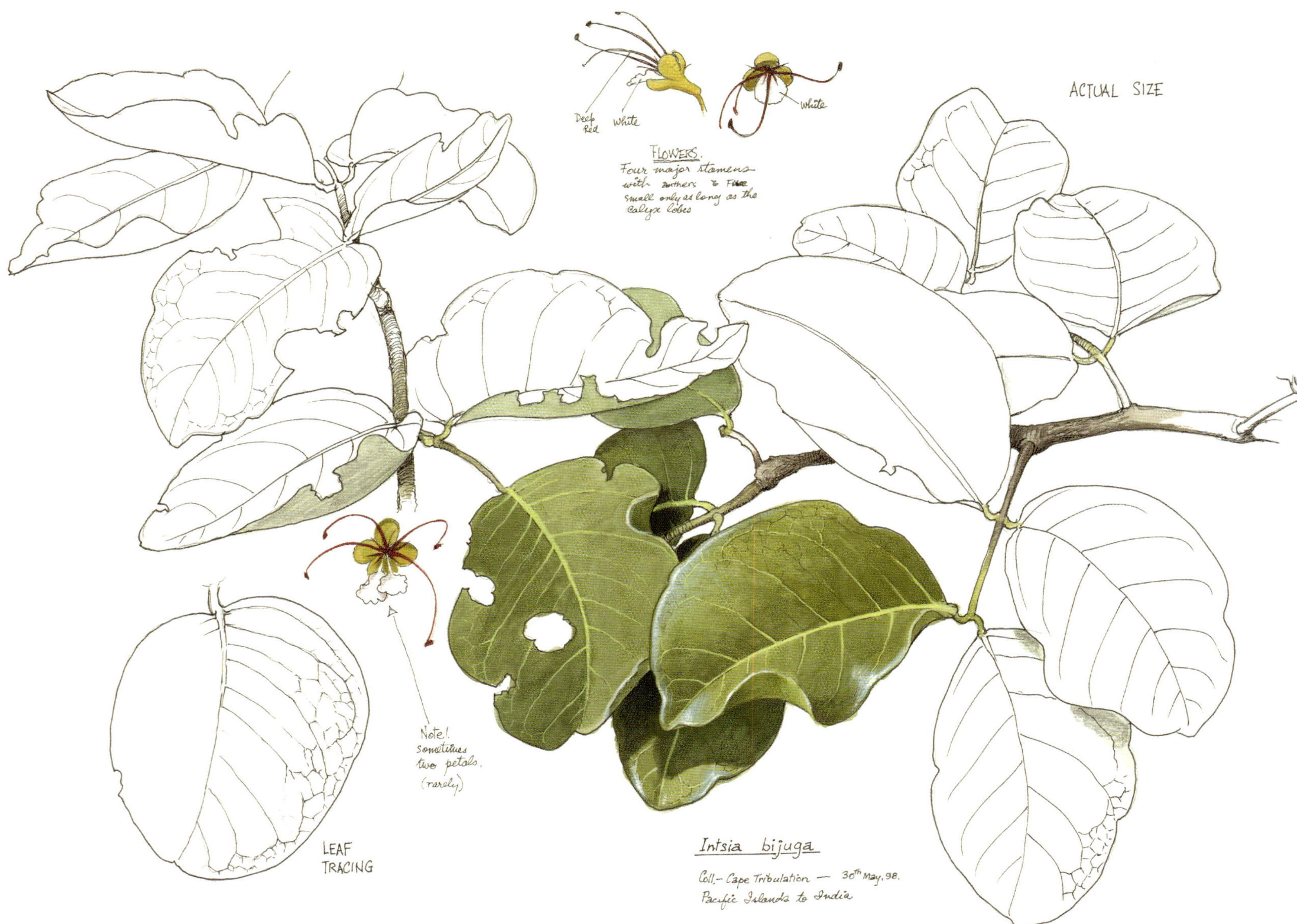

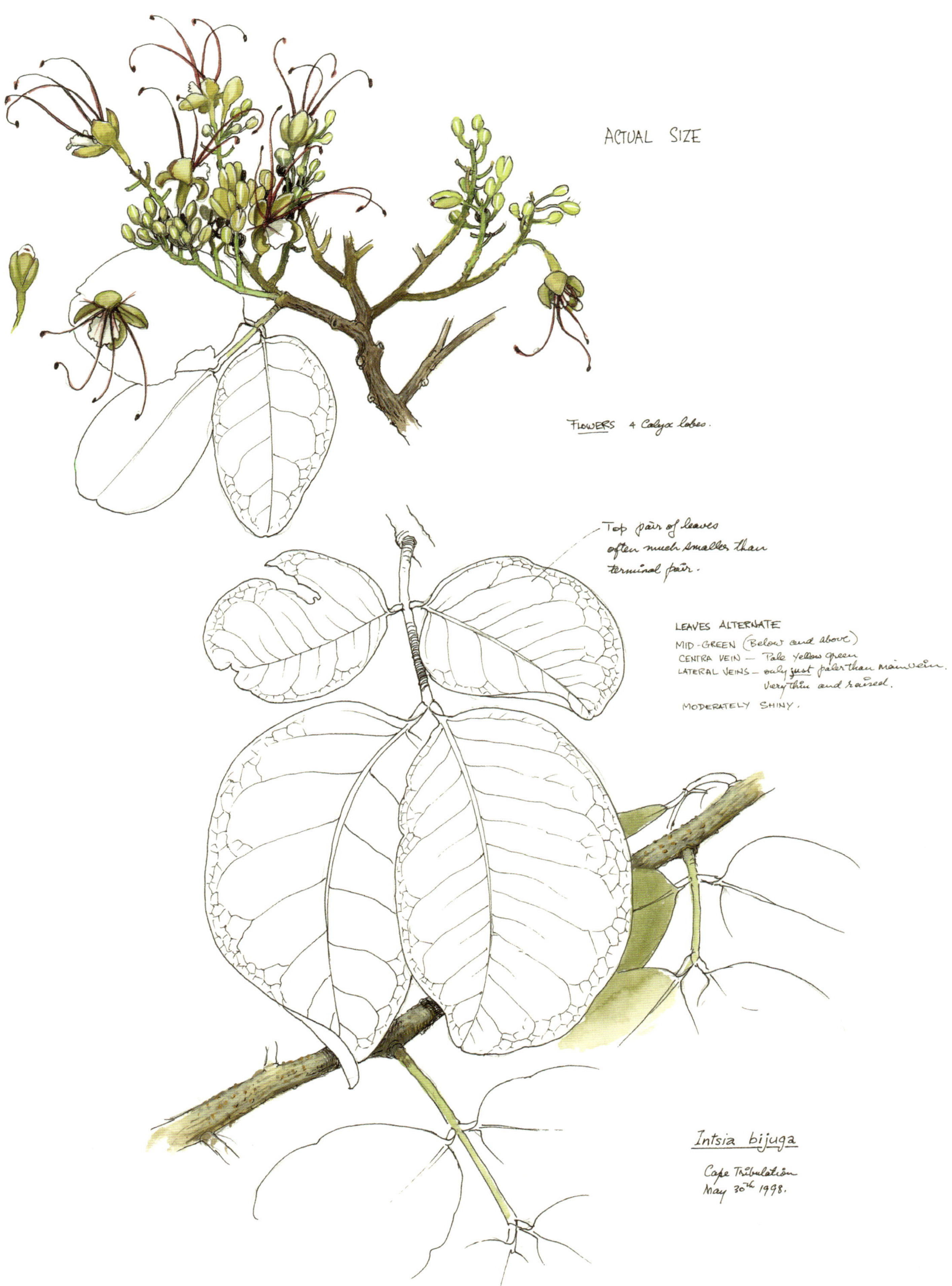
ACTUAL SIZE
FLOWERS 4 Calyx lobes.
Top pair of leaves
often much smaller than
terminal pair.
LEAVES ALTERNATE
MID-GREEN (Below and above)
CENTRA VEIN — Pale yellow green
LATERAL VEINS — only just paler than main vein.
Very thin and raised.
MODERATELY SHINY.
Intsia bijuga
Cape Tribulation
May 30th 1998.

Nymphaea gigantea

Giant Waterlily

NYMPHAEACEAE

Nymphaea is from the Latin *nymphaea* (waterlily).

gigantea is from *giganteus* (gigantic).

Waterlily is a favourite plant for many people. This species is native to northern Australia and New Guinea. While the large purple flowers are open during the day, they close each afternoon and reopen each morning over a few days.

These drawings were used for several paintings, including one of Green Pygmy Geese and another of a landscape with Jacanas at Keatings Lagoon near Cooktown. This painting is in the style of the Australian impressionists that Bill admired so much. It is quite small, about the size of the cigar box lids that were commonly used by the Australian impressionist artists (Arthur Streeton, Tom Roberts and Frederick McCubbin) as a cheap alternative to canvas for outdoor painting.

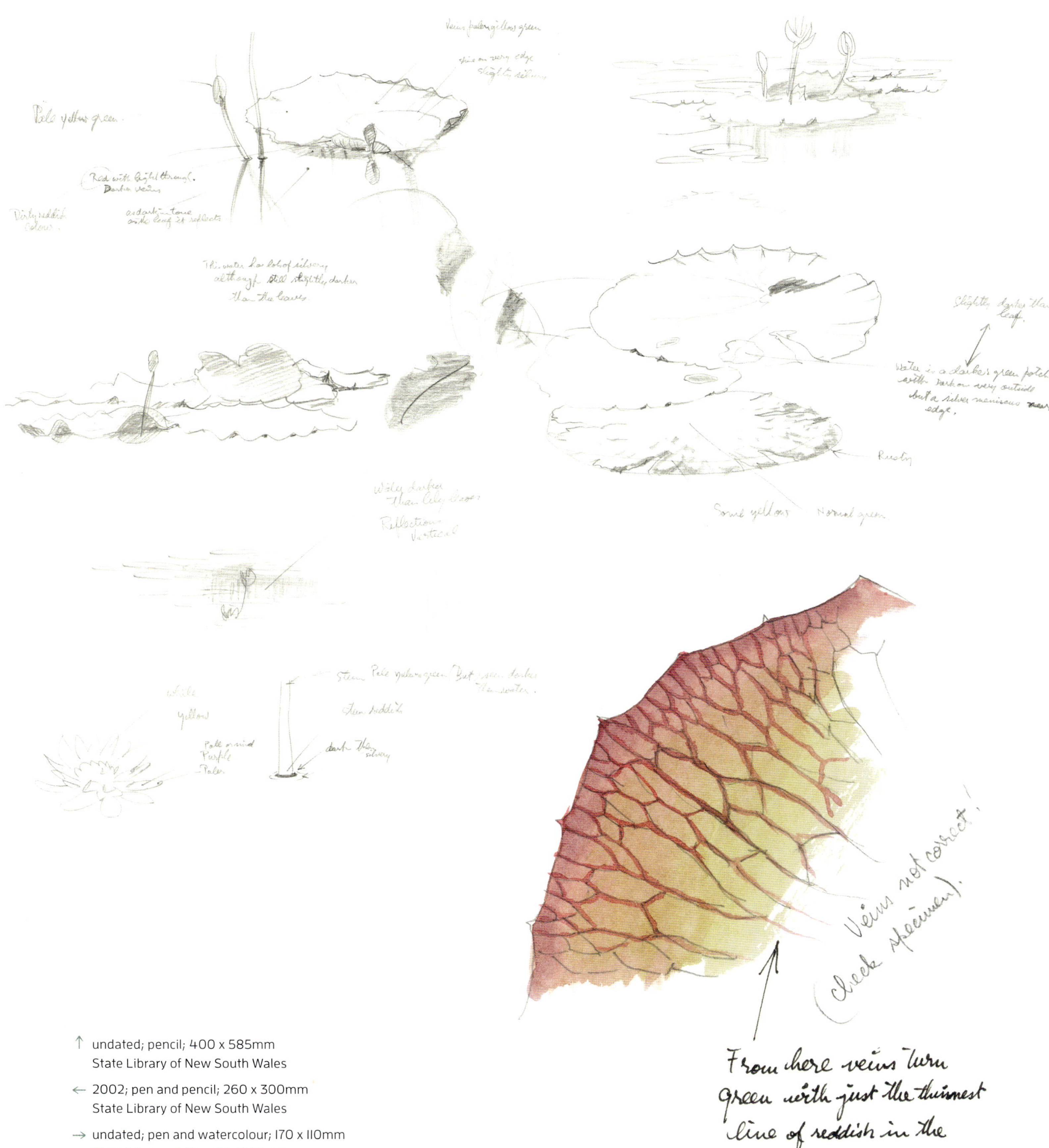

↑ undated; pencil; 400 x 585mm
State Library of New South Wales

← 2002; pen and pencil; 260 x 300mm
State Library of New South Wales

→ undated; pen and watercolour; 170 x 110mm
State Library of New South Wales

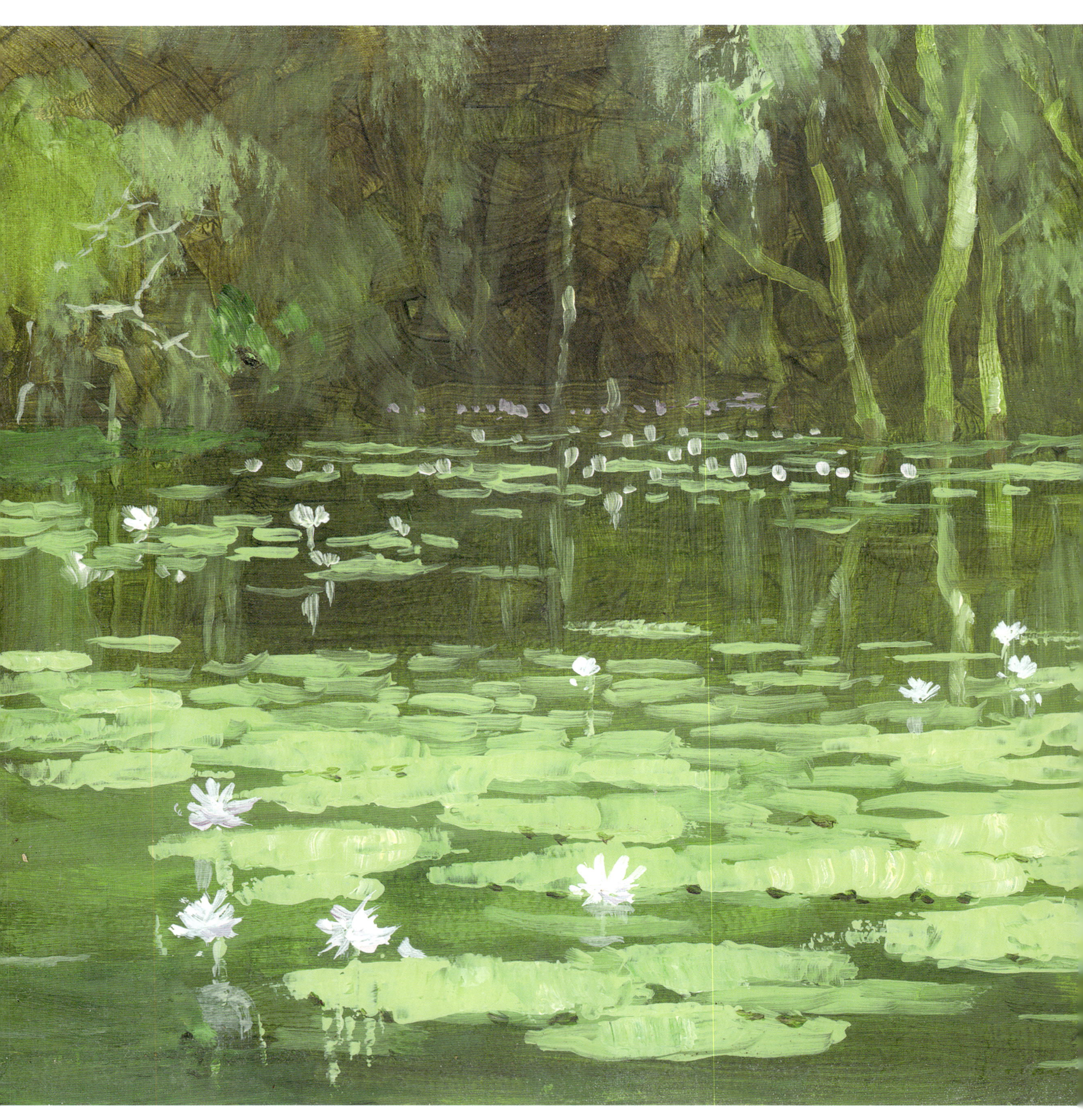

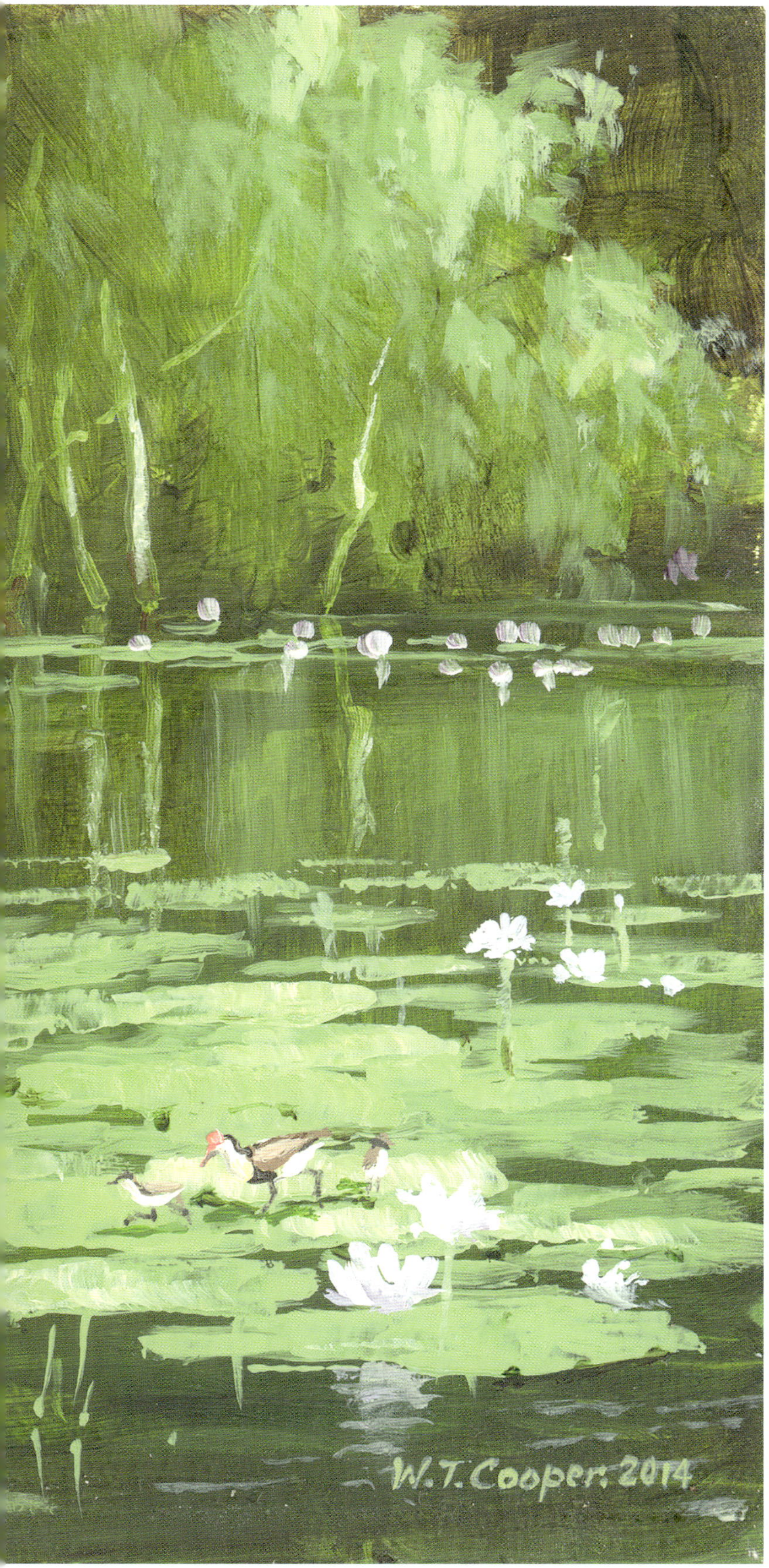

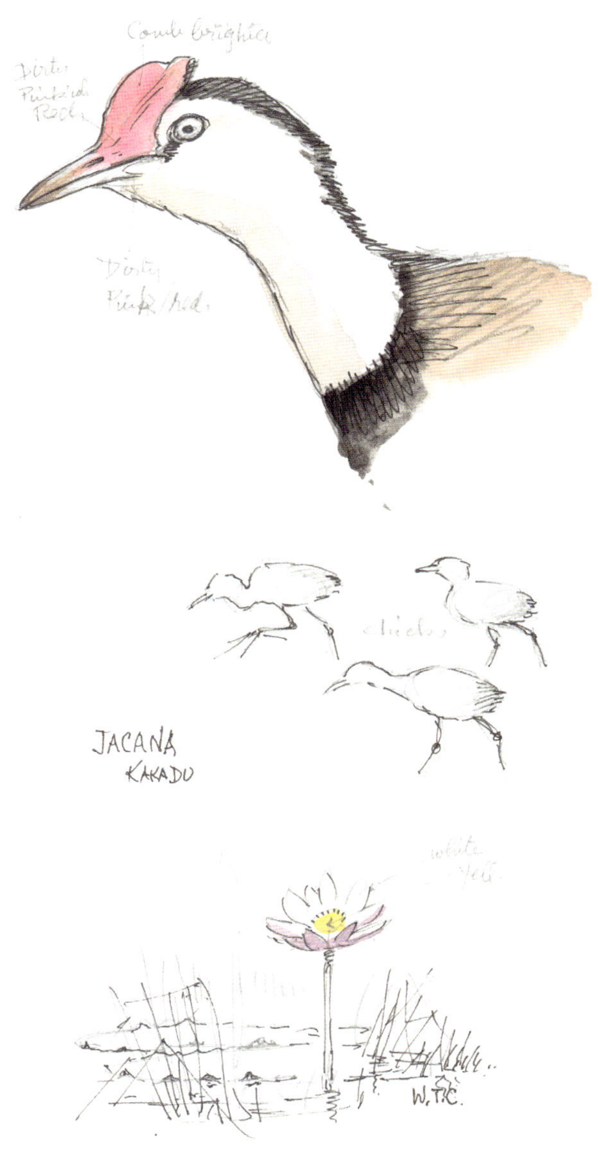

↑ undated; pen, pencil and watercolour; 135 x 210mm

← Keatings Lagoon
undated (c.2005); oil on board; 150 x 240mm
Private collection

Pandanus tectorius

Beach Pandan

PANDANACEAE

Pandanus is from the Malaysian word for conspicuous.

tectorious is from *tectus* (hidden) and *-orius* (capable), referring to the leaves forming a shelter or being used to construct a shelter.

Beach Pandan is a pandan with prop roots of up to 2 metres, occurring along beaches and headlands from northern New South Wales to Cape York, New Guinea and the Pacific. The large pineapple-like yellow fruit fall, breaking into as many as 50 segments. These segments are taken by Palm Cockatoos, which extract the long narrow seeds. Bill attempted to arrange these birds on this plant in his sketches, but he was never happy with the compositions he came up with.

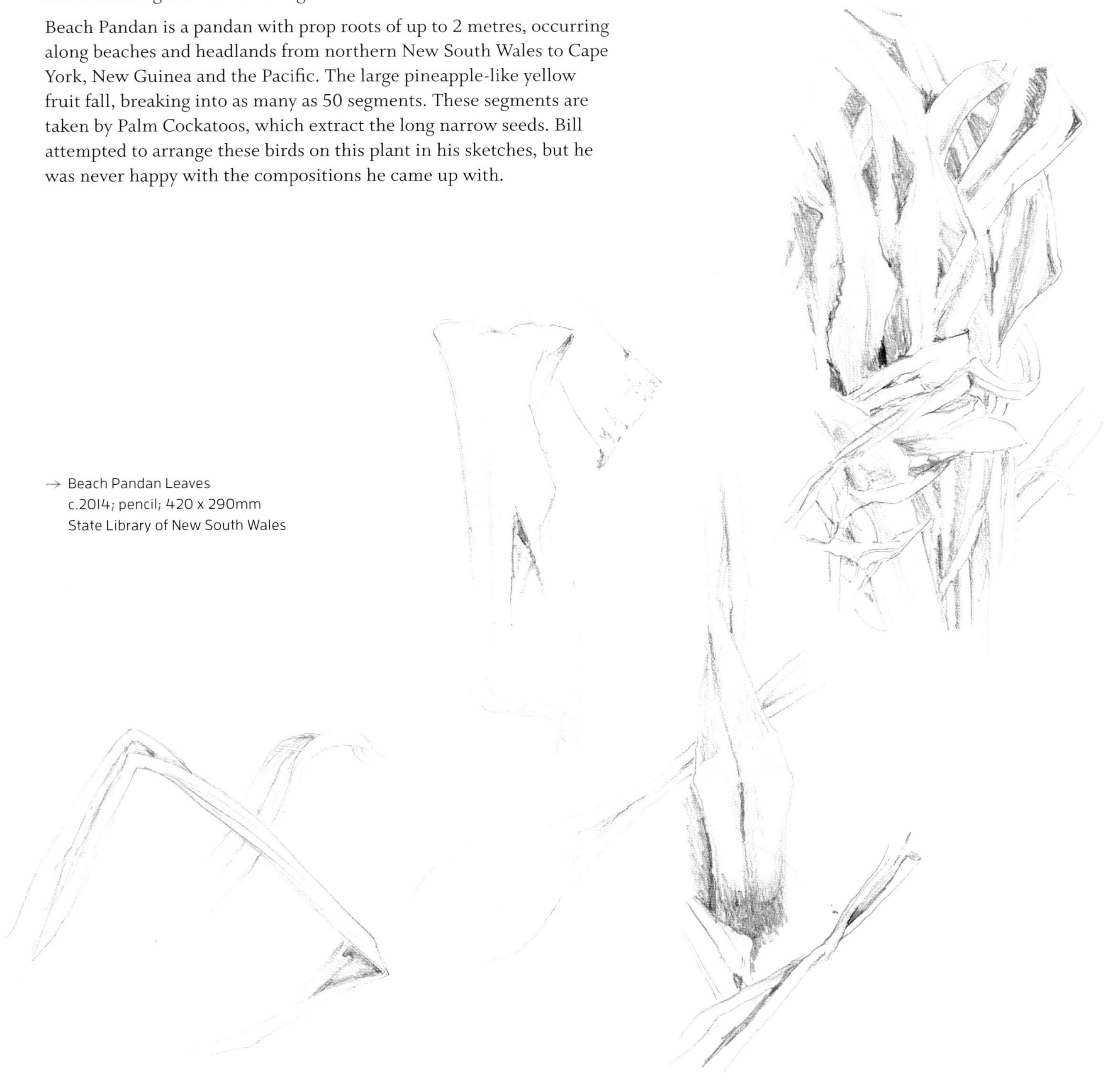

→ Beach Pandan Leaves
c.2014; pencil; 420 x 290mm
State Library of New South Wales

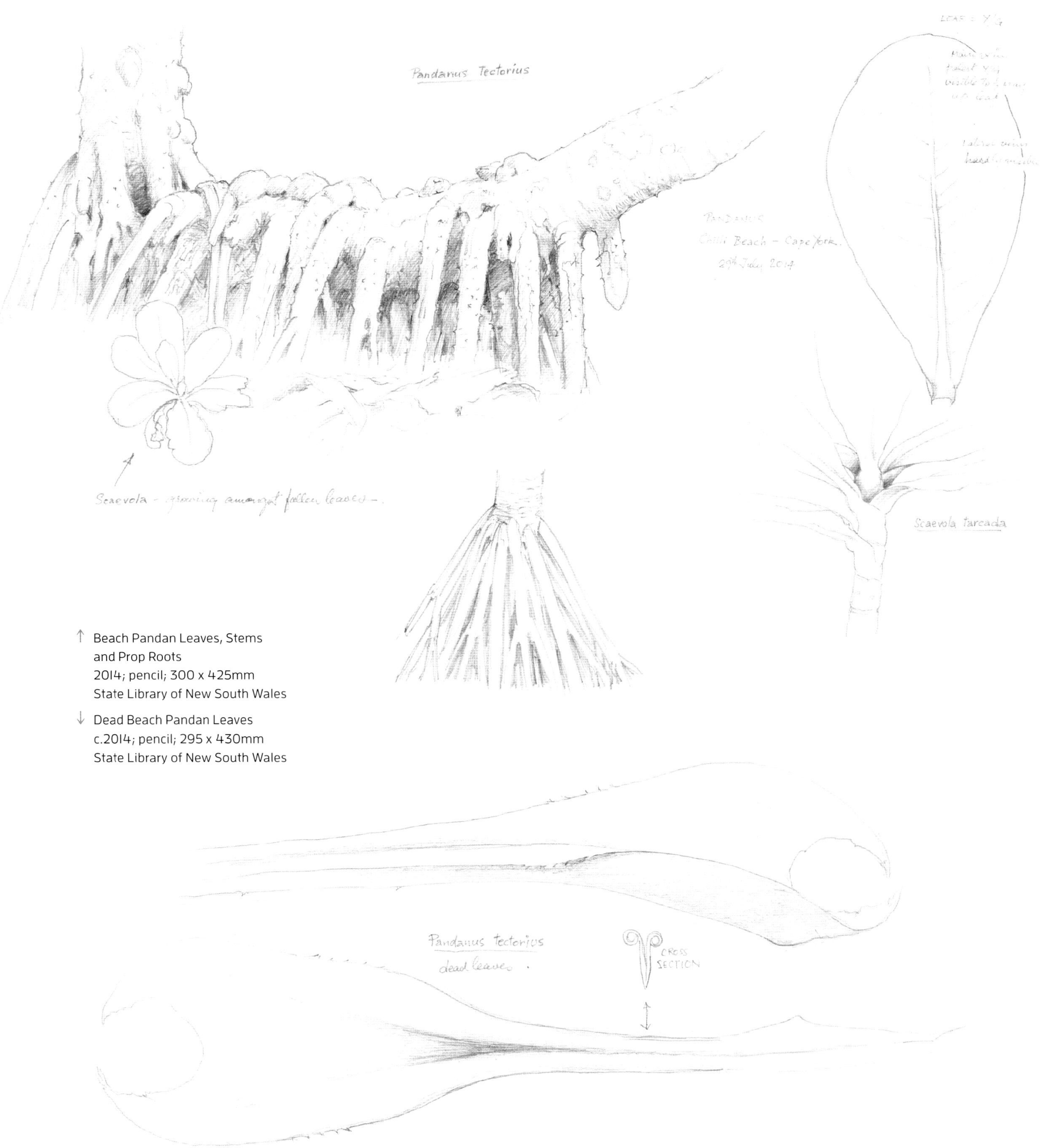

↑ Beach Pandan Leaves, Stems and Prop Roots
2014; pencil; 300 x 425mm
State Library of New South Wales

↓ Dead Beach Pandan Leaves
c.2014; pencil; 295 x 430mm
State Library of New South Wales

Terminalia catappa

Sea Almond or Indian Almond

COMBRETACEAE

Terminalia is from *terminalis* (terminal), referring to the many species with leaves clustered at the tips of the branches.

catappa is from *ketapang*, the Malaysian name for the tree.

Sea Almonds are dense, shade-giving trees occurring on tropical beaches in north Queensland, the Northern Territory and from the Pacific islands through to South-East Asia. They possess Bill's favourite pattern of large leaves in clusters. That the old leaves turn red before falling added warmth and colour to Bill's painting.

The seeds within the fruit are relished by Red-tailed Black-Cockatoos and Palm Cockatoos. We were able to collect these specimens from street trees in Cairns, and the sketches were used in a large painting of Palm Cockatoos.

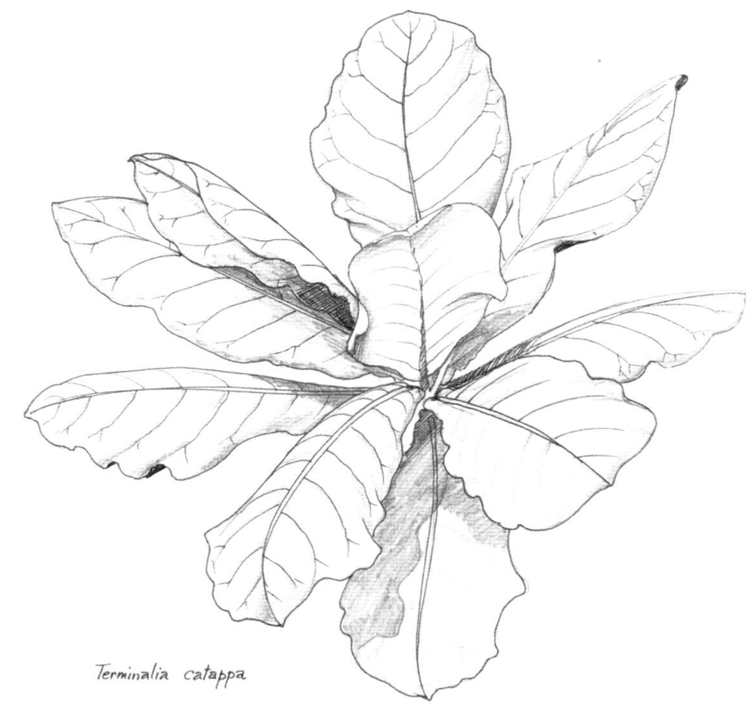

Terminalia catappa.

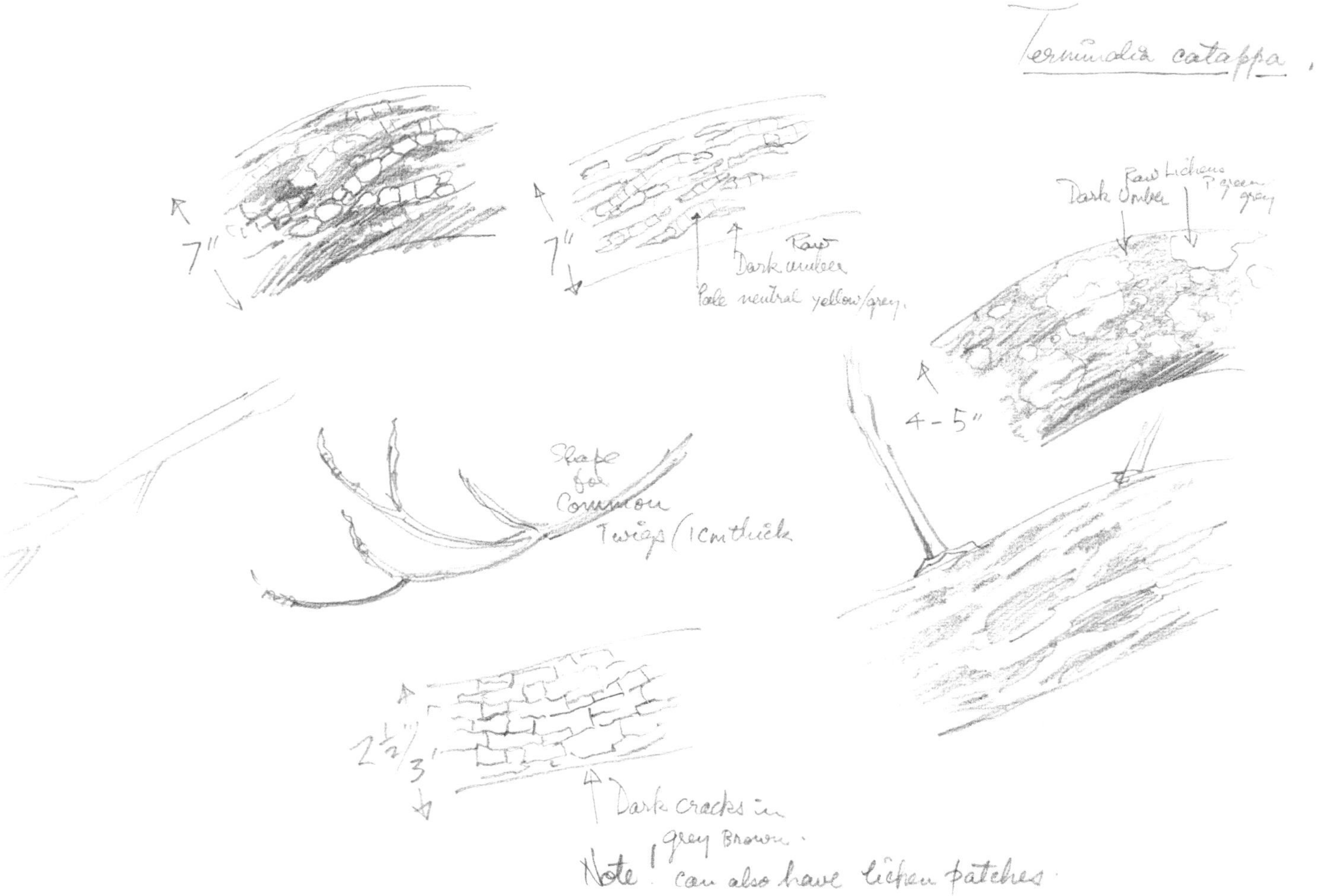

7"

7"

Raw
Dark umber

Pale neutral yellow/grey.

Dark Umber

Raw Lichens green grey

4-5"

Shape for Common Twigs (1cm thick

2½/3"

Dark cracks in grey Brown.

Note! can also have lichen patches.

← c.2002; pen and pencil; 295 x 325mm
State Library of New South Wales

↙ c.2002; pencil; 200 x 295mm
State Library of New South Wales

↓ 2002; pen, pencil and watercolour; 550 x 780mm
State Library of New South Wales

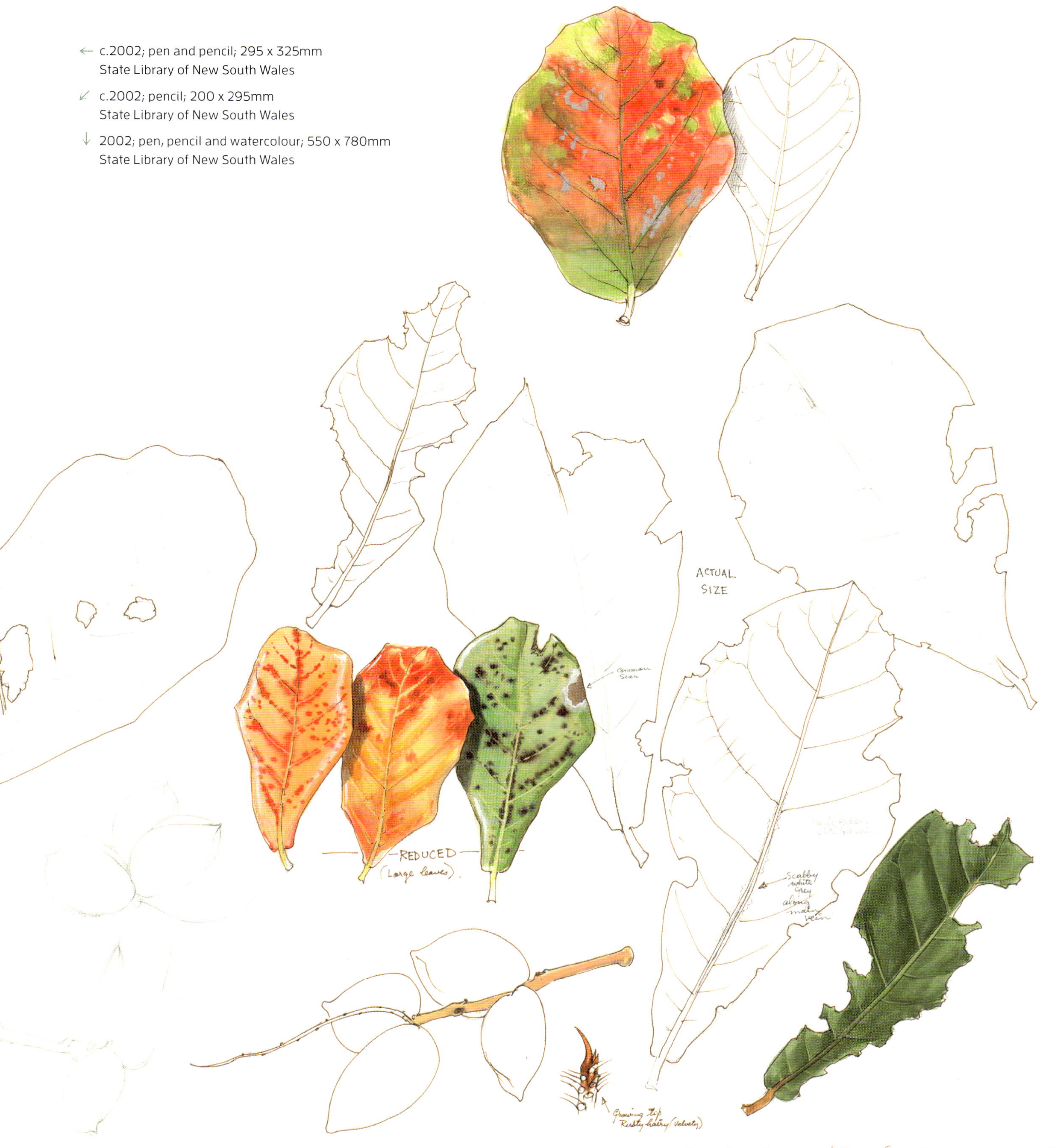

Xylocarpus moluccensis

Cedar Mangrove or Cannonball Mangrove

MELIACEAE

Xylocarpus is from *xylo-* (woody) and *-carpos* (fruited).

moluccensis is from the Maluku Islands in Indonesia, where the type specimen was collected.

Cedar Mangrove is a deciduous mangrove, occasionally growing up to 30 metres tall. It has small buttresses and rough, fissured, blackish bark. The remarkable cone-shaped pneumatophores (knee-roots) emerge from the underground roots, providing additional access to oxygen. This mangrove is distributed widely along Australia's tropical north coast, as well as in New Guinea and East Africa, where the timber is used for dhow masts.

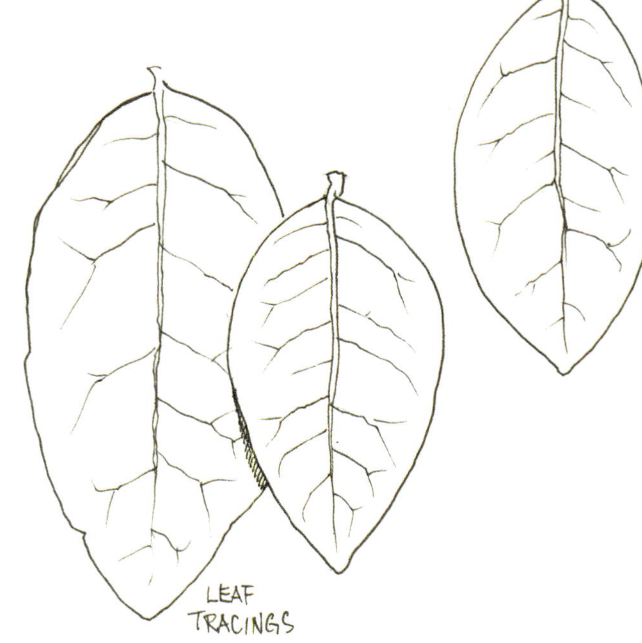

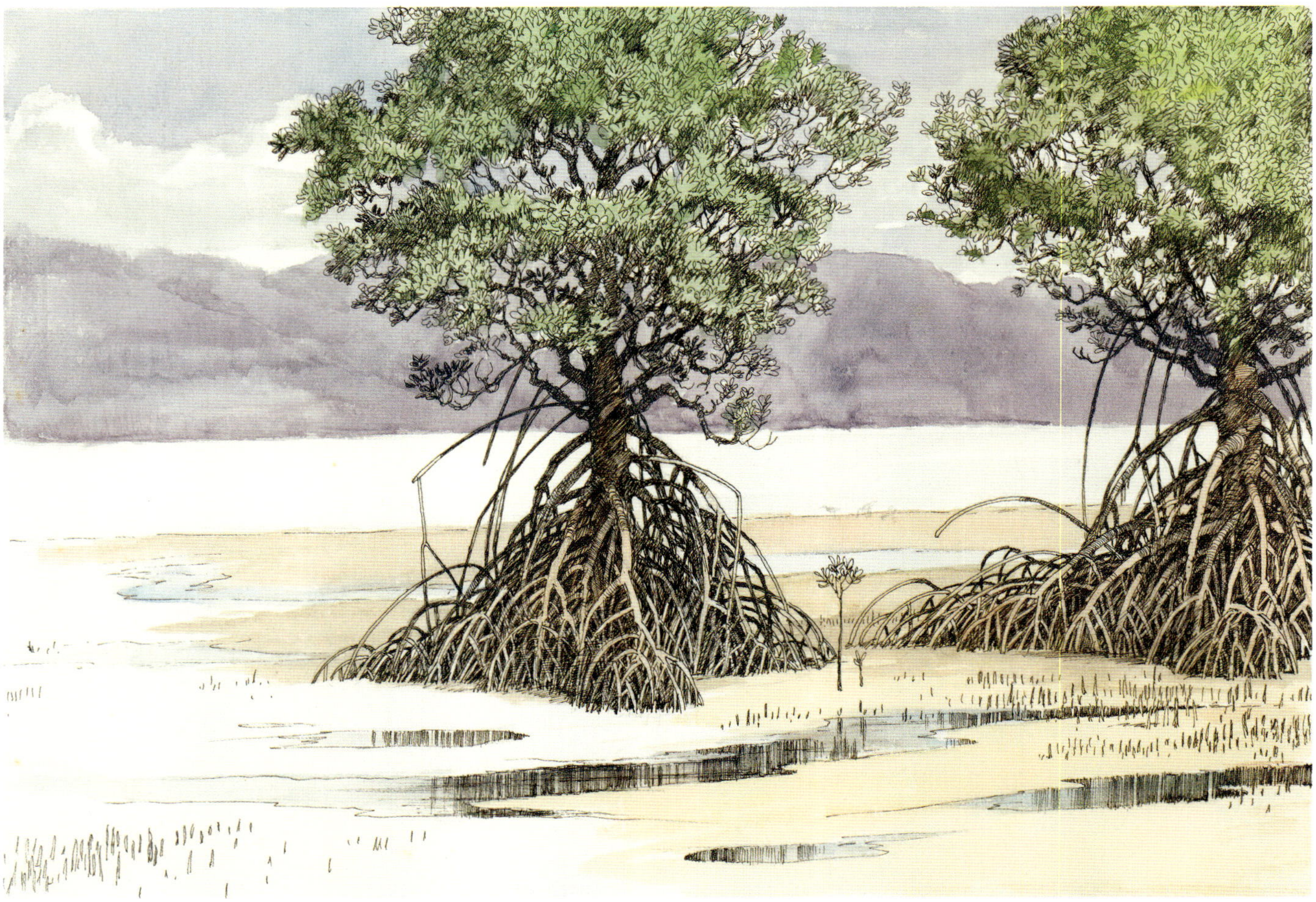

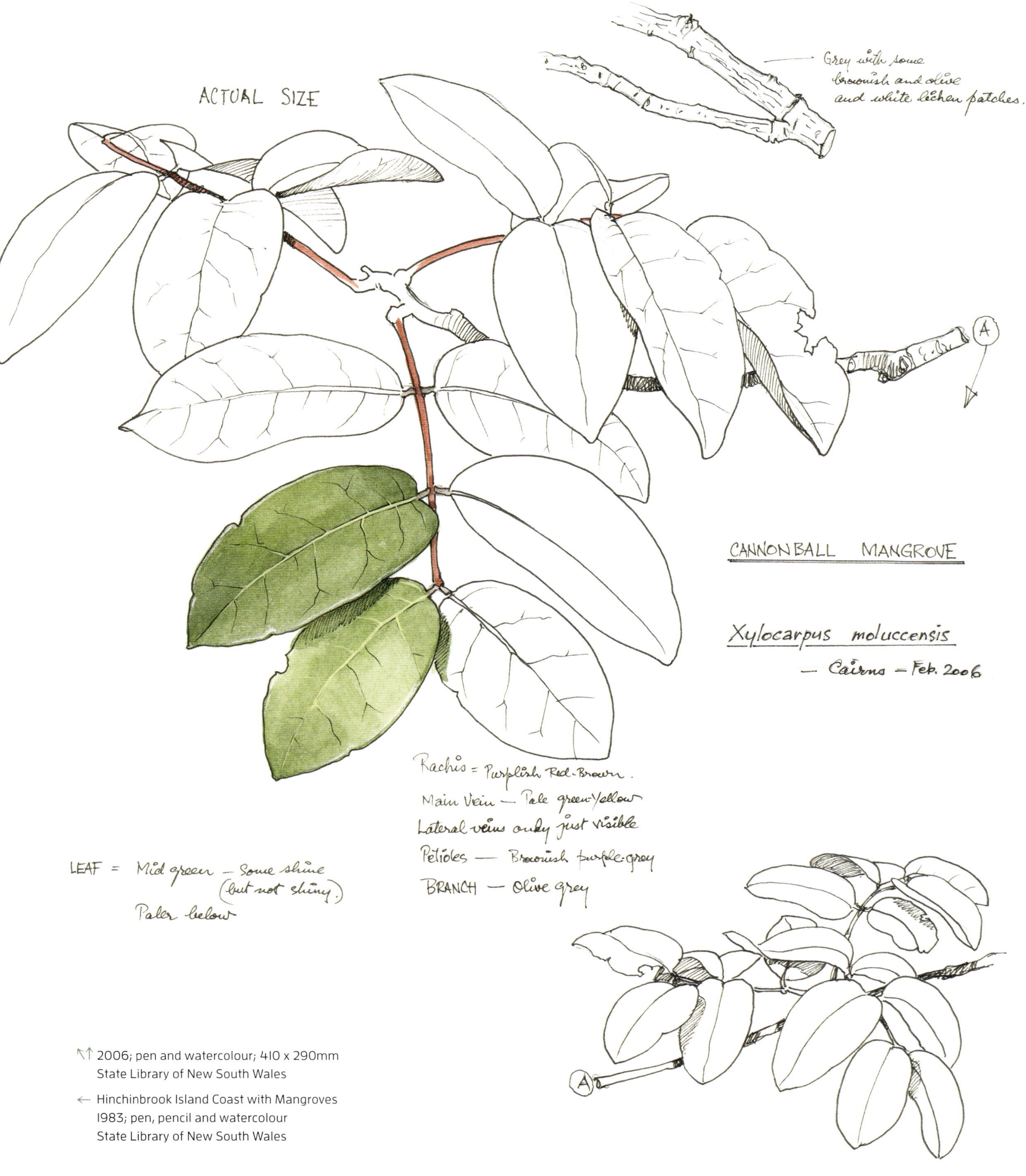

↖↑ 2006; pen and watercolour; 410 x 290mm
State Library of New South Wales

← Hinchinbrook Island Coast with Mangroves
1983; pen, pencil and watercolour
State Library of New South Wales

Chapter 5

Beyond Australia

→ An Unidentified Species of *Bulbophyllum* from Tomba, Papua New Guinea 1973; pen and gouache (in part) 400 x 275mm State Library of New South Wales

↙ Bill Sketching in Our Vehicle near Mwanza, Tanzania, 1995

All of Bill's trips overseas were work related, work meaning painting. He was very comfortable and confident on the ground and in the bush, but hated the claustrophobia of airports and flights. He accepted that part of travel, however, in exchange for the excitement of seeing the animals and forest in places he had dreamt about. When booking our tickets for a flight to Kenya, Bill requested a flight that would take us over the Indian Ocean in daylight hours as he couldn't bear the idea of drowning in the dark! He had a wonderful sense of humour.

From a young age, Bill was almost intoxicated by the romance of rainforest. He never tired of the sight of clouds smudging rainforested hillslopes and mountaintops. He considered a career as a Patrol Officer in Papua New Guinea as a means of spending time in those misty, lush forests enriched by the presence of birds of paradise. Bill travelled to Papua New Guinea on four occasions (1970, 1973, 1974 and 1976) before venturing to other countries. All of these trips were focused on becoming familiar with particular species and their habitats. As his diary shows, this was a stimulating and rewarding process:

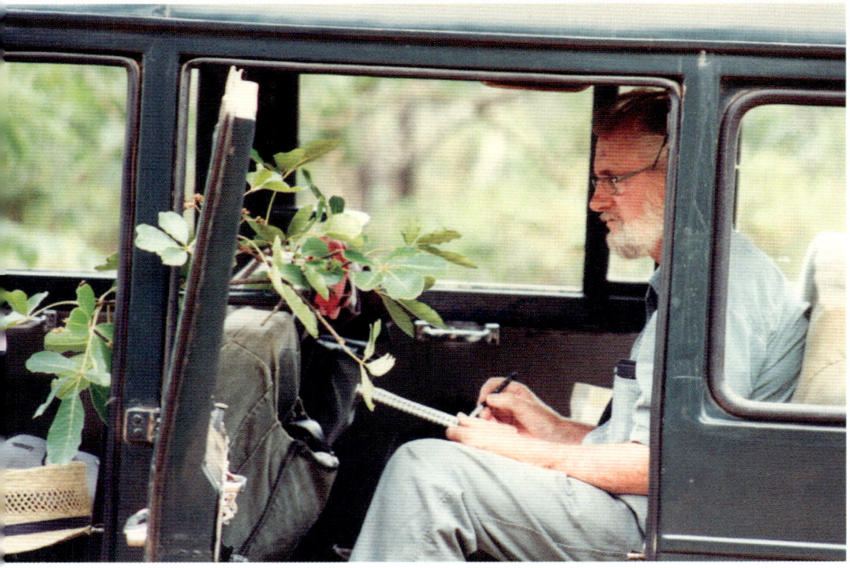

> *April 30th, 1973. Got up at 4:45 and went down the river by boat … Went a long way upstream into absolutely beautiful jungle, 'Garden of Eden type jungle' with Nipa Palms, mangroves, tree ferns (growing out of water) and tall jungle trees with almost silver white trunks. It was alive with birds: Eclectus Parrots male and female abundant, Shining Starlings, white cockatoos nesting, Red-cheeked Parrots, pigeons in great numbers … This morning's trip was one of the most interesting experiences in NG yet. I have never seen such picturesque jungle or such an abundance of birds. It was just like everybody's idea of a river in the tropics with colourful parrots screeching in the trees all around.*

During our field trips, Bill would enjoy the hunt to find the subject and then, if possible, he would edge closer and closer to be able to see it very well. Bill drew his subjects from life, and on the few occasions that he had to rely on photographs he disliked the experience intensely. He found that photos were misleading. He was used to twisting and turning the branches and leaves to suit his composition; trying to achieve this from a photograph, in which he could not always see the plant's structure, was difficult. The vital part of, for instance, a leaf attachment might be hidden. He could be creative with the living plant before him.

Trips were as much about the plants as the birds. Bill would often draw a bird several times until I would hear him say, with some satisfaction, 'Got it'. It usually took drawing a bird a few times to get to know it and to catch the bird's jizz. Then, if the bird hung around, he could confidently sketch various positions and aspects of behaviour. Plants would also be collected for drawing and painting. Bill delighted in finding nice branches adorned with various bits of moss and lichen. The plants would preferably be those that the birds had fed on, otherwise interesting plants from the area would work as authentic elements in the final painting of that species. Botanical paintings were mostly done back in camp or at a lodge, but sometimes Bill had to draw in the vehicle we were travelling in. The humid conditions in Papua New Guinea caused difficulties with watercolour paper, which would become damp and end up resembling blotting paper—a difficult surface to paint on. Watercolour paints were constantly growing mould on their wet surfaces.

The six volumes of *Kingfishers and Related Birds* involved trips for us to Singapore, Malaysia, India and Kenya, mostly to see hornbills, kingfishers, bee-eaters and rollers. India brought a new, unanticipated passion—tigers—which Bill painted several times. The first time we saw a tiger was breathtaking. Bill recorded the day in his diary:

> *Friday 15th Feb, 1985. Ranthambore. Fateh (our host) drove us down to the lake and showed us where he thought the tiger known as 'Links' was laying up. Sambar everywhere—very tame—Chital the same. Langurs in the trees. Parked where Fateh considered the tiger may come out to hunt or lay an ambush for the Sambar—Alarm calls. Langurs barking, Chital running and the Sambars barking and all looking towards the patch of grass—Tiger made a call not the full roar—we drove around to where we thought it was—monkeys barking loudly. Fateh decided it had moved back to where we had just come from—turned around and drove back—as we rounded a bend I saw the tiger standing facing us in the edge of the tall grass—only 40 m away. A weird feeling!*

A few days later, Bill recorded another sighting:

> *Monday 18th Feb, 1985 ... found the tiger. It was straight across [the lake] from our lodgings, so we sat on the verandah and watched as she stalked some Sambar Deer. They got past her into the water and were out of reach ... a little later some adult Sambar came down to the water in the same spot and the tiger charged but missed. The Sambar all made [it to] the water and swam across to our side. The tiger came out of the grass and stood watching the swimming deer, then turned back and disappeared into the tall grass. All this we were able to see from the verandah although we missed the actual charge as it must have gone on behind the tall grass. We could just hear the alarm calls and see the tiger come out and watch them. FABULOUS EXPERIENCE.*

Kenya introduced us to Turacos, beautiful birds, often with crests and exquisite 'eye makeup'. While in the country to collect material for hornbill paintings, Bill became obsessed with their beauty and character. Turacos are a bit like parrots, with a playful manner in the way they move through trees. Bill wrote in his diary of his first Turaco sighting:

> *Wed 1st Nov. 1989. Drove north from Sirikwa Farm and gradually dropped down in altitude as went down the escarpment to low warm country dry Acacia etc ... Got some good material for hornbills—Grey—Jacksons—Red billed—Crowned. The bird of the morning would have to be the White-crested Turaco—what a sight.*

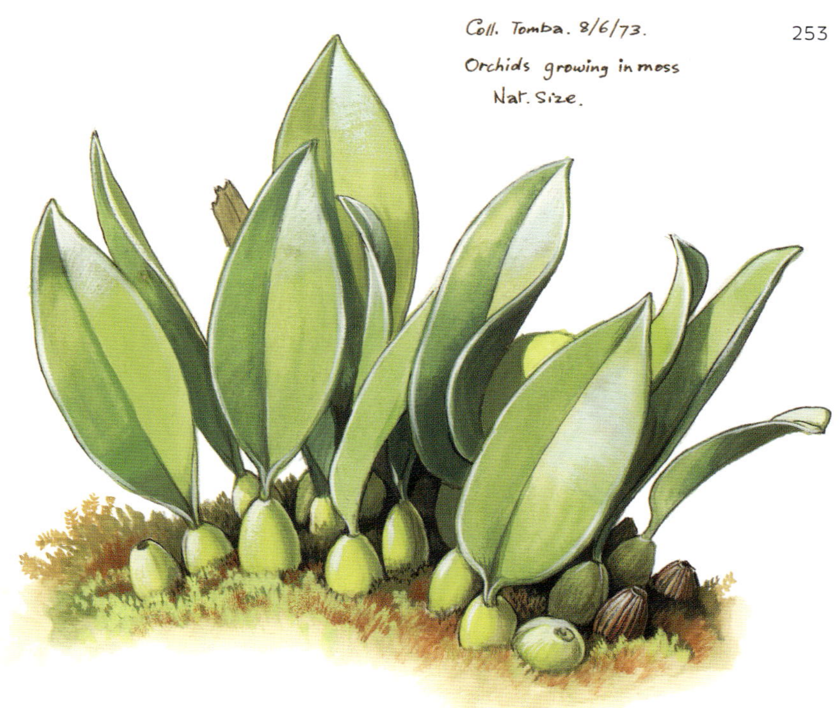

Bill decided that these birds would be the subject of his next monograph, *Turacos: A Portfolio of All Species*, and we would return to Kenya, Tanzania and Uganda to see as many species as possible. Schalow's Turaco is especially beguiling, with its tall tapered crest and long white line beneath its eye. Bill recorded seeing the species in his diary:

> *Kenya, 23rd Sept 1995. ... drove to Sekenani to meet the same Masai who said they would find Schalow's Turaco. As I write this there are 3 Giraffe walking past, amazing!, also Serval hunting!! The Masai explained that in the early morning the Turacos come down from the hills and move along the creek in the riverine forest but go back to the hills as the sun gets well up. So we walked right up to the base of the hills where the creek bed became a ravine clothed on both sides with riverine forest which had huge Euphorbias protruding from it ... Schalow's Turacos came into the trees nearby—they called but stayed high up in the taller trees ... Baboons were barking on the other side of the narrow gully. I was amazed to find Elephant dung and foot paths along the sides of the ravine—very steep ... I climbed a huge Euphorbia and managed to get some reasonable looks at the birds calling.*

In his later years, Bill preferred the comforts of home. He painted most days from his vast collection of drawings, with occasional field trips closer to home. Hunting for different rainforest fruits in tropical Queensland was a 20-year project, almost a hobby for Bill in between bird paintings. My role in Bill's bird books was minimal, but together Bill illustrated and I described more than 1,200 fruit specimens collected on our various field trips. Bill and I were lucky to have had a remarkable partnership.

Wendy Cooper

Artocarpus heterophyllus

Jakfruit

MORACEAE

Artocarpus from *artos* (bread) and *-carpos* (-fruited), referring to the Breadfruit Tree, which also belongs to this genus.

heterophyllus from *heteros* (different) and *-phyllus* (leaved), referring to the variously shaped leaves.

Jakfruit, the common name, is a Portuguese corruption of the Malayalam (a language in southern India) name for the tree, *Chakka*.

Jakfruit trees grow to 20 metres in the rainforests of India and Peninsular Malaysia. Flowering and fruiting are cauliflorous.

Alongside his birds, Bill also enjoyed painting many other subjects, including mammals. Giant Indian Squirrels held a particular fascination, partly because of their beautiful purplish-auburn-coloured fur. These squirrels live in the Western Ghats of India and are known to feed on Jakfruit. Bill took advantage of a fruiting Jakfruit growing on a friend's property to make these reference drawings in the hope that one day he would find time to paint the animal in its native forest, feeding on these intriguing fruit among the glossy dark-green leaves.

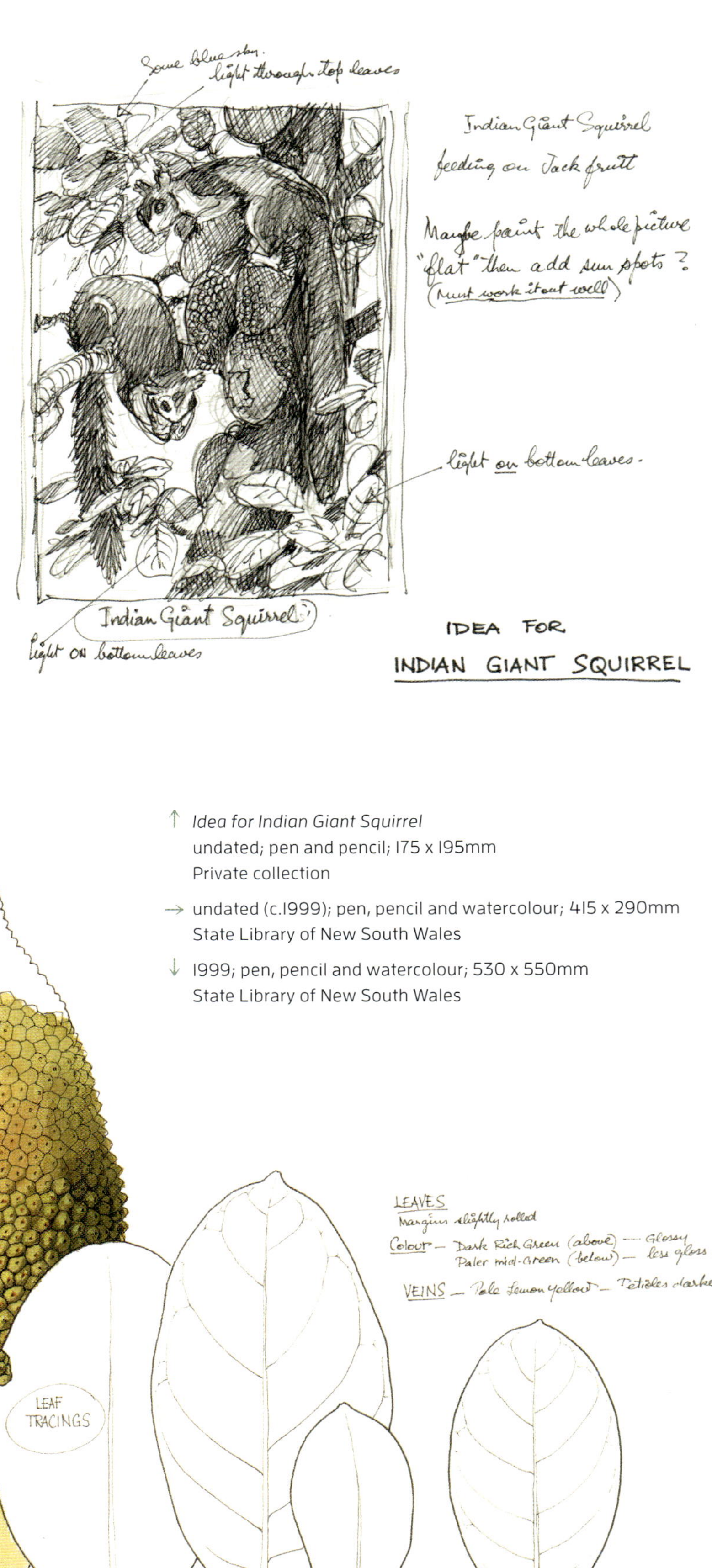

↑ *Idea for Indian Giant Squirrel*
undated; pen and pencil; 175 x 195mm
Private collection

→ undated (c.1999); pen, pencil and watercolour; 415 x 290mm
State Library of New South Wales

↓ 1999; pen, pencil and watercolour; 530 x 550mm
State Library of New South Wales

JACKFRUIT
Artocarpus heterophyllus
May 1999 – Topaz
(Food plant of Indian Giant Squirrel
Western Ghats)

Colour of segments
Note – many have rusty marks and the tip is burnt sienna but can appear darker.
These rusty marks and dark tips added to the lines separating the segments gives the fruit an overall darker appearance

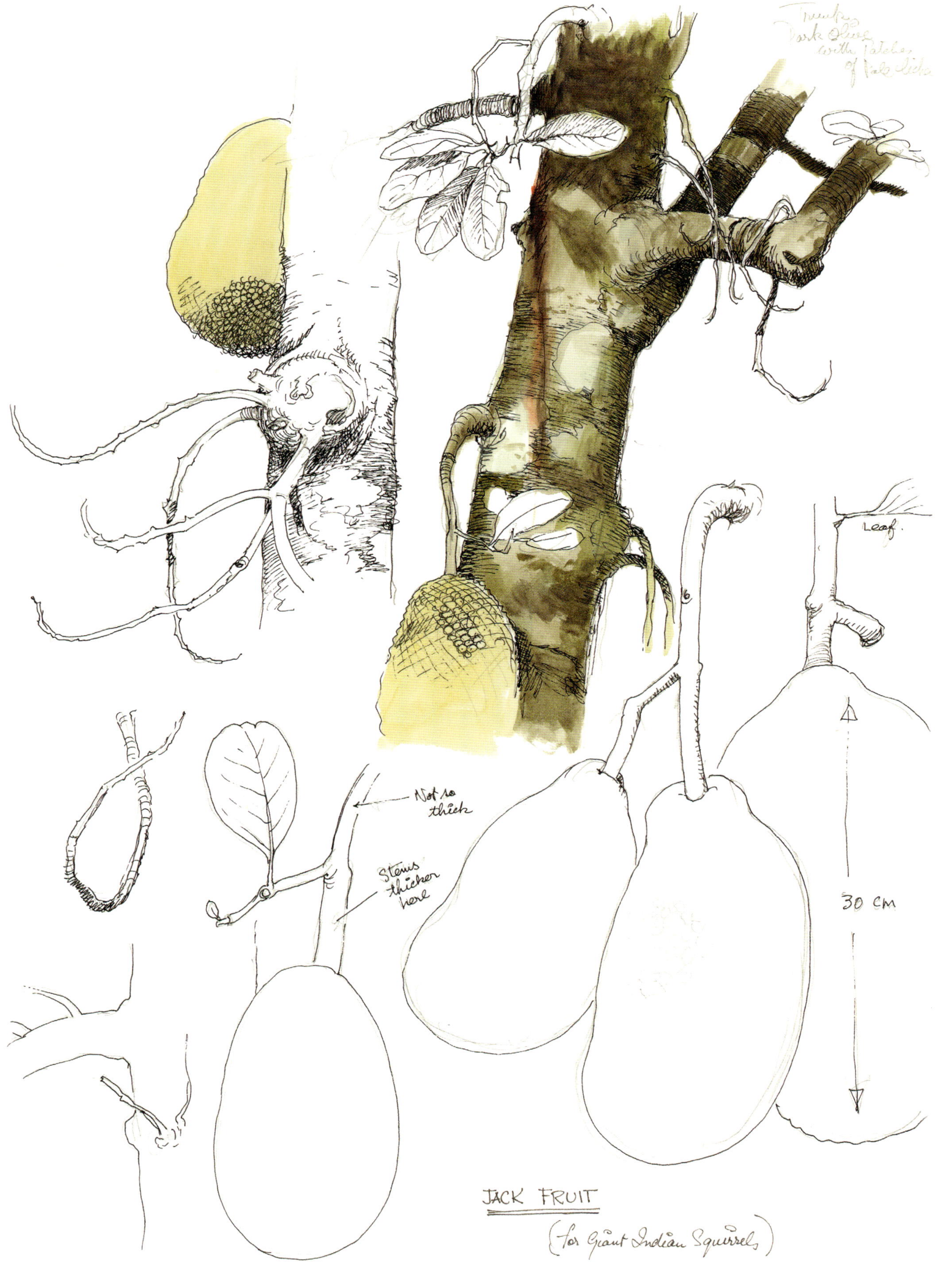
Leaf.
Not so thick
Stems thicker here
30 CM
JACK FRUIT
(for Giant Indian Squirrels)

Asplenium

Bird's Nest Fern

ASPLENIACEAE

Asplenium is from *a* (without) and spleen, referring to the plant's traditional use as a cure for spleen disorders.

There are numerous *Asplenium* species in the lush rainforest throughout New Guinea. They pepper branches, along with many other epiphytes. Bill loved the detail and lush exuberance of plants (orchids, mosses, ferns, lichens and so on), and he included them in all his paintings of birds in rainforest habitats. The illustrations on these pages have not been identified.

→ 1970; pen, pencil and gouache; 350 x 280mm
State Library of New South Wales

↙ 1973; pen, pencil and gouache; 355 x 328mm
National Library of Australia, 8070177

↓ Bird's Nest Fern on Trunk at Tanak Creek, Kubor Range, Papua New Guinea
1973; pen and pencil; (in part) 350 x 290mm
State Library of New South Wales

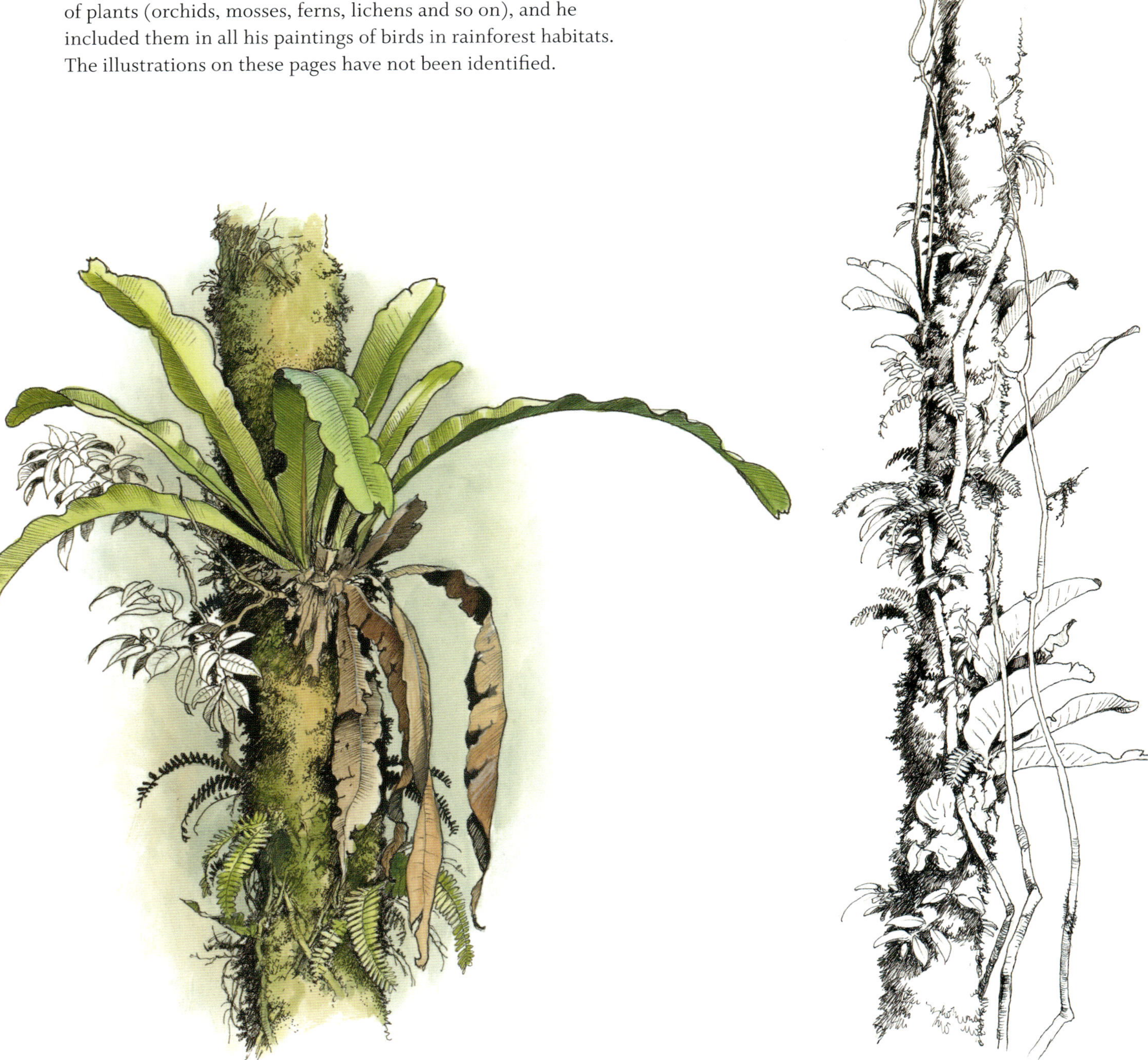

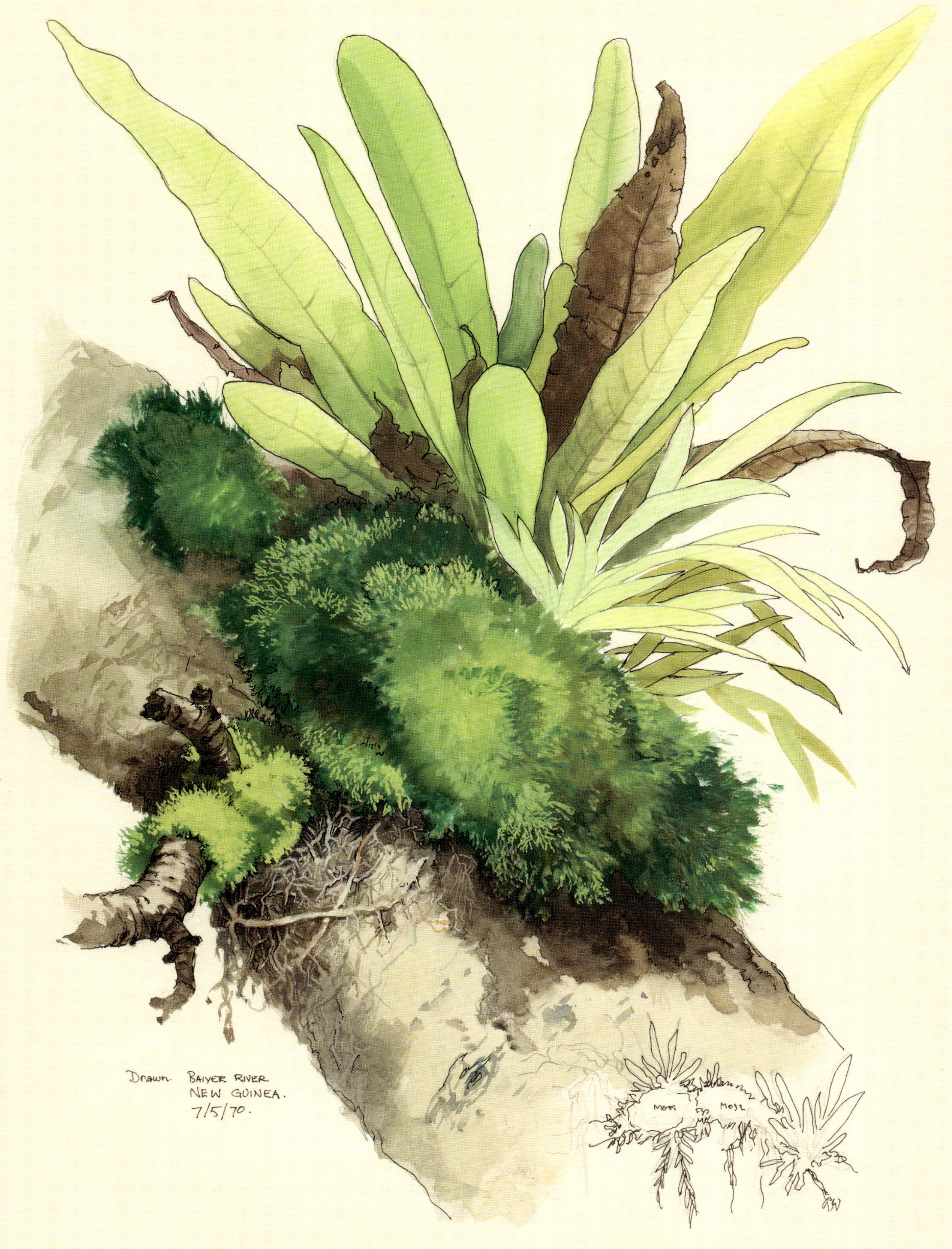
Drawn. Baiyer River
New Guinea.
7/5/70.

Bersama abyssinica

Winged Bersama

MELIANTHACEAE

→ 1989; pen, pencil and watercolour; 410 x 300mm
Private collection

↓ Crowned Hornbills, colour plate from *Kingfishers and Related Birds*, 1990
National Library of Australia, 2494045

Bersama is an Ethiopian name for the plant.

abyssinica is from Abyssinia (now Ethiopia).

Winged Bersama is a tree that grows up to 24 metres and is widespread in sub-Saharan Africa. All parts of the plant are poisonous, and are used medicinally.

In 1989, on an East African trip to gather information about hornbills, we became lost between Lake Baringo and Kitale in Kenya. However, we were unexpectedly rewarded by coming across a group of Crowned Hornbills in a Winged Bersama sapling beside the road. They were devouring the large *Nudaurelia* caterpillars feeding on the leaves of the tree. A young or female bird was begging and being fed by a male. This was exciting behaviour to witness and provided new material for a very different hornbill plate for *Kingfishers and Related Birds*.

Bill noted the incident in his diary:

> *Monday 30th Oct. 1989. Left Lake Baringo 0800 hrs. Drove up a range—down a wide valley then up again onto a plateau and to Kitale where we paid for Francis to go to a private dentist to have a tooth pulled. We then took the wrong road (horrific) but worth it when we found three Crowned Hornbills feeding on some small trees on the side of the road. They were getting huge caterpillars (see drawing), a female or young bird begged and was fed on the grubs.*

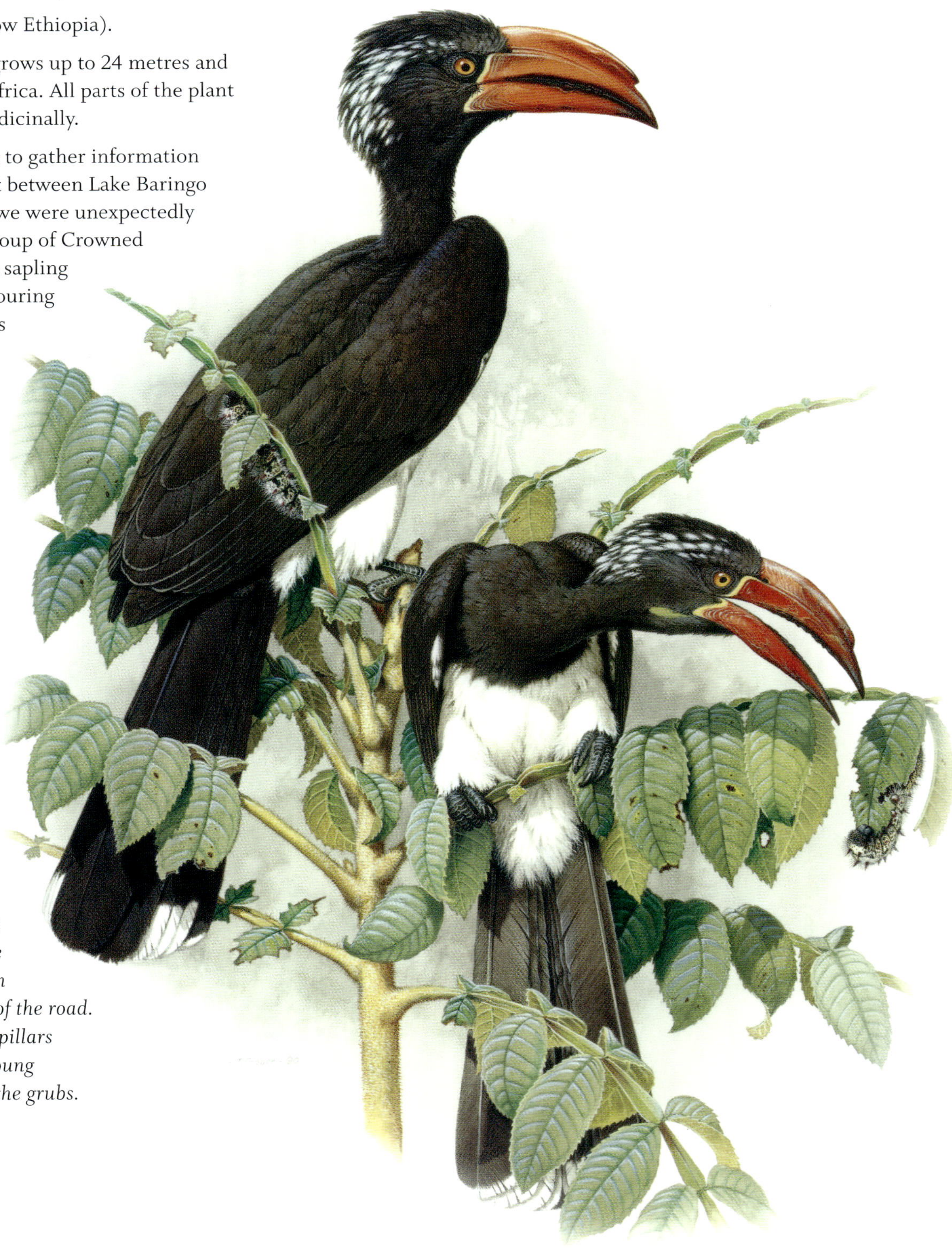

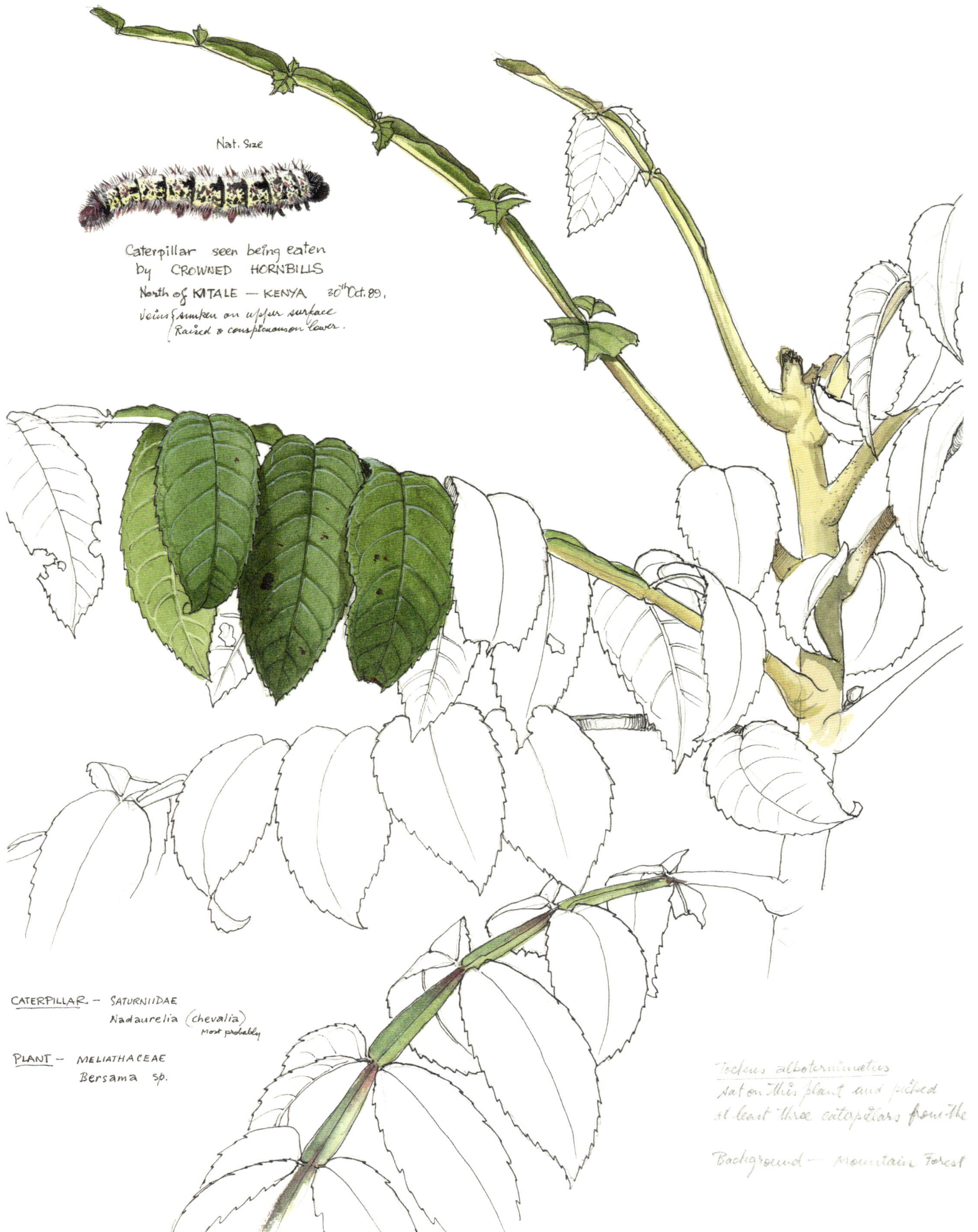
Nat. Size
Caterpillar seen being eaten
by CROWNED HORNBILLS
North of KITALE — KENYA 30th Oct. 89.
Veins sunken on upper surface
Raised & conspicuous on lower.
CATERPILLAR – SATURNIIDAE
Nadaurelia (chevalia)
Most probably
PLANT – MELIATHACEAE
Bersama sp.
Tockus alboterminatus
sat on this plant and picked
at least three caterpillars from the
Background — Mountain Forest

Billbergia pyramidalis

Flaming Torch

BROMELIACEAE

→ undated (c.1980); pen, pencil and watercolour
665 x 520mm
State Library of New South Wales

↓ undated; pen, pencil and watercolour
290 x 310mm
State Library of New South Wales

Billbergia is named in honour of Gustaf Johan Billberg (1772–1844), Swedish lawyer, botanist, zoologist and anatomist.

pyramidalis means pyramid-shaped, reference unknown.

The Flaming Torch bromeliad is a native epiphyte of South America widely cultivated in Australia.

Bill was able to take advantage of a cultivated Flaming Torch for use in a large acrylic painting on canvas (approximately 2 × 1 metres) of Blue and Yellow Macaws.

We were particular about finding plants near home that also occurred with overseas birds. However, it was often more difficult to confirm the habitats of South American subjects, and occasionally Bill was forced to use broadly correct plants with birds.

Bill painted the bromeliad below from a specimen of unknown identity in our garden. He described the activities of a small songbird on this plant in his diary on Christmas Day:

> *25th Dec. 1999. White-eared Monarch down on ground taking some prey from among Bromeliads in the garden. It was very quiet (tame), as was one we saw near the same spot a week or two ago.*

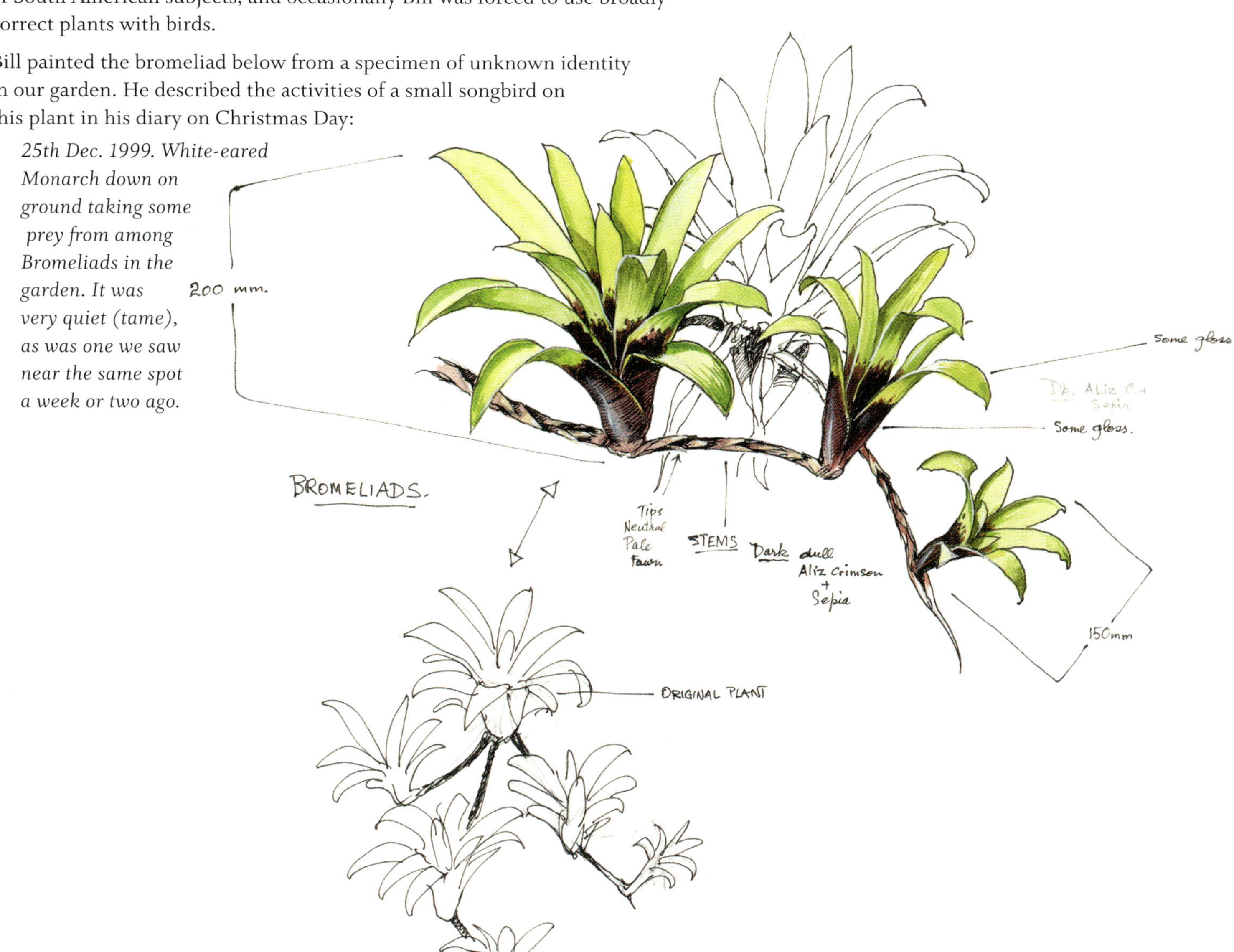

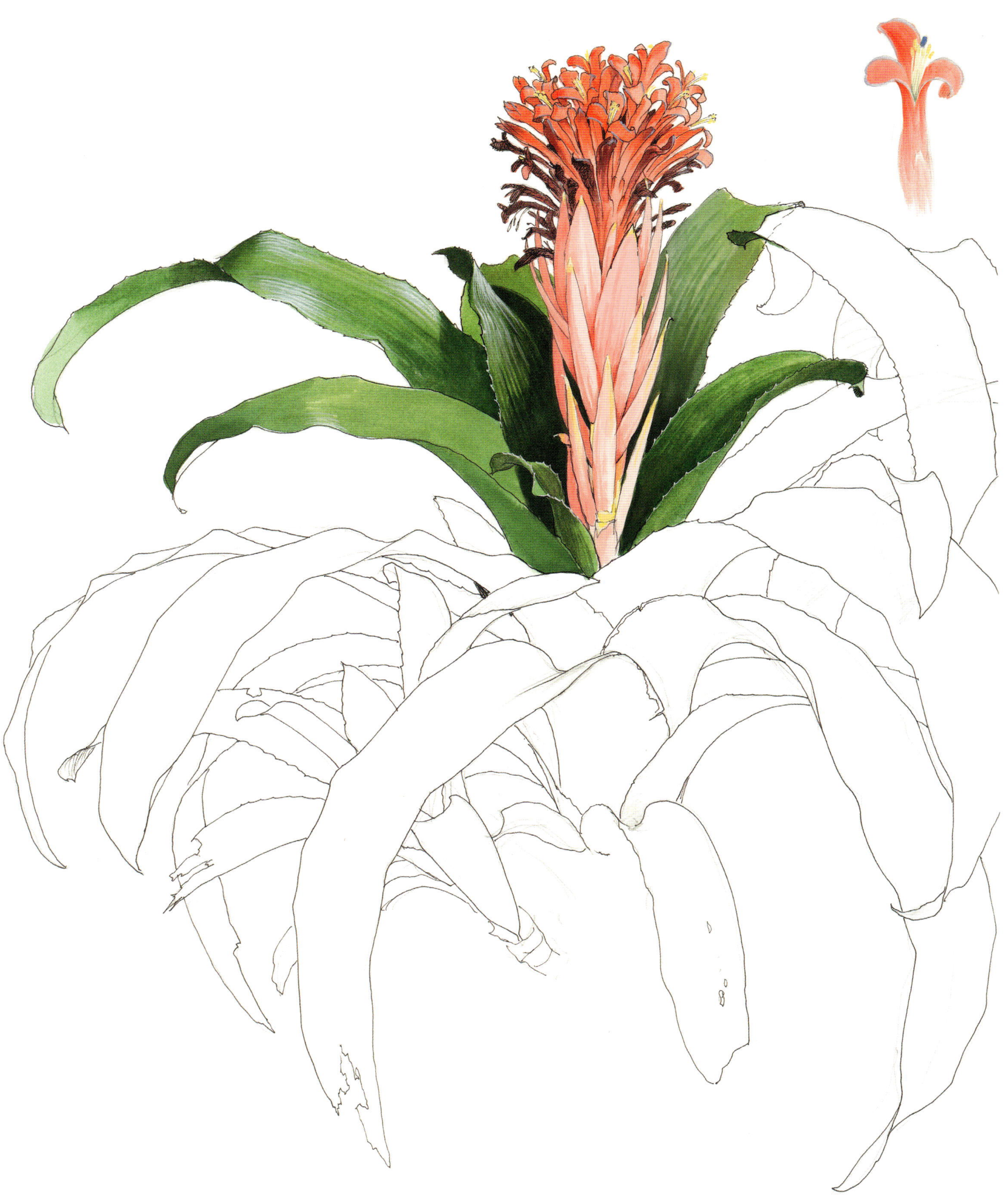

Bombax ceiba

Kapok Tree

MALVACEAE

Bombax is from *bombyx* (raw silk), referring to the fibre inside the fruit capsules.

ceiba is after *Ceiba*, a related genus.

The Kapok Tree is a deciduous species found in monsoon forests in Australia and throughout South-East Asia to India. Large waxy-red, nectar-rich flowers form on the leafless branches in the dry season and are visited by numerous bird species. The trunk is armed with conical thorns.

The painting of *Bombax malabaricum* (opposite) is now recognised as the widespread *Bombax ceiba*. It was done while we were in India to see tigers and hornbills. The sketch below, from north Queensland, is also of *Bombax ceiba*.

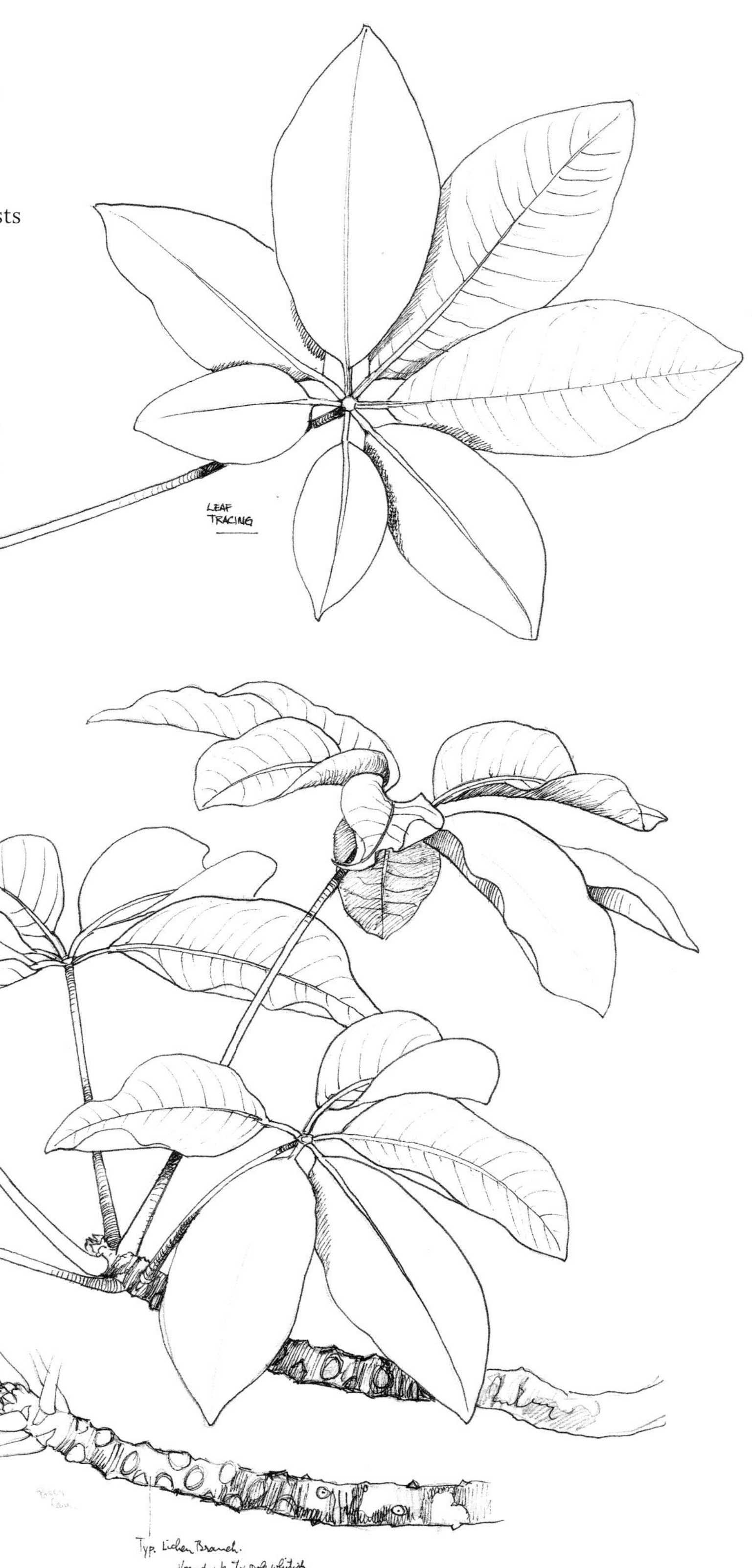

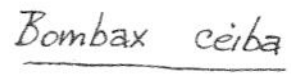

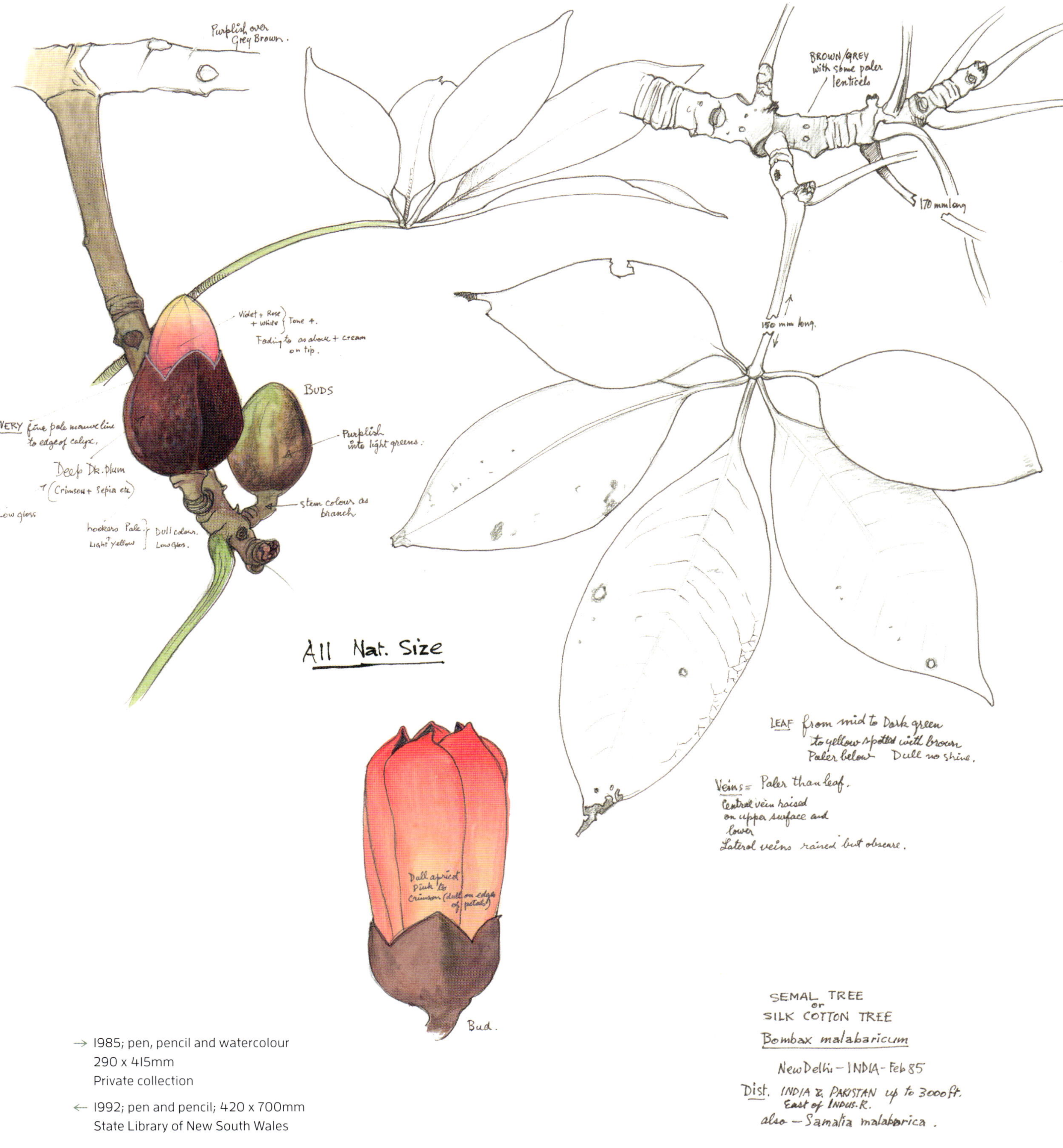

→ 1985; pen, pencil and watercolour
290 x 415mm
Private collection

← 1992; pen and pencil; 420 x 700mm
State Library of New South Wales

Capparis tomentosa

Woolly Caper Bush

CAPPARACEAE

→ 1995; pen, pencil and watercolour; 410 x 290mm
Private collection

↓ White-bellied Go-away Bird
1996; watercolour
Private collection

Capparis is after *kapar*, the Arabic name for *Capparis spinosa*.

tomentosa is from *tomentosus* (clothed in short, matted hairs), referring to the hair-like covering on new stems, spines and leaves.

Woolly Caper Bush is a thorny scrambler occurring in riverine forests, with a widespread distribution from South Africa and Namibia north to Senegal and Eritrea.

On a balmy morning after a stormy night in western Kenya, the riverine forest was alive with birds and mammals as a result of the previous night's rain.

Jackson's Hornbills were opening *Capparis* fruit and swallowing most of the seeds and pulp. White-bellied Go-away Birds then moved in to delicately extract the small bits of pulp remaining in cracked fruit hanging on the vine.

Bill wrote about the sighting in his diary:

> *Thurs. 5th Oct. 1995. Up at 0630 and off to West Pokot with our breakfast in a box … Went over to the Suam River which comes off Mt Elgon … Collected a species of Capparis which I drew on the spot. The White-bellied Go-aways eat the fruit—pecking a hole in the side of it exposing a lovely pink watermelon colour inside. Later watched Jackson's Hornbill squash a fruit with its bill and then extract the contents. I sat under a tree to draw the Capparis while the others went 'birding'.*

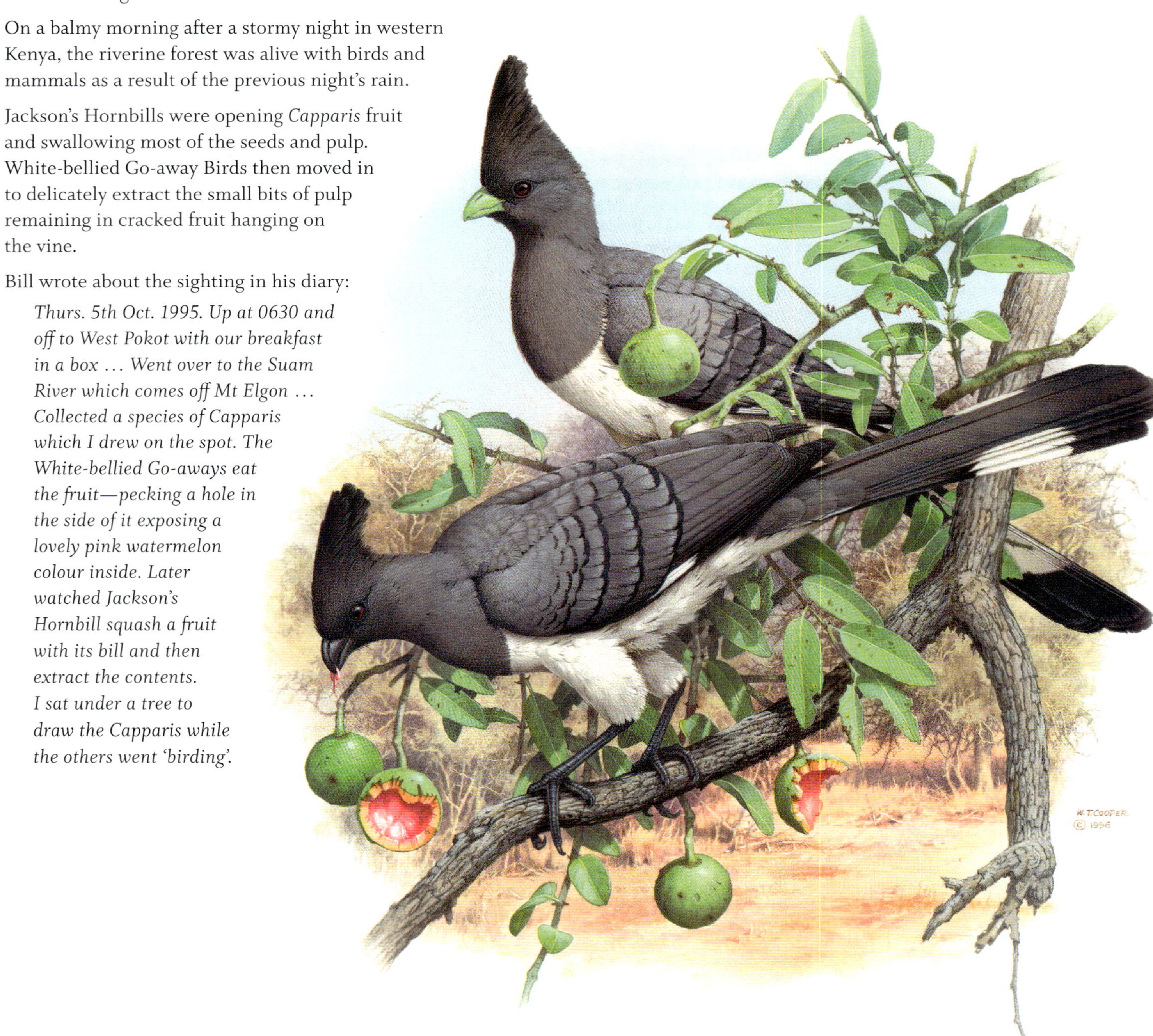

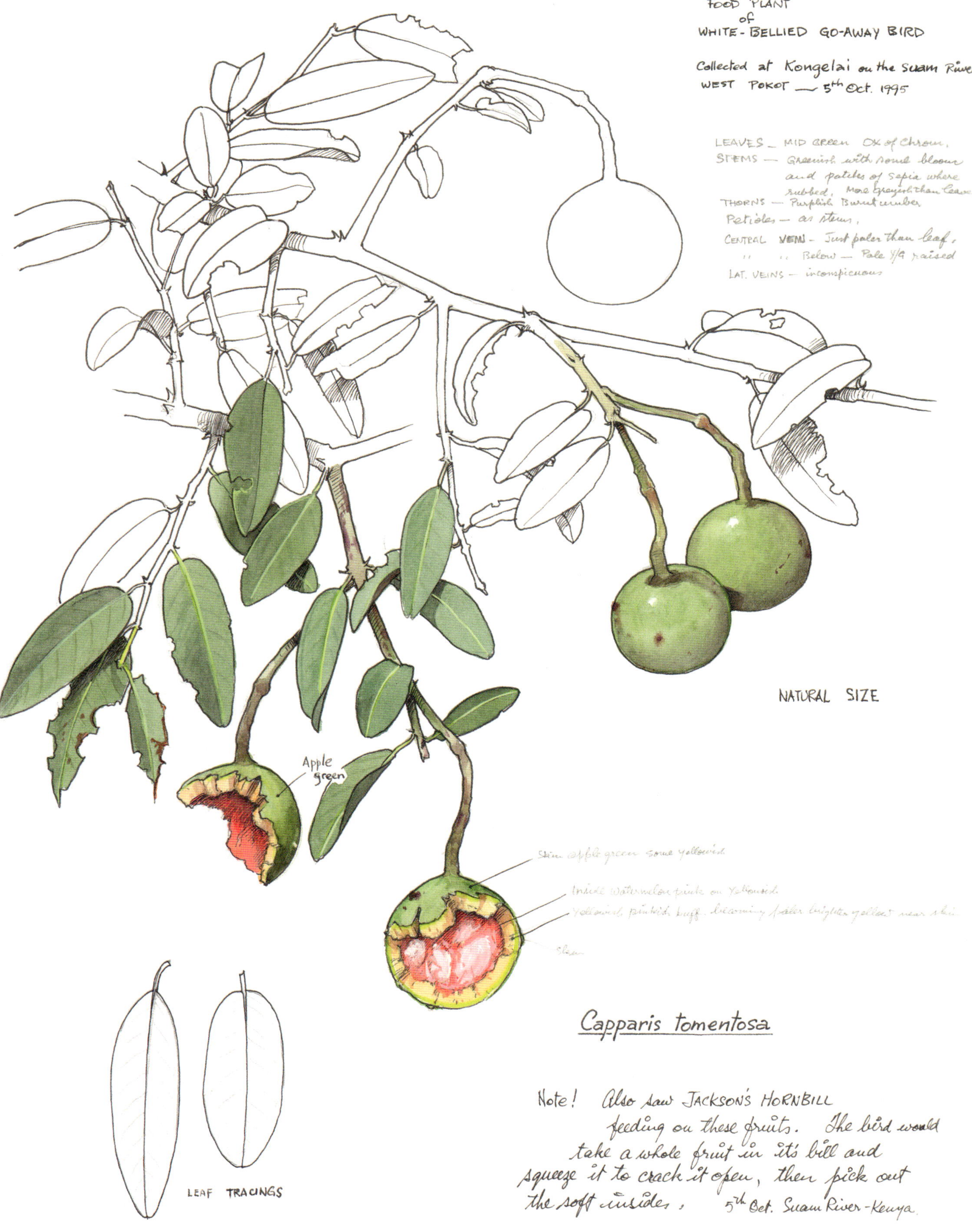
FOOD PLANT
of
WHITE-BELLIED GO-AWAY BIRD
Collected at Kongelai on the Suam River
WEST POKOT — 5th Oct. 1995
LEAVES – MID GREEN Ox of Chrom.
STEMS — Greenish with some bloom
and patches of sepia where
rubbed. More greyish than leave
THORNS — Purplish Burnt umber
Petioles — as stems.
CENTRAL VEIN – Just paler than leaf.
" " Below — Pale Y/G raised
LAT. VEINS – inconspicuous
NATURAL SIZE
Apple green
Skin apple green some yellowish
Inside watermelon pink on yellowish
Capparis tomentosa
Note! Also saw JACKSON'S HORNBILL
feeding on these fruits. The bird would
take a whole fruit in it's bill and
squeeze it to crack it open, then pick out
the soft insides. 5th Oct. Suam River-Kenya
LEAF TRACINGS

Cissus rotundifolia

Peruvian Grape Ivy

VITACEAE

Cissus is from the Greek name for Ivy, referring to these plants being climbers.

rotundifolia is from *rotundatus* (almost circular) and *-folius* (leaved), referring to the round fleshy leaves.

This tendril climbing vine occurs from India through Arabia down eastern Africa as far as South Africa. It has older stems that develop four or five corky wings.

While at the Arabuko-Sokoke Forest on the Kenyan coast, we watched Fischer's Turaco feeding on the delicious-looking berries. so the bird was painted with this vine in *Turacos: A Portfolio of All Species*. The sketch is labelled incorrectly as *Cissus quadrangularis*.

From Bill's diary:

> *Sunday 22nd Oct. 1995 0600 hrs. Drove to Arabuko-Sokoke forest … went looking for the birds (Fischer's Turaco). Got some good views of them but again found them to be very shy and wary. Collected some food plants. Back to the hotel for breakfast. Spent rest of day doing botanical drawings.*

→ 1995; pen, pencil and watercolour
420 x 290mm

↓ Fischer's Turaco
1996; watercolour
Private collection

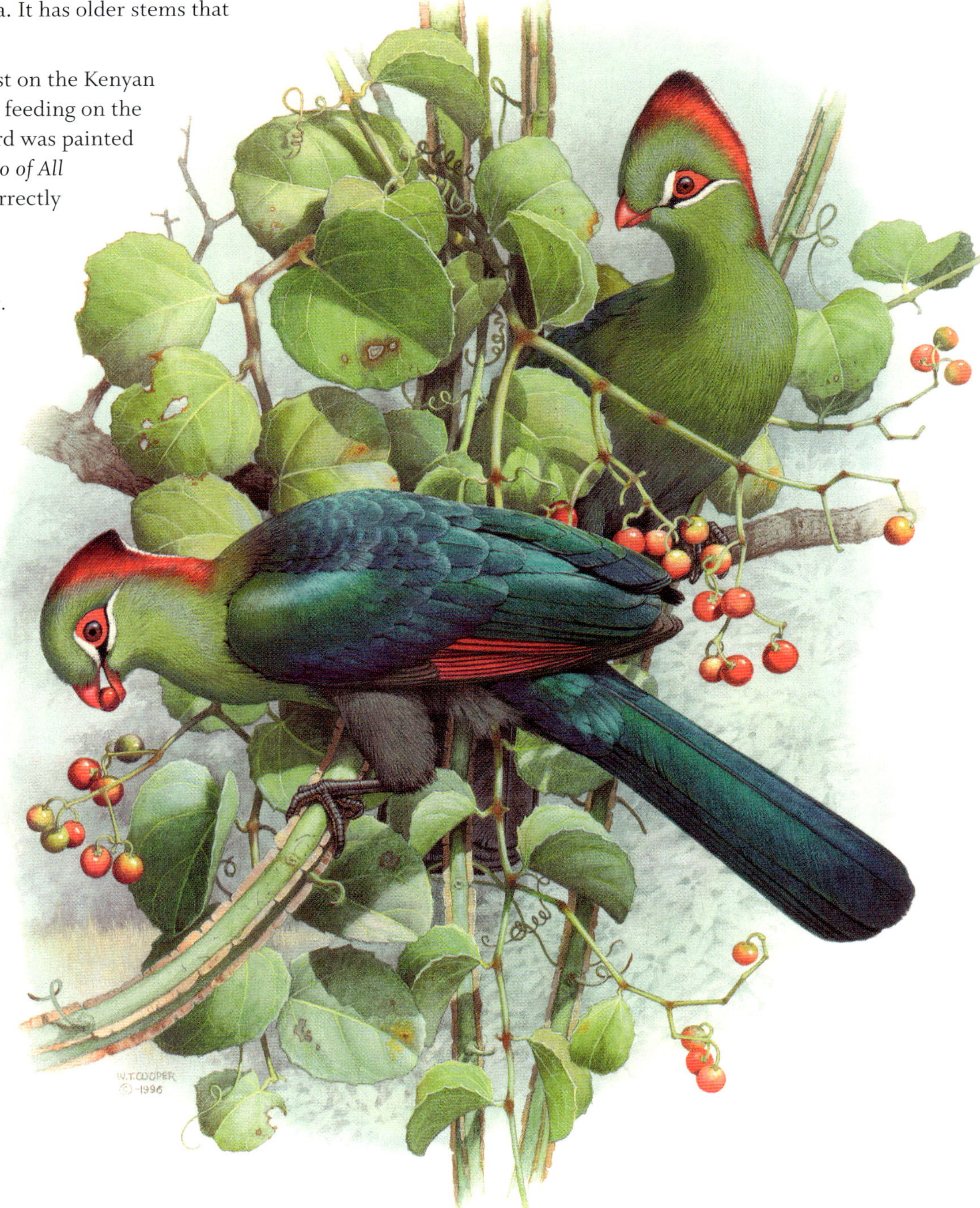

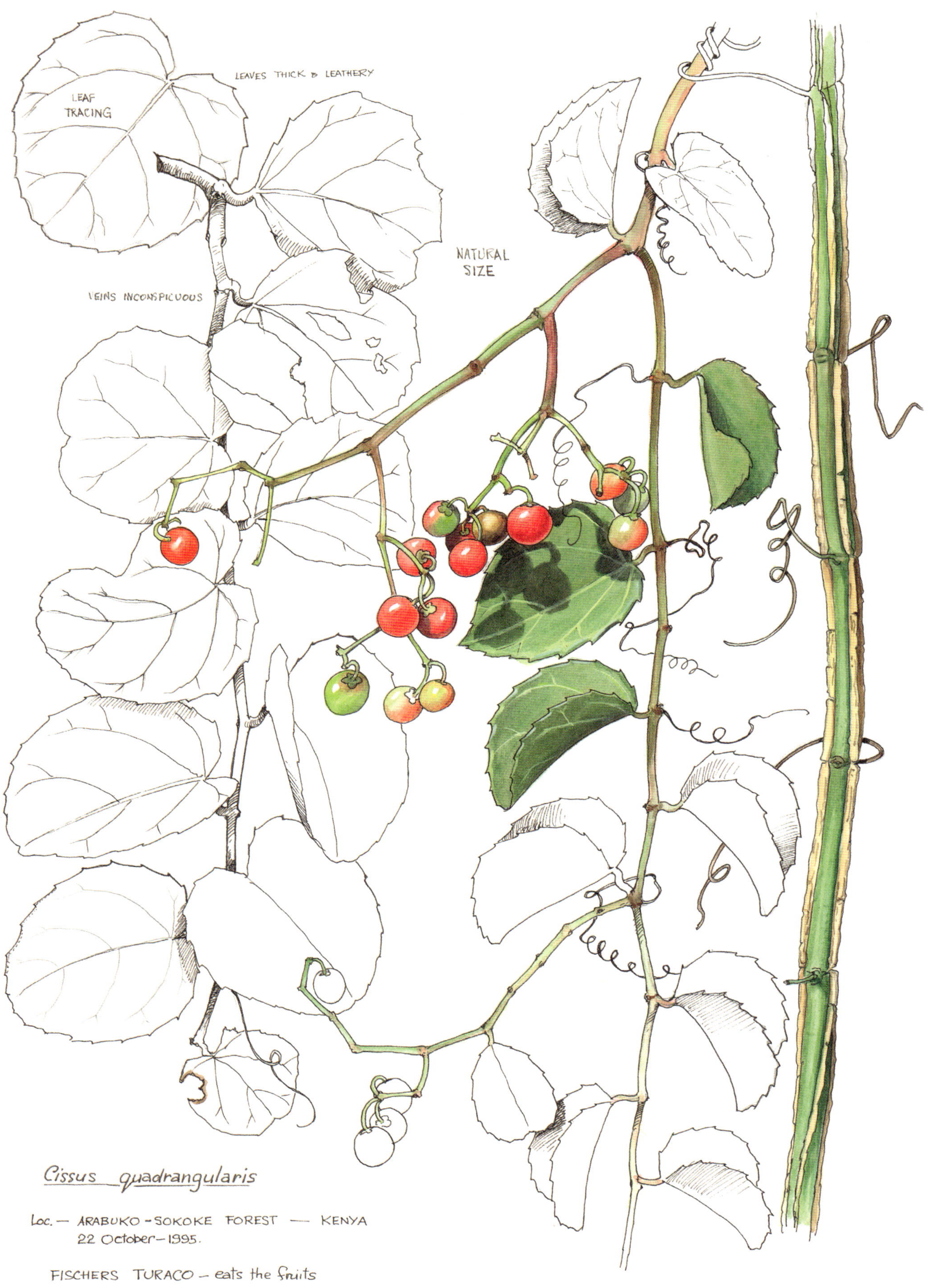
LEAVES THICK & LEATHERY
LEAF TRACING
NATURAL SIZE
VEINS INCONSPICUOUS
Cissus quadrangularis
Loc. — ARABUKO - SOKOKE FOREST — KENYA
22 October – 1995.
FISCHERS TURACO – eats the fruits

Dendrobium

ORCHIDACEAE

Dendrobium is from *dendron* (tree) and *bios* (life), referring to most *Dendrobium* orchids living on trees but not depending on them for life and nutrition.

This painting depicts a *Dendrobium* species with pinkish or orange pseudobulbs. The orchid in the foreground (opposite) may be a *Liparis* species. This piece was painted in Malaysia and used in the plate of the Banded Kingfisher for *Kingfishers and Related Birds*.

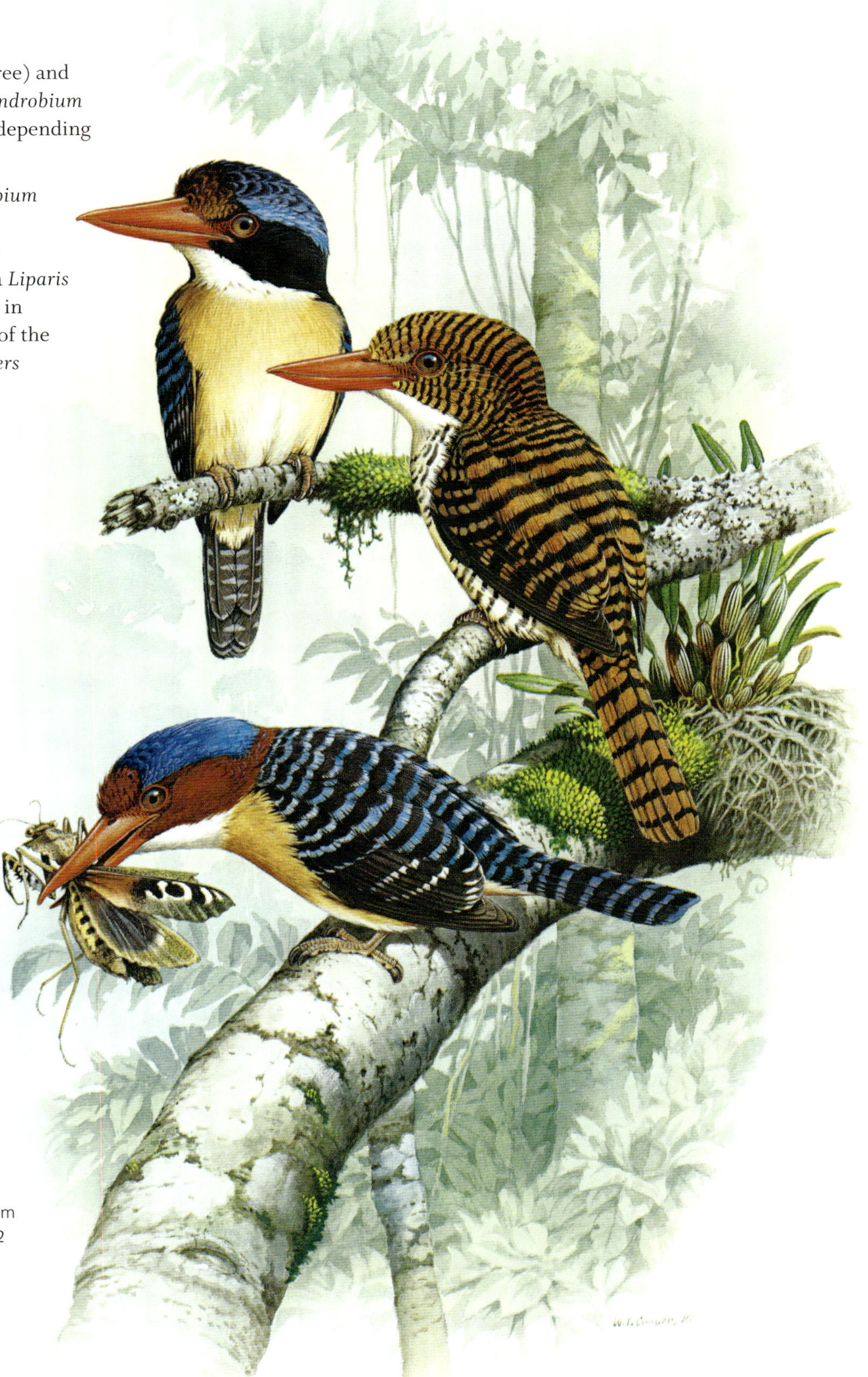

↗ Banded Kingfisher, colour plate from *Kingfishers and Related Birds*, 1982
National Library of Australia, 2494045

→ 1978; pen and watercolour
380 x 280mm
State Library of New South Wales

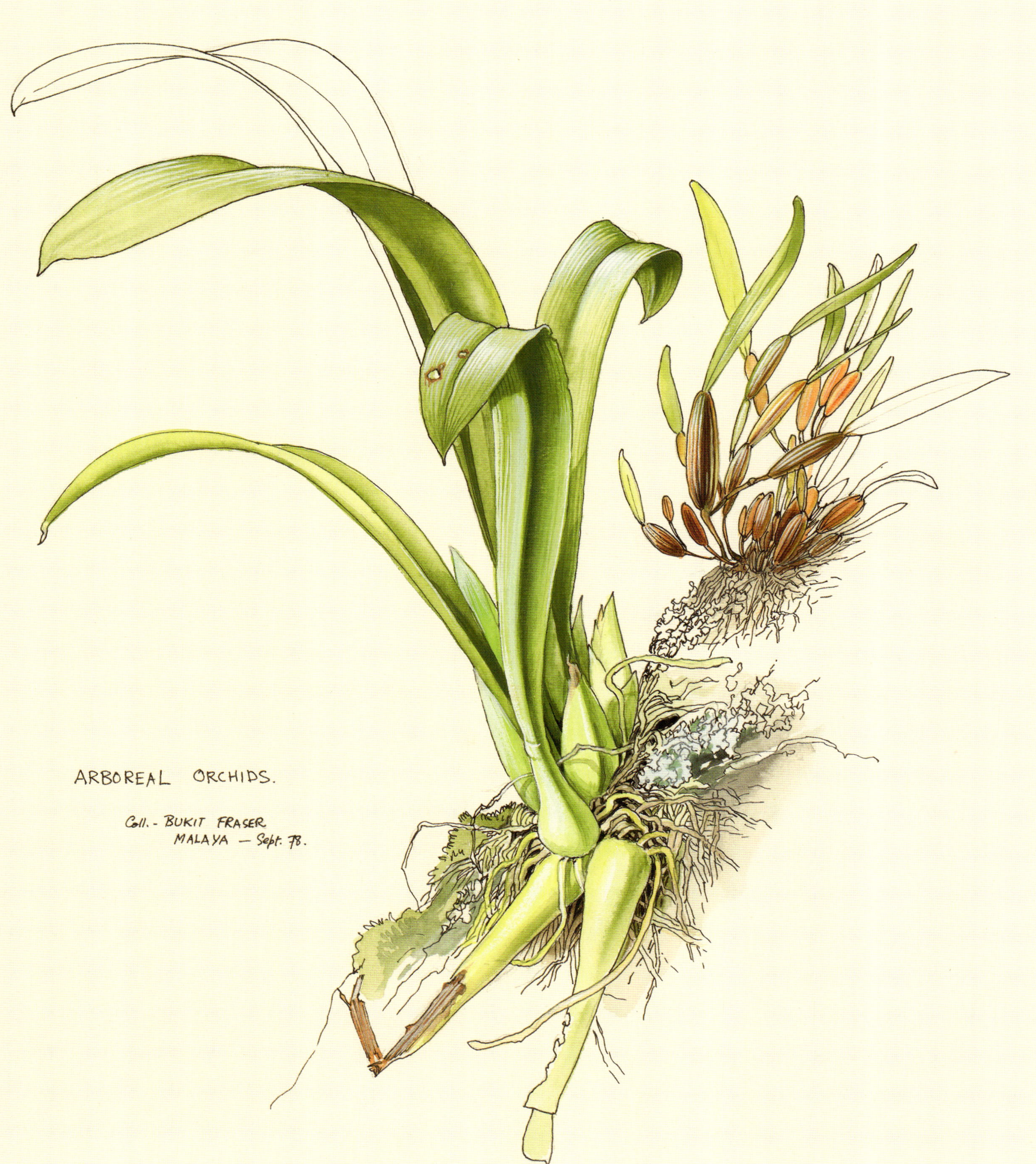
ARBOREAL ORCHIDS.
Coll. - BUKIT FRASER
MALAYA — Sept. 78.

Eichhornia crassipes

Water Hyacinth

PONTEDERIACEAE

Eichhornia is named in honour of J.A. Eichhorn (1779–1856), a Prussian Minister of Education.

crassipes is from *crassus* (thick) and *pes* (foot), referring to the swollen, bladder-like leaf stalks.

Water Hyacinth floats on still or moving water with the aid of its bladder-like leaf stalks. It is a native plant of the Amazon River area and has become prolific as an invasive species around the world.

This plant is now part of the habitat of many Australian bird species. Bill had planned a painting of Musk Ducks among its beautiful flowers. At Seppings Lake, near Albany in Western Australia, we watched Musk Ducks in display:

> *Wed. Sept. 1st 1999 … while it was displaying and making its bell-like note I could hear other birds 3 or 4 at different points around the lake making the same 'bell' call. Later we heard a fourth call—an even louder bell call, also a motor-like trrrrrrrrrr when the first female approached him.*

Opposite is an oil sketch of the proposed Musk Duck painting, with Bill's notes on using Hyacinth flowers.

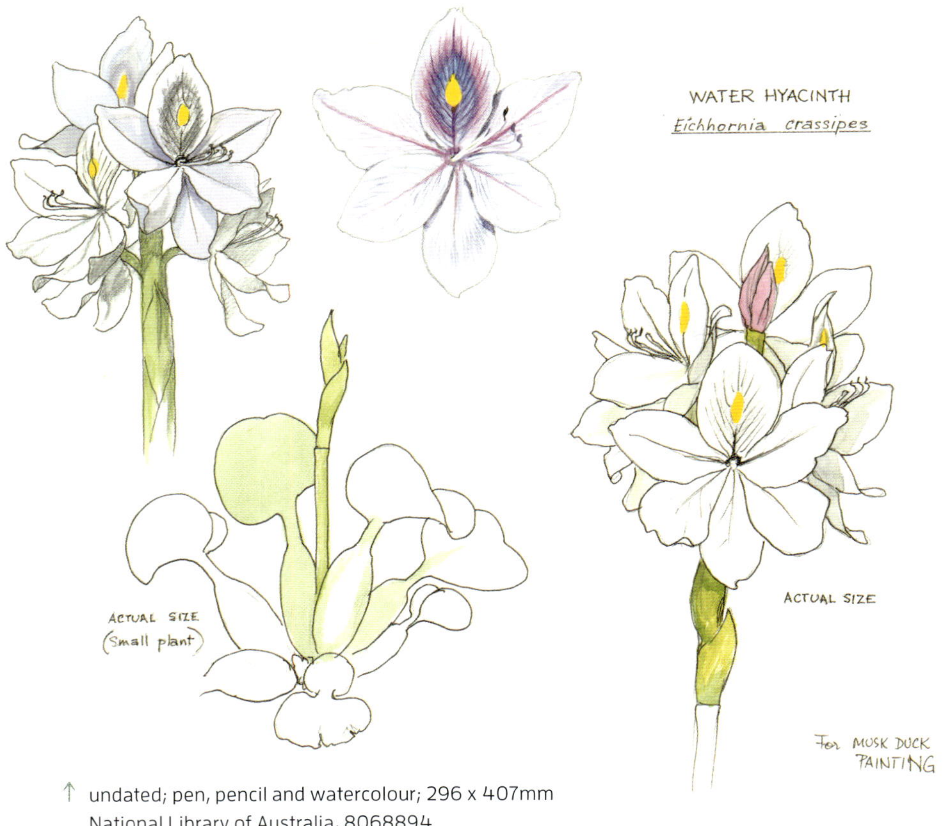

↑ undated; pen, pencil and watercolour; 296 x 407mm
National Library of Australia, 8068894

→ Musk Duck Sketch
undated; oil; 405 x 560mm
Private collection

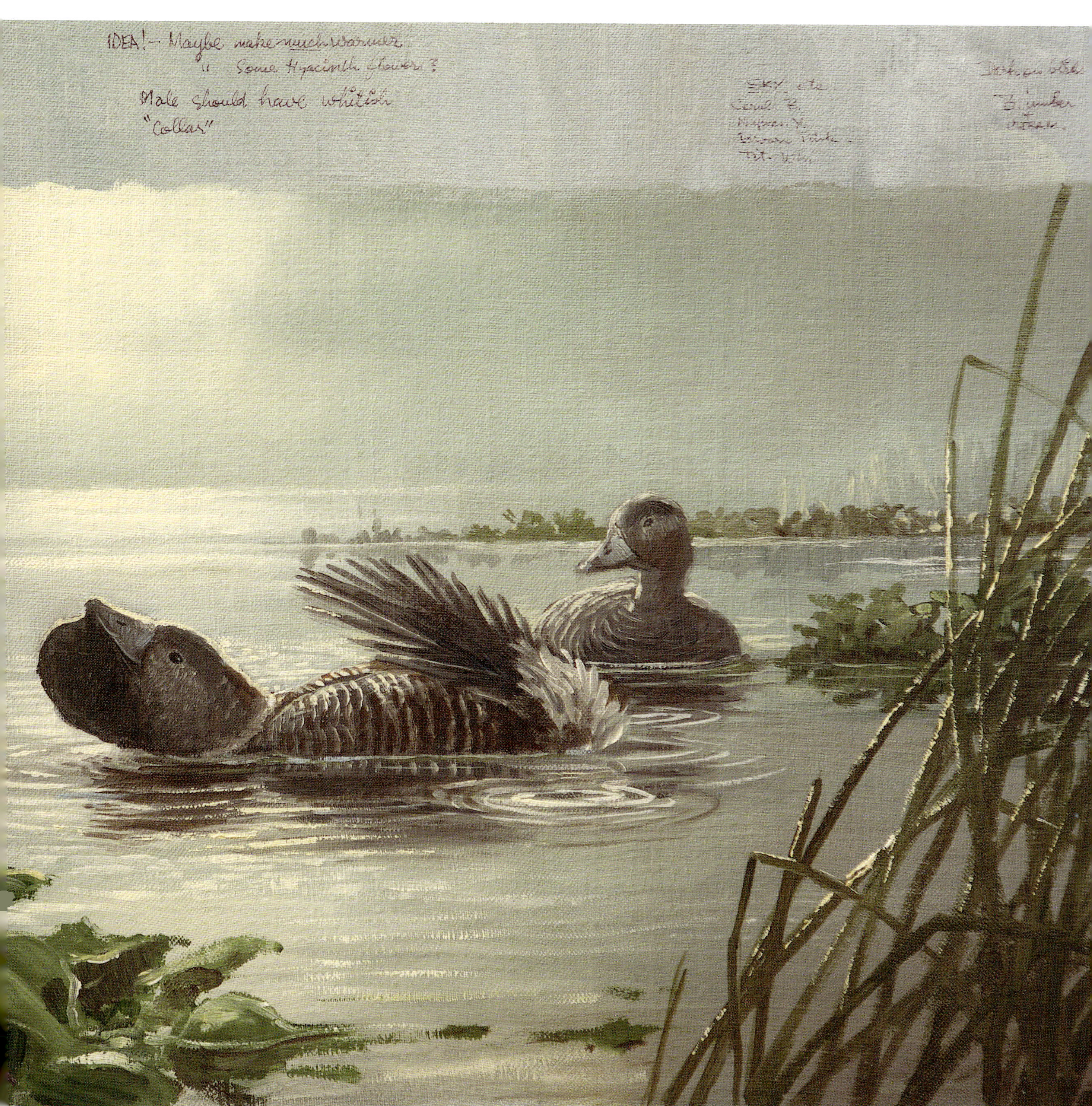
IDEA! – Maybe make much warmer
" Some Hyacinth flowers?
Male should have whitish
"collar"

Eurya tigang

Kilawa or Tigang

PENTAPHYLACACEAE

Eurya is from *Eurus* (the Greek god of the East Wind).

tigang is the local name for the plant in the area where the type specimen was collected by German Lutheran G. Bamler in the village of Sattelberg, near Finschhaffen, on the Huon Peninsula in Papua New Guinea.

Whilst the plant is known as Tigang at Sattelberg, Kilawa is the name used in the Mount Hagen area of the Papua New Guinea Highlands. The shrub is widespread in New Guinea, from the lowlands to the highlands.

Although Bill's notes state that the Ribbon-tailed Astrapia fed on these fruits, he chose to use different plants in his paintings of this bird, so this reference painting was never used. The painting below shows a lichen and moss embellished twig from Tomba in Papua New Guinea.

→ 1973; pen, pencil and watercolour
400 x 290mm
State Library of New South Wales

↓ 1973; pen and watercolour
(in part) 400 x 275mm
State Library of New South Wales

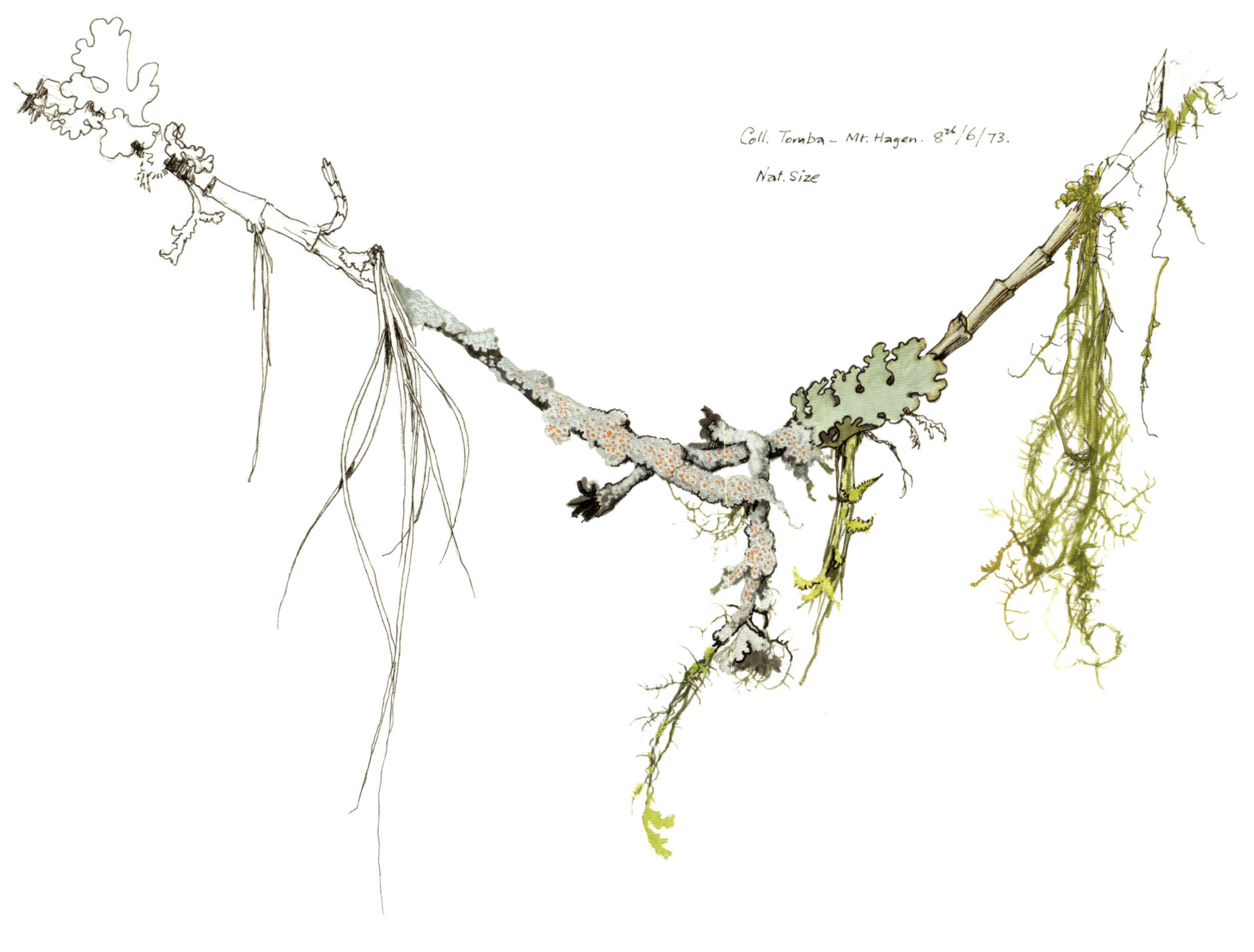

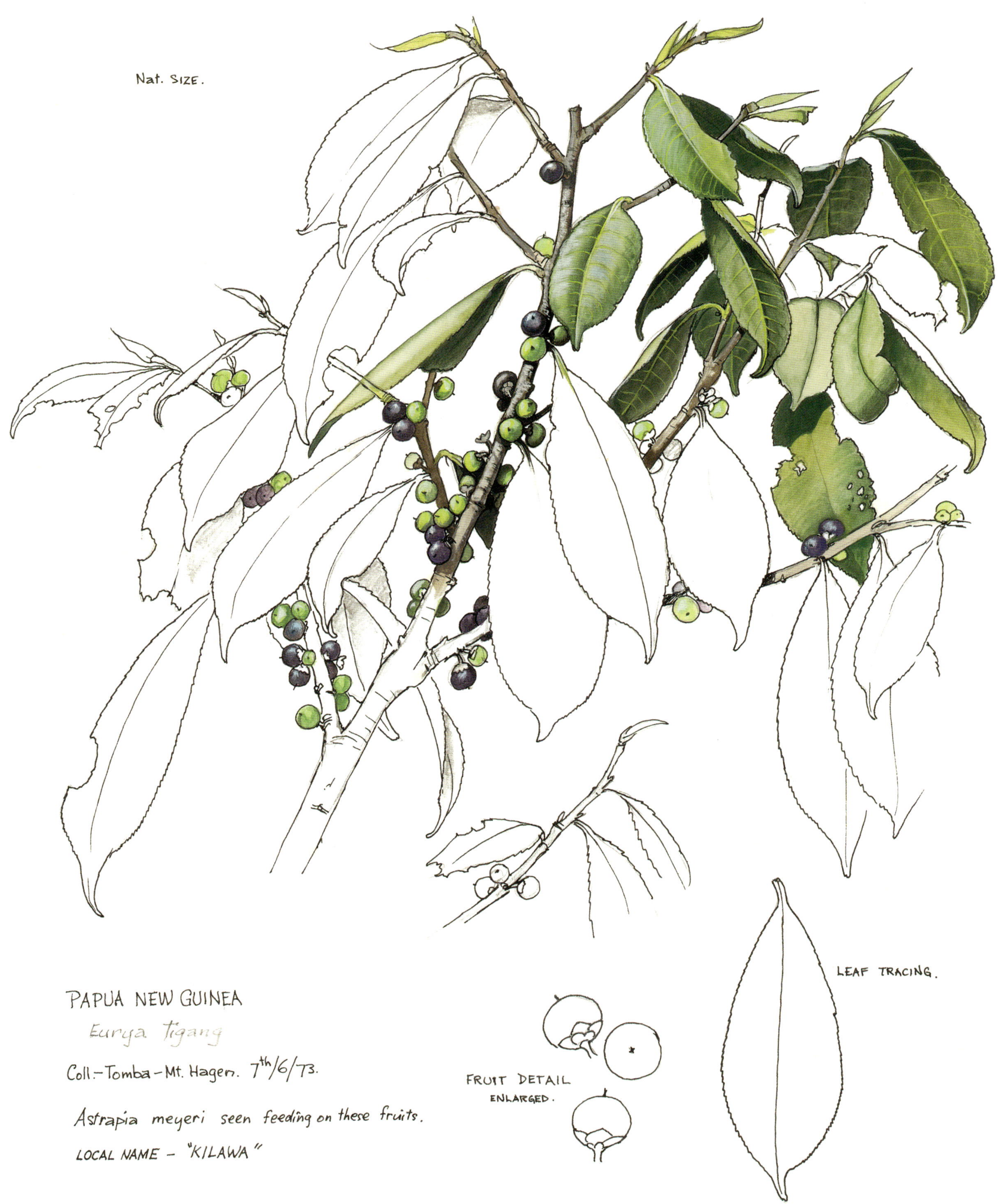
Nat. SIZE.
LEAF TRACING.
PAPUA NEW GUINEA
Eurya tigang
Coll.-Tomba-Mt. Hagen. 7th/6/73.
FRUIT DETAIL
ENLARGED.
Astrapia meyeri seen feeding on these fruits.
LOCAL NAME - "KILAWA"

Ficus

Fig

MORACEAE

Ficus is from the Greek name for the edible fig, *Ficus carica*.

The painting below is of an unidentified fig from Baiyer River in Papua New Guinea, where Bill was based while he gathered material relevant to parrots. He used this piece in the plate for Salvadori's and Desmarest's Fig Parrots for *Parrots of the World*.

→ Salvadori's Fig Parrot, Edward's Fig Parrot and Desmarest's Fig Parrot
1971; gouache; 455 x 363mm
National Library of Australia, 1812886

↓ 1970; pen, pencil and gouache; 380 x 275mm
National Library of Australia, 8069086

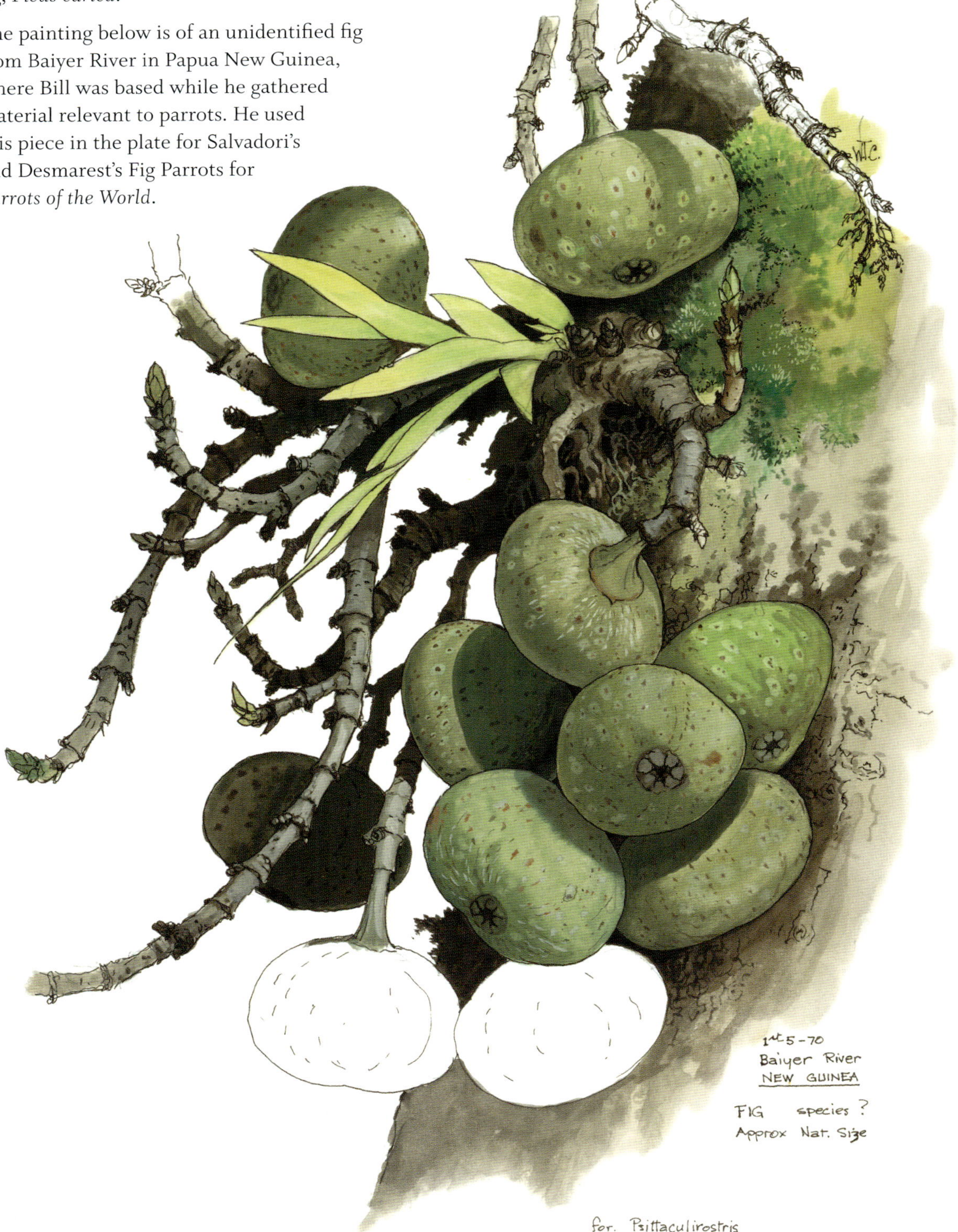

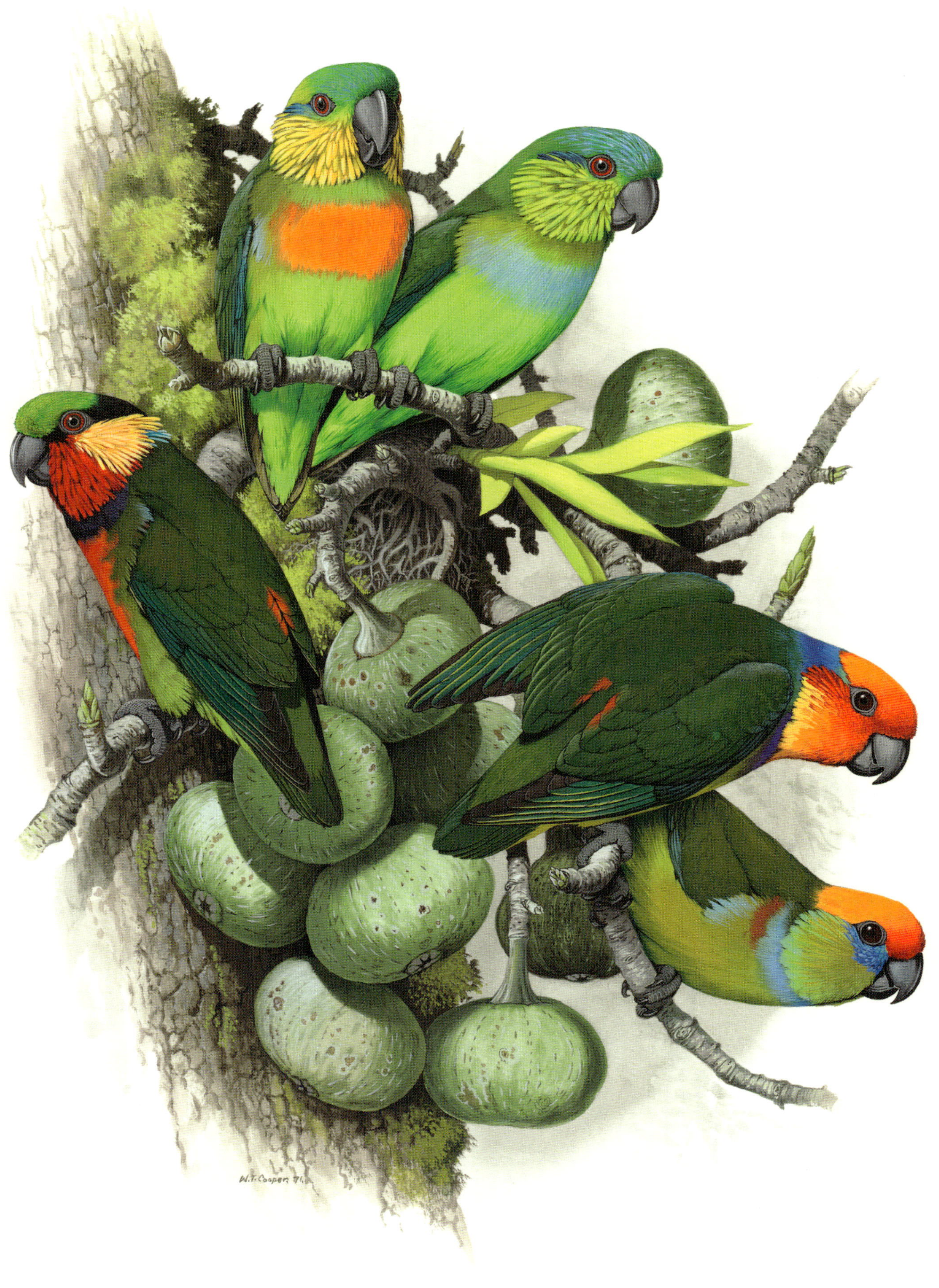
W.T.Cooper 74

Ficus benghalensis

Indian Banyan

benghalensis is from Bengal, north-eastern India.

Indian Banyan trees are considered to have the largest canopy of all trees in the world. They begin life as a strangler fig. Aerial roots grow from the branches creating drapes, which eventually coalesce to become woody and pillar-like, supporting the wide-spreading branches from where they emerged.

The Indian Banyan is the national tree of India, considered sacred by Hindus who often build temples in the shade beneath them.

The coloured sketch opposite was made while we were staying at Ranthambore National Park, in Rajasthan, to see hornbills and tigers in the wild.

→ 1985; pen, pencil and watercolour; 400 x 290mm
State Library of New South Wales

↓ Indian Grey Hornbill, colour plate from *Kingfishers and Related Birds*, 1989
National Library of Australia, 2494045

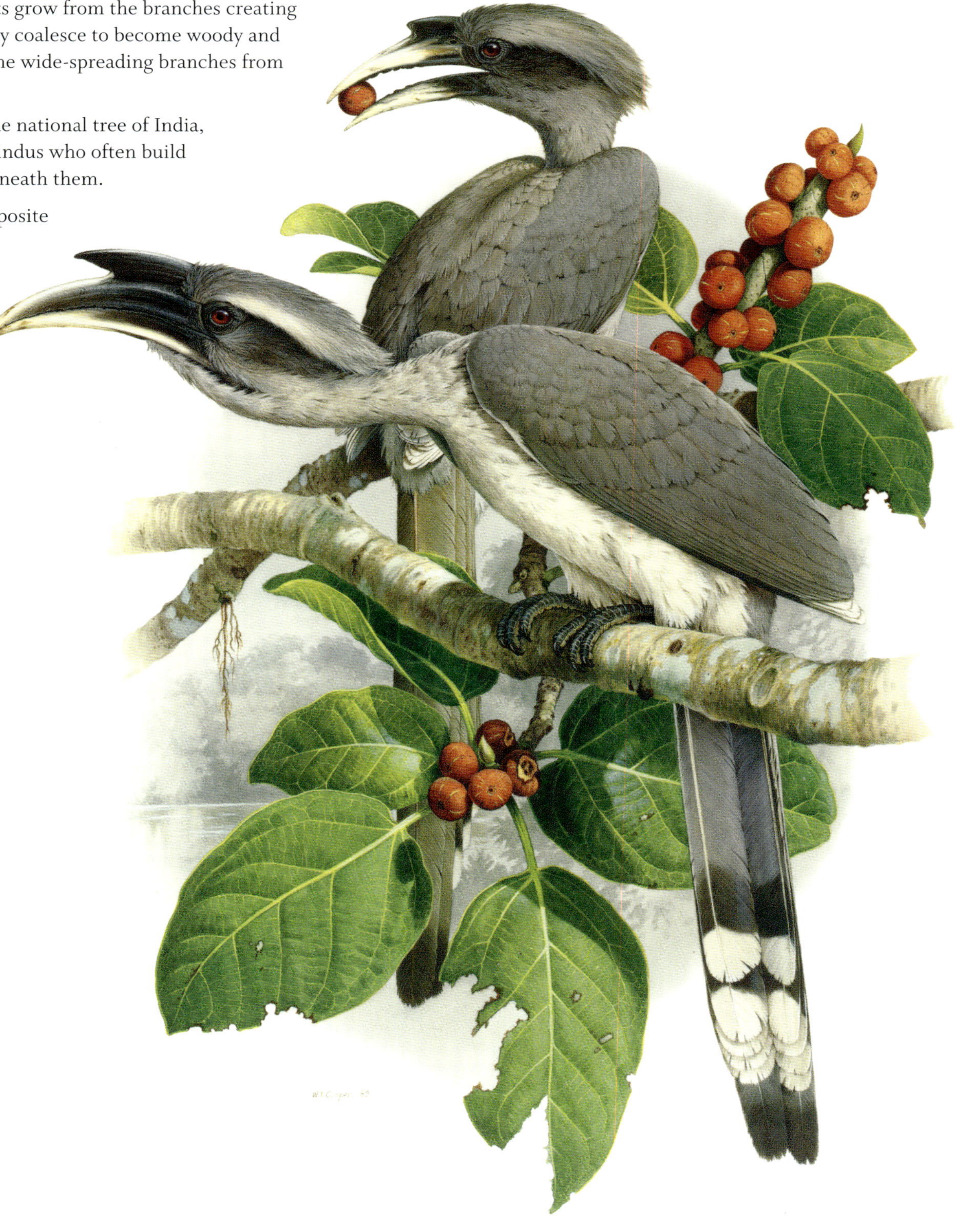

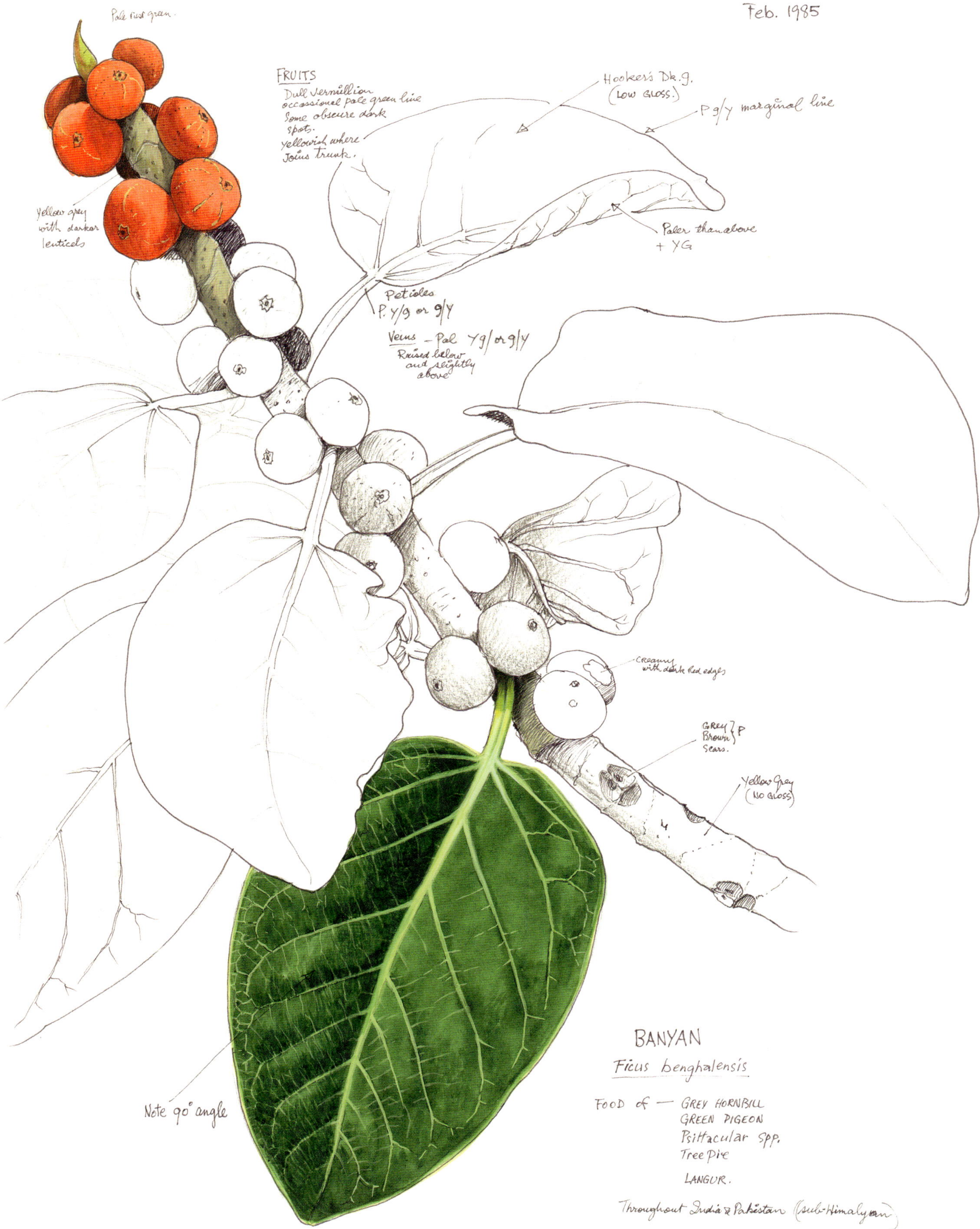
RANTHAMBHORE — INDIA
Feb. 1985
Pale rust green.
FRUITS
Dull vermillion
occassional pale green line
Some obscure dark
spots.
yellowish where
Joins trunk.
Hooker's Dk.g.
(LOW GLOSS.)
P g/y marginal line
Paler than above
+ YG
Yellow grey
with darker
lenticels
Petioles
P. y/g or g/y
Veins – Pal Yg/ or g/y
Raised below
and slightly
above
Creamy
with dark Red edges
Grey
Brown } P
Scars.
Yellow Grey
(NO GLOSS)
BANYAN
Ficus benghalensis
FOOD of — GREY HORNBILL
GREEN PIGEON
Psittacular spp.
Tree Pie
LANGUR.
Throughout India & Pakistan (sub-Himalayan)
Note 90° angle

Ficus schwarzii

Russet Stem-Fig

schwarzii is named in honour of Johannes Albert Traugott Schwarz (1836–1920), a teacher and missionary on the Indonesian island of Sulawesi (formerly the Celebes).

Russet Stem-Fig is a medium-sized fig tree that occurs in disturbed rainforest areas from the lowlands to the highlands in Malaysia, Thailand, Borneo and Myanmar. This reference sketch was used for the Sumba Hornbill plate in *Kingfishers and Related Birds*.

Whilst we did not observe birds feeding on these figs, we were confident that they would take them. All figs seem to be keenly sought after by fruit-eating birds. We once asked an African bird guide in Kenya whether anything ate the figs on the tree beside us. His answer was, 'It is a fig!'

→ 1978; pen, pencil and watercolour
State Library of New South Wales

↓ Sumba Hornbill, colour plate from *Kingfishers and Related Birds*, between 1985 and 1993
National Library of Australia, 2494045

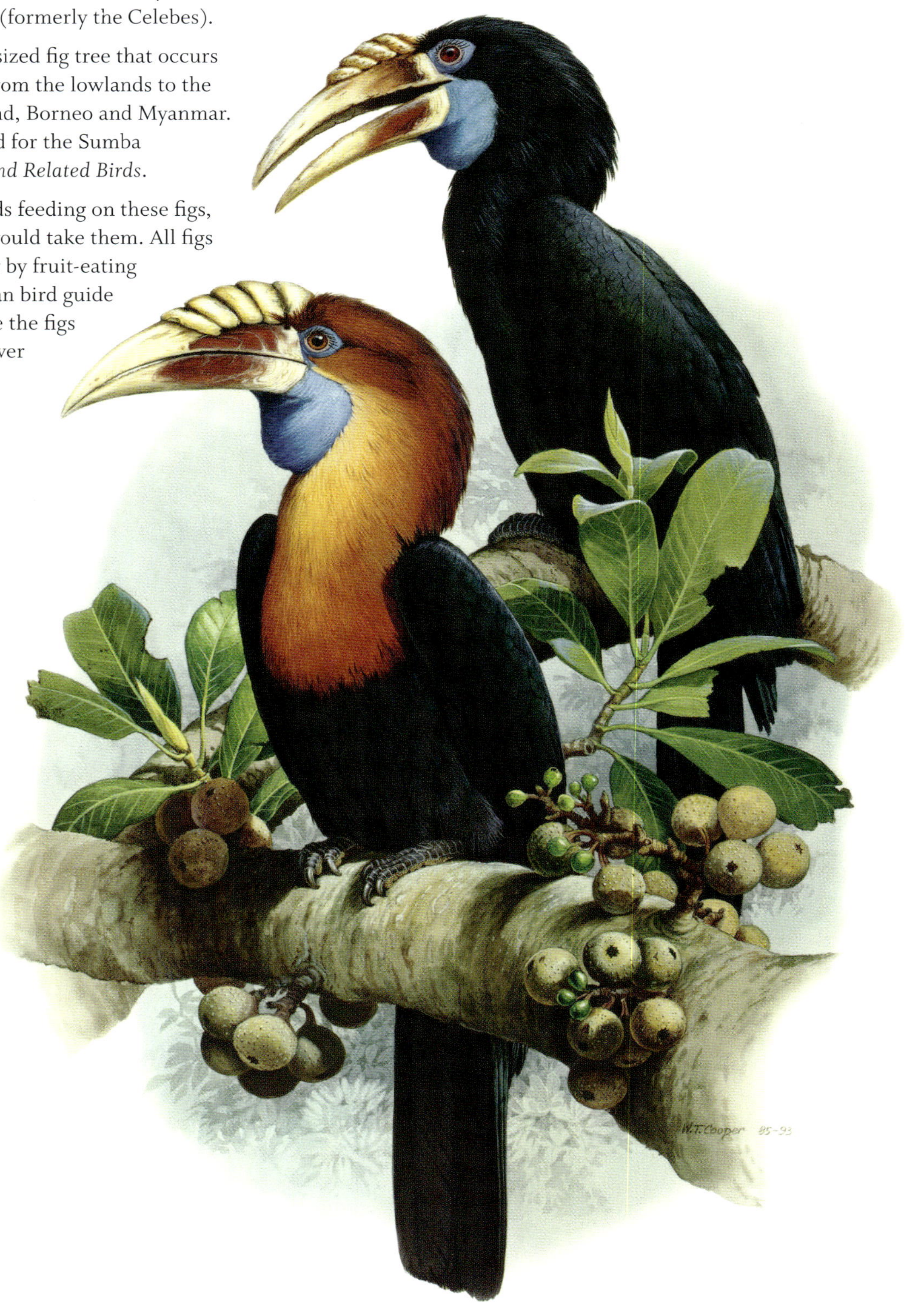

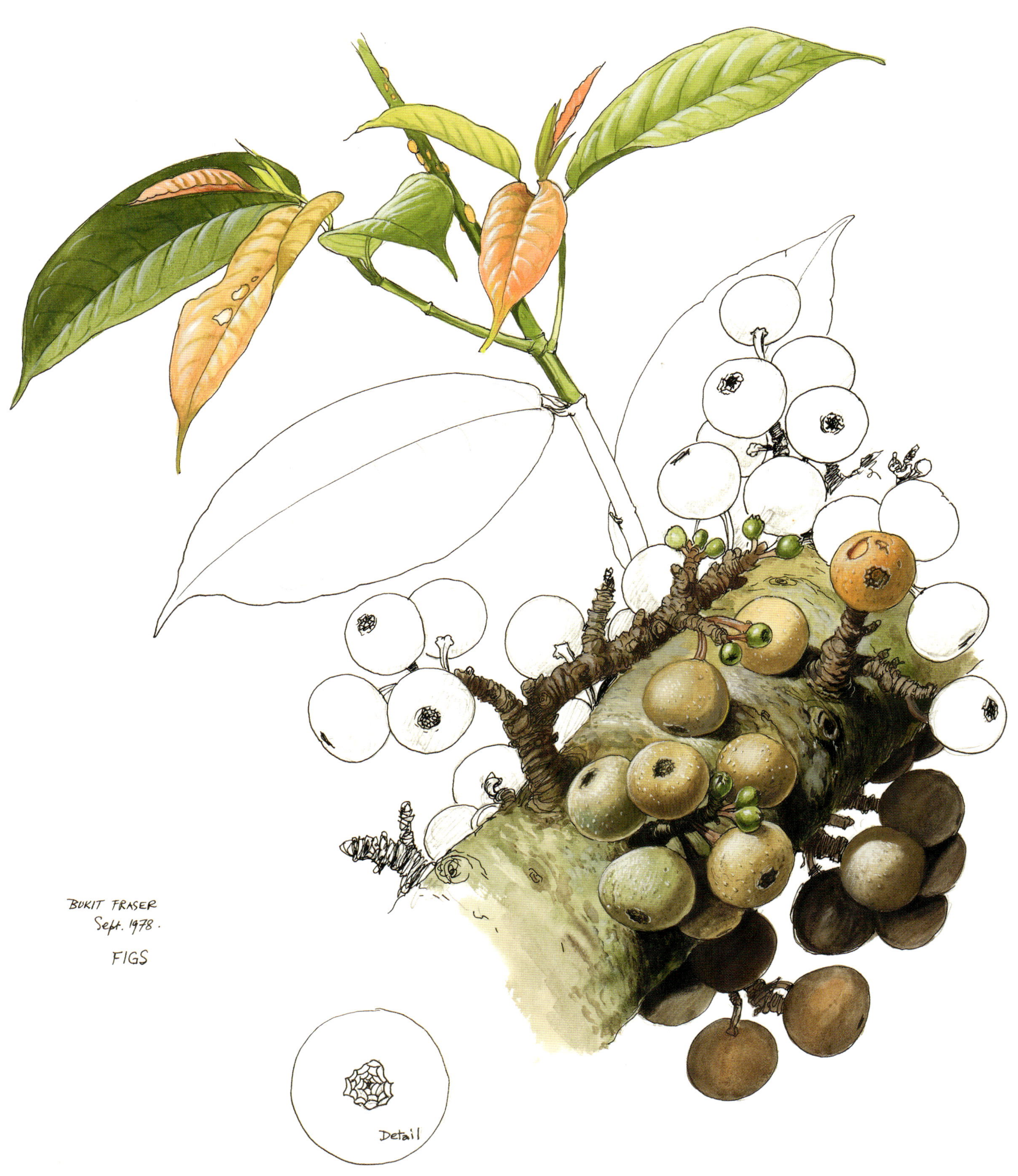
BUKIT FRASER
Sept. 1978.
FIGS
Detail

Lannea schweinfurthii

False Marula

ANACARDIACEAE

→ 1995; pen and watercolour; 410 x 285mm
Private collection

↓ Great Blue Turaco
1996; watercolour
Private collection

Lannea is from *lana* (wool), referring to the new leaf growth.

schweinfurthii is named in honour of Georg August Schweinfurth (1836–1925), a Latvian-born German botanist, ethnologist and explorer, working in Central Africa and the Libyan Desert.

False Marula is a beautifully shaped forest tree occurring in Kenya and beyond. The fruits are edible and the bark has many uses for local people.

At Entebbe, beside Lake Victoria in Uganda, we saw Great Blue Turacos feeding on these fruits. Bill wrote in his diary:

> *Sunday 8th Oct. 1995. Drove to Entebbe … Watched 5 Great Blue Turacos feeding in a huge buttressed tree. They were hanging almost upside down to pick fruit from the pendulous branches. We collected a piece … Lannea schweinfurthii.*

The birds are large and colourful and were animated, almost like parrots.

The reference sketch opposite was used in the painting of Great Blue Turacos in *Turacos: A Portfolio of All Species*.

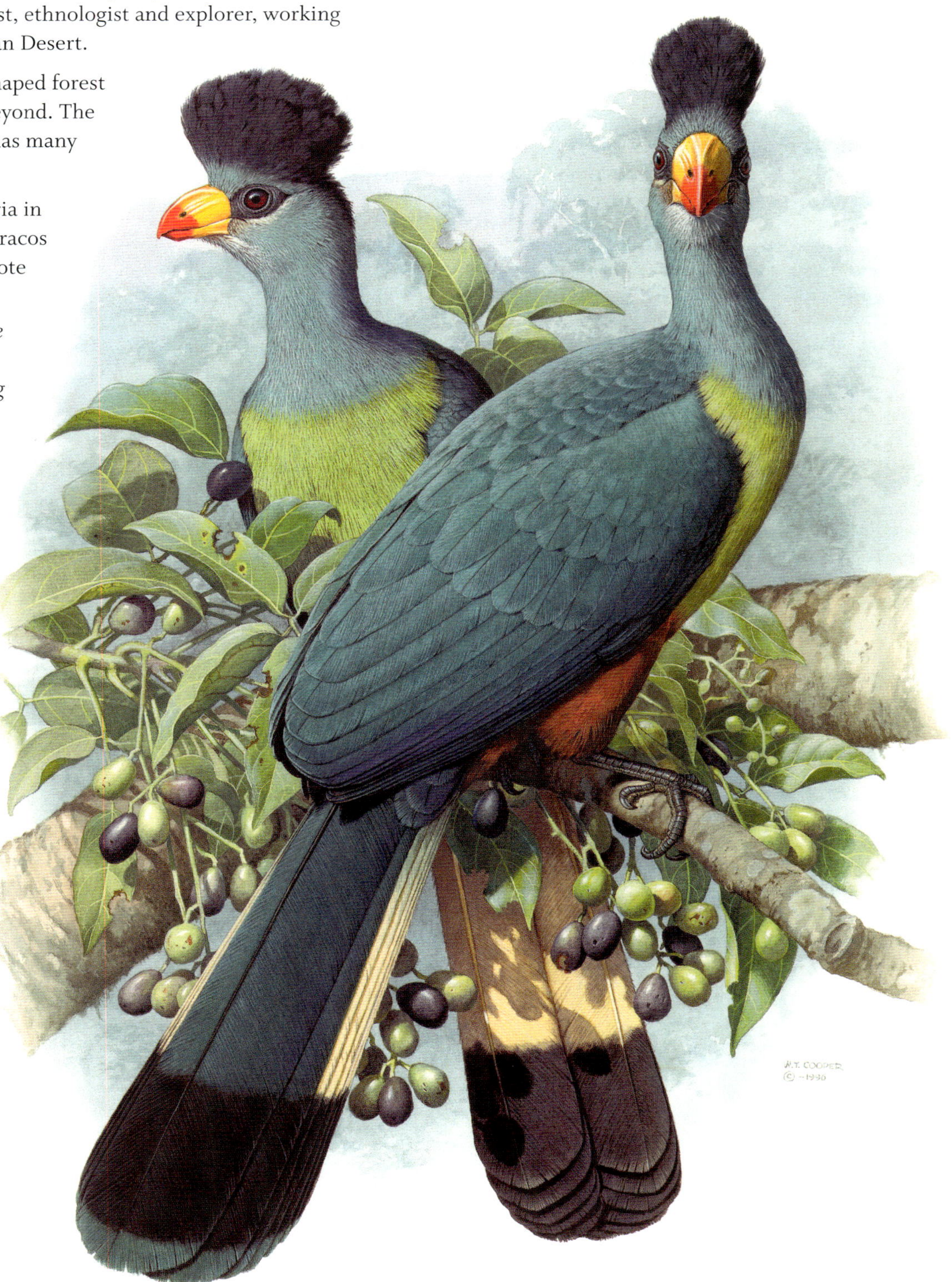

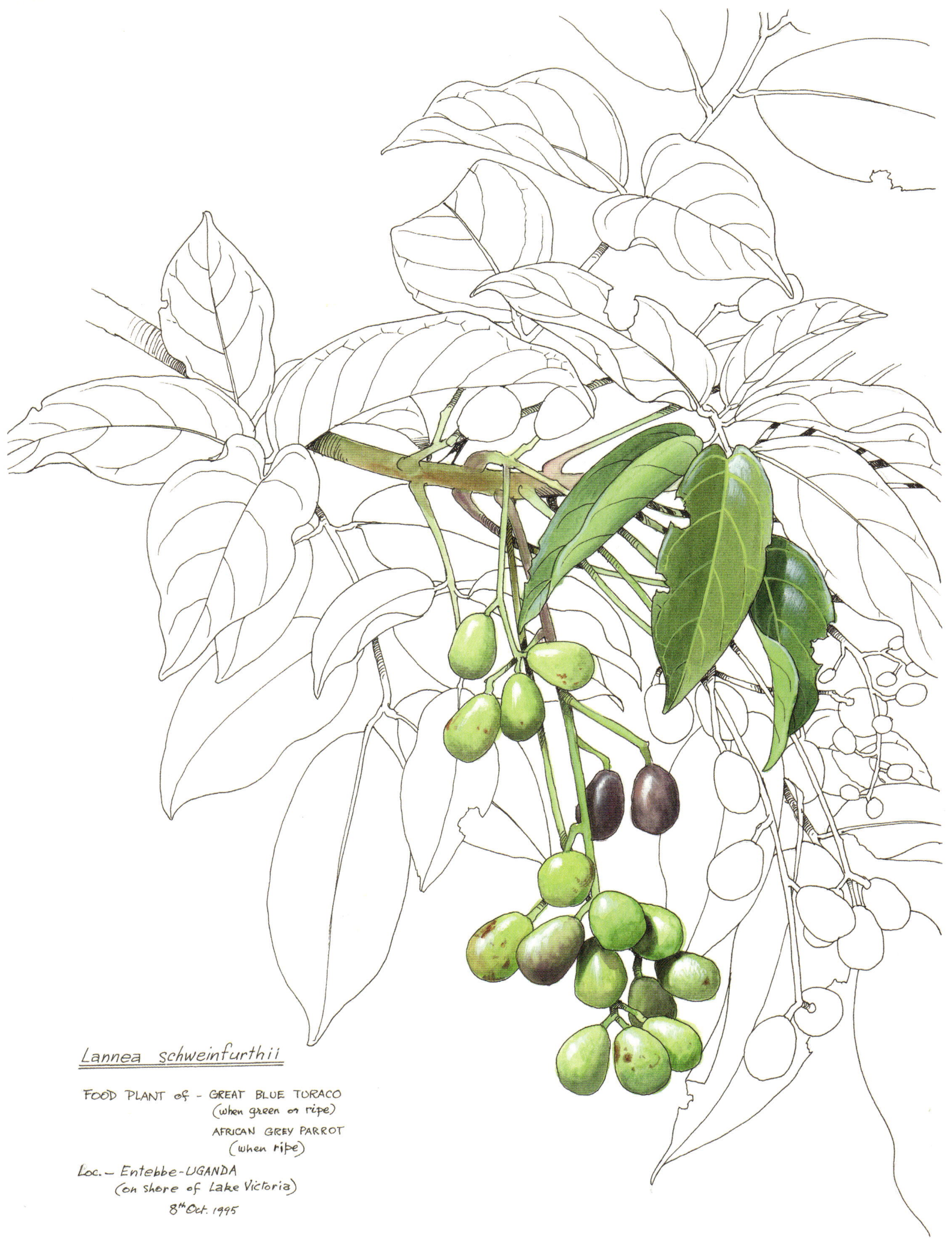
Lannea schweinfurthii
FOOD PLANT of - GREAT BLUE TURACO
(when green or ripe)
AFRICAN GREY PARROT
(when ripe)
Loc. – Entebbe-UGANDA
(on shore of Lake Victoria)
8th Oct. 1995

Lithocarpus rassa

Stone Oak

FAGACEAE

→ 1978; pen, pencil and watercolour
320 x 295mm
State Library of New South Wales

↓ Ampang Forest Canopy, Malaysia
1978; pencil; c.350 x 290mm

Lithocarpus from *litho-* (stone-) and *-carpus* (-fruited).

rassa is after a Malay name for the species.

Stone Oaks number about 334 species worldwide and occur only in South-East Asia. *Lithocarpus rassa* grows to about 25 metres, although the one illustrated opposite was shrubby, easily plucked from a branch beside the road at Bukit Fraser in Malaysia.

According to Bill's diaries, this sketch was done on what he called 'the day of the hornbills'. We were in Malaysia to see these large spectacular birds and, on this day, we watched an emergent fruiting tree, probably a fig, from above and higher up the valley. There were six species of hornbill noisily flapping as they came and went from the tree throughout the day. I really can't remember how Bill found the time to also collect and paint this botanical!

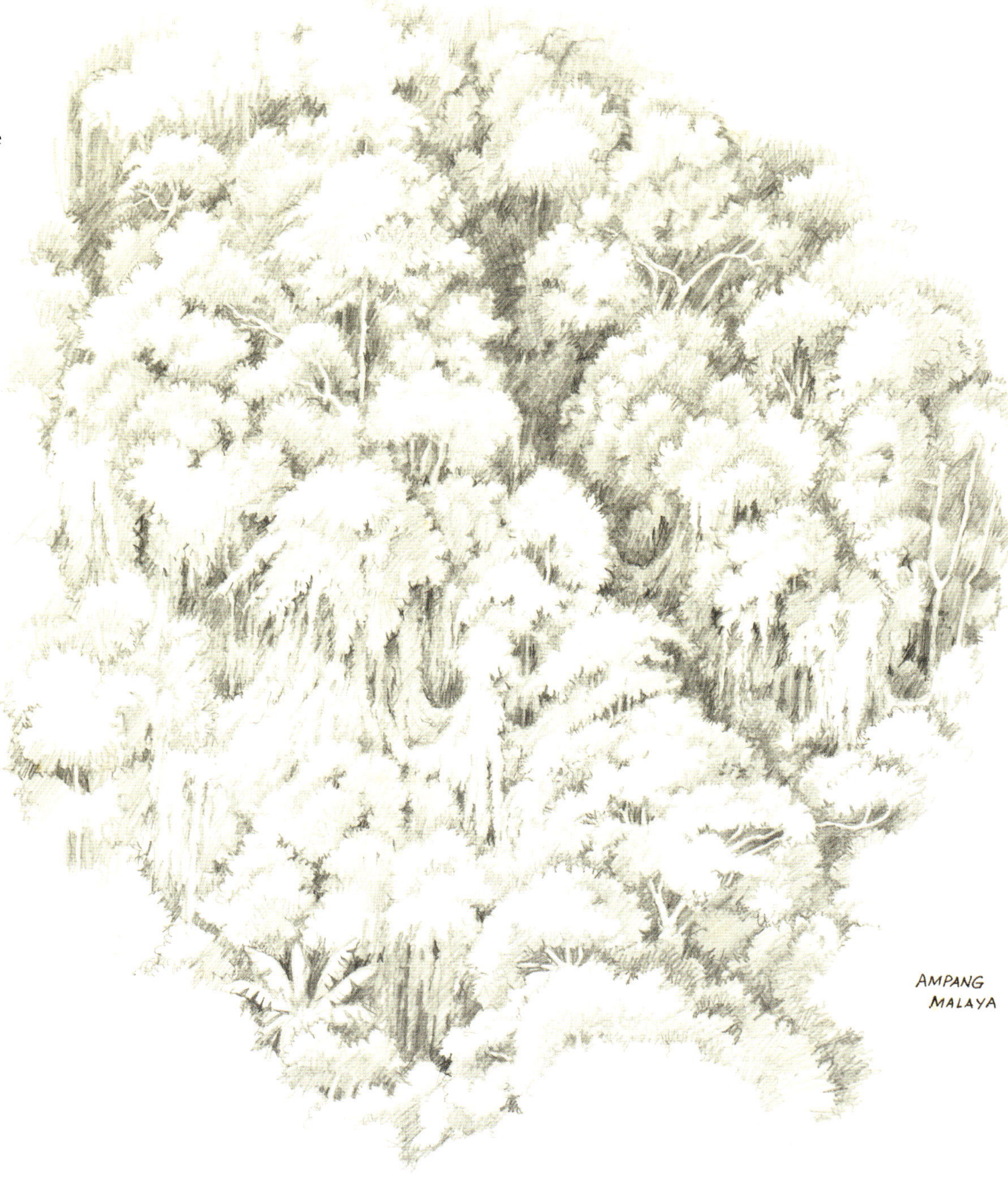

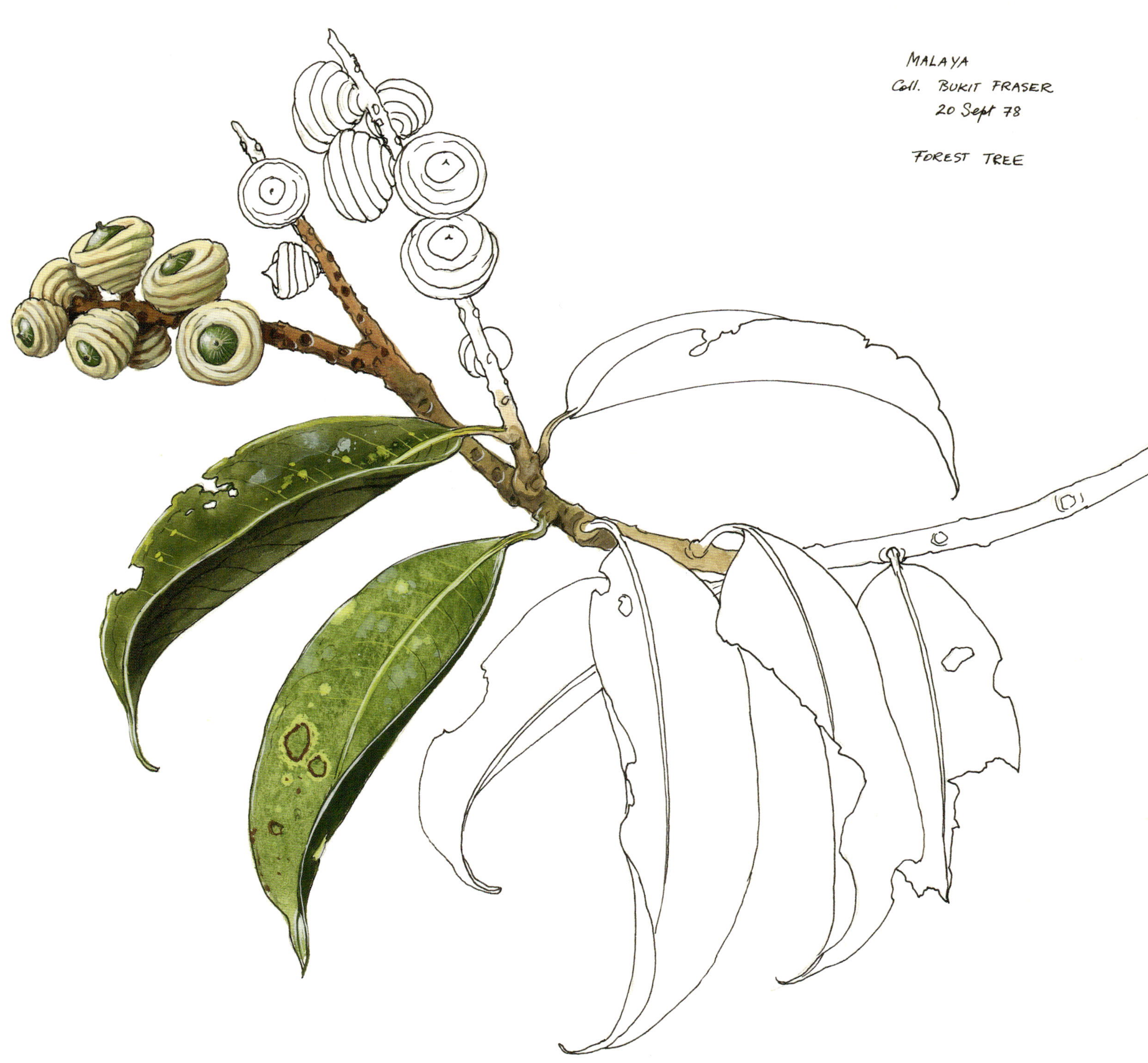
MALAYA
Coll. BUKIT FRASER
20 Sept 78
FOREST TREE

Maclurochloa montana

Buluh Padi

POACEAE

Maclurochloa is named in honour of Floyd Alonzo McClure (1897–1970), a bamboo specialist, and from *chloa* (grass), referring to the technicality that bamboos are actually large grasses.

montana is from *montanus* (montane), referring to their habitat.

This bamboo is a scrambler in the mountains of Indo-China and South-East Asia through to New Guinea.

This sketch was done at Bukit Fraser during our ill-fated trip to Malaysia in 1978 when we were involved in a car accident and had to return to Australia. Our driver went to sleep at the wheel, causing the car to roll several times down a hillslope.

The bamboo was obviously very appealing as a beautiful bird perch in a painting and was used in the plate for the Red-bearded Bee-eater in *Kingfishers and Related Birds*.

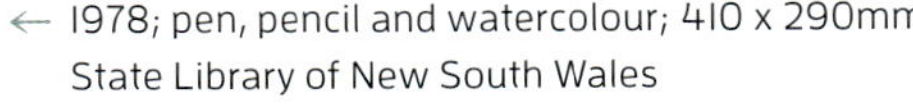

← 1978; pen, pencil and watercolour; 410 x 290mm
State Library of New South Wales

→ Red-bearded Bee-eater, colour plate from *Kingfishers and Related Birds*, 1986
National Library of Australia, 2494045

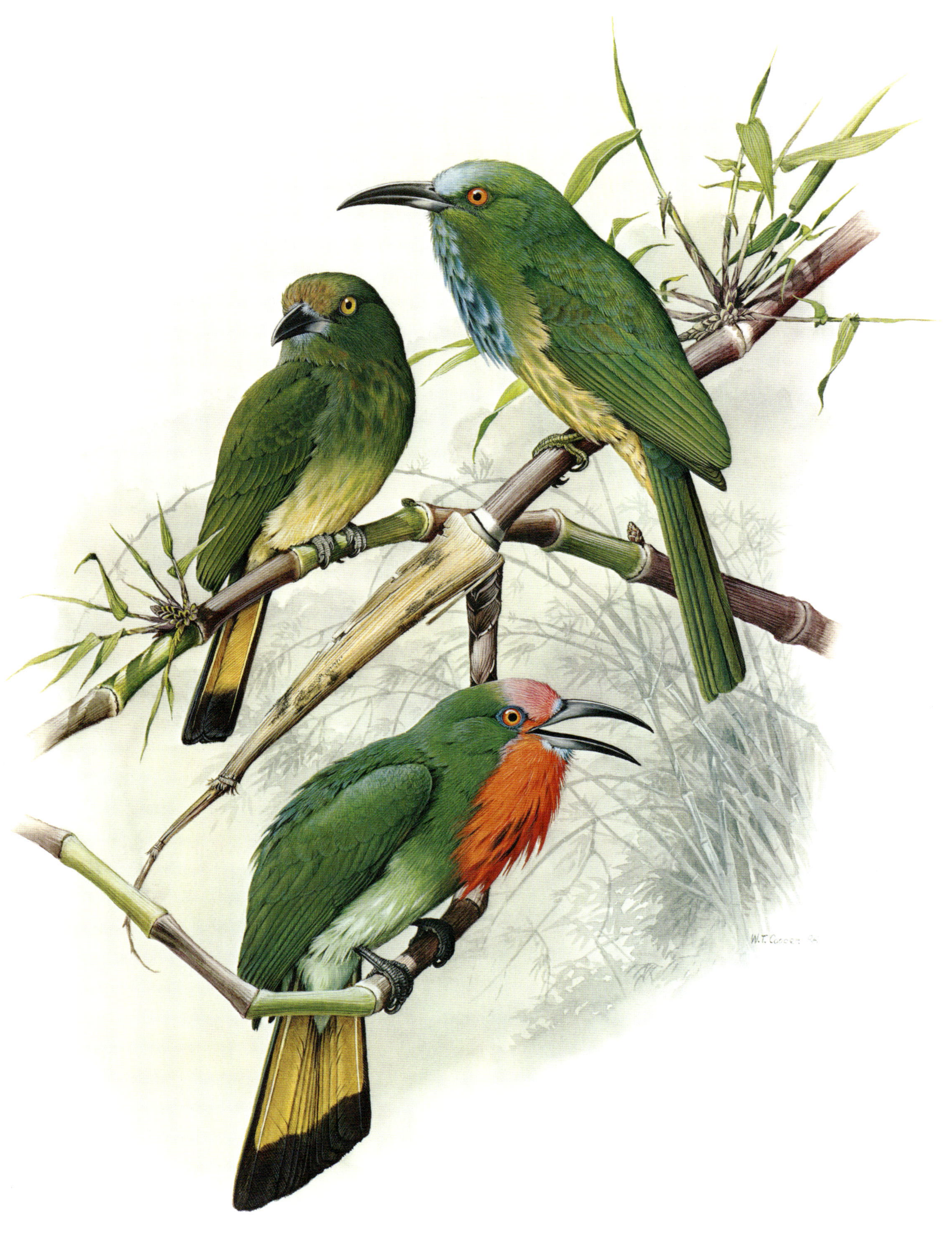

Melia azedarach

Persian Lilac or White Cedar

MELIACEAE

→ Cuban Macaw and Red-fronted Macaw
1972; gouache; 507 x 382mm
National Library of Australia, 1770290

↓ undated; pen, pencil and watercolour; 340 x 370mm
State Library of New South Wales

Melia is from the Greek name for the superficially similar tree *Fraxinus* sp.

azedarach is after *Azad-darakht*, the Persian name for the related Neem Tree.

White Cedar is a deciduous tree that grows to 45 metres in height and is widespread as a native from southern New South Wales to north Queensland, the Northern Territory, Western Australia, New Guinea, the Solomon Islands, Malesia and Asia. It has been cultivated for timber and fodder in Africa and the Americas. It is now naturalised in North and South America.

The name Persian Lilac is applied because of the fragrant blue or purple flowers, which emerge with the new flush of leaves. The Cuban Macaw is known to feed on the fruit, making it appropriate for inclusion in that bird's plate for *Parrots of the World*.

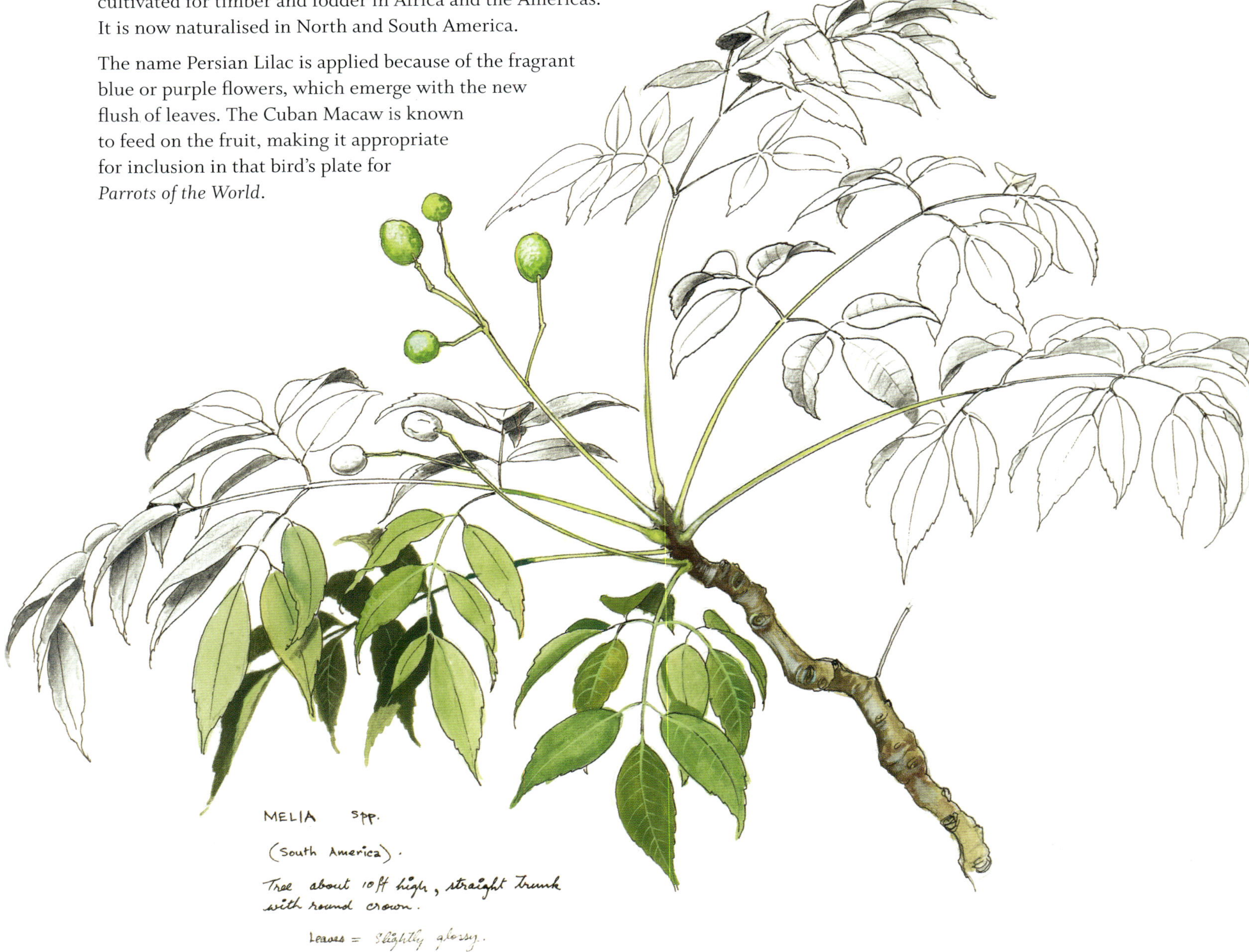

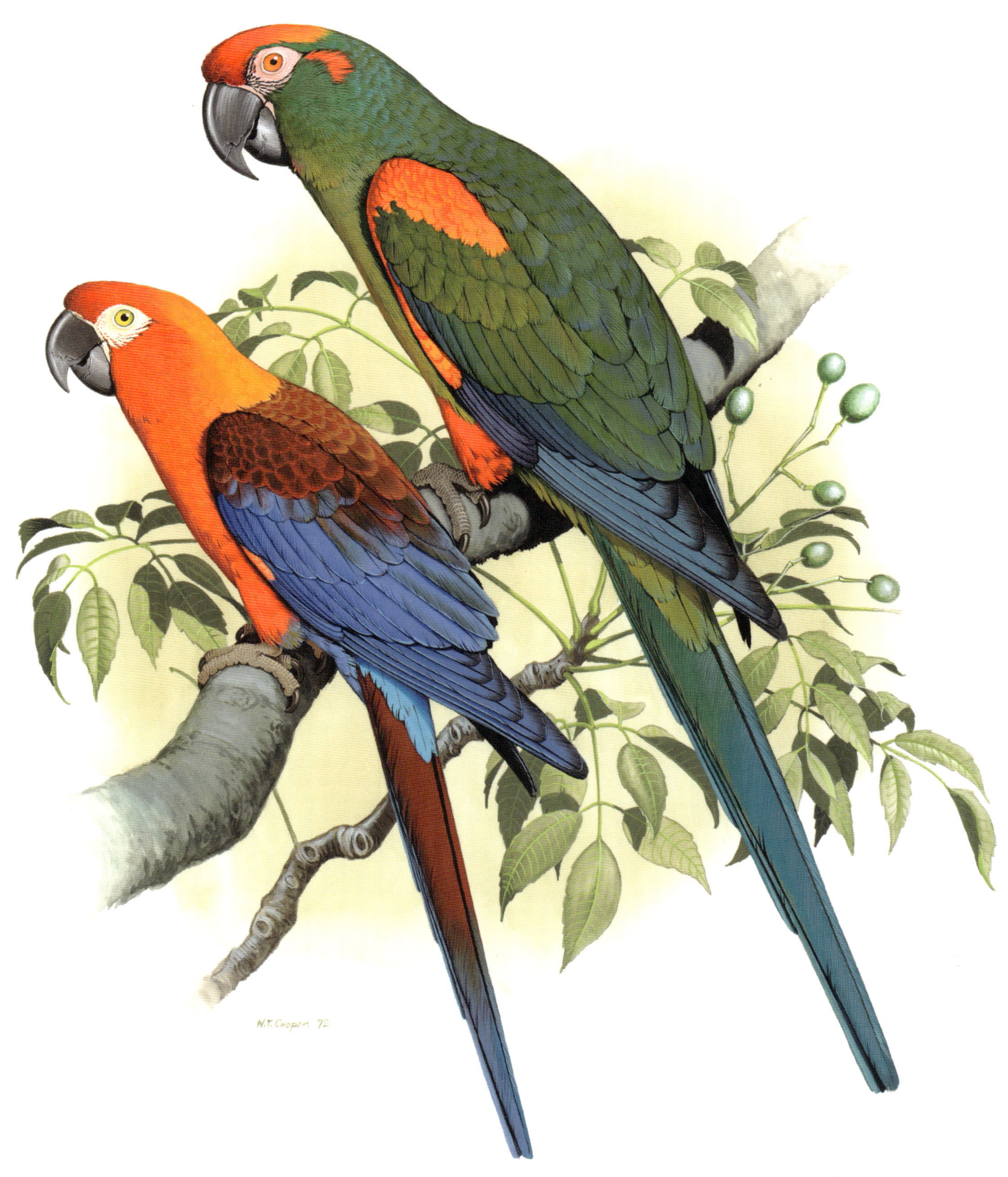
W.T. Cooper '72

Mitragyna parvifolia

Kadam

RUBIACEAE

Mitragyna is from *mitra* (headband or cap) and *gyne* (female), referring to the hat-shaped floral ovaries.

parvifolia is from *parvus* (small or puny) and *-folius* (-leaved), referring to the leaves that are smaller than some other *Mitragyna* species, even though these leaves are not especially small.

Kadam occurs in the monsoon forests of India and Sri Lanka. The flowers and fruit (painted by Bill here) are formed in balls.

We were at Ranthambore National Park in Rajasthan to see tigers, which became a passion of Bill's and a subject he returned to several times in his paintings. In between watching tigers, Bill took advantage of interesting local plants to sketch for future use.

The painting titled 'Ranthambore Tiger' was executed using the reference sketch below. Such was Bill's talent that, with an almost photographic memory and knowledge of his subjects, he did not need to be faithful to the original sketch and could twist and turn the plants (or anything else) to suit the composition that he had in mind.

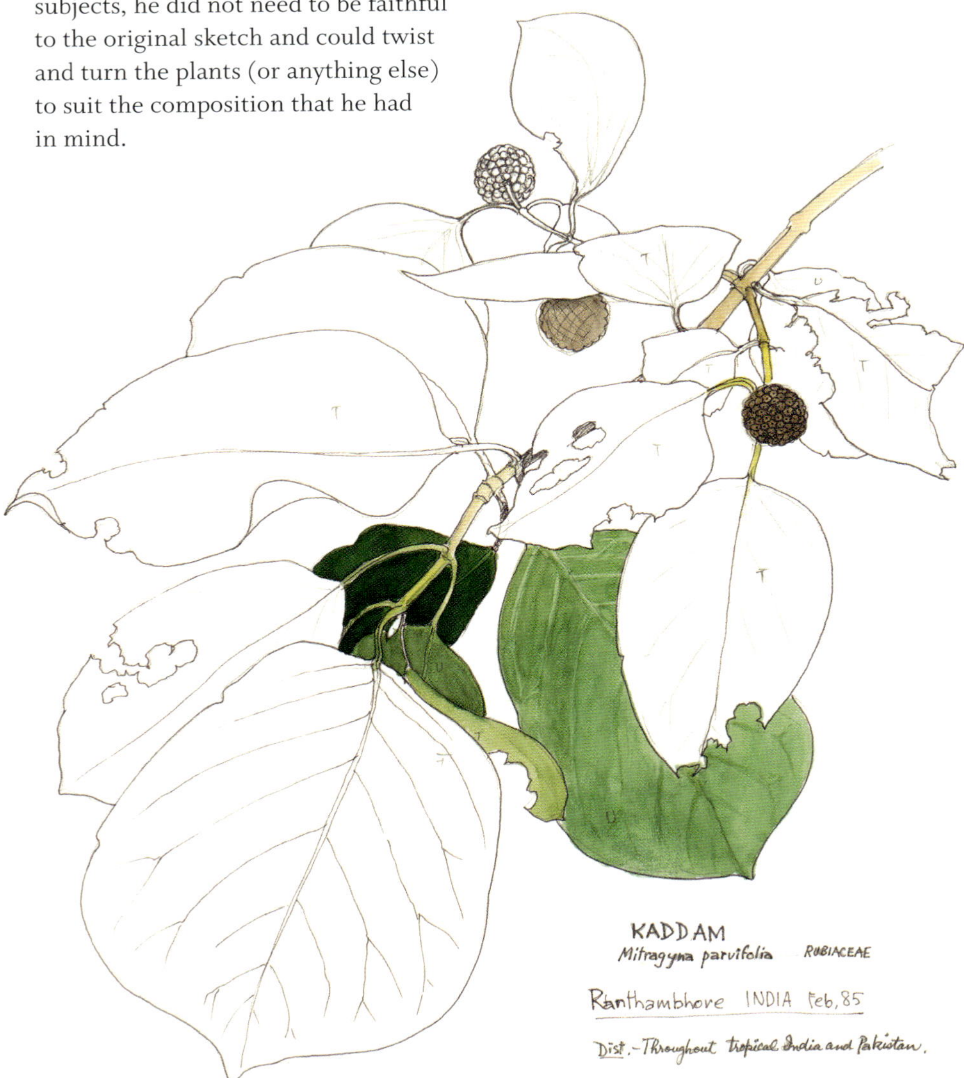

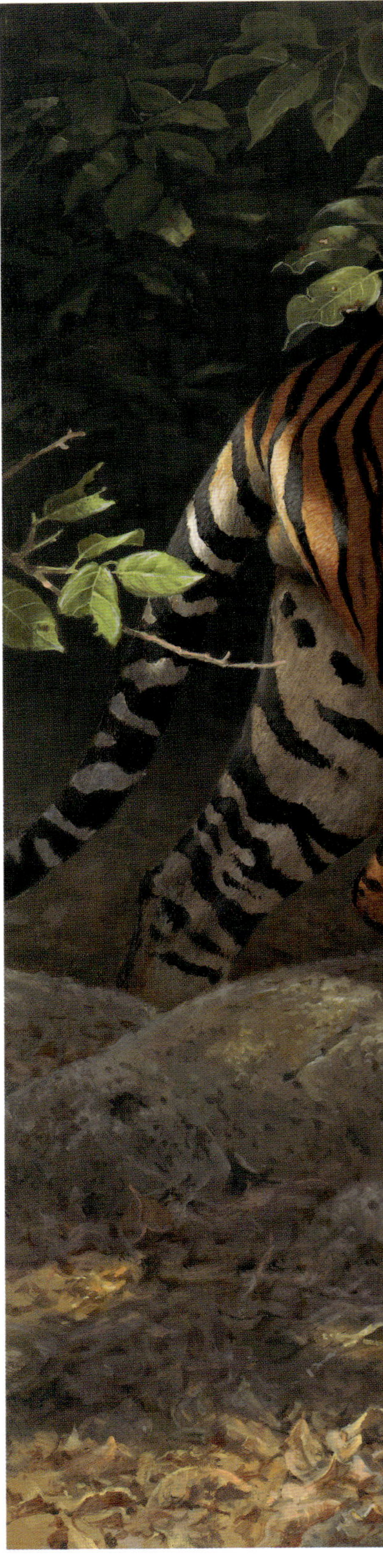

← 1985; pen, pencil and gouache
410 x 290mm

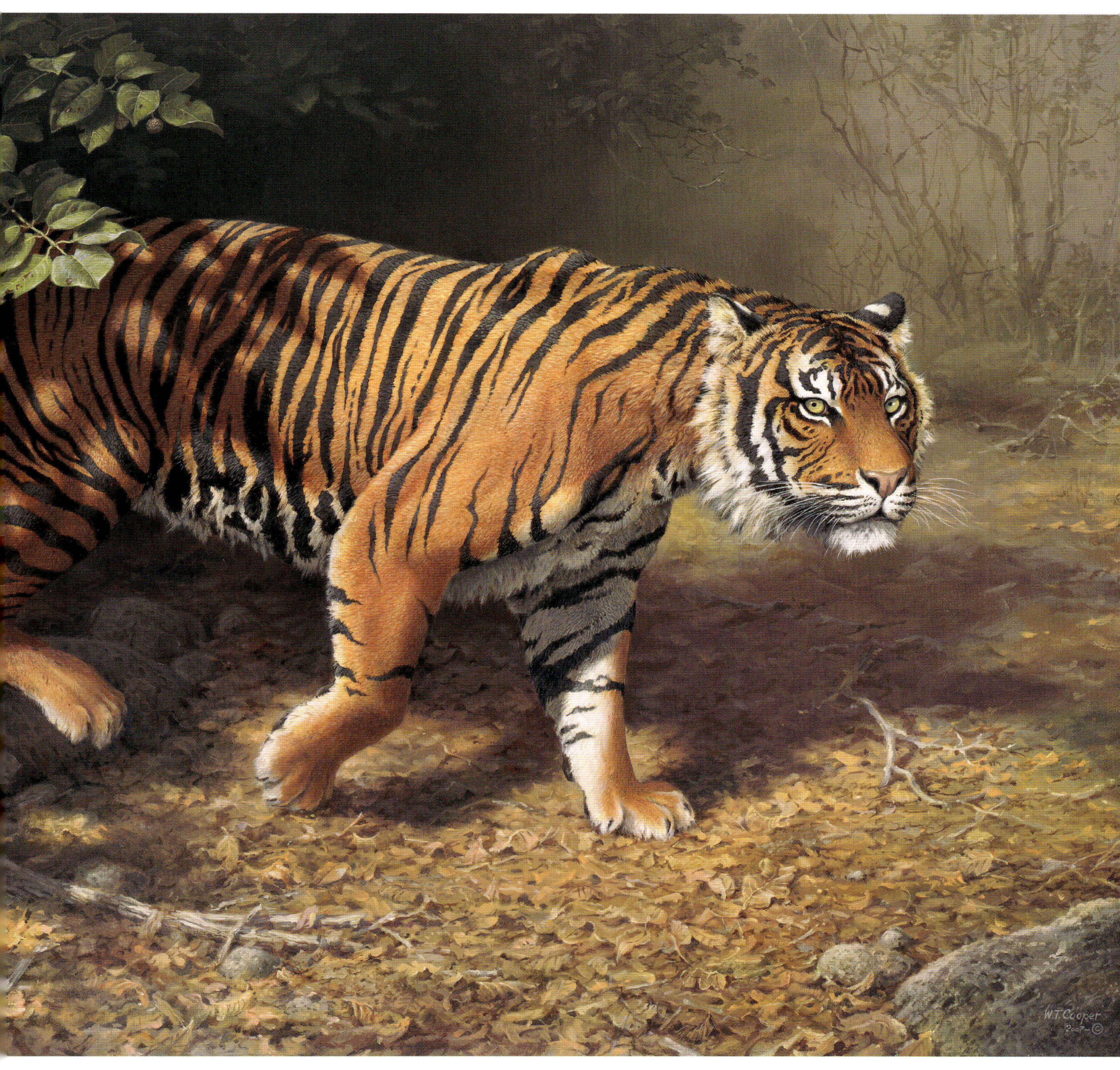

↑ Ranthambore Tiger
2007; oil; 610 x 770mm
Private collection

Muntingia calabura

Strawberry Tree or Panama Berry

MUNTINGIACEAE

Muntingia is named in honour of Abraham Munting (1626–1683), Dutch botanist and artist.

calabura, derivation unknown, is possibly from a local name in South America.

Strawberry Tree is a popular small ornamental tree from South America (Bolivia to Mexico). The fruits are edible for humans as well as being taken by birds, enabling it to become naturalised in many countries, including Australia.

The specimen opposite, taken from the forest edge in Cairns, was used for a painting of the Christmas Island Imperial Pigeon. The plant is naturalised on Christmas Island.

→ 2011; pen and watercolour
405 x 300mm
State Library of New South Wales

↓ Christmas Island Imperial Pigeon
2015; watercolour
State Library of New South Wales

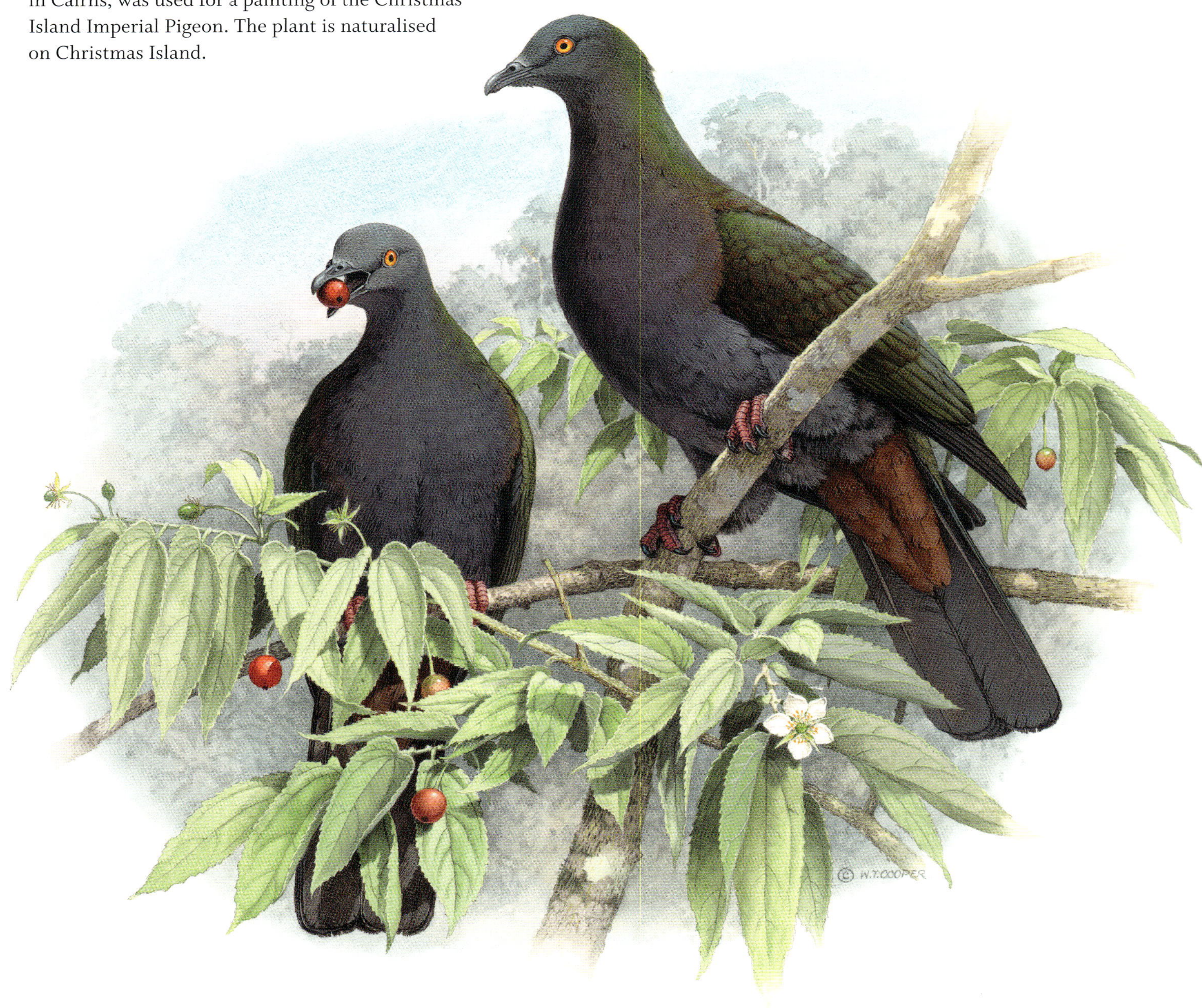

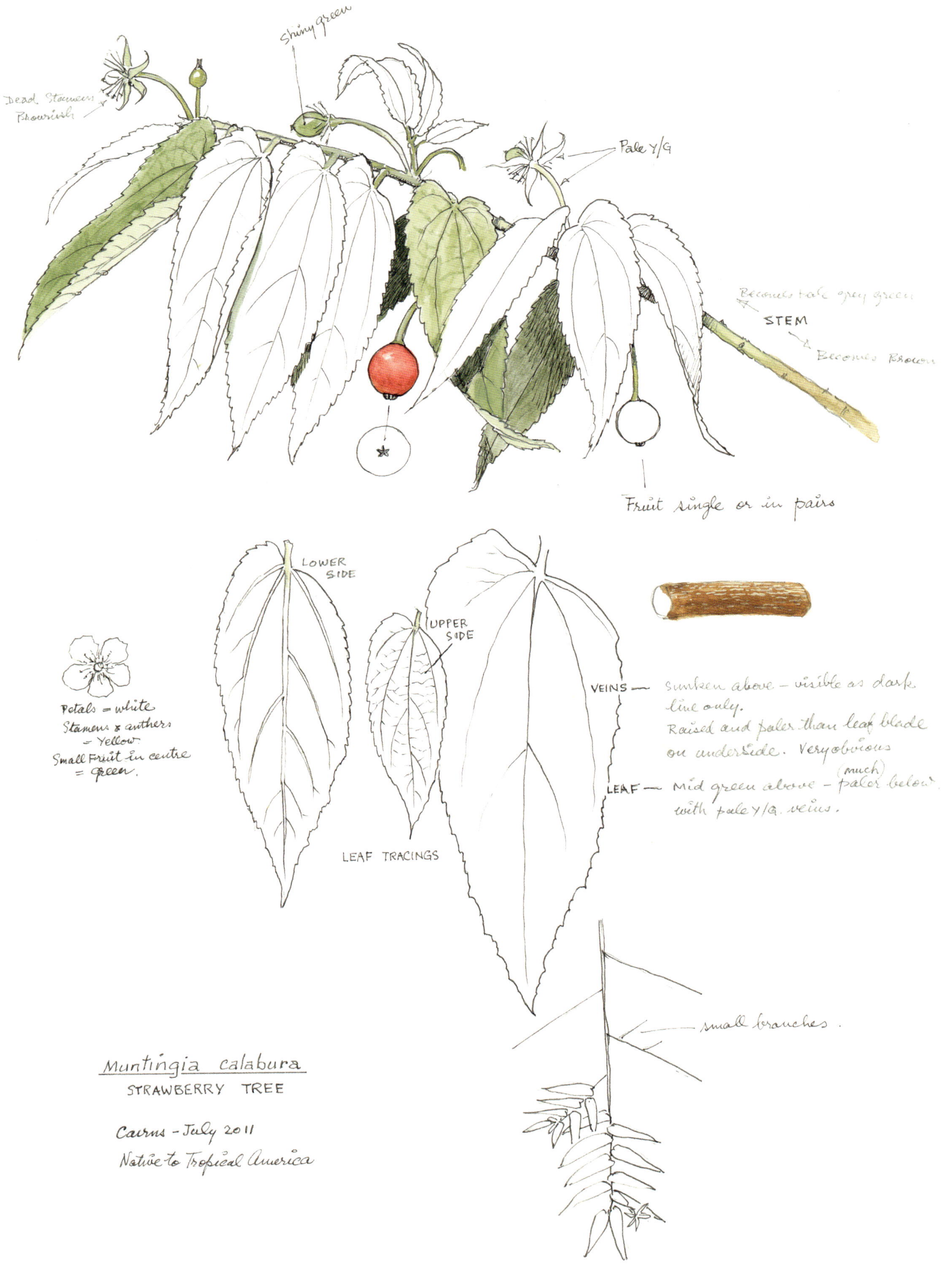

Muntingia calabura

STRAWBERRY TREE

Cairns – July 2011

Native to Tropical America

Nastus

Bamboo

POACEAE

Nastus is from a Greek name for a reed.

Nastus is a fine-leaved, scrambling or climbing bamboo with arching stems in New Guinea.

Bill described his experience of bamboo forest in a 1973 diary entry:

> *Mon. 7th May 1973. Left Avios hut this morning and walked 2½ hours to Murray Pass. Took 7 boys. Avi, Kova, and Patrick went back to Avios and will come up tomorrow afternoon to carry gear back. The walk up was really fantastic. The climbing bamboo draped all through the heavily moss laden trees and then all of a sudden to come out onto grassland dotted with strange tree ferns. The steepness of these hills is extreme, sometimes almost vertical and in other places the track was actually overhanging the drop of hundreds of feet into fast running creeks. Far below us Ero Creek was roaring on its twisted path to the Vanapa River.*

This unidentified *Nastus* species from a different bamboo forest was used in the Brehm's Parrot and Fairy Lorikeet plates in *Parrots of the World*.

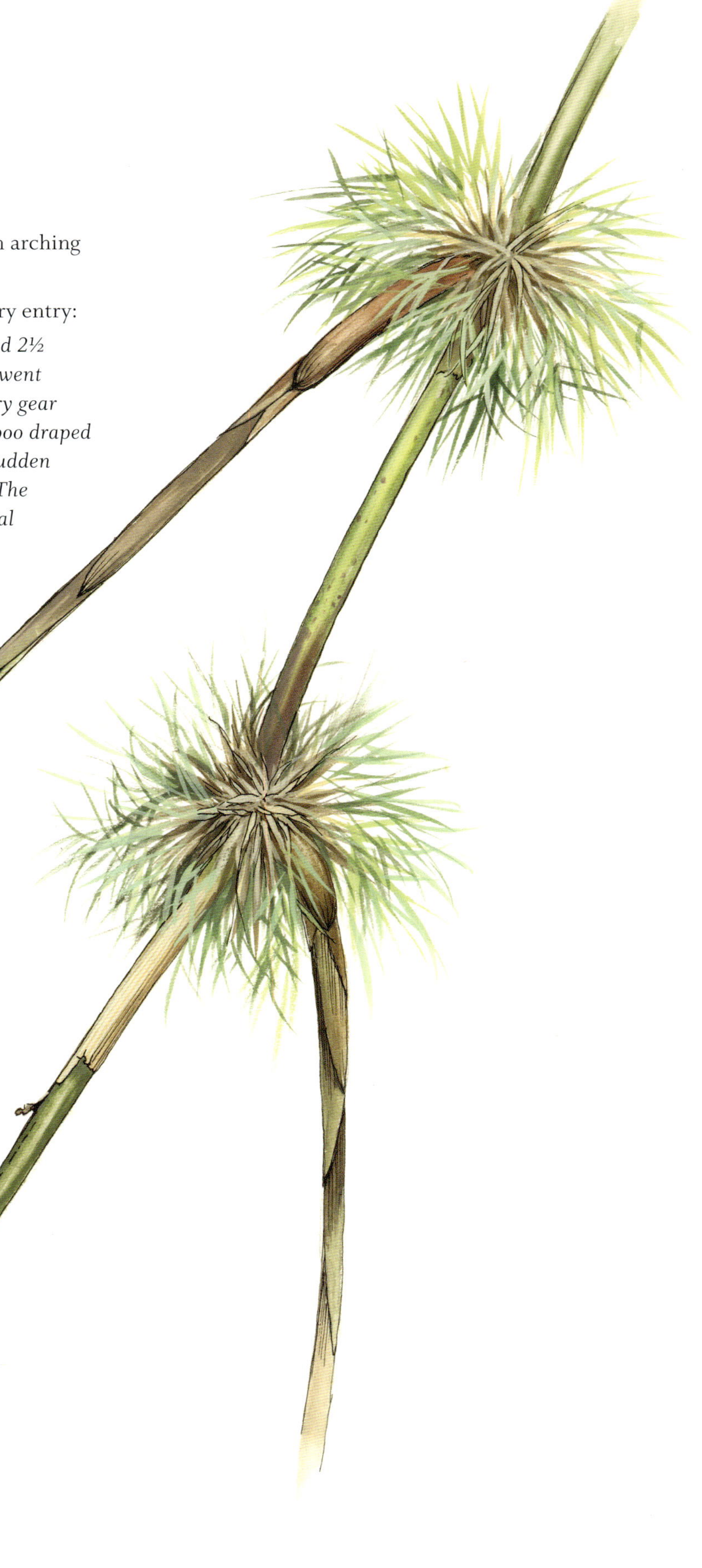

Type of bamboo
grows very thickly up trees and even
covers foliage of original tree in places.

Tufts of leaves grow much thicker and longer
than figured, so that the stem is not visible
among the leaves.

Saw Multi crested B. of Paradise in this
growth.

Coll. 9,000 ft. near TOMBA
NEW GUINEA.

← undated; pen and gouache; 400 x 280mm
State Library of New South Wales

↓ Brehm's Parrot
1971; gouache; 505 x 382mm
National Library of Australia, 1813114

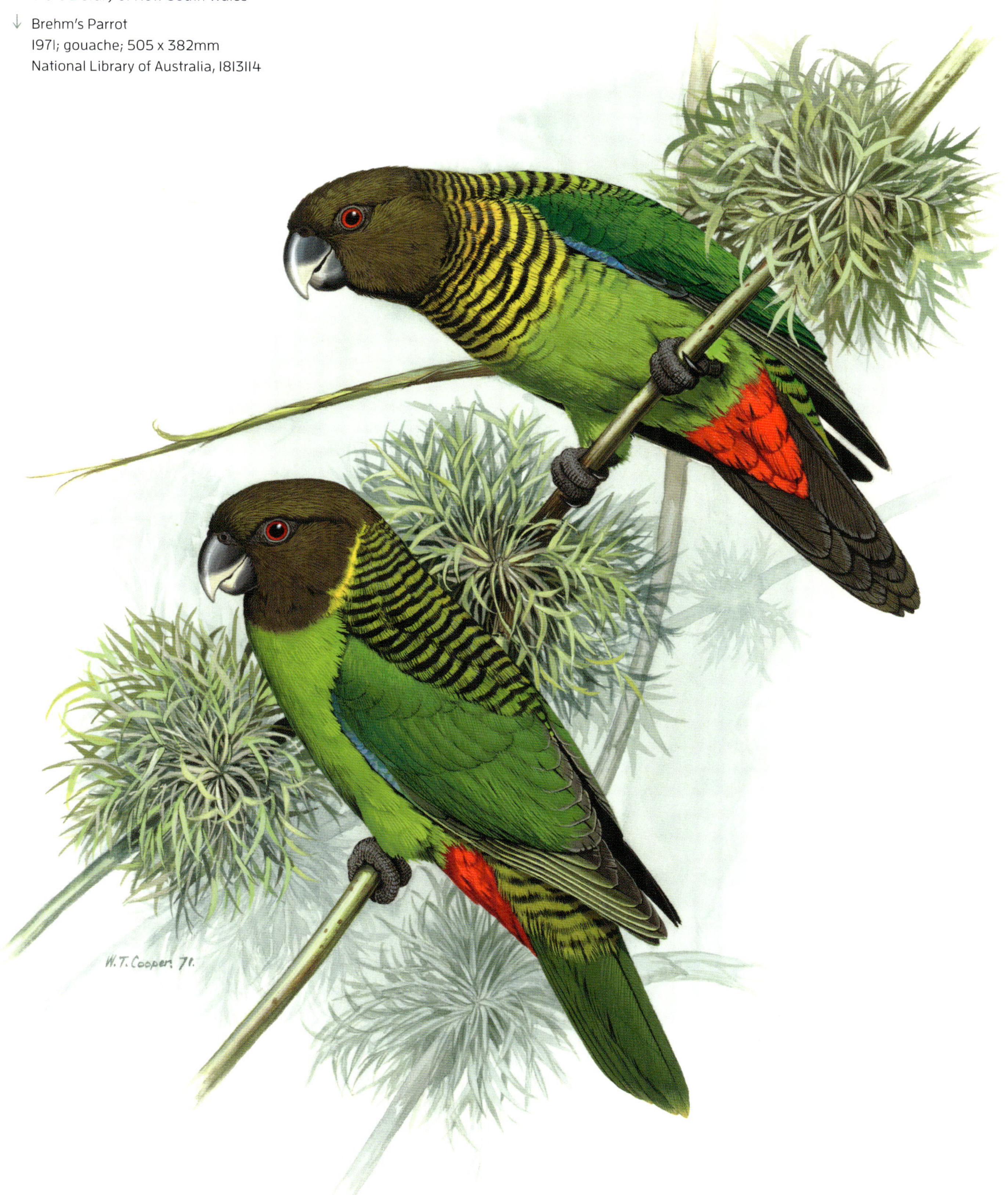

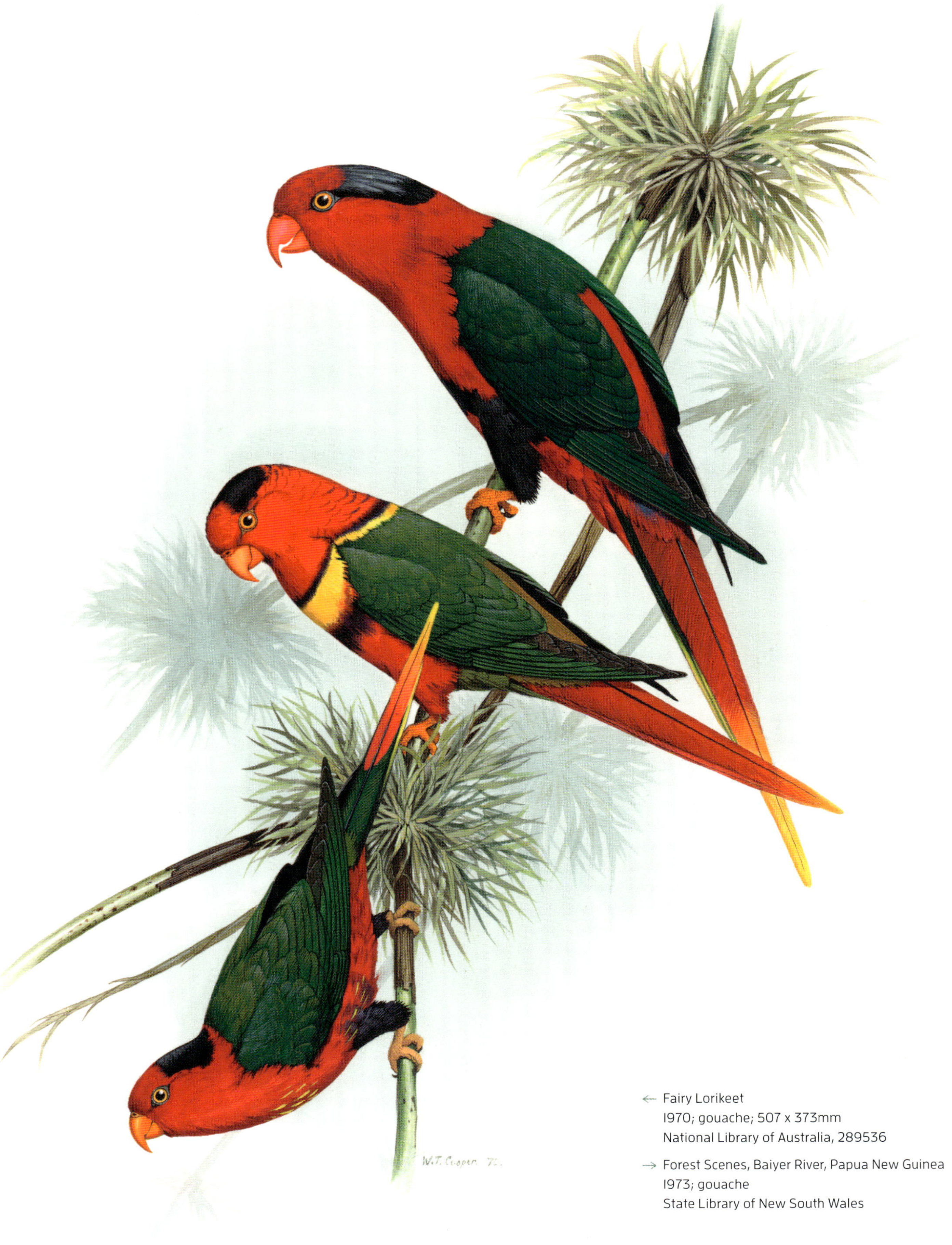

← Fairy Lorikeet
1970; gouache; 507 x 373mm
National Library of Australia, 289536

→ Forest Scenes, Baiyer River, Papua New Guinea
1973; gouache
State Library of New South Wales

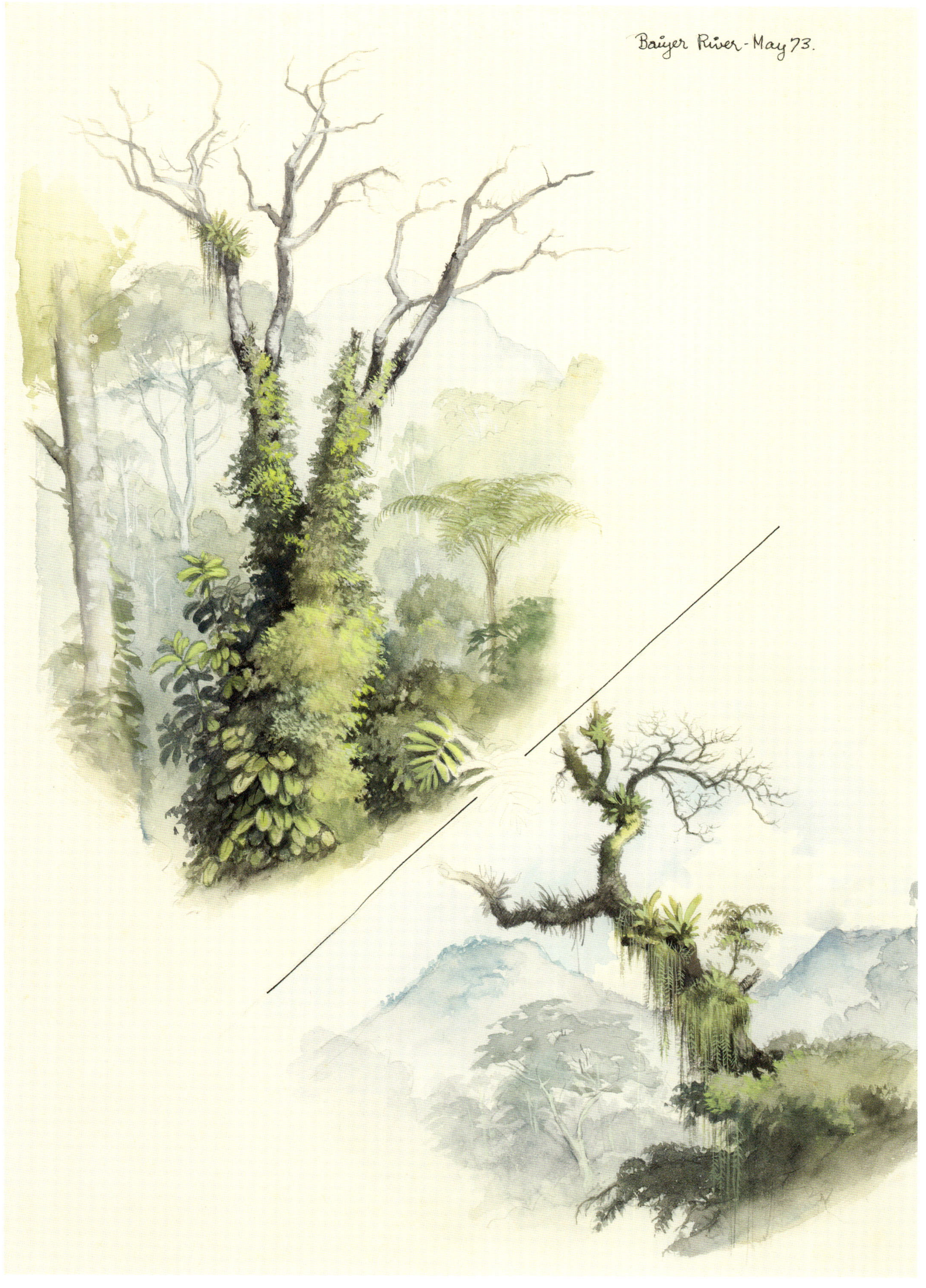
Baiyer River - May 73.

Passiflora vitifolia

Perfumed Passionflower or Crimson Passionflower

PASSIFLORACEAE

→ 1978; pen and watercolour
410 x 280mm
Private collection

↓ South American Mid Storey
undated; ink, pen and pencil
State Library of New South Wales

Passiflora is from *passio* (passion) and *-florus* (flowered), referring to the central parts of the flower and the crucifix.

vitifolia is after *Vitis* (grape vines) and *-folius* (-leaved), referring to the similarity of the leaves to those of grape vines.

With such large colourful flowers, Perfumed Passionflower is widely cultivated and has become naturalised in many countries, including Malaysia. It is a native vine of Central and South America, and because Bill was often commissioned to paint birds from places he hadn't visited, he was quick to capture this spectacular plant while he had the opportunity.

This specimen was hanging over the roadside at Bukit Fraser in Malaysia.

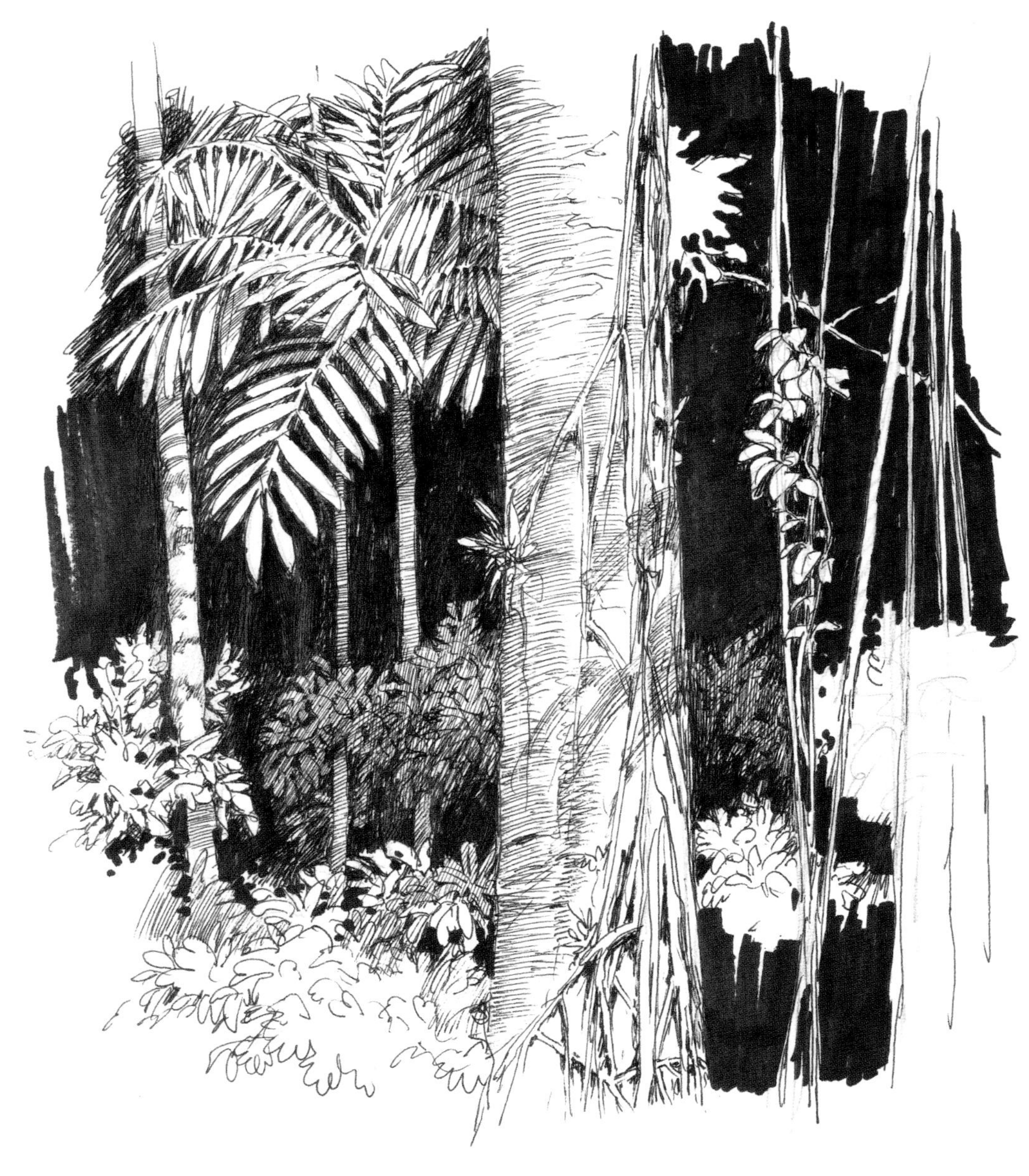

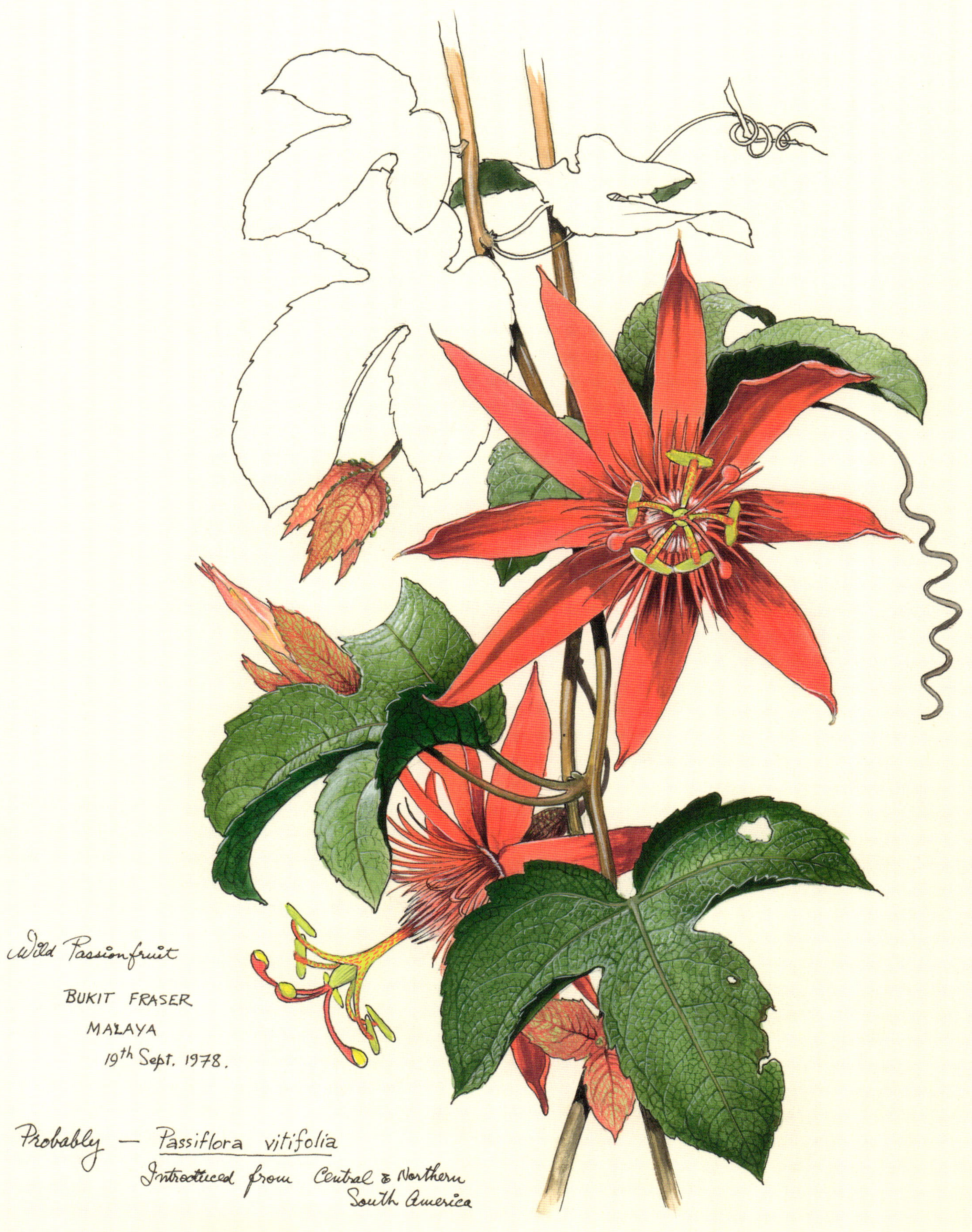
Wild Passionfruit
BUKIT FRASER
MALAYA
19th Sept. 1978.
Probably – Passiflora vitifolia
Introduced from Central & Northern
South America

Podocarpus archboldii

Yamiga

PODOCARPACEAE

→ 1973; pen and gouache; 400 x 300mm
State Library of New South Wales

↓ Splendid Astrapia, colour plate from *The Birds of Paradise and Bower Birds*, 1977
National Library of Australia, 69465

Podocarpus is from *podos-* (foot) and *-carpos* (fruited), referring to the fleshy receptacle at the fruit base.

archboldii is named in honour of Richard Archbold (1907–1976), an American zoologist and philanthropist who funded and led three biological expeditions to New Guinea in the 1930s.

This tall tree grows to 40 metres and is widespread on the island of New Guinea, east and west. The cone-like structures are actually leaf buds that haven't yet expanded like those at the top of the illustrated branch. The fleshy glaucous green fruit is 10 to 15 millimetres by 9 to 13 millimetres.

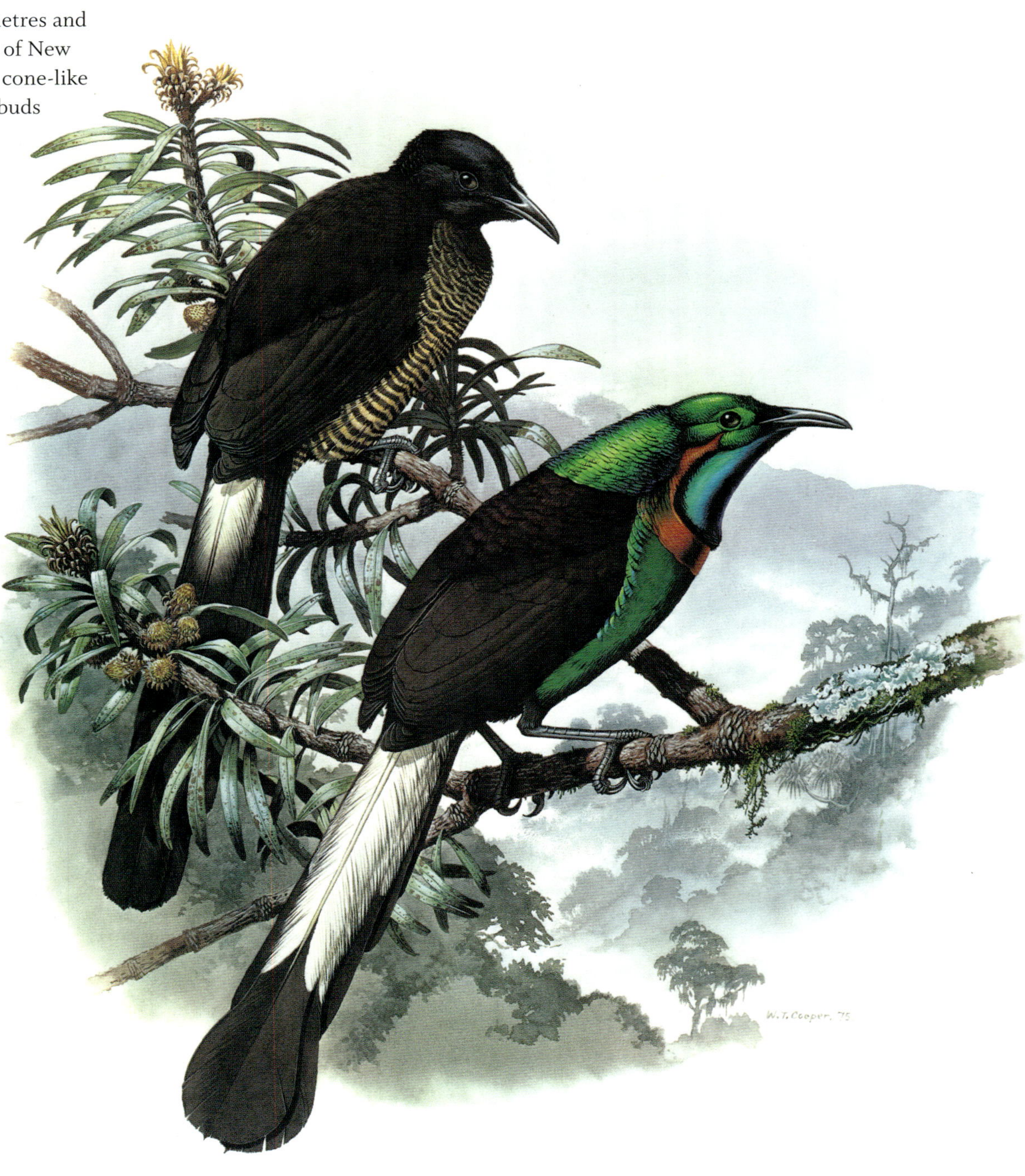

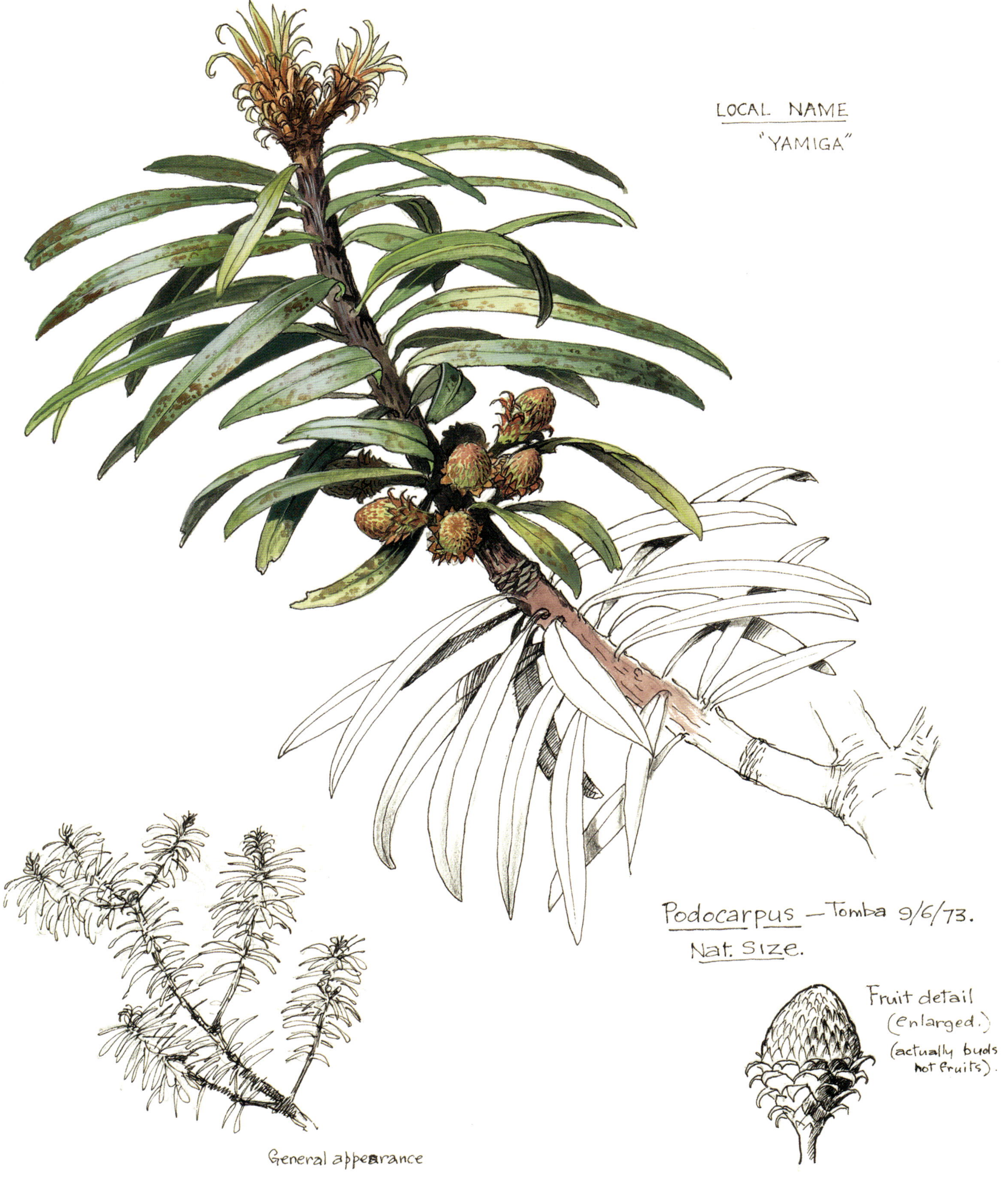
LOCAL NAME
"YAMIGA"
Podocarpus – Tomba 9/6/73.
Nat. Size.
Fruit detail
(enlarged.)
(actually buds
not fruits).
General appearance

Pseudobombax grandiflorum

Brazilian Shaving-brush Tree

MALVACEAE

Pseudobombax is from *pseudo-* (false-) and after the related genus *Bombax*.

grandiflorum is from *grandi* (grand) and *-florus* (-flowered), referring to the large flowers.

Brazilian Shaving-brush Tree is a deciduous tree from Brazil. The flowers are large with numerous long white stamens—hence the common name.

Bill was commissioned to paint two large macaw paintings (approximately 1 by 2 metres). With a need for South American vegetation, he sourced plants with known identification and distribution from the Royal Botanic Gardens in Sydney. This is one of those reference drawings. Parrots are known to eat the seeds.

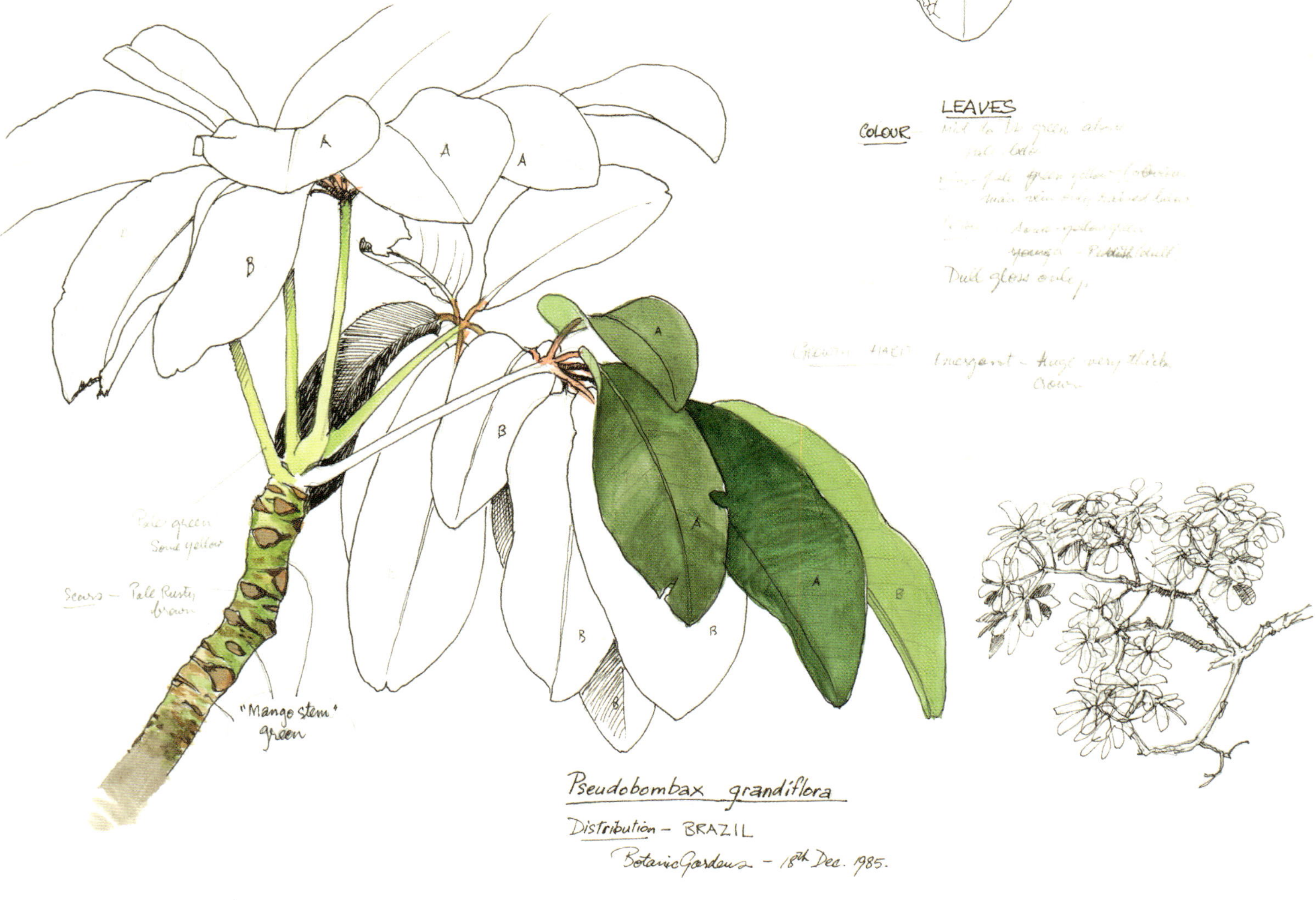

↓ 1985; pen, pencil and watercolour; 400 x 330mm
State Library of New South Wales

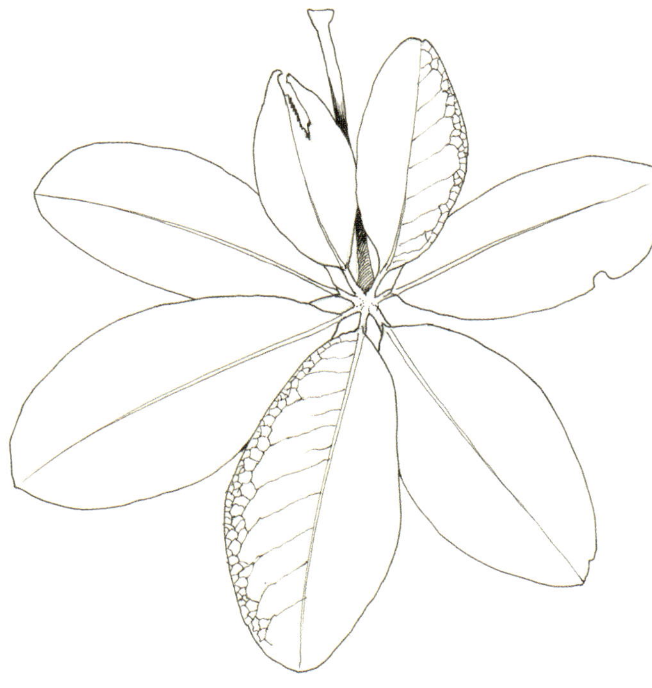

Psidium cattleianum

Cherry Guava

MYRTACEAE

Psidium is from *sidion* (Pomegranate), referring to the similar fruit.

cattleianum is named in honour of William Cattley (1788–1835), horticulturist and international merchant.

Cherry Guava is a Brazilian plant that has been cultivated for its tasty fruit. Unfortunately, the viable seeds are spread by birds, so the species has become naturalised in many areas, including closed-canopy rainforest.

Bill annotated this drawing for potential use in a hornbill painting, but it was not used. Although the plant does not naturally occur with hornbills, he did sometimes use cultivated plants in village gardens visited by birds.

↓ 1978; pen, pencil and watercolour; 270 x 320mm
State Library of New South Wales

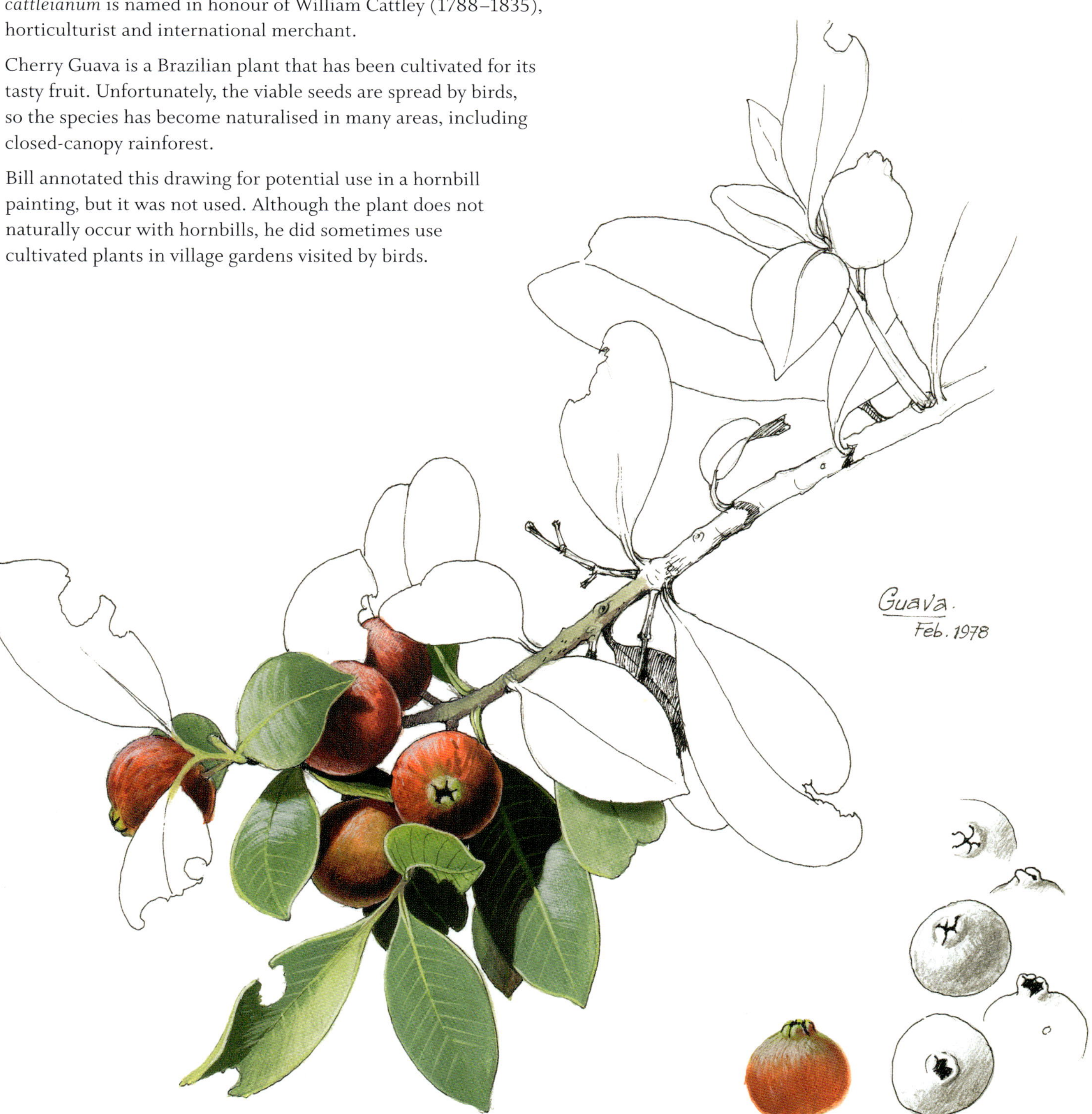

Schefflera

Umbrella Tree

ARALIACEAE

Schefflera is named in honour of Jacob Christian Scheffler.

Umbrella Trees seem to be a popular food source for many bird species in New Guinea and Australia.

This unidentified *Schefflera* was used for two paintings in *Parrots of the World*: the Papuan Lory and Musschenbroek's Lorikeet pictured opposite.

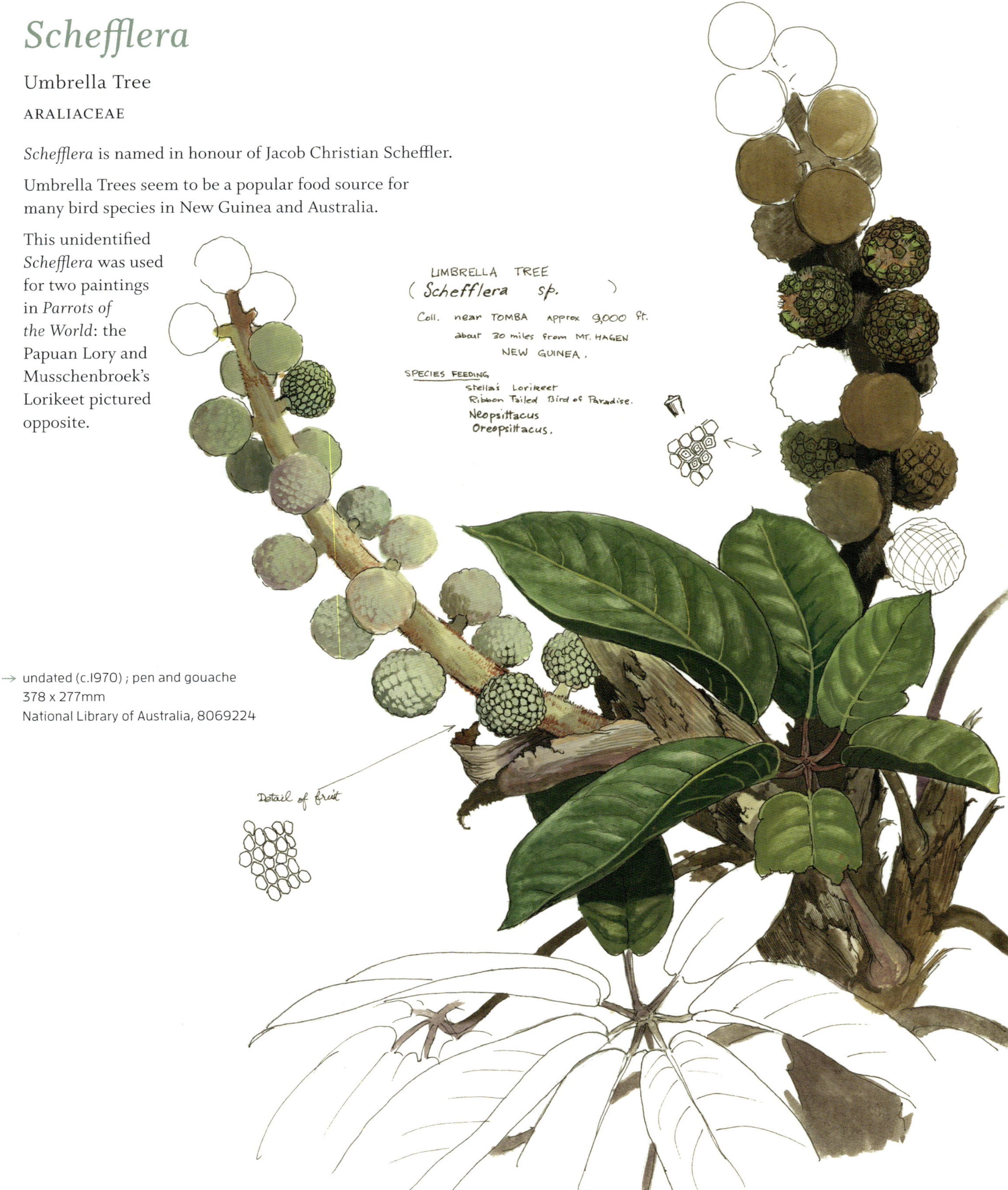

→ undated (c.1970) ; pen and gouache
378 x 277mm
National Library of Australia, 8069224

→ Papuan Lory and Stella's Lory
1970; gouache; 542 x 415mm
National Library of Australia, 289574

↓ Musschenbroek's Lory and Emerald Lory
1970; gouache; 510 x 377mm
National Library of Australia, 1812228

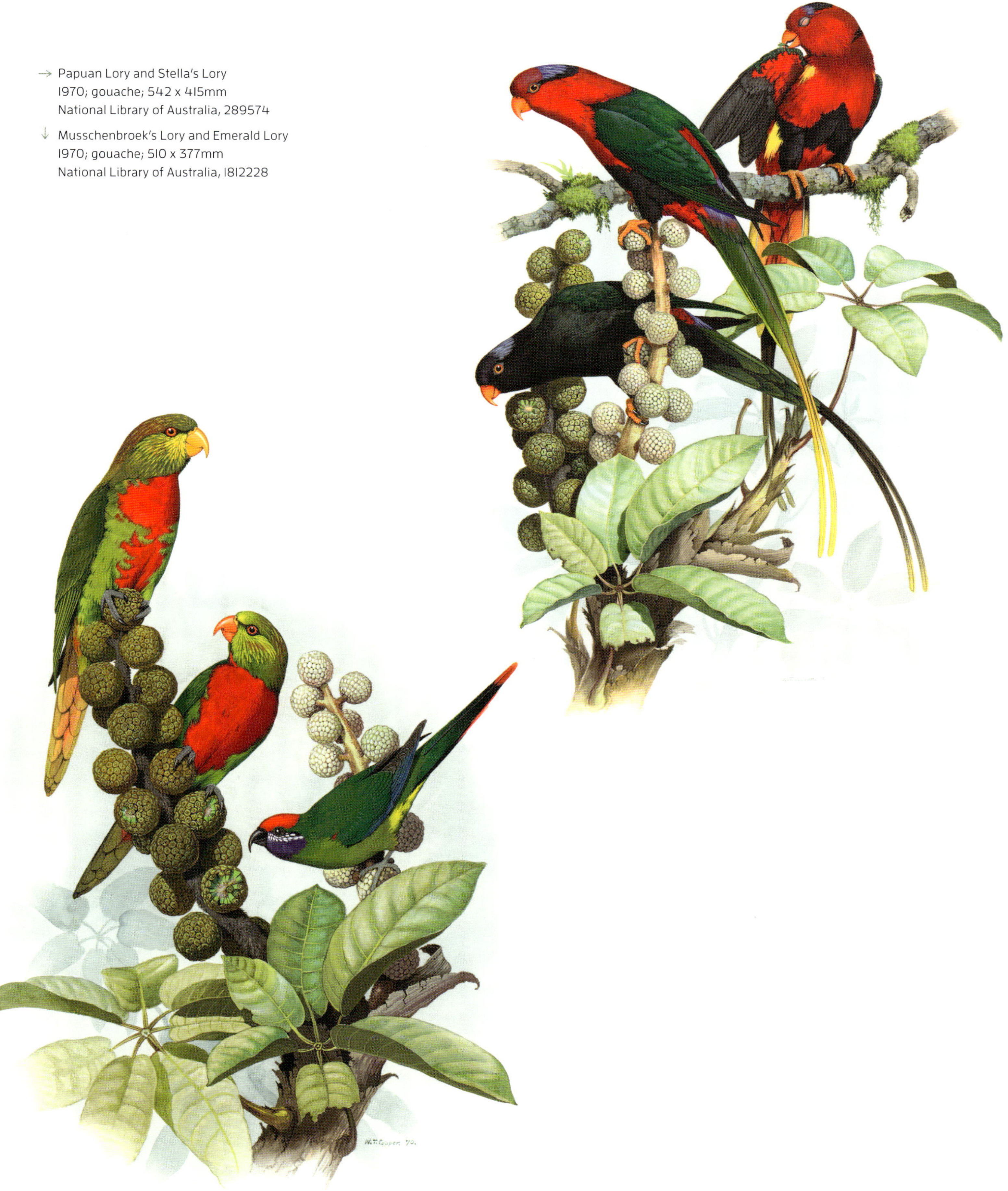

Spathodea campanulata

African Tulip Tree or Nandi Flame

BIGNONIACEAE

Spathodea is from *spathe* (broad, flat blade) and *-odes* (resembling), referring to the floral bracts.

campanulata is from *campanulatus* (bell-shaped), referring to the flowers.

African Tulip Trees are native to tropical Africa, and indeed they are spectacular in the forests of western Kenya, where they are known as Nandi Flame. Fruit with numerous winged seeds have made it a highly established weed in many tropical countries, including Australia.

These stunning flowers were cause for Bill to begin work on a large botanical watercolour painting of the plant, however it was never completed.

→ undated (c.2002); pen, pencil and watercolour; 290 x 420mm
National Library of Australia, 8069849

↘ 2002; pen, pencil and watercolour
290 x 420mm
National Library of Australia, 8069851

↓ c.2002; pen and pencil; c.290 x 420mm
State Library of New South Wales

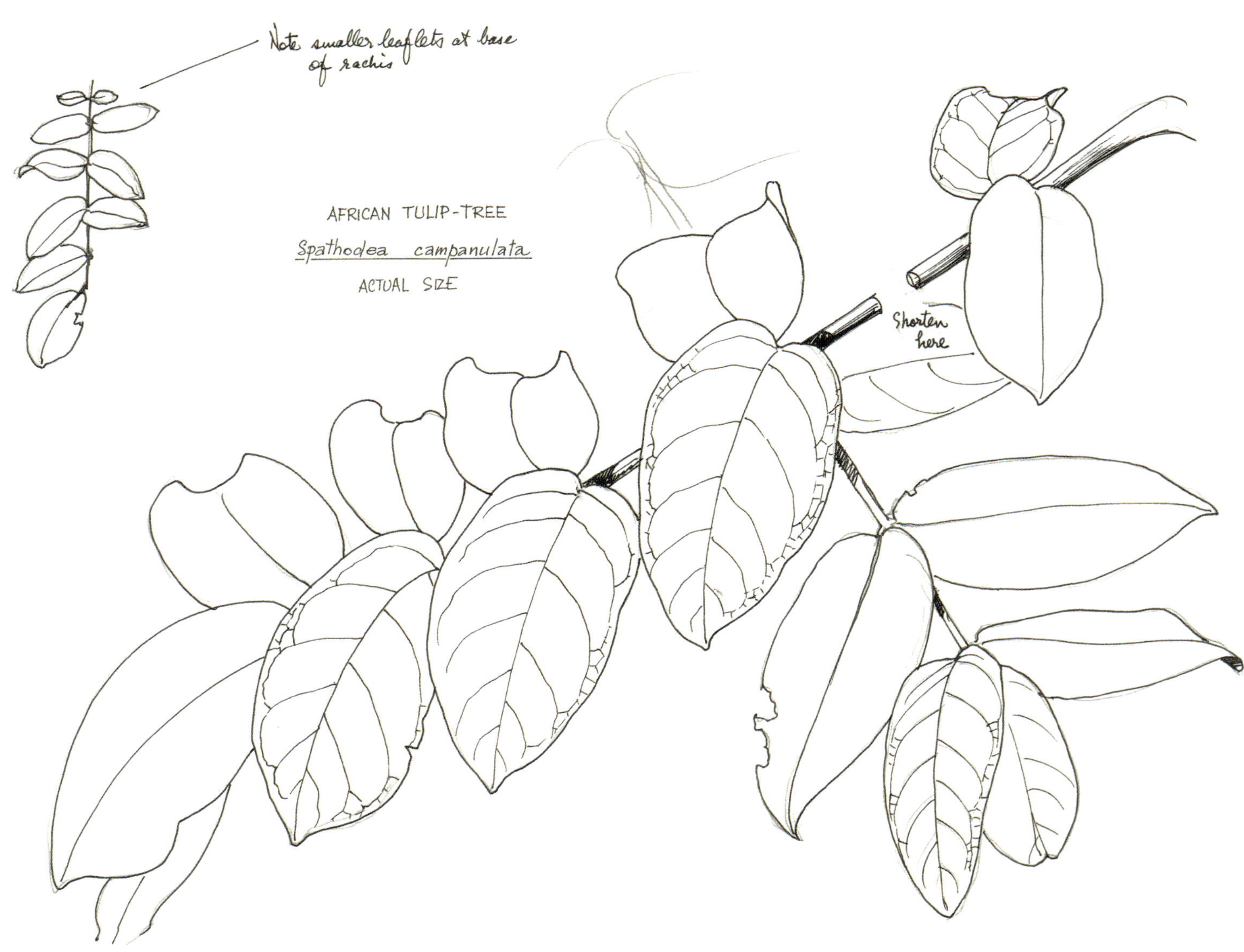

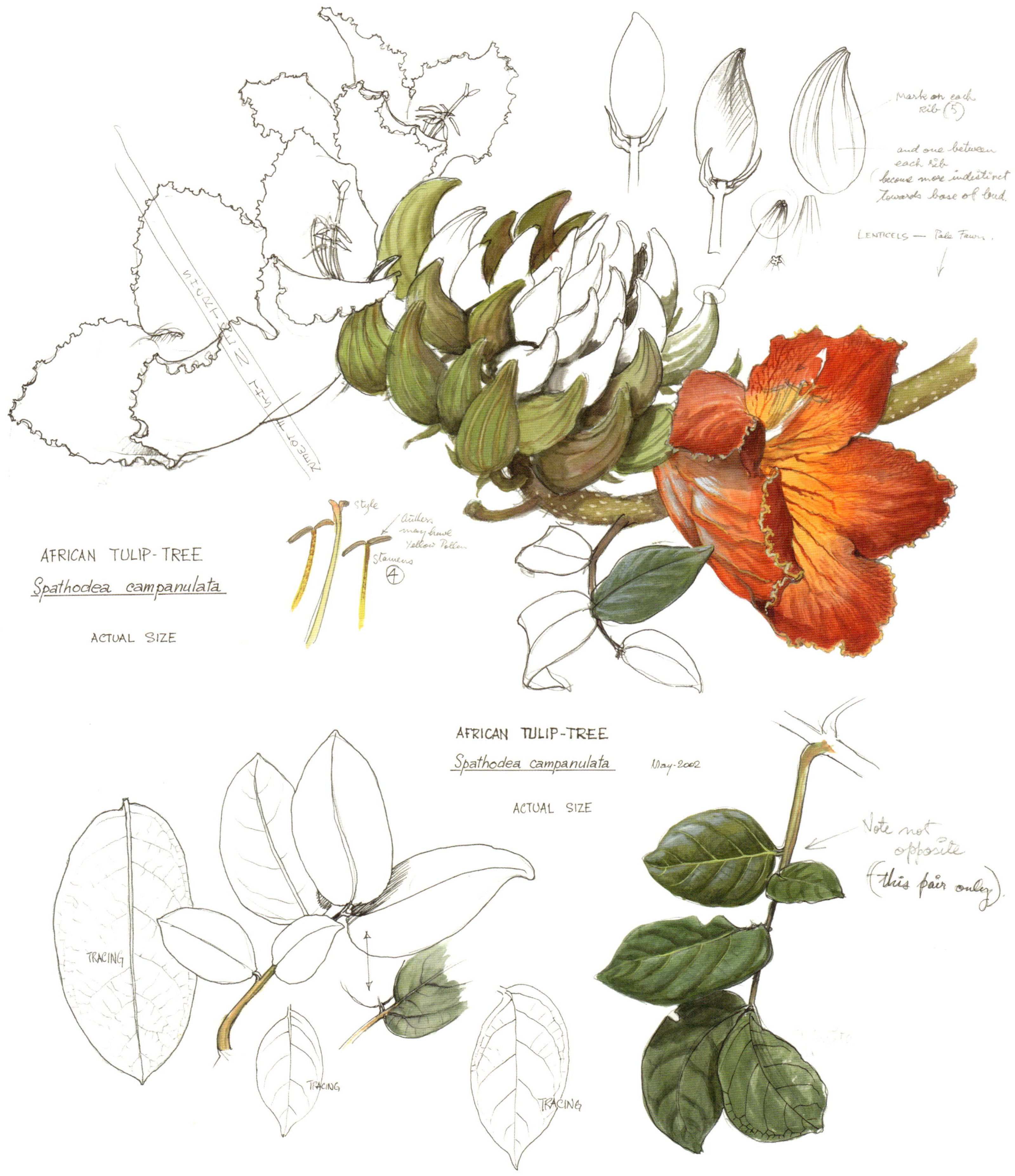
Mark on each rib (5)
and one between each rib
become more indistinct towards base of bud.
LENTICELS — Pale Fawn.
Style
Anthers may have Yellow Pollen
Stamens (4)
AFRICAN TULIP-TREE
Spathodea campanulata
ACTUAL SIZE
AFRICAN TULIP-TREE
Spathodea campanulata
May-2002
ACTUAL SIZE
Note not opposite
(this pair only).
TRACING
TRACING
TRACING

Thunbergia grandiflora

Bengal Clock Vine

ACANTHACEAE

→ undated (c.2005); watercolour
600 x 470mm
Private collection

↓ Unidentified Vine Stems
undated; pen and pencil
State Library of New South Wales

Thunbergia is named in honour of Carl Pehr Thunberg (1743–1828), naturalist and botanist.

grandiflora is from *grandis* (large) and *-florus* (-flowered).

Bengal Clock Vine is a rampant canopy climber from India. It has been widely cultivated for its spectacular blue or purple flowers and it has now become a serious weed in parts of Queensland, especially in the lowlands around Cairns. The plant grows from the smallest of bits and cuttings, as well as from a large persistent underground tuber.

The piece opposite is one of the few complete botanical paintings executed by Bill.

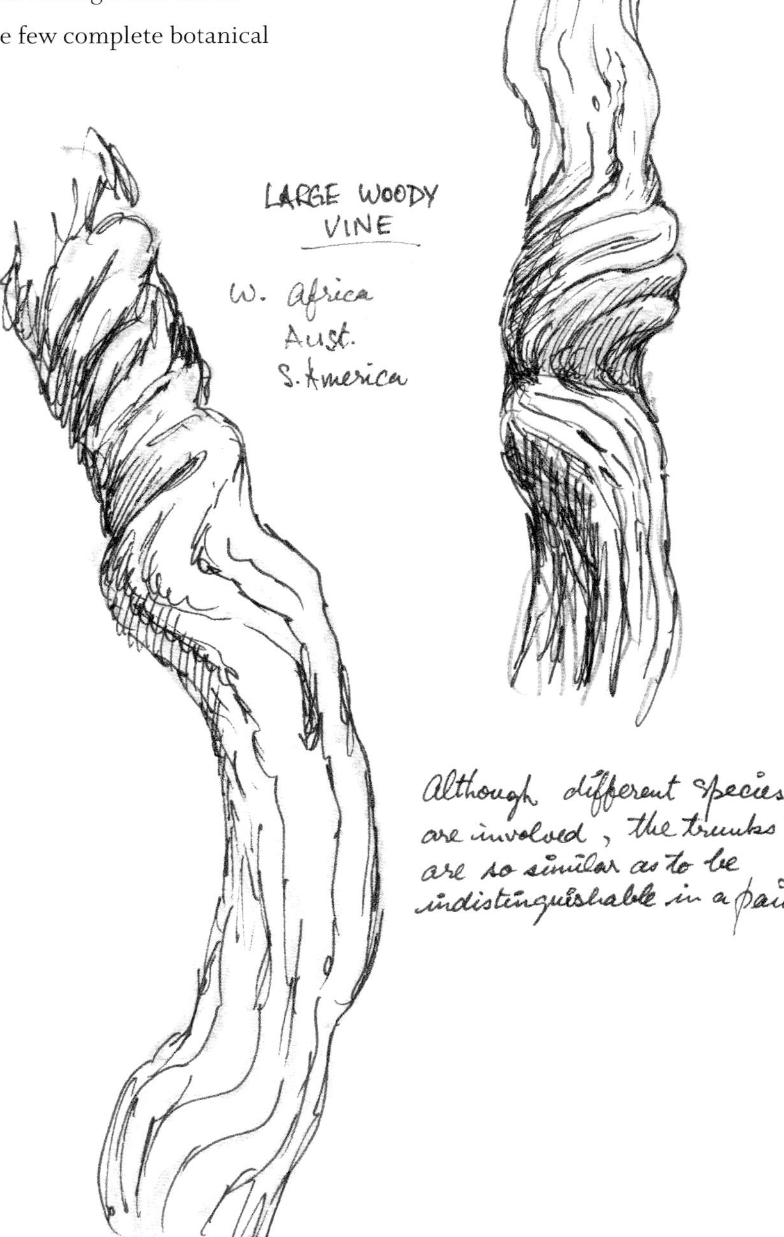

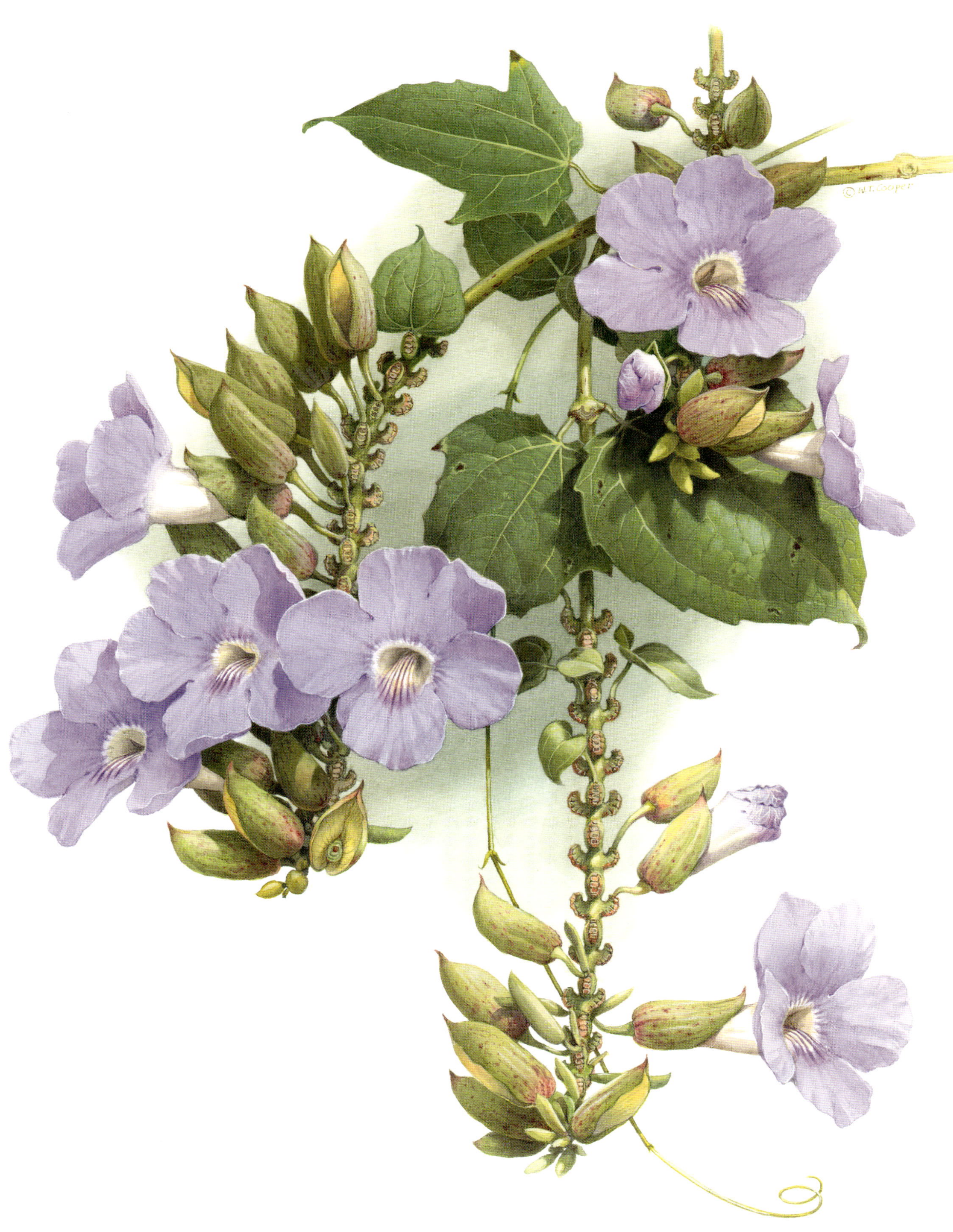
© W.T. Cooper

Vachellia karroo

Sweet Thorn

FABACEAE

Vachellia is named in honour of George Harvey Vachell (1799–1839), amateur botanist and chaplain to the British East India Company.

karroo is after Karroo, an old and incorrect spelling for the Karoo, an arid region in South Africa.

Sweet Thorn, once known as *Acacia karroo*, occurs naturally in various habitats from Angola and Zambia southwards, including the xerophytic Karoo region in South Africa. It is a tree that grows to about 12 metres, with spines mostly 70 millimetres in length but occasionally as long as 170 millimetres.

It is cherished by Africans for its uses and beauty. The bark exudes edible sweet gum and the plant has been variously used by Africans for its timber and thorns. The species has become naturalised beyond its natural range on grasslands in South Africa as well as coastal dunes and riparian areas in Spain and Portugal. Even though it is used as a source of fodder in Australia, it is considered a widespread weed.

Rüppell's Parrot (the upper two birds depicted here) occurs in Angola and Namibia. The reference sketch opposite was used as an appropriate plant to accompany this parrot for a plate in *Parrots of the World*.

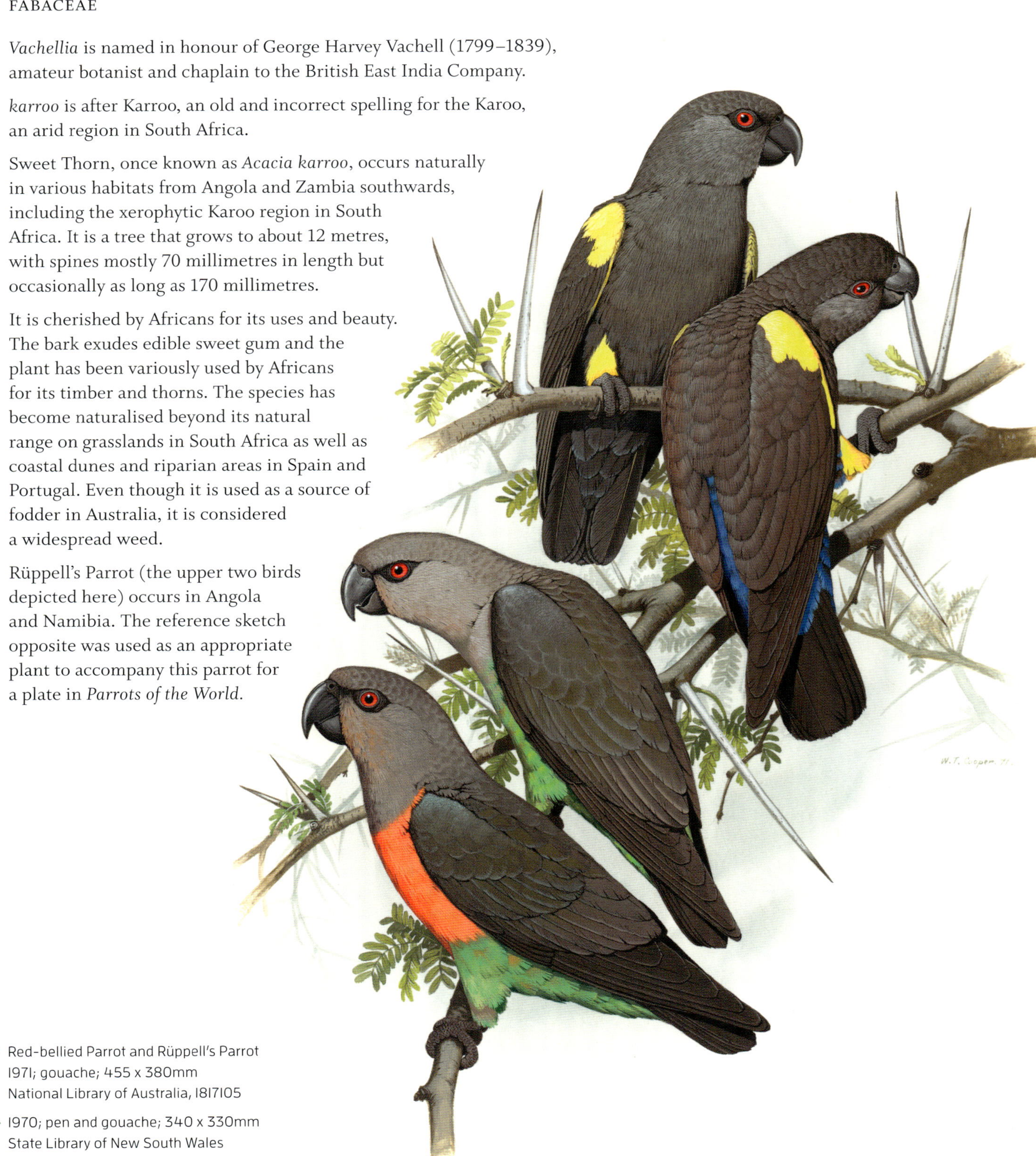

↗ Red-bellied Parrot and Rüppell's Parrot
1971; gouache; 455 x 380mm
National Library of Australia, 1817105

→ 1970; pen and gouache; 340 x 330mm
State Library of New South Wales

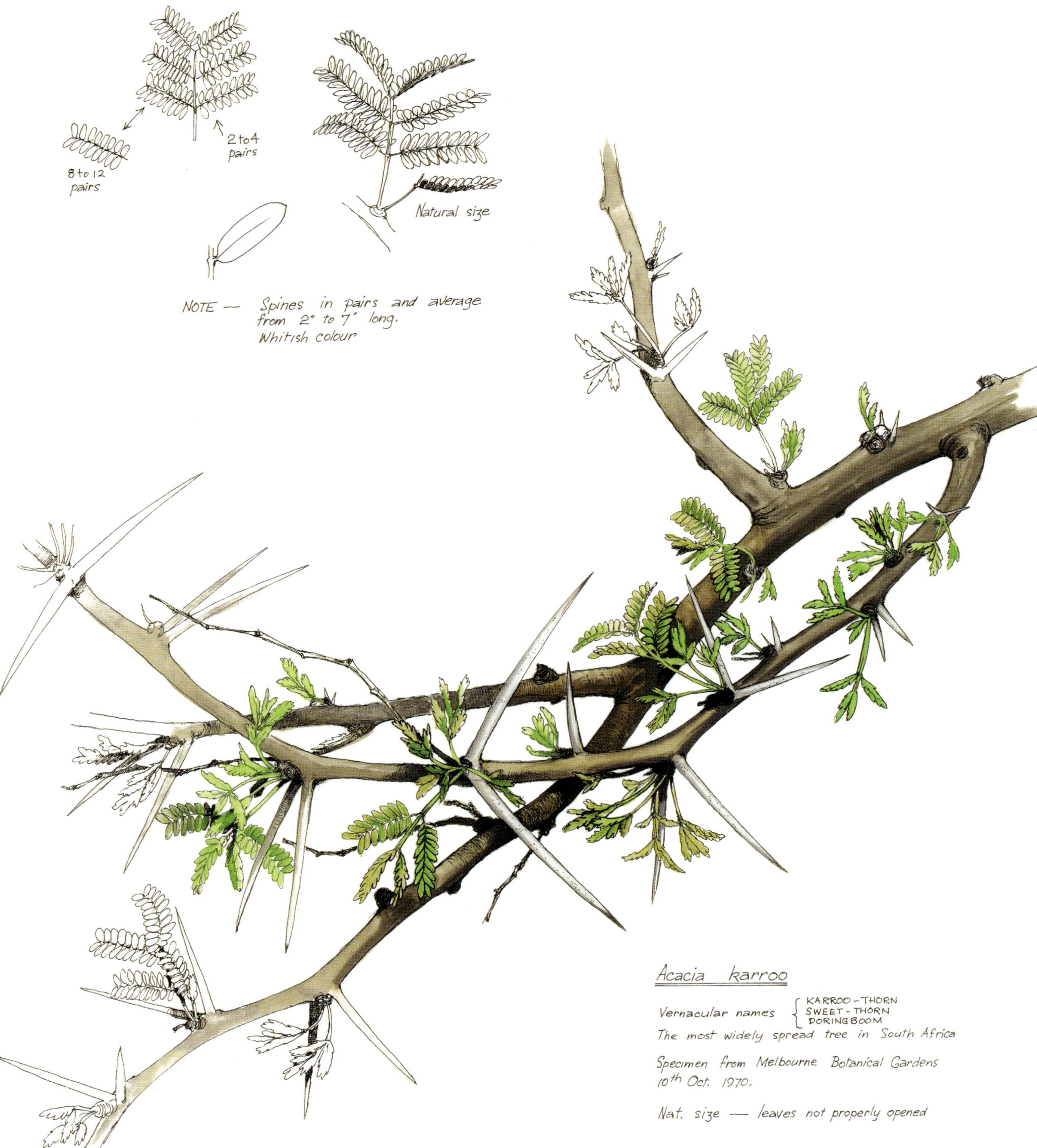
2 to 4 pairs
8 to 12 pairs
Natural size
NOTE — Spines in pairs and average from 2" to 7" long.
Whitish colour
Acacia karroo
Vernacular names { KARROO-THORN SWEET-THORN DORINGBOOM
The most widely spread tree in South Africa
Specimen from Melbourne Botanical Gardens 10th Oct. 1970.
Nat. size — leaves not properly opened

Zygogynum staufferianum

Kaik or Kup

WINTERACEAE

Zygogynum is from *zygo-* (joined) and *gyne* (female), referring to the carpels, which are often joined.

staufferianum is named in honour of Swiss botanist Hans Ulrich Stauffer (1929–1965).

Kaik is a small tree of the Western Highlands of Papua New Guinea above 2,400 metres. The forest is cold, wet, tall and mossy. Bill was there for an extended trip to study and draw birds of paradise. Beside his campfire at Tomba, on the slopes of Mount Hagen, Bill watched a male and two uncoloured Loria's Birds of Paradise feeding on these fruits.

Bill was camped in the Kubor Range when most of his companions left for a couple of days, leaving him with his local assistant, Tsu. Bill noted in his diary:

> *July 21st, 1973, 4:50pm Pouring rain again, this is a bloody wet place. Spent the afternoon drawing plants … 6:50pm Heard calls earlier from the hill. Tsu said it was 2 men who had gone up on Thurs morning. Now they just walked into camp. Carrying 2 huge Doria's Tree Kangaroos. The biggest one looked exactly like a bear. Both were dead. They (the men) were a real picture … with bows and arrows and all wet from the afternoon rains. These men are really part of the forest. They are now all sitting around the fire glistening wet. To see these men come in out of the inhospitable mountains with their catch was really something. I can't really describe the feeling or the atmosphere.*

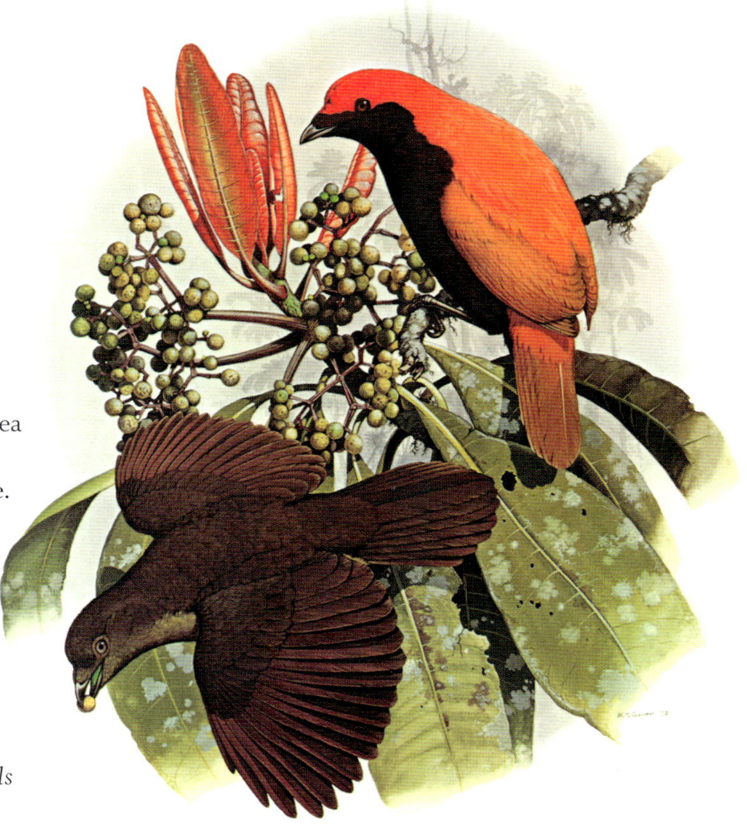

↑ Sickle-crested Bird of Paradise, colour plate from *Birds of Paradise and Bower Birds*, 1973
National Library of Australia, 69465

→ 1973; pen, pencil and gouache
408 x 295mm
National Library of Australia, 8070106

↓ Bill Painting Kaik in Camp at Tomba on the Kubor Range in Papua New Guinea, 1973

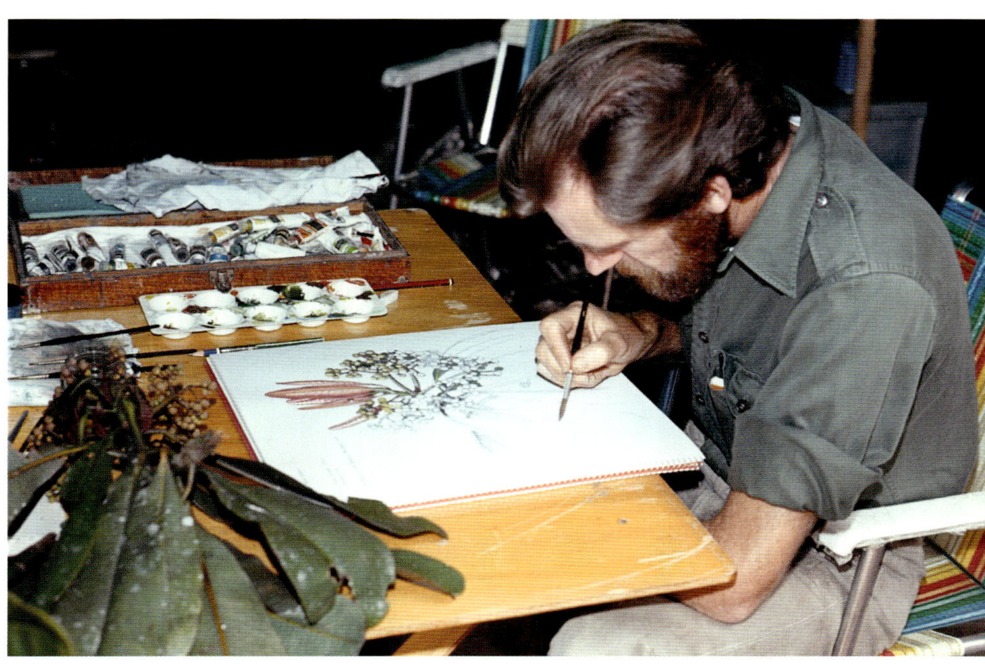

PAPUA NEW GUINEA

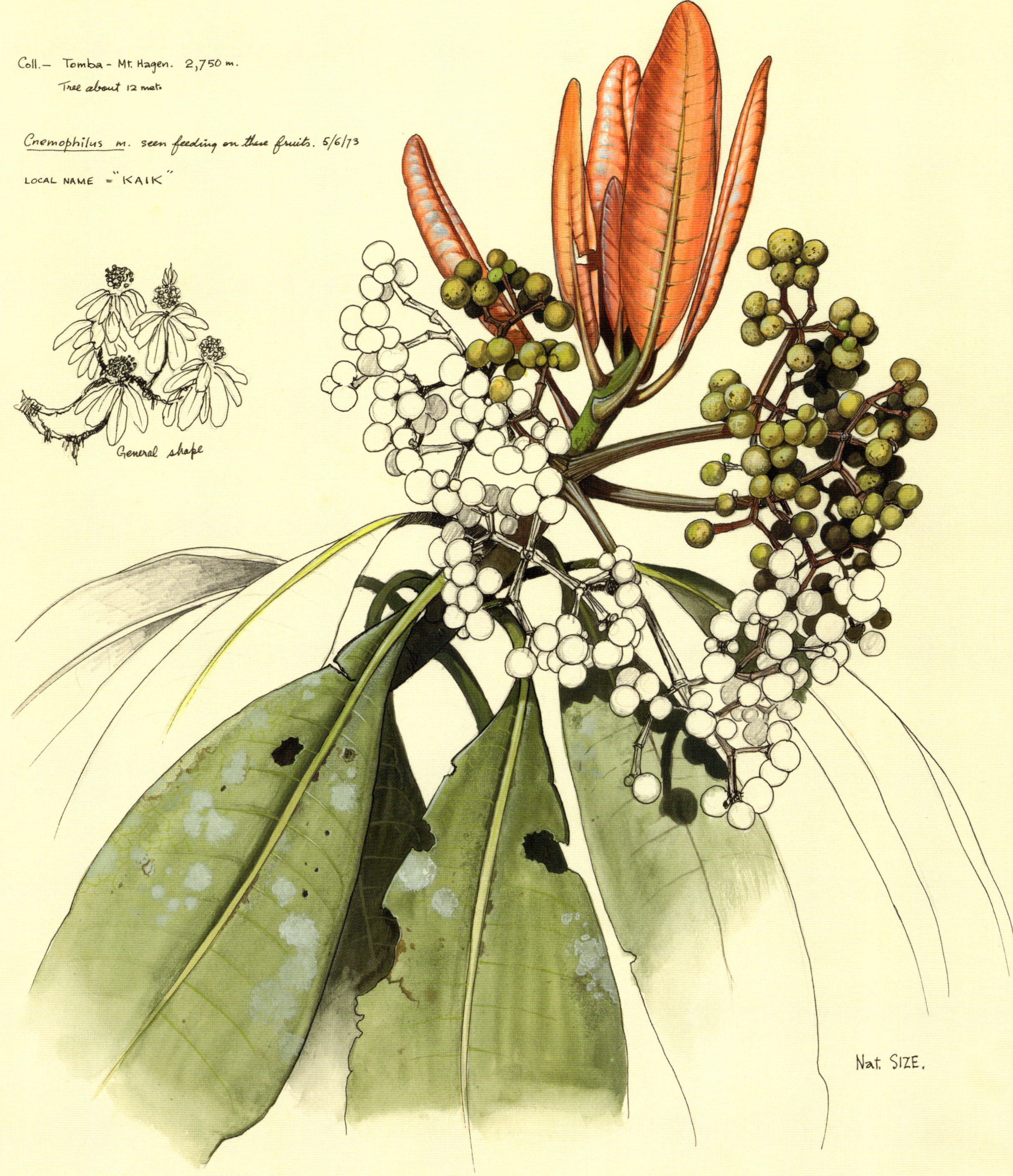

Glossary

adventitious – rising from stems above the ground, as in roots

aril – a fleshy growth from a seed base, partly or completely enclosing the seed

anther – the pollen-bearing section of a stamen

buttresses – extensions of tree trunks creating flared or prop supports at the base

calyx – the sepals collectively, which comprise the outer whorl of a flower

cauliflorous – bearing flowers and fruit on the main stem or trunk

deciduous – not evergreen, losing all leaves for part of the year

drupe – a fleshy fruit with one or more seeds inside a woody stone, such as plums and peaches

epiphyte – a plant that uses another plant as a support but does not take nutrients from that host

exudate – any substance excreted by a plant

funicle – a fleshy seed stalk, such as the coloured fleshy string attached to most wattle seeds

hemiparasitic – a plant that derives only some nourishment from its host

inflorescence – the flowering part of a plant

karst – the landform created by the dissolution of soluble rock; it usually has underground drainage and often forms caves

labellum – a distinct and often elaborate petal at the front of a flower, as in orchids and gingers

legume – the fruit of plants in the pea family

liana – woody vine

lignotuber – a woody swelling at plant base from which new shoots may emerge, especially after fire

littoral – growing near the sea

mallee –a plant with multiple stems emerging from near the base; landscape dominated by such plants

Malesia – botanical region including Malaysia, the Philippines, Indonesia, Bismarck Archipelago and New Guinea

node – the point on a twig where a leaf is or has been attached

pedicel – a flower or fruit stalk when it occurs within a multi-flowered inflorescence, not on solitary flowers

phyllode – a leaf-like stalk without true leaves, such as on most *Acacia* species

propagule – plant matter that is able to sprout as a new plant

riparian – growing on riverbanks

savanna – dry grassland sparsely populated with isolated trees

sclerophyll – plants with tough leathery leaves, such as *Eucalyptus*; a forest type that is dominated by such plants

senescent – too old to be reproductive

sepal – one lobe, or segment, of the calyx

spore – the asexual or sexual reproductive unit of plants such as ferns

staminode – modified sterile stamens that produce no pollen

stellate – star-shaped, radiating from a central point of attachment

symbiotic – different organisms living together, with or without mutual advantages

taxonomy – the classification of living things into relative groups

tuber – a storage organ, usually underground

type species – the plant that a genus is based on; for example, the first *Eucalyptus* species to be described is the basis for classifying all other *Eucalyptus* plants as the same genus

type specimen – the herbarium specimen from which a species is named and upon which descriptions are based

whorl – three or more organs (often leaves) rising from the same node or plane

Index of Families, Genera and Species

This is an index of plant families, genera and species referred to in this book, not a comprehensive index. It includes entries for illustrations of animal and bird species.

Page numbers in **bold** indicate illustrations.

A

Acacia aneura, 168, **168**, **169**
Acacia karroo, *see Vachellia karroo*
Acanthaceae, 224, **224**, **225**, 306, **307**
Actinotus helianthi, 170, **170**, **171**
Achariaceae, 76, **76**, **77**
Adenia heterophylla, 108, **108**, **109**
African Tulip Tree, 304, **304**, **305**
Aizoaceae, 230, **230**
Alexandra Palm, 54, **54–55**
Alloxylon flammeum, 4, **4**, **5**
Alphitonia petriei, 6, **6**, **7**
Amyema
  Amyema quandang var. *quandang*, 110, **111**
  Amyema queenslandica, 110, **110**
Anacardiaceae, vi, 68, **68–69**, 74, **74–75**, **82–83**, 280, **280**, **281**
Angophora, 178, 205, 216
  Angophora costata, 206, **206**
Ant-house Plant, 136, **136–137**
Antirhea tenuiflora, 8, **8–9**
Apocynaceae, 122, **122–123**, 124, **124**, 125, **125**, 134, **134**, **135**, 142, **142**, **143**, 222, 231, **231**
Apiaceae, 170, **170**, **171**
Araceae, 150, **150**, **151**
Araliaceae, 80, **80**, **81**, 302, **302**, **303**
Araucaria bidwillii, see Bunya Pine
Archontophoenix alexandrae, 54, **54–55**
Arecaceae, 54, **54–55**, 232, **232–233**
Argyle Apple, 196, **196**, **197**, **198–199**
Artocarpus heterophyllus, 254, **254**, **255**
Asparagaceae, 20, **20**, **21**, 70, **70**, **71**
Aspleniaceae, 78, 107, **107**, 112, **112**, **113**, 256, **256**, **257**
Asplenium, 78, 107, **107**, 112, **112**, **113**, 256, **256**, **257**
Australian King Parrot, **120**
Avicennia marina, 224, **224**, **225**

B

Bamboo, 292, **292**, **293**, **294**
Banana Fig, 32, **32**, **33**
Banded Kingfisher, **268**
Banksia, 216
  Banksia dentata, 172
  Banksia integrifolia, 166, 172, **172**, **173**
  Banksia menziesii, 174, **174**, **175**, **176–177**
  Banksia serrata, 166, 178, **178**, **179**, **180**, **181**
Barringtonia asiatica, 223, **223**
Bauhinia, 208, **208**, **209**
Basket Fern, 107, **115**
Beach Pandan, 246, **246**, **247**
Beach Tamarind, 22, **22**, **23**
Bean Tree, 208, **208**, **209**
Bengal Clock Vine, 306, **307**
Bersama abyssinica, 258, **258**, **259**
Bignoniaceae, 138, **138**, **139**, 158, **158**, 304, **304**, **305**
Billbergia pyramidalis, 260, **261**
Bird's Nest Fern, 78, 107, **107**, 112, **112–113**, **128**, 256, **256**, **257**
Black Bean, 10, **10–11**, **12–13**, **14–15**
Black Hornbill, **17**
Blandfordia grandiflora, **iv**
Blechnaceae, 114, **114**
Blechnum cartilagineum, 114, **114**, **115**
Blue-faced Honeyeaters, **212–213**
Blush Alder, 84, **84–85**
Bolbopsittacus, **225**
Bolwarra, 26, **26**
Bombax ceiba, 262, **262**
Bombax malabaricum, 262, **263**
Boraginaceae, 18, **18**, **19**
bowerbird, *see* Regent Bowerbird
Box Fruit, 223, **223**
Brazilian Shaving-brush Tree, 300, **300**
Brehm's Parrot, 292, **293**
Broad-leaved Carbeen, 167, 186, **186**, **187**
Broad-leaved Paperbark, 210, **210**, **211**
Broad-leaved Tea Tree, 212, **212–213**
Bromeliaceae, 260, **260**, **261**
Brown Kurrajong, 16, **16**, **17**
Brown Pine, *see Podocarpus grayae*
Bruguiera gymnorhiza, 223, 226, **226**, **227**, **228**, **229**
Brush Box, 206, **207**
Bulbophyllum longiflorum, 116, **117**
Buluh Padi, 284, **284**, **285**
Bumpy Satinash, 92, **92**, **93**
Bunya Pine, 3
Burdekin Plum, vi, 68, **68–69**
Button Orchid, 124, **124**

C

Cabbage Gum, 167, 186, **186**, **187**
Callitris glaucophylla, 182, **182**, **183**
Cannonball Mangrove, 250, **250–251**
Cape Bamboo, 62, **62**, **63**
Capparaceae, 264, **264**, **265**
Capparis tomentosa, 264, **264**, **265**
Carnaby's Black-Cockatoos, 174, **176–177**
Carpobrotus glaucescens, 230, **230**
Castanospermum australe, 10, **10–11**, **12–13**, **14–15**
Casuarina, 222
Casuarinaceae, 222
Cedar Mangrove, 250, **250–251**
Celastraceae, 156, **156–157**
Cerbera manghas, 222, 231, **231**
Cherry Guava, 301, **301**
Christmas Bells, **iv**
Christmas Island Imperial Pigeon, **290**

Cissus, 3
Cissus hypoglauca, 118, **118**, **119**
Cissus rotundifolia, 266, **266**, **267**
Cissus sterculiifolia, 120, **120**, **121**
Climbing Guinea Flower, 130, **130**, **131**
Climbing Pandan, 128, **128**
Coast Banksia, 166, 172, **172**, **173**
Cocky Apple, 217, **217**
Coconut Palm, 232, **232–233**
Cocos nucifera, 232, **232–233**
Combretaceae, 3, 98, 99, 221, **221**, **248**, **249**
Commersonia bartramia, 16, **16**, **17**
Common Silkpod, 134, **134**
Cook's Pandan, 214, **214–215**
Cordia subcordata, 18, **18**, **19**
Cordyline stricta, 20, **20**, **21**
Corymbia
Corymbia calophylla, 184, **184–185**
Corymbia confertiflora, 167, 186, **186**, **187**
Corymbia eximia, 188
Corymbia intermedia, 188, **189**
Corymbia gummifera, 188, **189**
Corymbia ptychocarpa, **ii–iii**, 190, **190–191**
Crimson Berry, 8, **8**, **9**
Crimson Passionflower, 296, **297**
Crinkle Bush, 205, **205**
Crow's Nest Fern, 107
Crowned Hornbill, **258**
Cuban Macaw, **287**
Cucurbitaceae, 160, **160**, **161**, 162, **162**, **163**, **164–165**
Cunoniaceae, 72, **72–73**
Cupaniopsis anacardioides, 22, **22**, **23**
Cupressaceae, 182, **182**, **183**

D

Daintree Hickory, 40, **40**, **41**
Damson, 98, **98–99**, **100–101**
Dendrobium, 268, **268**, **269**
Desmarest's Fig Parrot, **275**
Dillenia alata, 223, 234, **234–235**
Dilleniaceae, 130, **130**, **131**, 223, 234, **234–235**
Dischidia
Dischidia major, 122, **122–123**
Dischidia nummularia, 124, **124**
Dischidia ovata, 125, **125**
Dog Bane, 222, 231, **231**
Double-eyed Fig Parrot, **34–35**
Drynaria rigidula, 107, **115**
Dysoxylum arborescens, 24, **24**, **25**

E

Edward's Fig Parrot, **275**
Eichhornia crassipes, 270, **270–271**
Elaeagnus, 106
Elaeocarpaceae, 84, **84–85**
Emerald Lory, **303**
Entada phaseoloides, 126, **126**, **127**
Erythrina variegata, 236, **236**, **237**
Eucalyptus, 184, 186, 218
Eucalyptus camaldulensis, 166, 192, **192**, **193**
Eucalyptus chartaboma, 194, **194**, **195**
Eucalyptus cinerea, 196, **196**, **197**, **198–199**
Eucalyptus melanophloia, 200
Eucalyptus shirleyi, 166–167, 200, **200**, **201**
Euphorbiaceae, 48, **48**, **49**, 56, **56**, **57**, 140, **140–141**
Eupomatia laurina, 26, **26**
Eupomatiaceae, 26, **26**
Eurya tigang, 272, **272**, **273**

F

Fabaceae, 10, **10–11**, **12–13**, **14–15**, 64, **64**, **65**, 126, **126**, **127**, 132, **132**, **133**, 168, **168**, **169**, 208, **208**, **209**, 236, **236**, **237**, 240, **240**, **241**, 308, **308**, **309**
Fagaceae, 282, **283**
Fairy Lorikeet, **294**
False Euodia, 58, **58**, **59**
False Marula, 280, **280**, **281**
Fan Palm, 54, **54–55**
Fern-leaved Grevillea, 202, **202**, **203**
Ficus, 28, 274, **275**,
Ficus benghalensis, 276, **276**, **277**
Ficus benjamina, 3
Ficus carica, 274
Ficus copiosa, 28, **28**
Ficus fraseri, 30, **30**, **31**
Ficus pleurocarpa, 32, **32**, **33**
Ficus schwarzii, 278, **278**, **279**
Ficus variegata, 34, **34–35**
Ficus virens, 36, **36**, **37**
Ficus watkinsiana, 38, **38**, **39**
Fig, 28, 274, **274**, **275**, 282
Banana Fig, 32, **32**, **33**
Green Fig, 36, **36**, **37**
Indian Banyan, 276, **276**, **277**
Plentiful Fig, 28, **28**
Russet Stem-Fig, 278, **278**, **279**
Sandpaper Fig, 30, **30**, **31**
Variegated Fig, 34, **34–35**
Watkins Fig, 38, **38**, **39**
Weeping Fig, 3
White Fig, 36, **36**, **37**
Fischer's Turaco, **266**
Five-leaved Grape, 118, **118**, **119**
Flaming Torch, 260, **261**
Flannel Flower, 170, **170**, **171**
Forest Kingfisher, **137**
Freycinetia excelsa, 128, **129**
Freycinetia scandens, 128, **128**

G

Gang-gang Cockatoo, **198–199**
Ganophyllum falcatum, 40, **40**, **41**
Gardenia scabrella, 42, **42**, **43**
Ghittoe, 46, **46**, **47**
Giant Pepper Vine, 144, **144–145**, **146**, **147**, **148–149**
Giant Waterlily, 242, **242**, **243**, **244–245**
Glochidion harveyanum, 44, **44**, **45**
Golden Guinea Tree, 223, 234, **234–235**
Great Blue Turaco, **280**
Green Fig, 36, **36**, **37**
Green Possum, **33**
Green Rosella, **95**
Grevillea pteridifolia, 202, **202**, **203**
Grey Mangrove, 224, **224**, **225**
Grey Mistletoe, 110, **111**
Gristle Fern, 114, **114**, **115**

H

Halfordia scleroxyla, 46, **46**, **47**
Hard Alder, 72, **72–73**
Harvey's Buttonwood, 44, **45**
Hernandia nymphaeifolia, 222, 238, **238–239**
Hernandiaceae, 222, 238, **238–239**

Hibbertia scandens, 130, **130**, **131**
Homalanthus novo-guineensis, 48, **48**, **49**
Hope's Cycad, 50, **50**, **51**

I

Ibatiria furfuracea, 132, **132**, **133**
Indian Almond, 248, **248–249**
Indian Banyan, 276, **276**, **277**
Indian Coral Tree, 236, **236**, **237**
Indian Grey Hornbill, **276**
Intsia bijuga, 240, **240**, **241**

J

Jacana, **244–245**, **245**
Jakfruit, 254, **254**, **255**
Jitta, 46, **46**, **47**
Jungle Vine, 138, **138**, **139**

K

Kadam, 288, **288–289**
Kaik, 310, **310**, **311**
Kapok Tree, 262, **262**
Kilawa, 272, **272**, **273**
King Parrot, **120**
Kup, 310, **310**, **311**
Kwila, 240, **240**, **241**

L

Lacewing Vine, 108, **108**, **109**
Lambertia formosa, 204, **204**
Lannea schweinfurthii, 280, **280**, **281**
Large-flowered Milk Vine, 134, **135**
Lawyer Vine, 106
Lecythidaceae, 217, **217**, 223, **223**
Lepidozamia hopei, 50, **50**, **51**
Levieria acuminata, 52, **52–53**
Licuala ramsayi, 54, **54–55**
Linospadix microcaryus, 54, **54**
Lithocarpus rassa, 282, **283**
Little Kingfisher, **226**
Lolly Berry, 156, **156–157**
Loranthaceae, 110, **110**, **111**
Lomatia silaifolia, 205, **205**
Long-leaved Grape, 120, **120**, **121**
Lophostemon confertus, 206, **207**
Lysiphyllum cunninghamii, 208, **208**, **209**

M

Maclurochloa montana, 284, **284**, **285**
Maiden's Blush, 84, **84–85**
Malvaceae, 16, **16**, **17**, 262, **262**, **263**, 300, **300**
Mallotus philippensis, 56, **56**, **57**
Marri, 184, **184**, **185**
Marsdenia liisae, 134, **135**
Matchbox Bean, 126, **126**, **127**
Melaleuca
Melaleuca quinquenervia, 210, **210**, **211**
Melaleuca viridiflora, 212, **212–213**
Melia azedarach, 286, **286**, **287**
Meliaceae, 24, **24**, **25**, 90, **90**, **91**, 250, **250**, **251**, 286, **286**, **287**
Melianthaceae, 258, **258**, **259**
Melicope broadbentiana, 58, **58**, **59**
Menzies' Banksia, 174, **174**, **175**, **176–177**
Milky Silkpod, 142, **143**
Mistletoe, 110, **110–111**
Mitragyna parvifolia, 288, **288–289**
Monimiaceae, 52, **52–53**, **86–87**
Monkey's Earrings, 64, **64**, **65**
Moraceae, 28, **28**, **29**, 30, **30**, **31**, 32, **32**, **33**, 34, **34–35**, 36, **36**, **37**, 38, **38**, **39**, 254, **254**, **255**, 274, **274**, **275**, 276, **276**, **277**, 278, **278**, **279**
Mossman Mahogany, 24, **24**, **25**
Mountain Devil, 204, **204**
Mulga, 168, **168**, **169**
Muntingia calabura, 290, **290**, **291**
Muntingiaceae, 290, **290**, **291**
Musa banksii, 60, **60**, **61**
Musaceae, 60, **60**, **61**
Musk Duck, **270–271**
Musk Lorikeet, **210**
Musschenbroek's Lory, **303**
Myrmecodia beccarii, 136, **136**, **137**
Myrtaceae, 92, **92**, **93**, 166, 184, **184**, **185**, 186, **186**, **187**, 188, **188**, 190, **190–191**, 192, **192**, **193**, 194, **194**, **195**, 196, **196**, **197**, **198–199**, 200, **200**, **201**, 206, **206**, **207**, 210, **210**, **211**, 212, **212**, **213**, 218, **218**, **219**, 220, 301, **301**

N

Narrow-leaved Palm Lily, 20, **21**
Nastus, 292, **292**, **293**, **294**, **295**
Native Banana, 60, **60**, **61**
Native Cashew, 82, **82–83**
Native Dracaena, 70, **70–71**
Neololeba atra, 62, **62**, **63**
Neosepicaea jucunda, 138, **138**, **139**
Nudaurelia, 258, **259**
Nymphaea gigantea, 242, **242**, **243**, **244–245**
Nymphaeaceae, 242, **242**, **243**, **244–245**

O

Old Man Banksia, 166, 178, **178–181**
Omphalea queenslandiae (omphalea), 140, **140–141**
Orchidaceae, 116, **117**, 268, **268**, **269**

P

Pale Umbrella Orchid, 116, **117**
Palm Cockatoo, **100–101**, **164–165**
Panama Berry, 290, **290**, **291**
Pandanaceae, 3, 128, **128**, **129**, 214, **214**, **215**, 246, **246**, **247**
Pandanus, 3
Pandanus cookii, 214, **214**, **215**
Pandanus tectorius, 246, **246**, **247**
Papuan Lory, **303**
Paradise Crow, **225**
Pararchidendron pruinosum, 64, **64**, **65**
Parsonsia latifolia, 142, **143**
Parsonsia straminea, 134, **134**, 142, **142**
Passiflora vitifolia, 296, **297**
Passifloraceae, 108, **108**, **109**, 296, **297**
Pentaphylacaceae, 272, **272**, **273**
Perfumed Passionflower, 296, **297**
Persian Lilac, 286, **286**, **287**
Persoonia linearis, 216, **216**
Peruvian Grape Ivy, 266, **266**, **267**
Phyllanthaceae, 44, **45**
Pigface, 230, **230**
Pink Bloodwood, 188, **188–189**
Pink Silky Oak, 4, **4**, **5**
Pink-eyed Cerbera, 222, 231, **231**
Piper hederaceum, 144, **144–145**, **146**, **147**, **148–149**

Piper novae-hollandiae, *see Piper hederaceum*
Piperaceae, 144, **144–145**, **146**, **147**, **148–149**
Pittosporaceae, 66, **66**
Pittosporum undulatum, 66, **66**
Planchonia careya, 217, **217**
Platycerium, **115**
Pleiogynium timorense, vi, 68, **68–69**
Plentiful Fig, 28, **28**
Pleomele angustifolia, 70, **70–71**
Poaceae, 62, **62**, **63**, 284, **284**, **285**, 292, **292**, **293**, **294**
Podocarpaceae, 3, 298, **298**, **299**
Podocarpus archboldii, 298, **298**, **299**
Podocarpus grayae, **3**
Podocarpus neriifolius, *see Podocarpus grayae*
Polygalaceae, 104, **104**, **105**
Polypodiaceae, 152, **152**, **153**
Polyscias, 28, **29**
Pontederiaceae, 270, **270–271**
Pothos, 106
Pothos longipes, 150, **150**, **151**
Princess Parrot, **168**
Proteaceae, 4, **4**, **5**, 94, **94**, **95**, **96–97**, 102, **102**, 166, 172, **172**, **173**, 174, **174**, **175**, **176–177**, 178, **178**, **179**, **180**, **181**, 202, **202**, **203**, 204, **204**, 205, **205**, 216, **216**
Pseudobombax grandiflorum, 300, **300**
Psidium cattleianum, 301, **301**
Psittinus, **225**
Pullea stutzeri, 72, **72–73**
Purple-headed Bee-eater, **56**
Pyrrosia confluens, 152, **152**, **153**

R

Rainbow Lorikeet, **14–15**, **195**, **203**
Rainforest Gourd, 160, **161**, **162**
Rattle Skulls, 122, **122–123**
Red Beech, 234, **234–235**
Red Gourd, 162, **162**, **163**, **164–165**
Red Kamala, 56, **56**, **57**
Red Lacewing Butterfly, **109**
Red Mangrove, 223, 226, **226**, **227**, **228**, **229**
Red Silky Oak, 4, **4**, **5**
Red-bearded Bee-eater, **285**
Red-bellied Parrot, **308**
Red-capped Parrot, **184**
Red-fronted Macaw, **287**
Red-winged Parrot, **201**
Regent Bowerbird, **84–85**, **90**
Rhamnaceae, 6, **6**, **7**
Rhaphidophora, 106
Rhizophoraceae, 223, 226, **226**, **227**, **228**, **229**
Rhus taitensis, 74, **74–75**
riflebird, *see* Victoria's Riflebird
Ripogonaceae, 154, **154**, 155, **155**
Ripogonum
Ripogonum album, 154, **154**
Ripogonum fawcettianum, 155, **155**
Ripogonum papuanum, *see Ripogonum album*
River Red Gum, 166, 192, **192**, **193**
Rubber Tree, 231, **231**
Rubiaceae, 8, **8**, **9**, 42, **43**, 136, **136**, **137**, 288, **288–289**
Rüppell's Parrot, **308**
Russet Stem-Fig, 278, **278**, **279**
Rutaceae, 46, **46**, **47**, 58, **58**, **59**
Ryparosa javanica, *see Ryparosa kurrangii*
Ryparosa kurrangii, 76, **76**

S

Salacia
Salacia chinensis, 156, **156–157**
Salacia disepala, 156, **156**
Salvadori's Fig Parrot, **275**
Sandpaper Fig, 30, **30**, **31**
Sapindaceae, 22, **22**, **23**, 40, **40**, **41**, 78, **78–79**, 88, **88**, **89**
Sarcotoechia protracta, 78, **78–79**
Sarsaparilla, 6, **6**, **7**
Scaly-breasted Lorikeet, **181**
Scentless Rosewood, 90, **90**, **91**
Schefflera, 302, **302**, **303**
Schefflera actinophylla, 80, **80**, **81**
Sea Almond, 222, 248, **248–249**
Sea Hearse, 222, 238, **238–239**
Sea Trumpet, 18, **18**, **19**
Semecarpus australiensis, 82, **82–83**
Senna artemisioides subsp. *helmsii*, **167**
Silver-leaved Ironbark, 166–167, 200, **200**, **201**
Silvery Felt Fern, 152, **152**, **153**
Slender Climbing Pandan, 128, **128**, **129**
Sloanea, 3
Sloanea australis subsp. *parviflora*, 84, **84–85**
Small Supplejack, 155, **155**
Snowwood, 64, **64**, **65**
Soft Ghittoe, 104, **104**, **105**
Sovereignwood, 98, **98–99**, **100–101**
Spathodea campanulata, 304, **304**, **305**
Spectacled Flying-fox, **33**
Spice Bush, 102, **102**, **103**
Splendid Astrapia, **298**
Star Flower, 42, **43**
Staghorn, **115**
Steganthera laxiflora, 86, **86–87**
Stella's Lory, **303**
Stone Oak, 282, **283**
Straw Beech, 52, **52–53**
Strawberry Tree, 290, **290**, **291**
Sumac, 74, **74–75**
Sumba Hornbill, **278**
Superb Fruit-Dove, **119**
Swamp Bloodwood, 190, **190–191**
Sweet Pittosporum, 66, **66**
Sweet Thorn, 308, **308**, **309**
Syncarpia glomulifera, 218, **218**, **219**, **220**
Synima macrophylla, 88, **88**, **89**
Synoum glandulosum, 90, **90**, **91**
Syzygium cormiflorum, 92, **92**, **93**

T

Tar Tree, 82, **82–83**
Tarictic Hornbill, **19**
Tasmanian Waratah, 94, **94**, **95**, **96–97**
Tecomanthe burungu, 158, **158**
Telopea truncata, 94, **94**, **95**, **96–97**
Terminalia, 3
Terminalia catappa, 222, 248, **248–249**
Terminalia platyphylla, 221, **221**
Terminalia sericocarpa, 98, **98–99**, **100–101**
Tetra Beech, 86, **86–87**
Tetrasynandra laxiflora see *Steganthera laxiflora*
Thunbergia, 80
Thunbergia grandiflora, 306, **307**
Thylacine, **96–97**
Tigang, 272, **272**, **273**
Tiger, **288–289**

Topaz Tamarind, 88, **88**, **89**
Topknot Pigeon, **46**
Trichosanthes
Trichosanthes odontosperma, 160, **161**, **162**
Trichosanthes pentaphylla, 162, **162**, **163**, **164–165**
Triunia erythrocarpa, 102, **102**, **103**
Tropical Bleeding Heart, 48, **48**, **49**
Trumpeter Hornbill, **229**
Tuckeroo, 22, **23**
Turpentine, 218, **218**, **219**, **220**

U

Ulysses Swallowtail, **139**
Umbrella Tree, 80, **80**, **81**, 302, **302**, **303**

V

Vachellia karroo, 308, **308**, **309**
Variegated Fig, 34, **34**, **35**
Victoria's Riflebird, **6**, **89**, **147**, **148–149**
Vitaceae, 118, **118**, **119**, 120, **120**, **121**

W

Walking-stick Palm, 54, **54**
Water Hyacinth, 270, **270–271**
Watermelon Dischidia, 125, **125**
Watkins' Fig, 38, **38**, **39**
Weeping Fig, 3
White Cedar, 286, **286**, **287**
White Cypress, 182, **182**, **183**
White Fig, 36, **36**, **37**
White Mangrove, 224, **224**, **225**
White Supplejack, 154, **154**
White-bellied Go-away Bird, **264**
Wild Parsley, 205, **205**
Wild Plum, 221, **221**
Winged Bersama, 258, **258**, **259**
Winteraceae, 310, **310**, **311**
Woolly Caper Bush, 264, **264**, **265**
Woollybutt, 194, **194**, **195**
Wreathed Hornbill, **49**

X

Xanthophyllum octandrum, 104, **104**, **105**
Xylocarpus moluccensis, 250, **250–251**

Y

Yamiga, 298, **298**, **299**
Yellow Boxwood, 104, **104**, **105**
Yellow-billed Turaco, **41**
Yellow-tailed Black-Cockatoo, **220**

Z

Zamia Palm, 50, **50**, **51**
Zamiaceae, 50, **50**, **51**
Zodiac Moth, **140–141**
Zygogynum staufferianum, 310, **310**, **311**

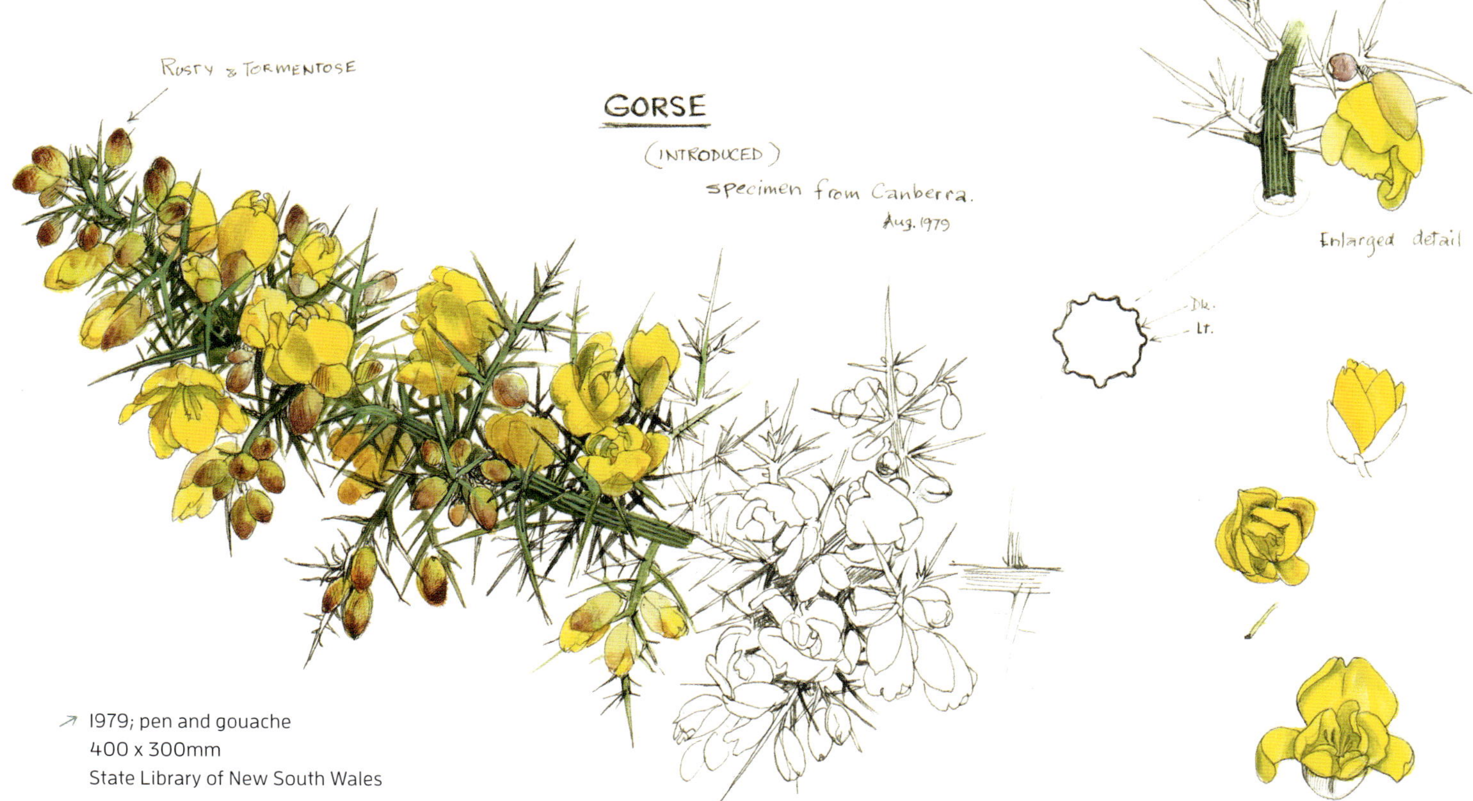

↗ 1979; pen and gouache
400 x 300mm
State Library of New South Wales

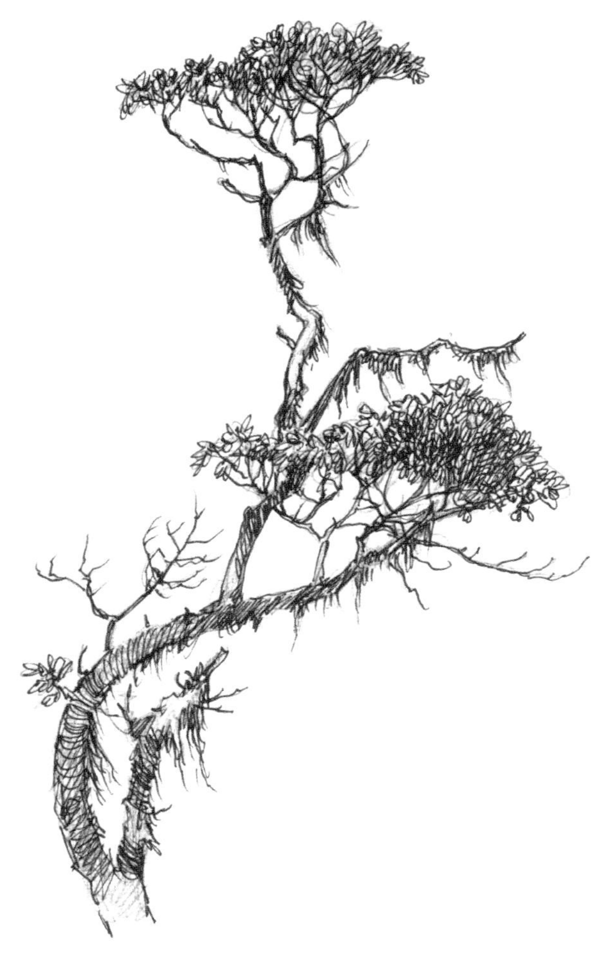

undated; pen and pencil; (in part) 210 x 270mm
State Library of New South Wales